W0112354

ELECTRONS AND PHONONS IN LAYERED CRYSTAL STRUCTURES

PHYSICS AND CHEMISTRY OF MATERIALS WITH LAYERED STRUCTURES

VOLUME 3

ELECTRONS AND PHONONS IN LAYERED CRYSTAL STRUCTURES

Edited by

T. J. WIETING

U.S. Naval Research Laboratory, Washington, D.C. 20375, U.S.A.

and

M. SCHLÜTER

Bell Laboratories, Murray Hill, N.J. 07974, U.S.A.

D. REIDEL PUBLISHING COMPANY

DORDRECHT: HOLLAND / BOSTON: U.S.A.
LONDON: ENGLAND

Library of Congress Cataloging in Publication Data

Main entry under title:

Electrons and phonons in layered crystal structures.

(Physics and chemistry of materials with layered structures; v. 3)
Includes bibliographical references and index.
1. Layer structure (Solids). 2. Electrons. 3. Phonons. I. Wieting. T. J.,
1935-   II. Schlüter, Michael, 1945-   III. Series.
QD478.P47 vol. 3 [QD921] 530.4'1s [548'.81] 78-14733
ISBN 13 978-94-009-9372-3   ISBN 978-94-009-9370-9 (eBook)
DOI 10.1007/978-94-009-9370-9

Published by D. Reidel Publishing Company
P.O. Box 17, Dordrecht, Holland

Sold and distributed in the U.S.A., Canada, and Mexico
by D. Reidel Publishing Company, Inc.
Lincoln Building, 160 Old Derby Street, Hingham,
Mass. 02043, U.S.A.

TABLE OF CONTENTS

CHAPTER C

PART II: PHONONS

T. J. WIETING and J. L. VERBLE / Infrared and Raman Investigations of Long-Wavelength Phonons in Layered Materials   321

FOREWORD

This volume is devoted to the electron and phonon energy states of inorganic layered crystals. The distinctive feature of these low-dimensional materials is their easy mechanical cleavage along planes parallel to the layers. This feature implies that the chemical binding within each layer is much stronger than the binding between layers and that some, but not necessarily all, physical properties of layered crystals have two-dimensional character. In Wyckoff's *Crystal Structures*, SiC and related compounds are regarded as layered structures, because their atomic layers are alternately stacked according to the requirements of cubic and hexagonal close-packing. However, the uniform (tetrahedral) coordination of the atoms in these compounds excludes the kind of structural anisotropy that is fundamental to the materials discussed in this volume. An individual layer of a layered crystal may be composed of either a single sheet of atoms, as in graphite, or a set of up to five atomic sheets, as in Bi_2Te_3. A layer may also have more complicated arrangements of the atoms, as we find for example in Sb_2S_3. But the unique feature common to all these materials is the structural anisotropy, which directly affects their electronic and vibrational properties.

The nature of the weak interlayer coupling is not very well understood, despite the frequent attribution of the coupling in the literature to van der Waals forces. Two main facts, however, have emerged from all studies. The first is the extraordinary variation in the interlayer coupling strength from material to material. By comparison with the forces acting within the layers, the interlayer forces are smaller by factors ranging between twenty and a thousand. These ratios have been determined from phonon studies, and their values are a convenient measure of the structural anisotropy of a particular material. The second fact is the nearly three-dimensional character of certain electronic energy states in materials that otherwise show large structural anisotropy. This combination of two- and three-dimensional effects in the same material is an intriguing aspect of the physical properties of layered solids. Since most studies have focussed on the stronger intralayer forces, few quantitative models of the interlayer interaction have thus far been proposed, and much important work remains to be done in this area.

In organizing this review the editors have chosen to restrict the number of contributions, so that a broader and more synthetic treatment of electron and phonon energy states could be achieved. The first three articles in this volume, contributed by Schlüter and Fong, review the calculations of energy bands in layered materials and the supporting optical data. After discussing the group theoretical tools, the authors consider the various theoretical methods used to calculate the band structures of layered crystals. They then discuss specific materials and their bonding models and compare theoretical results with relevant experiments. The two articles on phonons, which make up the second part of the volume, divide along the lines of the experimental techniques employed in studying lattice vibrations. Long-wavelength phonons are treated in the article by Wieting and Verble on infrared absorption and Raman scattering, and phonon dispersion is taken up in the article by Wakabayashi and Nicklow on inelastic neutron scattering. Each article in turn attempts to evaluate all the published work on electron or phonon energy states in inorganic layered crystals.

In a rapidly growing field of research, the most recent results are inevitably omitted from consideration. Since completing the manuscript, a number of papers have appeared that supplement the discussion in this review or provide examples of new families of layered crystals, or in some cases describe novel physical effects. Perhaps the most interesting recent development is the discovery of electron charge density waves in the tantalum dichalcogenides and the periodic structural distortions that are associated with the formation of these waves. Many investigators have already entered into this new area. An assessment of this and other work, however, must remain for future reviews of electrons and phonons in layered crystals.

U.S. Naval Research Laboratory   T. J. WIETING AND M. SCHLÜTER
Bell Laboratories
July 1977

PART I

ELECTRONS

SYMMETRY CONSIDERATIONS

M. SCHLÜTER

Bell Laboratories, Murray Hill, N.J. 07974, U.S.A.

1. Group Theoretical Basis

1.1. SYMMETRY OF THE SCHRÖDINGER EQUATION

Many problems in solid state theory consist in solving the time-independent Schrödinger equation

$$H\psi_n = E_n\psi_n \tag{A1}$$

to obtain the energy eigenvalues E_n together with the corresponding eigenfunctions ψ_n. Since one is dealing with a large number of particles ($\sim 10^{23}$ electrons + nuclei) a solution of this many-particle problem can only be obtained after various approximations and simplifications. Without going into detail we shall simply mention the usual approximations made in solving Schrödinger's equation:

(i) Since our main interest is in the electronic system, we separate the motion of electrons and nuclei and consider the latter as being fixed (Born-Oppenheimer approximation). This approximation of course is invalid if we are interested in finite temperature effects. It is, however, in most cases sufficient to treat finite temperature effects, like electron-phonon coupling by perturbation.

(ii) The remaining many-electron problem can be reduced to a one-electron problem by defining an averaged potential, like the Hartree or Hartree-Fock potential. Exchange and correlation can be treated in several local or non-local approximations according to their importance in the particular problem. The resulting one-electron potential can be iterated in a self-consistent way until convergence is reached.

(iii) Spin and other relativistic effects are usually introduced through the standard two-component Pauli matrix formalism. This approximation of Dirac's treatment is valid in the low energy region we are concerned with.

After all these simplifications we are left with a one-electron Hamiltonian $H(\mathbf{r}, \mathbf{p}, \mathbf{s})$ which is a function of one electron's coordinate $\mathbf{r}$, momentum $\mathbf{p}$ and spin $\mathbf{s}$. Even with a Hamiltonian of this simple form it remains a difficult task to determine the eigensolutions of the problem. It is very convenient at this point to consider the symmetry of the problem and to find any simplification or reduction which can be made rigorously on the basis of symmetry.

The symmetry properties of the system are described by the symmetry of the Hamiltonian operator. (In some cases we might consider a more complicated Schrödinger equation to find some symmetry properties and their full implications, as e.g., the time-dependent Schrödinger equation to describe time reversal symmetry). We shall use the tools provided by group theory for exploiting all symmetry properties to their full extent. Most of this can be done by very elementary group theory; thus a profound knowledge of group theory is not required for our purposes. We shall introduce and use group theoretical concepts in a natural way without any formal proofs. For more details we refer the reader to the standard textbooks on group theory [1–6]. An excellent description of the use of group theory in solid state physics is also given in [7, 8].

To discuss the symmetry of the Hamiltonian we introduce its symmetry transformations R as linear coordinate transformations leaving the Hamiltonian invariant. Later we shall add time reversal symmetry to these operations. The common coordinate transformations are translations, rotations, inversion, reflections and permutations and combinations thereof. The set of all symmetry transformations of a Hamiltonian forms the group of the Schrödinger equation. Since the considered symmetry operations R leave the Hamiltonian invariant, they commute with it: $RH = HR$. It follows from this that there are no matrix elements of H between states which have different eigenvalues for the operator R i.e., which differ in their classification according to any of the symmetry operators.

For example, considering a problem with inversion symmetry, we find that all eigenfunctions of H are either even or odd under inversion. For the case of several symmetry operations R which all commute with H but not necessarily among themselves (which is usually the case in crystal symmetry groups) we cannot find simultaneous eigenfunctions of H and all operations R anymore. However, finite sets of functions can be found such that the effect of any symmetry operation R on any one of the functions results in a linear combination of functions within the same set. This set of functions spans an invariant subspace under the crystal symmetry operations of the total space of eigenfunctions of H.

The linear combinations can be expressed by matrices $\Gamma(R)_{ji}$:

$$R\psi_i = \sum_j \Gamma(R)_{ji}\psi_j \tag{A2}$$

which 'represent' the group of symmetry operations. The dimensions of the invariant subspaces which equal the dimensions of the representation matrices give the degeneracies of the Hamiltonian eigenfunctions. We shall discuss these properties in more detail in the next section when we consider some important points of the theory of group representations.

Applying results of this theory to our quantum mechanical problem will enable us to largely simplify the calculations and to characterize its solutions. We shall also see that the influence of external perturbations which lead to transitions between the system's eigenstates is mainly determined by symmetry and that the corresponding selection rules can be obtained by group theoretical arguments.

We now present the most important results of the theory of representations and illustrate their meanings in several examples.

1.2. GROUP REPRESENTATIONS

An abstract group $\{R\}$ can be 'represented' by a number of square matrices $\{\Gamma(R)\}$ associated with each group element and fulfilling the group multiplication table by matrix multiplication. The dimension of the matrices $\Gamma(R)$ gives the 'dimensionality' of the representation. Any similarity transformation applied to these matrices will leave the multiplication table and thus the 'character' of this representation unchanged. The infinity of representations related to each other in this way for various similarity transformations are called equivalent. An invariant characterization of a representation is obtained by considering the traces of the matrices. This defines the character of the lth representation as being the set of numbers $\{X^l(R)\}$ with

$$X^l(R) = \text{Tr}\, \Gamma^{(l)}(R). \tag{A3}$$

This set of numbers can be reduced by collecting the group elements R according to classes φ_k within which the $X^l(R)$ are the same. We are then left with a set of numbers $\{X^l(\varphi_k)\}$ characterizing the lth representation of the group. Most of the discussion of the group theory related to the electronic structure of a system is based on these characters rather than on an actual set of representation matrices. There are however, exceptions like the construction of symmetry adapted basis functions for a Hamiltonian, for which representation matrices are explicitly needed. We shall therefore continue discussing some important properties of representation matrices.

Two or more representations can be combined into bigger matrices still representing the group. However, such an artificially enlarged matrix representation is said to be reducible. This fact might be concealed and not be obvious for a given representation. The criterion for reducibility of a representation is that it be possible to reduce all the matrices representing the group to block form by the same similarity transformation. Or conversely a representation is said to be *irreducible* if it cannot be expressed in terms of representations of lower dimensionality. It is the structure of these *irreducible* representations which is important in our discussion of the symmetry of a system.

Many of the symmetry properties are based on the *great orthogonality theorem* which states for all inequivalent, irreducible, unitary representations of a group:

$$\sum_R \Gamma^{(i)}(R)^*_{\mu\nu} \Gamma^{(j)}(R)_{\alpha\beta} = \frac{h}{d_i} \delta_{ij} \delta_{\mu\alpha} \delta_{\nu\beta}, \tag{A4}$$

where the sum runs over all group elements, h is the dimension of the group and d_i is the dimensionality of the representation $\Gamma^{(i)}$. This theorem states the orthogonality of different elements of the representation matrices in a space spanned by the group operations.

A similar theorem can be derived for the characters of representations. As already mentioned, the table of characters of a group presents in many cases sufficient information for a group theoretical discussion. We shall therefore mention some simple rules which follow from the orthogonality theorem and which enable us to construct the character table of a given abstract group:

(i) The number of irreducible representations equals the number of classes of group elements.

(ii) The dimensionalities d_i of the irreducible representations are connected to the dimension h of the group by $\sum_i d_i^2 = h$.

(iii) The identity element E must be represented by a unit matrix, hence $X^i(E) = d_i$.

(iv) There always exists the totally symmetric representation with $X^l(\varphi_k) = 1$, i.e, each group element is represented by unity.

(v) All rows and columns respectively of the character table must be orthogonal.

We now discuss the relation of the group-theoretical arguments mentioned so far to the quantum mechanical problem we want to solve. As already mentioned, the group of interest is the group of symmetry operations which leave the Hamiltonian of the problem invariant. Each of these operations is usually defined as a real orthogonal transformation of coordinates eventually combined with a translation of coordinates. Following Wigner [1] one can introduce a new group which is isomorphic to this group of coordinate transformations, in which the group elements are now operators which operate on *functions* rather than on *coordinates*. These operators then change the contours of a function in such a way as to compensate for the change of coordinates:

$$Rf(\mathbf{r}) = f(R^{-1}\mathbf{r}). \tag{A5}$$

Since all these operations leave the Hamiltonian invariant, they commute with H and we have

$$RH\psi_n = RE_n\psi_n = HR\psi_n = E_nR\psi_n. \tag{A6}$$

We conclude from this that any function $R\psi_n$ obtained by operating on an eigenfunction ψ_n by a symmetry operator of the group of H will also be an eigenfunction to the same eigenenergy. Thus, given any eigenfunction, we can generate by application of all group operations other eigenfunctions which are all degenerate with the original eigenfunction. If this procedure yields *all* degenerate functions, the degeneracy is called *normal* or symmetry induced; any further degeneracy is called *accidental*. Let us assume that the eigenvalue E_n is l_n-fold symmetry-degenerate and let us choose a set of l_n orthonormal eigenfunctions to this energy. The operation of any R on any ψ_n of these l_n functions produces another function having the same energy, which therefore can be expressed as a linear combination of the complete set of l_n degenerate functions.

$$R\psi_v^{(n)} = \sum_{\mu=1}^{l_n} \Gamma^{(n)}(R)_{\mu v}\psi_\mu^{(n)}. \tag{A7}$$

The l_n-dimensional matrices $\Gamma^{(n)}(R)$ form an l_n-dimensional irreducible representation of the group of the Hamiltonian. The degeneracy of E_n is simply given by the dimensionality l_n of the representation. Thus, by finding the dimensionalities of all irreducible representations of the group of the Hamiltonian, which we do to set up the character table, we are able to determine the different degrees of (*non*-accidental) degeneracies of the system. Moreover, if the representation matrices are worked out, they contain according to Equation (A7) the transformation properties of all eigenfunctions of the system under all symmetry operations of the group of the Hamiltonian.

In many cases the inclusion of perturbation terms into the Hamiltonian reduces the size of the symmetry group. A typical example for this is e.g., the application of an external electric field to a crystal with inversion symmetry. In this and similar cases the decrease in size of the symmetry group may cause some irreducible representations to become reducible. This is equivalent to decreasing or removing degeneracies in the eigenfunction system. The decomposition of a reducible representation Γ (which e.g., was an irreducible representation of the full symmetry group) into irreducible representations $\Gamma^{(i)}$ of the reduced symmetry group can be achieved by knowing the character tables of both groups. If $X(R)$ and $X^i(R)$ denote the characters of Γ and $\Gamma^{(i)}$ respectively we can write

$$X(R) = \sum_i a_i X^i(R), \tag{A8}$$

where $a_i = (1/h) \sum_R X^i(R)^* X(R)$ denotes the number of times the various irreducible representations $\Gamma^{(i)}$ of the reduced group appear in the representation Γ of the full group and where the sum runs over all operations R of the reduced group.

We summarize that a perturbation can lift degeneracies if its inclusion in the Hamiltonian reduces the symmetry group and hence changes the possible irreducible representations.

Let us now suppose that we introduce a perturbation term H_1 into the Hamiltonian H which lowers its symmetry. We then are interested in matrix elements of H_1 between eigenfunctions of the unperturbed Hamiltonian. Using group-theoretical considerations we can restrict the variety of possible non-zero matrix elements. In other words we can find selection rules for matrix elements of the kind $\langle \phi_l^i | H_1 | \phi_{l'}^j \rangle$ to be nonvanishing. To obtain these selection rules, we first determine the irreducible representation $\Gamma^{(H_1)}$, according to which H_1 transforms (i.e., for which H_1 is a basis function). In a general case H_1 may contain parts of several symmetries; it can then be decomposed into several irreducible representations. Having found $\Gamma^{(H_1)}$ we determine the direct product of representations $\Gamma^{(H_1)} \Gamma^{(j)} = \Gamma^{(p)}$ according to the rule

$$X^{\Gamma^{(p)}}(R) = X^{\Gamma^{(H_1)}}(R) X^{\Gamma^{(j)}}(R). \tag{A9}$$

This direct product representation may be reducible; it can be decomposed according to Equation (A8) into irreducible representations. The selection rule then states that the matrix element $\langle \phi_l^i | H_1 | \phi_{l'}^i \rangle$ must be zero, unless $\Gamma^{(i)}$ is found in the decomposition of the direct product representation of $\Gamma^{(H_1)} \Gamma^{(j)}$.

An equivalent formulation of this statement is that the matrix element must vanish unless the direct product representation of $\Gamma^{(i)*} \Gamma^{(H_1)} \Gamma^{(j)}$ includes $\Gamma^{(1)}$, the identical representation. This statement does not guarantee the existence of any matrix element: it rather excludes the possibility of some kinds and thus helps considerably to reduce calculations based on perturbation theory.

The selection rules given above have to be distinguished from the general matrix element theorem applying for matrix elements of operators O which are invariant under *all* group operations. This theorem

$$\langle \varphi_l^i | O | \varphi_{l'}^j \rangle = \delta_{ij} \delta_{ll'} M \tag{A10}$$

states that matrix elements of an operator O which is invariant under all operations of a group vanish between functions belonging to different irreducible representations or to different columns of the same unitary representation.

This general matrix element theorem is of great help whenever variational solutions of the Hamiltonian are desired. In that case O is replaced by H and the resulting Hamiltonian matrix can be decomposed into noninteracting blocks, provided symmetry adapted basis functions (i.e., basis functions which transform according to one column of one of the irreducible representations) are used. Basis functions of this kind can easily be obtained by the so called group projection operators which are expressed in terms of the irreducible representation matrices of the group. We shall discuss the functioning of these operators and their explicit application to plane wave basis functions in a later chapter. We will describe more specifically the nature of groups met in crystalline systems. We will also introduce the concepts of point and space groups, of symmorphic and non-symmorphic groups and discuss their particularities using several examples.

1.3. CRYSTAL POINT AND SPACE GROUPS

The objective of this section is to discuss the irreducible representations of the symmetry groups of solids. We then shall use these tools to describe the motion of an electron in a perfect crystal. In that case the Hamiltonian has perfect translational symmetry which means that there exists a specific set of vectors $\mathbf{a}_1$, $\mathbf{a}_2$, and $\mathbf{a}_3$, called 'primitive vectors', such that the Hamiltonian is left invariant when it is expressed as a functional of $\mathbf{r} + \mathbf{t}$ instead of $\mathbf{r}$, where

$$\mathbf{t} = n_1\mathbf{a}_1 + n_2\mathbf{a}_2 + n_3\mathbf{a}_3 \tag{A11}$$

and n_1, n_2, n_3 are any set of integers.

All translations of the form (A11) define the crystal lattice. It can be shown [2] that there exist exactly 32 point groups of rotations and reflections which are consistent with this lattice translational symmetry. Conversely, the different existing lattices may then be classified according to these point groups of rotations and reflections. Thus, inspecting all the 32 point groups one finds that there exist 14 distinct types of lattices, the 'Bravais lattices' which may further be classified into the 7 different 'crystal systems'. Enumerations of the point groups, Bravais lattices and crystal systems can be found in the group theoretical literature [6]. In the following discussion of the electronic properties of layered compounds we only shall meet hexagonal and rhombohedral crystal systems. All point groups associated with these systems derive from the 'full hexagonal' point group D_{6h}. A detailed discussion of this group and its various subgroups will be given later. The classification into point groups, Bravais lattices and crystal systems does not complete the description of all symmetries possible for a perfect crystal. A systematic enumeration of all these symmetry properties yields 230 distinct 'space groups'. They involve, in addition to the point group operations – rotations, reflections and combinations thereof – and in addition to translations, new elements which are combinations of rotations and translations – screw axes – or combinations of reflections and translations – glide planes.

To describe the properties of space groups it is convenient to introduce a systematic notation for all the operations involved. A general space group operation is denoted by $R = \{\alpha \mid \mathbf{t} + \tau\}$, where α is a point group operation, $\mathbf{t}$ is a primitive translation of the lattice and τ any fractional translation different from any $\mathbf{t}$. Operationally $\{\alpha \mid \mathbf{t} + \tau\}$ is defined by

$$\{\alpha \mid \mathbf{t} + \tau\}\mathbf{r} = \alpha\mathbf{r} - \mathbf{t} - \tau \tag{A12}$$

or operating on a function $f(\mathbf{r})$ by

$$\begin{aligned}\{\alpha \mid \mathbf{t} + \tau\}f(\mathbf{r}) &= f(\{\alpha \mid \mathbf{t} + \tau\}^{-1}\mathbf{r}) \\ &= f(\alpha^{-1}\mathbf{r} + \alpha^{-1}(\mathbf{t} + \tau)). \end{aligned} \tag{A13}$$

In Equation (A13) we used the definition of the inverse element $\{\alpha \mid \mathbf{t} + \tau\}^{-1} = \{\alpha^{-1} \mid -\alpha^{-1}(\mathbf{t} + \tau)\}$. The identity element is given by $\{\varepsilon \mid 0\}$. The space group G consists of all operations $R = \{\alpha \mid \mathbf{t} + \tau\}$ which leave a given lattice invariant. The associated point group G^0 which not necessarily is a symmetry group of the crystal, is obtained by setting $\mathbf{t} + \boldsymbol{\tau} = 0$ in all elements of G. If all elements of G^0 are contained in G (i.e., if $\boldsymbol{\tau} = 0$ for all elements), then G is called *symmorphic*; otherwise, G is called *non-symmorphic*. In the discussion of layered compounds most of the space groups which we will consider are symmorphic. Typical exceptions and examples for *non-symmorphic* space groups are: D_{6h}^4, the group of the hexagonal close packed lattice and C_{6v}^4, the group of the wurtzite structure.

All elements of G which are of the form $\{\varepsilon \mid \mathbf{t}\}$ constitute the translation group T which is an invariant subgroup of G. Since real, *finite* crystals are not exactly invariant under the operations of T (i.e., the crystal-surface is not translational invariant) one has to modify the problem so that it has T as an exact symmetry group. One possible modification is called the application of 'periodic boundary conditions', namely

$$\{\varepsilon \mid 0\} = \{\varepsilon \mid N_1 \mathbf{a}_1\} = \{\varepsilon \mid N_2 \mathbf{a}_2\} = \{\varepsilon \mid N_3 \mathbf{a}_3\}, \tag{A14}$$

where N_1, N_2, and N_3 are large numbers so that $\mathbf{L}_1 = N_1 \mathbf{a}_1$, $\mathbf{L}_2 = N_2 \mathbf{a}_2$ and $\mathbf{L}_3 = N_3 \mathbf{a}_3$ are vectors connecting opposite boundaries of the crystal. With these restrictions T becomes a finite, Abelian group of order $N = N_1 \times N_2 \times N_3$. Its irreducible representations are given by

$$\Gamma^{(k)}(\{\varepsilon \mid \mathbf{t}\}) = \exp(i\mathbf{k} \cdot \mathbf{t}). \tag{A15}$$

Each of these one-dimensional representations is labeled by a wave vector $\mathbf{k}$. The group T of order $N_1 \times N_2 \times N_3$ has only $N_1 \times N_2 \times N_3$ distinct representations; we can obtain them from Equation (A15) by putting restrictions on $\mathbf{k}$. First we introduce in $\mathbf{k}$-space a lattice of vectors

$$\mathbf{G} = m_1 \cdot \mathbf{b}_1 + m_2 \cdot \mathbf{b}_2 + m_3 \cdot \mathbf{b}_3 \tag{A16}$$

defining the *reciprocal lattice*, each satisfying the condition

$$\exp(i\mathbf{G} \cdot \mathbf{t}) = 1 \quad \text{for all } \mathbf{t} \text{ and } \mathbf{G}. \tag{A17}$$

It follows from this definition that $\mathbf{k}$ and $\mathbf{k} + \mathbf{G}$ give identical representations. To label the representations uniquely, it is therefore necessary to restrict $\mathbf{k}$ to some region which is a unit cell of the reciprocal lattice. The particular unit cell which has been adopted by convention is called *Brillouin zone* and is obtained by bisecting the lines connecting $\mathbf{G} = 0$ to the nearest reciprocal lattice points. Within the 'reduced' zone $\mathbf{k}$ is restricted to the allowed discrete values

$$\mathbf{k} = \frac{r_1}{N_1} \mathbf{b}_1 + \frac{r_2}{N_2} \mathbf{b}_2 + \frac{r_3}{N_3} \mathbf{b}_3, \tag{A18}$$

where r_1, r_2, r_3 are integers.

By these conventions the $N_1 \times N_2 \times N_3$ irreducible representations of T are labeled uniquely. Based on *Bloch's theorem*, all functions $\psi_{\mathbf{k}}(\mathbf{r})$ which transform under the group T according to the $\mathbf{k}$th irreducible representation can be written in the form

$$\psi_{\mathbf{k}}(\mathbf{r}) = e^{i\mathbf{k}\cdot\mathbf{r}} u_{\mathbf{k}}(\mathbf{r}), \tag{A19}$$

where $u_{\mathbf{k}}(\mathbf{r})$ has full translational symmetry, i.e., $u_{\mathbf{k}}(\mathbf{r} + \mathbf{t}) = u_{\mathbf{k}}(\mathbf{r})$.

In discussing solutions of the crystal Hamiltonian we shall only use eigenfunctions which are Bloch functions. Moreover, each energy level itself may be labelled with the same **k**-vector which labels the irreducible representation of T according to which its eigenfunctions transform. We thus are left with the problem to solve a Schrödinger equation of the form

$$H\psi_{\mathbf{k}}(\mathbf{r}) = E(\mathbf{k})\psi_{\mathbf{k}}(\mathbf{r}) \qquad (A20)$$

with **k** lying inside the Brillouin zone.

Usually, with the exception of the triclinic crystal system, the space group of a crystal contains more operations than the pure translations. However, by construction, the translation group T is always a subgroup of the crystal space group G. It follows from this that in setting up the irreducible representations of any space group G, we can restrict ourselves to basis functions which are in Bloch form. We first shall consider symmorphic space groups.

1.3.1. *Symmorphic Space Groups*

Even in the case of symmorphic space groups when $\tau = 0$ for any operation, G is *not* a direct product of the translation group T and the point group G^0 as it can be seen from the non-commutativity of the product elements. However, T is an invariant subgroup of G. It is therefore convenient to study the factor group G/T, whose 'abstract' elements are 'complexes' or sets of elements $\{\{\alpha \mid \mathbf{t}\}\}$ of G formed by giving **t** all the possible translation vector values. Introducing this factor group means to deal with the rotational aspects of the space group and to divide out or remove the translational effects. The factor group describes the space group multiplication modulo T. The factor group G/T is isomorphic to the point group G^0. This is true for symmorphic or non-symmorphic groups, in the latter case however G^0 is *not* contained in G which leads to consequences we shall discuss later. Since G/T is isomorphic to G^0 it has only h elements and not N, where $N = N_1 \times N_2 \times N_3 = 10^{23}$ is the order of T and h is the order of G^0. Representations of G/T can therefore easily be constructed. The irreducible representations Γ of a symmorphic space group G are then simply related to the irreducible representations Γ^0 of the appropriate point group G^0 and to the irreducible representations of the translation group T by

$$\Gamma(\{\alpha \mid \mathbf{t}\}) = e^{i\mathbf{k}\cdot\mathbf{t}} \, \Gamma^0(\{\alpha \mid 0\}). \qquad (A21)$$

In particular, if we use Bloch functions the correct translational properties are automatically given and we only have to find functions which transform like Γ^0.

For this let us consider a specific Bloch function $\psi_{\mathbf{k}}(\mathbf{r})$ and operate upon it with all the point group elements $\{\alpha \mid 0\}$. We then create new Bloch functions $\{\psi'_{\alpha\mathbf{k}}(\mathbf{r})\}$ of **k**-vectors $\alpha\mathbf{k}$. The set of distinct vectors in **k**-space, $\alpha\mathbf{k}$, generated from one **k**-vector and all possible $\{\alpha \mid 0\}$ is called the 'star of **k**'. For a general **k**-point in the Brillouin zone we generate exactly h distinct **k**-vectors by this procedure, where h is the dimension of the point group G^0. This set of h distinct Bloch functions $\{\psi'_{\alpha\mathbf{k}}(\mathbf{r})\}$ for a general **k** and all $\{\alpha \mid 0\} \in G^0$ generates an h-dimensional irreducible representation of the space group G. Consequently, this set of h functions is degenerate, i.e., $E(\mathbf{k}) = E(\alpha\mathbf{k})$ for all $\{\alpha \mid 0\} \in G^0$, or in other words $E(\mathbf{k})$ has the full symmetry of G^0. Conventionally, however, these wave functions are *not* considered to be degenerate unless there are two or more eigenfunctions with the same energy *and* the same **k**-vector. This latter case can

happen at 'symmetry points' in the Brillouin zone, where the application of all operations $\{\alpha \mid 0\}$ yields less than h distinct Bloch functions. It is this case which is of practical
interest, since it reflects the essential degeneracies of the system.

A point $\mathbf{k}$ in the Brillouin zone is defined to be a symmetry point if there exists at
least one operation $R = \{\alpha \mid 0\}$ with $\alpha \neq \varepsilon$ of the point group such that $e^{i\alpha\mathbf{k}\cdot\mathbf{t}} = e^{i\mathbf{k}\cdot\mathbf{t}}$,
i.e., an operation R which leaves $\mathbf{k}$ unchanged (modulo any reciprocal lattice vector $\mathbf{K}$)
$\alpha\mathbf{k} = \mathbf{k} + \mathbf{K}$. For symmetry points inside the Brillouin zone $\mathbf{K} = 0$ and the points lie on
a rotation axis or a mirror plane or both. For symmetry points on the surface of the
Brillouin zone we might find $\mathbf{K} \neq 0$. The set of operations $R = \{\alpha \mid \mathbf{t}\}$ of the crystal
space group G with the property of leaving $\mathbf{k}$ invariant forms a group itself – 'the group
of $\mathbf{k}$', which is symbolized by $G_\mathbf{k}$. As in the total crystal space group G, the translational
group T is contained in $G_\mathbf{k}$ as an invariant subgroup too. Consequently one may construct the factor group $G_\mathbf{k}/T$ which is isomorphic with a corresponding point group
$G_\mathbf{k}^0$ consisting of all operations of the form $\{\alpha \mid 0\}$ which are contained in $G_\mathbf{k}$. The
irreducible representations of $G_\mathbf{k}$ are therefore given by the irreducible representations
of $G_\mathbf{k}^0$ and of T according to Equation (A21). We thus need to know the representations
and character tables of $G_\mathbf{k}^0$ for every $\mathbf{k}$ in the Brillouin zone. The eigenfunctions $\psi_{\mathbf{k}, \lambda}^i$
transforming according to these irreducible representations are labeled by:

(i) the wave vector $\mathbf{k}$, indicating their translational transformation properties,

(ii) the superscript i denoting the ith irreducible representation of $G_\mathbf{k}^0$ if this representation is of dimension $l_i \geq 2$,

(iii) the subscript λ, denoting the particular column of the representation matrix,
according to which $\psi_{\mathbf{k}, \lambda}^i$ transforms.

If $l_i \geq 2$, there are degenerate eigenfunctions with the same $\mathbf{k}$-vector. This situation
can occur only at symmetry points. For a general $\mathbf{k}$-point $G_\mathbf{k}^0 = \{\varepsilon \mid 0\}$ and the non-
degenerate eigenfunctions are completely labeled by $\mathbf{k}$.

Knowing the irreducible representations of T (i.e., considering Bloch functions) and
of any $G_\mathbf{k}^0$ the irreducible representations of the full group G can in principle be obtained.
In practice, however, there is little need for the representations of the full group G. All
the important information is contained in the various 'groups of $\mathbf{k}$', $G_\mathbf{k}$.

To illustrate these ideas, we now consider an example of a symmorphic crystal space
group in full detail.

1.3.2. *The Symmorphic Crystal Space Group D_{3d}^3*

A convenient example is the hexagonal group D_{3d}^3, which is the symmetry group of the
CdI_2 crystal-structure in which many layered materials crystallize. The CdI_2 structure
corresponds to an octahedral coordination of the cations while the sequence in which
the anions are arranged derives from hexagonal close packing. A perspective view of
this structure is given in Figure 1. The hexagonal crystal lattice is spanned by two
vectors $\mathbf{a}_1$ and $\mathbf{a}_2$ of equal length and having an angle of $120°$ to each other and by a
third vector $\mathbf{a}_z$ of length c which is perpendicular to the basal plane formed by $\mathbf{a}_1$ and
$\mathbf{a}_2$. In the case of the CdI_2 structure the unit cell contains one molecule. To facilitate the
enumeration of the possible space group operations we give in Figure 2 a projection
of the structure on the $\mathbf{a}_1$, $\mathbf{a}_2$ basal plane.

With the usual periodic boundary conditions, any lattice translation $\{\varepsilon \mid \mathbf{t}\}$, where

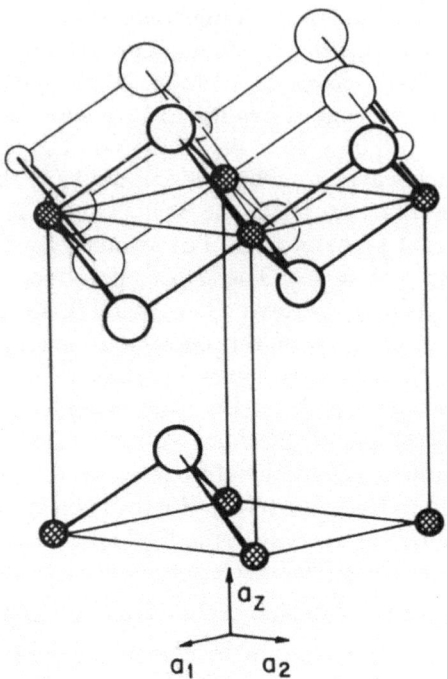

Fig. 1.   Perspective view of the atomic arrangement in the CdI_2 structure. The large circles represent the anions, the small shaded circles represent the cations.

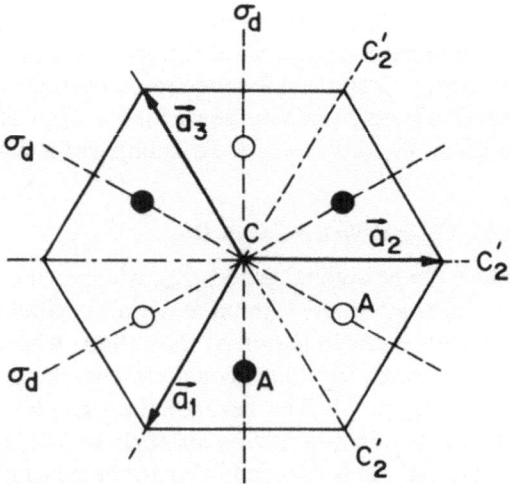

Fig. 2.   Projection of the CdI_2 crystal upon the xy-plane. The primitive translation vectors a_1, a_2, $a_3 = -(a_1 + a_2)$ and some of the symmetry operations are indicated together with the atomic positions of anions (A) and cation (C).

$\mathbf{t} = n_1\mathbf{a}_1 + n_2\mathbf{a}_2 + n_z\mathbf{a}_z$, is a symmetry operation which leaves the crystal structure unchanged. To enumerate the point group operations we conveniently choose the origin at the cation site and we find the twelve symmetry operations of D_{3d} listed in Table 1.

The last six operations of Table 1 can be obtained by multiplying the first six elements by the inversion element $\{I \mid 0\}$. The choice of the cation site as origin is convenient but not necessary. It can be shown that any symmetry operation which leaves the crystal

TABLE 1

Point group operations of D_{3d}. $\mathbf{a}_1$, $\mathbf{a}_2$, $\mathbf{a}_z$ are the basis vectors of the hexagonal unit cell, $\mathbf{a}_3 = -(\mathbf{a}_1 + \mathbf{a}_2)$ as shown in Figure 2.

Symbol	Operation
E	identity
$2C_3$	$\pm\, 2\pi/3$ rotation about $\mathbf{a}_z$
$3C_{2'}$	π rotation about $\mathbf{a}_1$, $\mathbf{a}_2$, $\mathbf{a}_3$
I	inversion
$2S_6$	$\pm\, \pi/3$ rotation about $\mathbf{a}_z$
	$+$ mirror plane $\perp \mathbf{a}_z$
$3\sigma_d$	mirror planes $\perp \mathbf{a}_1$, $\mathbf{a}_2$, $\mathbf{a}_3$ and $\parallel \mathbf{a}_z$

invariant can be written in the form $\{\alpha \mid \mathbf{t}\}$, where α denotes any of the twelve point group operations listed in Table 1. Since the space group operations $\{\alpha \mid \mathbf{t}\}$ do not contain fractional translations, the group D_{3d}^3 is symmorphic. It is interesting to note the existence of other space groups that are based on the same *point* group D_{3d}. In discussing several layer crystals we shall meet two more groups, such as D_{3d}^5, describing the $CdCl_2$ structure and D_{3d}^1 found for e.g., BiI_3. While D_{3d}^5 is connected to a rhombohedral lattice, D_{3d}^1 like D_{3d}^3 is based on a hexagonal lattice. As already mentioned D_{3d}^1 derives from the same abstract point group D_{3d} as does D_{3d}^3. The difference appears in the position of the $\{C_2 \mid 0\}$ and $\{\sigma_d \mid 0\}$-like operations with respect to the primitive translation axes $\mathbf{a}_1$, $\mathbf{a}_2$, and $\mathbf{a}_3$.

While in D_{3d}^3 the $\{C_2 \mid 0\}$-like axes lie parallel and the $\{\sigma_d \mid 0\}$-like planes perpendicular to the three vectors $\mathbf{a}_1$, $\mathbf{a}_2$, and $\mathbf{a}_3$, the situation is interchanged in D_{3d}^1. The origins for this are of course the different positions of the basis atoms in CdI_2 and BiI_3 respectively. These differences are *not* manifested in the crystal point groups, however, they may lead to different 'groups of $\mathbf{k}$' for particular high symmetry $\mathbf{k}$-points and therefore to different degeneracies of $E(\mathbf{k})$ at these points. In discussing the different groups of $\mathbf{k}$ in the following we shall mention these differences.

Let us now construct the Brillouin zone. The reciprocal lattice of the hexagonal lattice is also a hexagonal lattice, spanned by the three basis vectors $\mathbf{b}_1(\perp\mathbf{a}_2)$, $\mathbf{b}_2(\perp\mathbf{a}_1)$ and $\mathbf{b}_z(\perp\mathbf{a}_1, \mathbf{a}_2$ and $\parallel\mathbf{a}_z)$. The two vectors $\mathbf{b}_1$ and $\mathbf{b}_2$ now form a $60°$ angle. The symmetric unit cell (Brillouin zone) of this reciprocal lattice is shown in Figure 3.

There are 14 classes of special $\mathbf{k}$-points in this zone, if we define them using the operations of D_{3d}. They are listed together with the corresponding point groups $G_\mathbf{k}^0$ in Table 2. Corresponding $\mathbf{k}$-points on the top face of the Brillouin zone (i.e., $k_z = +\pi/c$) and in the middle plane (i.e., $k_z = 0$) belong to the same small point groups $G_\mathbf{k}^0$. The

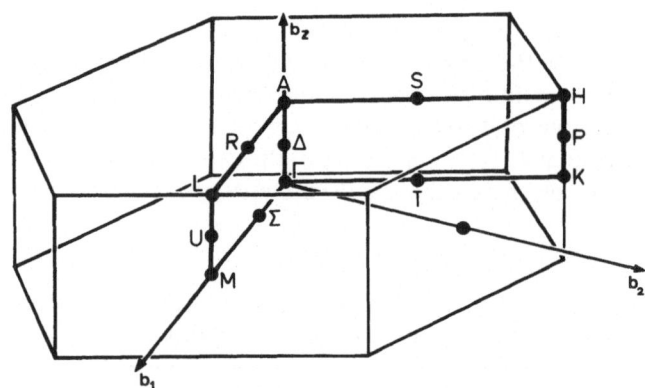

Fig. 3.   First Brillouin zone of the hexagonal lattice. The basis vectors $\mathbf{b}_1$, $\mathbf{b}_2$, $\mathbf{b}_z$ of the reciprocal lattice and high symmetry $\mathbf{k}$-points are shown.

TABLE 2

High-symmetry $\mathbf{k}$-points and $\mathbf{k}$-lines as indicated in Figure 3 and their associated point groups.

$\mathbf{k}$-points of high symmetry	Corresponding point-groups	Number of group elements	Elements of G_k^0
Γ	D_{3d}	12	E, $2C_3$, $3C_{2'}$, I, $2S_6$, $3\sigma_d$
A	D_{3d}	12	as Γ
Δ	C_{3v}	6	E, $2C_3$, $3\sigma_d$
K	D_3	6	E, $2C_3$, $3C_{2'}$
H	D_3	6	as K
L	C_{2h}	4	E, $C_{2'}$, I, σ_d
M	C_{2h}	4	as L
P	C_3	3	E, $2C_3$
Σ	C_s	2	E, σ_d
U	C_s	2	as Σ
R	C_s	2	as Σ
T	C_2	2	E, $C_{2'}$
S	C_2	2	as T
F	C_s	2	as Σ

bottom face ($k_z = -\pi/c$) is considered to belong to the adjacent Brillouin zone. More-over, the point group D_3 of K and H is isomorphic with the point group C_{3v} of Δ and the point group C_2 of T and S is isomorphic with the point group C_s of Σ, R, U and F. In the case of D_{3d}^1, Δ, K, H and P all belong to the same point group C_{3v}. Since this group is isomorphic with D_3, the only real difference between D_{3d}^3 and D_{3d}^1 appears for the line P. Inversion symmetry is present only in the small groups of Γ, A, M and L.

With the selection of these symmetry points we also have selected the smallest part of the Brillouin zone (irreducible part) whose interior contains only general $\mathbf{k}$-points and from which the entire Brillouin zone can be generated by applying to all its $\mathbf{k}$-vectors the h operations of G^0. Therefore any wavefunction $\psi_{\mathbf{k}}$ or energy $E(\mathbf{k})$ needs only to be calculated within the irreducible part; the solutions for the rest of k-space can be obtained from them by applying symmetry operations. This irreducible part in the present case of D_{3d} is the triangular prism bounded by the points A, L, H in the two planes $\mathbf{k} = \pm \mathbf{b}_z/2$ on top and bottom of the Brillouin zone. The special points are only indicated for $k_z \geq 0$ in Figure 3. The corresponding special points in the lower half of the irreducible part ($k_z < 0$) are connected to the shown special points by symmetry operations. This is *not* the case for general k-points inside the upper or lower half of the irreducible part. As a consequence all calculations which require a full k-space integration have to be carried out over the whole irreducible part, any calculation for high symmetry points only (like usual band structures) can be restricted to $k_z \geq 0$.

Obviously, the irreducible part for the present case of D_{3d} is *not* the smallest irreducible volume compatible with the full hexagonal symmetry. In the present case the irreducible part fills $\frac{1}{12}$ of the Brillouin zone according to the 12 symmetry operations of D_{3d}. If we would instead have to consider the full hexagonal symmetry group D_{6h} which has 24 operations (i.e., the 12 operations of D_{3d} plus 12 more operations obtained by multiplying the 12 operations of D_{3d} by the reflection in the $\mathbf{a}_1, \mathbf{a}_2$-plane $\{\sigma_h \mid 0\}$, the irreducible part would shrink to the upper part ($k_z \geq 0$) of the triangular prism. On the other hand, if we would have to consider the crystal point group D_3 (as e.g., for trigonal Se and Te), the irreducible part in principle would only cover $\frac{1}{6}$ of the entire Brillouin zone. In this case, however, where the inversion symmetry is missing among the space group operations, time reversal symmetry can be considered. For many purposes it acts like inversion as it e.g., always assumes $E(\mathbf{k}) = E(-\mathbf{k})$. We thus would end up in D_3 with an effective irreducible part which extends over $\frac{1}{12}$ of the Brillouin zone only. The effect of time reversal symmetry on solutions of the Schrödinger equation will be discussed in a later section in more detail.

In the following tables (Tables 3a–3e) we present the characters of the various point groups $G_\mathbf{k}^0$ of k which differ from each other [9]. We here only show the 'single-valued' representations which are valid only if spin-orbit interaction is neglected. The effect of the inclusion of spin will be discussed in a later chapter. In a separate column we list a letter describing the behaviour under time inversion, this also will be discussed in a later chapter. By looking at the character tables, we see from $X(E) = d_j$ that only Γ and A with Γ_3^+, Γ_3^- (A_3^+, A_3^-), Δ, K and H with $\Delta_3, (K_3, H_3)$ have degenerate representations of order $d_j = 2$. There is, however, one more concealed degeneracy: for the line P the

TABLE 3a

Character table for the group D_{3d}.

Γ, A	E	$2S_6$	$2C_3$	I	$3\sigma_d$	$3C_{2'}$	Time inversion
Γ_1^+	1	1	1	1	1	1	a
Γ_2^+	1	1	1	1	-1	-1	a
Γ_3^+	2	-1	-1	2	0	0	a
Γ_1^-	1	-1	1	-1	-1	1	a
Γ_2^-	1	-1	1	-1	1	-1	a
Γ_3^-	2	1	-1	-2	0	0	a

TABLE 3b

Character tables for the groups C_{3v} and D_3.

Δ	C_{3v}			
	E	$2C_3$	$3\sigma_d$	
	D_3			Time inversion
K, H	E	$2C_3$	$3C_{2'}$	
$\Delta_1\ K_1$	1	1	1	a
$\Delta_2\ K_2$	1	1	-1	a
$\Delta_3\ K_3$	2	-1	0	a

TABLE 3c

Character table for the group C_{2h}.

M, L	C_{2h}				Time inversion
	E	I	σ_d	$C_{2'}$	
M_1^+	1	1	1	1	a
M_2^+	1	1	-1	-1	a
M_1^-	1	-1	-1	1	a
M_2^-	1	-1	1	-1	a

TABLE 3d

Character table for the group C_3, ω stands for $\exp(i\pi/3)$.

P	C_3			Time inversion
	E	C_3	C_3^{-1}	
P_1	1	1	1	a
P_2	1	ω^2	$-\omega$	b
P_3	1	$-\omega$	ω^2	b

TABLE 3e

Character tables for the groups C_2 and C_s.

	C_2		
T, S	E	$C_{2'}$	
	C_s		Time inversion
U, Σ, R, F	E	σ_d	
$T_1\ U_1$	1	1	a
$T_2\ U_2$	1	-1	a

two one-dimensional representations P_2 and P_3 of C_3 are degenerate by time reversal symmetry.

All other representations are one-dimensional, thus indicating that the energy $E(\mathbf{k})$ in CdI_2-type crystals will not be degenerate by symmetry at $\mathbf{k}$-points except at the points and lines mentioned above.

We shall now discuss the effect on the energy $E(\mathbf{k})$ of going from one $\mathbf{k}$-point of high symmetry to one of lower symmetry and vice versa. Let us take as an example the line Σ. All states $\psi_\mathbf{k}(\mathbf{r})$ with $\mathbf{k}$-vectors ending on this line have according to Table 3e to transform either like Σ_1 or like Σ_2 depending on whether the function is even or odd with respect to the vertical reflection plane σ_d. Approaching the point Γ these symmetries do not change and it is reasonable that at Γ itself the eigenfunctions have the same symmetry under σ_d. By inspecting the Tables 3a and 3e we thus would conclude that Σ_1 either turns into a state Γ_1^+ or into a state Γ_2^-. Correspondingly Σ_2 turns either into Γ_2^+ or into Γ_1^-. Finally Σ_1 and Σ_2 can merge and turn into one of the two two-fold degenerate states Γ_3^+ or Γ_3^-. Without giving a rigorous proof of this hypothesis we here present the recipe how to obtain the so-called compatibility-relations which connect neighboring $\mathbf{k}$-points of different degrees of symmetry.

In going from a point of high symmetry to another of lower symmetry, we go from one group of $\mathbf{k}$ to another group of $\mathbf{k}'$ which is a subgroup of the first. This corresponds to a loss of symmetry; the correct compatibility relations are obtained by inspecting the character tables of the two groups of $\mathbf{k}$ and $\mathbf{k}'$ and by fulfilling the relationship

$$\sum_i \alpha_{ij} \Gamma^{(i)}(\mathbf{k}') = \Gamma^{(j)}(\mathbf{k}) \tag{A22}$$

for each irreducible representation $\Gamma^{(j)}(\mathbf{k})$. We thus find for the example of the space group D_{3d}^3 the relations summarized in Table 4. These compatibility relations are of great help in establishing a 'continuous' band structure $E(\mathbf{k})$ from only a small number of $\mathbf{k}$-points.

We now turn to the case where we perturb the system (which is invariant under the space group D_{3d}^3) by some external probe. This perturbation will cause transitions between the original eigenstates of the unperturbed system; the transitions are described by matrix elements

$$M_{fi} = \langle \psi_f | H_1 | \psi_i \rangle \tag{A23}$$

TABLE 4

Compatibility relations for various symmetry points in the Brillouin zone for a crystal with D_{3d}^3 space group symmetry.

$\Gamma(A)$	Γ_1^+	Γ_2^+	Γ_3^+	Γ_1^-	Γ_2^-	Γ_3^-
$\Sigma(R)$	Σ_1	Σ_2	$\Sigma_1 + \Sigma_2$	Σ_2	Σ_1	$\Sigma_1 + \Sigma_2$
$T(S)$	T_1	T_2	$T_1 + T_2$	T_2	T_1	$T_1 + T_2$
Δ	Δ_1	Δ_2	Δ_3	Δ_2	Δ_1	Δ_3

$M(L)$	M_1^+	M_2^+	M_1^-	M_2^-
$\Sigma(R)$	Σ_1	Σ_2	Σ_2	Σ_1
U	U_1	U_2	U_2	U_1

$K(H)$	K_1	K_2	K_3
$T(S)$	T_1	T_2	$T_1 + T_2$
P	P_1	P_1	$P_2 + P_3$

between initial and final states. As pointed out in Section 1.2 these transition matrix-elements may vanish because of the symmetry properties of ψ_i, ψ_f, and H_1. The statement we formulated was that M_{fi} must vanish by symmetry unless the direct product representation $\Gamma^{(f)*}\Gamma^{(H_1)}\Gamma^{(i)}$ includes the identical representation $\Gamma^{(1)}$. If ψ_i, ψ_f, and H_1 are described by different groups, the analysis has to be done in the largest subgroup contained in the three groups. Let us consider the most common example of external perturbation in solid state spectroscopy: to shine light onto a crystal. The interaction between the solid and the electromagnetic field is in lowest order described by its strongest component: the interaction of electric dipoles in the solid with the electric field. The matrix elements determining the selection rules are therefore $M_{fi} = \langle \psi_f | \mathbf{r} | \psi_i \rangle$.

Because of the uniaxial symmetry of D_{3d}^3 the selection rules for dipole transitions are different for light polarized parallel ($\mathbf{r} = (0, 0, z)$) and perpendicular ($\mathbf{r} = (x, y, 0)$) to the crystal c-axis. Their derivation according to Equation (A10) is straightforward and the results for transitions between states $\psi_\mathbf{k}^n(\mathbf{r})$ and $\psi_\mathbf{k}^{n'}(\mathbf{r})$ at various high symmetry $\mathbf{k}$-points are summarized in Table 5. The inclusion of spin-orbit coupling which alters these selection rules will be discussed later. An interesting result appears for the selection rules at the line P. The representations P_2 and P_3 transform like $(x - iy)$ and $(x + iy)$ respectively. The transition probabilities therefore do not depend on the direction of the linear polarization of the incident light, but rather on the sense of its circular polarization. Since P_2 and P_3 are degenerate by time reversal symmetry this effect is unobservable unless the degeneracy is possibly lifted by a perturbation which is not time reversible like a magnetic field. This example shall end our discussion on symmorphic crystal space groups and we now turn to the more involved (but frequent) case of *non*-symmorphic space groups. Again, we shall choose a convenient example to illustrate the various rules describing non-symmorphic space groups.

TABLE 5

Single group selection rules for optical dipole transitions in a crystal with D_{3d}^3 space group symmetry.

Γ, A	$\Gamma_1^\pm$	$\Gamma_2^\pm$	$\Gamma_3^\pm$	P	P_1	P_2	P_3
z	$\Gamma_2^\mp$	$\Gamma_1^\mp$	$\Gamma_3^\mp$	z	P_1	P_2	P_3
x, y	$\Gamma_3^\mp$	$\Gamma_3^\mp$	$\Gamma_1^\mp, \Gamma_2^\pm, \Gamma_3^\mp$	$x + iy$	P_2	P_3	P_1
				$x - iy$	P_3	P_1	P_2

M, L	$M_1^\pm$	$M_2^\pm$
z, x	$M_2^\mp$	$M_1^\mp$
y	$M_1^\mp$	$M_2^\mp$

K, H	K_1	K_2	K_3	Σ, U, R, F	Σ_1	Σ_2
z	K_2	K_1	K_3	z, x	Σ_1	Σ_2
x, y	K_3	K_3	K_1, K_2, K_3	y	Σ_2	Σ_1

Δ	Δ_1	Δ_2	Δ_3	T, S	T_1	T_2
z	Δ_1	Δ_2	Δ_3	z, x	T_2	T_1
x, y	Δ_3	Δ_3	$\Delta_1, \Delta_2, \Delta_3$	y	T_1	T_2

1.3.3. *Non-Symmorphic Space Groups* [10, 11]

If a crystal is left invariant under an ensemble of operations $\{\alpha \mid \tau_\alpha + \mathbf{t}\}$ and if no origin exists such that $\tau_\alpha \equiv 0$ for all elements, the space group defining the symmetry properties of the crystal is called *non*-symmorphic. Before we discuss possibilities of how to construct irreducible representations for *non*-symmorphic space groups, let us recall the corresponding procedure we applied for symmorphic space groups. There the essential fact was that the factor group G/T was isomorphic with the point group G^0. Moreover, this also was true for a particular high symmetry point $\mathbf{k}$: the factor group $G_{\mathbf{k}}/T$ was isomorphic with the point group $G_{\mathbf{k}}^0$ which was a subgroup of $G_{\mathbf{k}}$. We then found that it was sufficient to construct Bloch functions which transformed according to an irreducible representation of $G_{\mathbf{k}}^0$.

In the case of *non*-symmorphic groups $G_{\mathbf{k}}/T$ is still isomorphic with $G_{\mathbf{k}}^0$, but in contrast to the symmorphic case the elements of $G_{\mathbf{k}}^0$ which are all of the form $\{\alpha \mid 0\}$ are *not* all elements of $G_{\mathbf{k}}$ or G. In other words the ensemble of elements $\{\alpha \mid \tau_\alpha\}$ does *not* form a subgroup of $G_{\mathbf{k}}$. This can be seen from the simple fact that the product of two such elements $\{\alpha_1 \mid \tau_{\alpha_1}\} \cdot \{\alpha_2 \mid \tau_{\alpha_2}\}$ might give the result $\{\varepsilon \mid \mathbf{t}\}$ which itself is *not* among the set $\{\alpha \mid \tau_\alpha\}$. We see that the ensemble $\{\alpha \mid \tau_\alpha\}$ does not form a group by itself.

We can get out of this dilemma by enlarging the factor group $G_{\mathbf{k}}/T$ to a new factor group $G_{\mathbf{k}}/T_{\mathbf{k}}$ where $T_{\mathbf{k}}$ is an invariant subgroup of G which includes only certain translations $\{\varepsilon \mid \mathbf{t_k}\}$ such that $e^{i\mathbf{k} \cdot \mathbf{t_k}} = 1$ or equivalently $\mathbf{k} \cdot \mathbf{t_k} = 2\pi n$. The (abstract) elements of the new factor group G_k/T_k are now 'complexes' of elements $\{\alpha \mid \tau_\alpha + t\}$ either with t in T_k or with t not in T_k.

For $k = 0$, $T_k = T$ and G_k/T_k is isomorphic with G_k^0. The irreducible representations can be obtained from G^0 as in the case of symmorphic space groups according to Equation (A21).

If k is a general vector ending inside the Brillouin zone, G_k consist only of the translation group and the construction of its representations is trivial.

If k ends inside the Brillouin zone, but lies on an axis of symmetry or in a plane of symmetry, G_k contains in addition rotations and/or reflections. Again G_k/T_k is isomorphic with G_k^0 and the irreducible representations can be obtained accordingly.

The only k-vectors for which the construction of character tables and irreducible representations can be non-trivial are vectors terminating on the boundary of the Brillouin zone. The procedure there is to form the factor group G_k/T_k and to find its character table and its irreducible representations in the usual way, using the relation between class multiplication and character multiplication. The character table thus obtained will be square, i.e., the number of irreducible representations is equal to the number of classes in G_k/T_k. There is, however, one problem with this method: some of the irreducible representations of G_k/T_k may not be appropriate to classify the Bloch states for the given particular k-vector.

The reason for this artifact can be seen as follows: The definition of T was given by the requirement that for each $\{\varepsilon \mid t\}$, $\exp(ik \cdot t_k) = 1$. This relation, however, also implies $(\exp ik \cdot t_k)^n = 1$ or $\exp(i(nk) \cdot t_k) = 1$. Thus the subgroup T_k not only belongs to k but also to nk, where n is any integer. Consequently only some of the irreducible representations of G_k/T_k can be used to label k, others may belong to nk. To find out the representations which are appropriate to k we consider the characters of the pure translations. In irreducible representations of G_k appropriate to a wave function with k the character of the lattice translations $\{\varepsilon \mid t\}$ is equal to $e^{ik \cdot t}$ times the dimension of the irreducible representations and *not* to $e^{ink \cdot t}$ times the dimension. In the latter case the representation would be appropriate for $k' = nk$. In general there is some small integer n_0 for which $n_0 k$ differs from k by a reciprocal lattice vector so that only a finite number of different k-vectors k' can occur. This of course is consistent with the fact that the dimension of G_k/T_k is finite. Considering only the irreducible representations appropriate for k, the corresponding character table may *not* be square any more.

We are now in a position to construct the representations of *non*-symmorphic space groups. To obtain a particular representation of G we consider a set of Bloch functions ψ_k which transform under all operations of G_k according to an irreducible representation of its factor group G_k/T_k. (This representation has to be appropriate to k in the sense of the preceding discussion.)

We may then generate from this set of ψ_k new functions by using operations of G which are *not* in G_k. These functions complete the star of k and form an irreducible representation of G. The *essential* degeneracy of this representation, however, is equal to the dimension of the irreducible representation of G_k/T_k we started with.

Before illustrating these ideas by a particular example we shall note that a different procedure to construct irreducible representations of *non*-symmorphic space groups has been developed by Zak [12] and Streitwolf [13]. We refer the reader to the literature for a discussion of this procedure.

1.3.4. *The Non-Symmorphic Space Group D_{6h}^4* [14]

The space group D_{6h}^4 is known to be the symmetry group of hexagonal close packed structures. Its associated point group D_{6h} has the full symmetry of the hexagonal Bravais lattice. In addition to the hexagonal close packed structures a number of very 'open-packed' layer structures like β-GaSe, GaS, or 2H–MoS_2, $MoSe_2$ have D_{6h}^4 symmetry.

We now discuss the space group D_{6h}^4 and its associated groups of **k** by considering a crystal of β-GaSe. In contrast to the CdI_2 structure in which the anions were found in an octahedral coordination, here the anions sit in a trigonal prismatic arrangement, i.e., the two anion sheets on the outside of each layer sit on top of each other. The hexagonal unit cell of β-GaSe which extends over two layers is shown in Figure 4. It contains four anions and four cations. A projection of the structure on the $\mathbf{a}_1$, $\mathbf{a}_2$ basal plane together with the location of some rotation axes and reflection planes is presented in Figure 5.

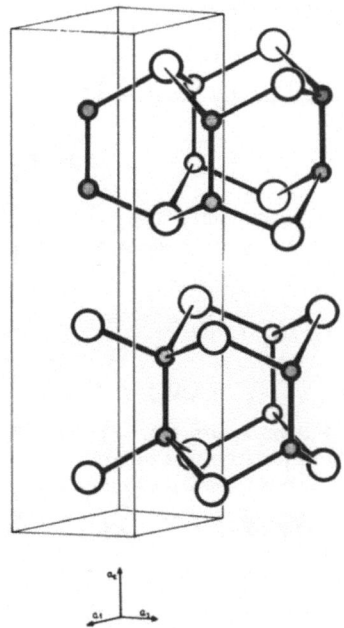

Fig. 4.   Unit cell of β-GaSe. The Ga atoms are represented by the small shaded circles, the Se atoms by the large open circles.

To enumerate the group operations we conveniently choose the origin half way between two layers. A set of 24 symmetry operations which leave the structure unchanged is listed in Table 6. The non-primitive translation $\tau = \frac{1}{2}\mathbf{a}_z$. The 24 operations are obtained by ignoring all primitive lattice translations in the various space group elements. The 24 operations do *not* form a group by themselves as one can see by forming the product $\{C_2 \mid \tau\}\{C_2 \mid \tau\} = \{\varepsilon \mid \mathbf{a}_z\}$ where $\{\varepsilon \mid \mathbf{a}_z\}$ is an operation of the space group D_{6h}^4 but *not* among the 24 operations we listed.

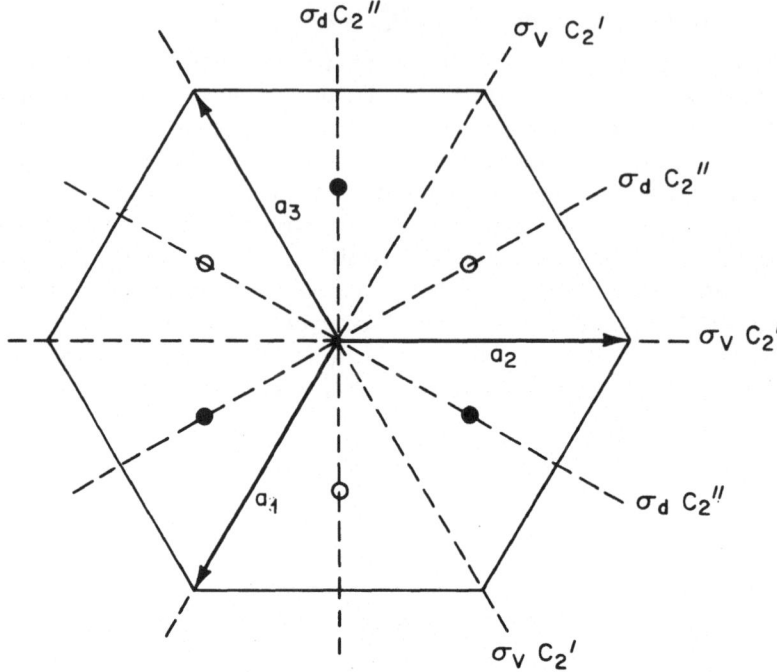

Fig. 5. Projection of the β-GaSe crystal upon the xy plane. The primitive translation vectors $\mathbf{a}_1$, $\mathbf{a}_2$, $\mathbf{a}_3 = -(\mathbf{a}_1 + \mathbf{a}_2)$ and some of the symmetry operations are indicated.

TABLE 6

Point group operations of D_{6h}. $\mathbf{a}_1$, $\mathbf{a}_2$, $\mathbf{a}_z$ are the basis vectors of the hexagonal unit cell. $\mathbf{a}_3 = -(\mathbf{a}_1 + \mathbf{a}_2)$ as shown in Figure 5. The associated non-primitive translations τ for the space group elements of D_{6h} are indicated.

Number	Symbol	Non-primitive translation	Operation
1	E	–	identity
2, 3	$2C_6$	τ	$\pm \pi/3$ rotation about $\mathbf{a}_z$
4, 5	$2C_3$	–	$\pm 2\pi/3$ rotation about $\mathbf{a}_z$
6	C_2	τ	π rotation about $\mathbf{a}_z$
7, 8, 9	$3C_{2'}$	–	π rotation about $\mathbf{a}_1$, $\mathbf{a}_2$, $\mathbf{a}_3$
10, 11, 12	$3C_{2''}$	τ	π rotation about axes $\perp$ to $\mathbf{a}_1$, $\mathbf{a}_2$, $\mathbf{a}_3$
13	I	–	inversion
14, 15	$2S_3$	τ	$\pm 2\pi/3$ rotation about $\mathbf{a}_z$ followed by a mirror plane $\perp \mathbf{a}_z$
16, 17	$2S_6$	–	$\pm \pi/3$ rotation about $\mathbf{a}_z$ followed by a mirror plane $\perp \mathbf{a}_z$
18	σ_h	τ	mirror plane $\perp \mathbf{a}_z$
19, 20, 21	$3\sigma_d$	–	mirror plane $\perp \mathbf{a}_1$, $\mathbf{a}_2$, $\mathbf{a}_3$ and $\parallel \mathbf{a}_z$
22, 23, 24	$3\sigma_v$	τ	mirror plane $\parallel \mathbf{a}_1$, $\mathbf{a}_2$, $\mathbf{a}_3$ and $\parallel \mathbf{a}_z$

The last 12 operations in Table 6 can be obtained by multiplying the first 12 elements by the inversion element $\{I \mid 0\}$. The 12 operations which are *not* associated with a non-primitive translation τ are those found for the group D_{3d}^3 of the CdI$_2$ structure (see example for symmorphic space groups). The first Brillouin zone of the hexagonal lattice with the various high symmetry points has already been shown in Figure 3. The 17 different types of special **k**-points together with the corresponding point groups $G_{\mathbf{k}}^0$ (which are *not* necessarily subgroups of $G_{\mathbf{k}}$) are listed in Table 7. As in the case of the

TABLE 7

High symmetry **k**-points and **k**-lines as indicated in Figure 3 and their associated point groups.

k-points of high symmetry	Corresponding point groups	Number of group elements	Elements of $G_{\mathbf{k}}^0$
Γ	D_{6h}	24	$E, 2C_6, 2C_3, C_2, 3C_{2'}, 3C_{2''}, I, 2S_3, 2S_6, \sigma_h, 3\sigma_d, 3\sigma_v$
A	D_{6h}	24	as Γ
Δ	C_{6v}	12	$E, 2C_6, 2C_3, C_2, 3\sigma_d, 3\sigma_v$
K	D_{3h}	12	$E, 2C_3, 3C_{2'}, \sigma_h, 2S_3, 3\sigma_v$
H	D_{3h}	12	as K
L	D_{2h}	8	$E, C_2, C_{2'}, C_{2''}, I, \sigma_h, \sigma_d, \sigma_v$
M	D_{2h}	8	as L
P	C_{3v}	6	$E, 2C_3, 3\sigma_v$
Σ	C_{2v}	4	$E, C_{2''}, \sigma_h, \sigma_d$
U	C_{2v}	4	$E, C_2, \sigma_d, \sigma_v$
R	C_{2v}	4	as Σ
T	C_{2v}	4	$E, C_{2'}, \sigma_h, \sigma_v$
S	C_{2v}	4	as T
F	C_s	2	E, σ_d
B	C_s	2	E, σ_v
N	C_s	2	E, σ_h
G	C_s	2	as N

symmorphic group D_{3d}^3, corresponding **k**-points in the top-face and in the middle plane of the Brillouin zone belong to the same point groups $G_{\mathbf{k}}^0$. However, in contrast to D_{3d}^3 their factor groups $G_{\mathbf{k}}/T_{\mathbf{k}}$ are in general different.

As already pointed out in the discussion of the group D_{3d}^3, the irreducible part of the Brillouin zone for D_{6h}^4 is a triangular prism with $k_z \geq 0$ which extends over $\frac{1}{24}$ of the Brillouin zone.

We now construct the character tables for the various factor groups $G_{\mathbf{k}}/T_{\mathbf{k}}$ which differ from each other. Let us begin with **k**-points inside the Brillouin zone. No particular difficulties arise there; the various space group representations $D(\{\alpha \mid \tau + \mathbf{t}\})$ can be obtained from the corresponding point group representations $\Gamma(\{\alpha \mid 0\})$ by

$$D(\{\alpha \mid \tau + \mathbf{t}\}) = e^{i\mathbf{k}\cdot\tau} e^{i\mathbf{k}\cdot\mathbf{t}} \Gamma(\{\alpha \mid 0\}). \tag{A24}$$

The same simple relation holds for surface **k**-points which do *not* lie on the top surface of the Brillouin zone. The reason for this simplification is as follows: for surface **k**-points with $k_z \neq 0$ like P or U, $T_{\mathbf{k}} = T$ and the factor group $G_{\mathbf{k}}/T_{\mathbf{k}}$ is isomorphic with

$G_{\mathbf{k}}^{0}$; for surface **k**-point with $k_z = 0$ like K or M, no additional phase factors due to non-primitive translations can arise, since $\mathbf{k} \cdot \boldsymbol{\tau} = 0$, and the space group representations can be simply obtained from the point group representations.

In the following tables we shall include, whenever present, the phase factors $e^{i\mathbf{k}\cdot\boldsymbol{\tau}}$ arising from the non-primitive translations $\boldsymbol{\tau}$ but no phase factors arising from the primitive lattice translations **t**. The latter are automatically obtained by using the appropriate Bloch functions. Also, in enumerating the factor group elements we omitted the primitive lattice translation for simplicity.

The factor group $G_{\mathbf{k}}/T_{\mathbf{k}}$ for the point $\Gamma(\mathbf{k} = 0)$ contains 24 elements. There are 8 one-dimensional and 4 two-dimensional representations. The superscripts $\pm$ refer to the parity with respect to inversion (Table 8a). A classification into cases (a), (b), and (c)

TABLE 8a

Character table for the group D_{6h} which is isomorphic with the factor group $G_{\mathbf{k}}/T_{\mathbf{k}}$ for $\mathbf{k} = 0$ (Γ). The associated non-primitive translations $\boldsymbol{\tau}$ are indicated. Since $\mathbf{k} = 0$ they do not result in additional phase factors.

Γ	E	C_2	$2C_3$	$2C_6$	$3C_{2'}$	$3C_{2''}$	I	σ_h	$2S_6$	$2S_3$	$3\sigma_d$	$3\sigma_v$	Time inversion
		τ		τ		τ		τ		τ		τ	
Γ_1^+	1	1	1	1	1	1	1	1	1	1	1	1	a
Γ_2^+	1	1	1	1	-1	-1	1	1	1	1	-1	-1	a
Γ_3^+	1	-1	1	-1	1	-1	1	-1	1	-1	1	-1	a
Γ_4^+	1	-1	1	-1	-1	1	1	-1	1	-1	-1	1	a
Γ_5^+	2	2	-1	-1	0	0	2	2	-1	-1	0	0	a
Γ_6^+	2	-2	-1	1	0	0	2	-2	-1	1	0	0	a
Γ_1^-	1	1	1	1	1	1	-1	-1	-1	-1	-1	-1	a
Γ_2^-	1	1	1	1	-1	-1	-1	-1	-1	-1	1	1	a
Γ_3^-	1	-1	1	-1	1	-1	-1	1	-1	1	-1	1	a
Γ_4^-	1	-1	1	-1	-1	1	-1	1	-1	1	1	-1	a
Γ_5^-	2	2	-1	-1	0	0	-2	-2	1	1	0	0	a
Γ_6^-	2	-2	-1	1	0	0	-2	2	1	-1	0	0	a

according to the effect of time reversal symmetry which will be discussed later is also indicated in the following tables.

For the line Δ, which connects Γ and A the point group is C_{6v}. Only those twelve operations of D_{6h} which leave $\mathbf{a}_z$ unchanged are symmetry operations of C_{6v}. The characters have to be multiplied by the phase factors $\alpha = e^{i\mathbf{k}\cdot\boldsymbol{\tau}} = e^{i\pi k_z}$ for those operations which contain $\boldsymbol{\tau}$. For K the corresponding point group is D_{3h} which is isomorphic to C_{6v}. No phase factors arise in this case since $k_z = 0$ (Table 8b).

The point M is left invariant under eight operations including inversion. The corresponding 8 one-dimensional representations of D_{2h} are therefore labeled by superscripts $\pm$ (Table 8c).

For the line P which connects K and H, only those six operations of C_{6v} are symmetry operations which represent the *three*-fold symmetry axis. The point group is C_{3v} and has 2 one-dimensional and 1 two-dimensional representations (Table 8d).

For Σ, U, and T the point groups are C_{2v} though they contain different sets of opera-

TABLE 8b

Character table for the groups C_{6v} (Δ) and D_{3h} (K) which are isomorphic with the corresponding factor groups $G_{\mathbf{k}}/T_{\mathbf{k}}$. The associated non-primitive translations τ are indicated. For the line $\Delta \mathbf{k} = (0, 0, k_z)$ all characters of operations containing τ have to be multiplied by the phase factor $\alpha = e^{i\pi k_z}$. For K, $k_z = 0$ and no phase factors arise.

Δ	E	C_2	$2C_3$	$2C_6$	$3\sigma_d$	$3\sigma_v$	
		τ		τ		τ	
K	E	σ_h	$2C_3$	$2S_3$	$3C_{2'}$	$3\sigma_v$	Time
		τ		τ		τ	inversion
Δ_1	1	1	1	1	1	1	a
Δ_2	1	-1	1	-1	1	-1	a
Δ_3	1	1	1	1	-1	-1	a
Δ_4	1	-1	1	-1	-1	1	a
Δ_5	2	2	-1	-1	0	0	a
Δ_6	2	-2	-1	1	0	0	a

TABLE 8c

Character table for the group D_{2h} which is isomorphic with the factor group $G_{\mathbf{k}}/T_{\mathbf{k}}$ at point M. The associate non-primitive translations τ are indicated. Since $k_z = 0$ no additional phase factors arise.

M	E	C_2	$C_{2'}$	$C_{2''}$	I	σ_h	σ_d	σ_v	Time
		τ		τ		τ		τ	inversion
M_1^+	1	1	1	1	1	1	1	1	a
M_2^+	1	-1	-1	1	1	-1	-1	1	a
M_3^+	1	-1	1	-1	1	-1	1	-1	a
M_4^+	1	1	-1	-1	1	1	-1	-1	a
M_1^-	1	1	1	1	-1	-1	-1	-1	a
M_2^-	1	-1	-1	1	-1	1	1	-1	a
M_3^-	1	-1	1	-1	-1	1	-1	1	a
M_4^-	1	1	-1	-1	-1	-1	1	1	a

TABLE 8d

Character table for the group C_{3v} which is isomorphic with the factor group $G_{\mathbf{k}}/T_{\mathbf{k}}$ for the line P. The associated non-primitive translations τ are indicated. For the line P $\mathbf{k} = (\frac{1}{3}, \frac{1}{3}, k_z)$ and all characters of operations containing τ have to be multiplied by the phase factor $\alpha = e^{i\pi k_z}$.

P	E	$2C_3$	$3\sigma_v$	Time
			τ	inversion
P_1	1	1	1	a
P_2	1	1	-1	a
P_3	2	-1	0	a

TABLE 8e

Character table for the group C_{2v} which is isomorphic with the factor groups $G_{\mathbf{k}}/T_{\mathbf{k}}$ of the lines Σ, T and U. The associated non-primitive translations τ are indicated. For the line U $\mathbf{k} = (\frac{1}{2}, 0, k_z)$ and the characters of operations containing τ have to be multiplied by the phase factor $\alpha = e^{i\pi k_z}$. No additional phase factors arise for the lines Σ and T.

Σ	E	σ_d	C_2''	σ_h	
T	E	σ_v τ	$C_{2'}$	σ_h τ	
U	E	σ_v τ	σ_d	C_2 τ	Time inversion
Σ_1	1	1	1	1	a
Σ_2	1	-1	1	-1	a
Σ_3	1	1	-1	-1	a
Σ_4	1	-1	-1	1	a

tions. For U the characters have to be multiplied by the phase factors $\alpha = e^{i\pi k_z}$ for operations containing τ (Table 8e).

For F, B and N the point groups are C_s. Only in the case of B, the characters of $\{\sigma_v \mid \tau + \mathbf{t}\}$ have to be multiplied by the phase factor α (Table 8f).

TABLE 8f

Character table for the group C_s which is isomorphic with the factor groups $G_{\mathbf{k}}/T_{\mathbf{k}}$ for the planes F, B and N. The associated non-primitive translations τ are indicated. For the plane B $\mathbf{k} = (k_1, k_1, k_z)$ the characters of the operations $\{\sigma_v \mid \tau + \mathbf{t}\}$ have to be multiplied by the phase factor $\alpha = e^{i\pi k_z}$. No additional phase factors arise for the planes F and N.

F	E	σ_d	
B	E	σ_v τ	
N	E	σ_h τ	Time inversion
F_1	1	1	a
F_2	1	-1	a

The remaining $\mathbf{k}$-points A, H, L, R, S and G lie on the top face of the Brillouin zone ($k_z = \pi/c$) and the enlarged factor groups $G_{\mathbf{k}}/T_{\mathbf{k}}$ which contain twice as many elements as the corresponding point groups $G_{\mathbf{k}}^0$ have to be considered. Since $k_z = \pi/c$ for all these $\mathbf{k}$-points the set of translation vectors $T_{\mathbf{k}}$ contains only vectors with *even* z-components t_z and different first and second components according to the $\mathbf{k}$-point. This set which we shall call $\{\mathbf{t}_{even}\}$ is complimented by the set $\{\mathbf{t}_{odd}\}$ to span the entire translation group T.

The different factor group elements are then formed by the various 'complexes' $\{\alpha \mid \tau_\alpha + t_{even}\}$ and $\{\alpha \mid \tau_\alpha + t_{odd}\}$ with the appropriate point group transformations α and with $\tau_\alpha = 0$ or τ.

As already mentioned these enlarged factor groups G_k/T_k contain representations appropriate for states with wave vectors $\mathbf{k}$, $2\mathbf{k}$, $3\mathbf{k}$ etc. Since originally $k_z = \pi/c$ for all these points, the additional new points $\mathbf{k}' = 2n\mathbf{k}$ lie in the plane $k_z = 0$ and the points $\mathbf{k}' = (2n + 1)\mathbf{k}$ again have $k_z = \pi/c$ and therefore are with respect to their symmetry behavior equivalent to the original points $\mathbf{k}$. Thus each factor group contains some representations appropriate for $\mathbf{k}$ and some representations appropriate for $\mathbf{k}' = 2n\mathbf{k}$.

These 'pairs' of symmetry points, lines or planes are the following: (A, Γ), (H, K), (L, M), (S, T), (R, Σ) and (G, N). Let us now explicitly construct the enlarged factor group G_k/T_k for the line S which contains representations appropriate for both lines S and T.

The four symmetry elements for S (see Table 7) are $\{\varepsilon \mid 0\}$, $\{C_{2'} \mid 0\}$, $\{\sigma_h \mid \tau\}$, $\{\sigma_v \mid \tau\}$. From these elements we construct the following eight 'complexes' which span the factor group G_k/T_k for S and which can be decomposed into five classes:

$$
\begin{aligned}
&C_1 \quad \{\varepsilon \mid t_{even}\} \\
&C_2 \quad \{\varepsilon \mid t_{odd}\} \\
&C_3 \quad \{C_{2'} \mid t_{even}\}, \{C_{2'} \mid t_{odd}\} \qquad\qquad\qquad (A25) \\
&C_4 \quad \{\sigma_h \mid \tau + t_{even}\}, \{\sigma_h \mid \tau + t_{odd}\} \\
&C_5 \quad \{\sigma_v \mid \tau + t_{even}\}, \{\sigma_v \mid \tau + t_{odd}\}.
\end{aligned}
$$

The abstract group of eight elements which form five classes is either D_4, C_{4v} or D_{2d} which are all isomorphic with each other. The characters of this group are given in Table 9. There are 4 one-dimensional representations and 1 two-dimensional representation. To find out, which representations are appropriate to $\mathbf{k}$ and which of them are appropriate to $\mathbf{k}'$ we consider the characters of the pure translations $\{\varepsilon \mid t_{even}\}$ and $\{\varepsilon \mid t_{odd}\}$ which have to be equal to $e^{i\mathbf{k}\cdot\tau}$ or to $e^{i\mathbf{k}'\cdot\tau}$ times the dimension of the representation. With $\{t_{even}\} = \{2n\mathbf{a}_z\}$ and $\{t_{odd}\} = \{(2n + 1)\mathbf{a}_z\}$ we find for S, $e^{i\mathbf{k}\cdot t_{even}} = 1$ and $e^{i\mathbf{k}'\cdot t_{odd}} = -1$ and for T, $e^{i\mathbf{k}\cdot t_{even}} = e^{i\mathbf{k}'\cdot t_{odd}} = 1$. Thus only the two-dimensional representation S_1 is appropriate for S; the remaining 4 one-dimensional representations

TABLE 9

Characters for the factor group G_k/T_k of the line S. The non-primitive translations τ and the two sets of primitive lattice translations $\{t_{even}\}$ and $\{t_{odd}\}$ are indicated. S_1 is an appropriate representation for S while T_1 to T_4 are appropriate representations for T.

S, T	E t_{even}	E t_{odd}	$2C_{2'}$ $t_{even,\,odd}$	$2\sigma_h$ $\tau + t_{even,\,odd}$	$2\sigma_v$ $\tau + t_{even,\,odd}$	Time inversion
S_1	2	-2	0	0	0	a
T_1	1	1	1	1	1	a
T_2	1	1	1	-1	-1	a
T_3	1	1	-1	-1	1	a
T_4	1	1	-1	1	-1	a

T_1 to T_4 are appropriate for T and are identical to those given in Table 8e, where we omitted the redundant distinction between t_{even} and t_{odd}. By similar analysis the results for the 'pairs' (A, Γ), (H, K), (L, M), (R, Σ) and (G, N) can be obtained straightforwardly. We shall present only the final character tables retaining only those representations which are appropriate for A, H, L, R and G.

The factor group G_k/T_k for A contains forty-eight elements which form fifteen classes. There are 2 two-dimensional and 1 four-dimensional representations appropriate for A and there are 8 one-dimensional and 4 two-dimensional representations appropriate for Γ. The twelve representations appropriate for Γ are of course identical with those given in Table 8a. In Table 10a we list the characters of the three representations which are

TABLE 10a

Characters for representations of the factor group G_k/T_k which are appropriate for A. The non-primitive translations τ are indicated. The phase factors ± 1 arising from 'even' or 'odd' primitive translations are omitted.

A	E	C_2 τ	$2C_3$	$2C_6$ τ	$3C_{2'}$	$3C_{2''}$ τ	I	σ_h τ	$2S_6$	$2S_3$ τ	$3\sigma_d$	$3\sigma_v$ τ	Time inversion
A_1	2	0	2	0	0	0	0	0	0	0	2	0	a
A_2	2	0	2	0	0	0	0	0	0	0	-2	0	a
A_3	4	0	-2	0	0	0	0	0	0	0	0	0	a

appropriate for A. The phase factors ± 1 arising from $e^{ik \cdot t_{even}}$ or $e^{ik \cdot t_{odd}}$ have been omitted which leaves twelve distinct classes.

The factor group G_k/T_k for H contains twenty-four elements which form nine classes. There are 3 two-dimensional representations appropriate for H and 2 two-dimensional and 4 one-dimensional representations appropriate for K which are identical with those given in Table 8b. In Table 10b we present the characters of the three representations appropriate for H. Omitting the phase factors ± 1 which arise from 'even' and 'odd' primitive translations leaves seven distinct classes. This particular result comes from the fact that $\{S_3 \mid \tau + t_{even}\}$ and $\{S_3^{-1} \mid \tau + t_{odd}\}$ form one class C_I which differs from the class C_{II} formed by $\{S_3 \mid \tau + t_{odd}\}$ and $\{S_3^{-1} \mid \tau + t_{even}\}$. Thus for the same

TABLE 10b

Characters for representations of the factor group G_k/T_k which are appropriate for H. The non-primitive translations τ are indicated. The phase factors ± 1 arising from 'even' or 'odd' primitive translations are omitted.

H	E	$2C_3$	$3C_{2'}$ τ	σ_h τ	S_3 τ	S_3^{-1} τ	$3\sigma_v$ τ	Time inversion
H_1	2	2	0	0	0	0	0	a
H_2	2	-1	0	0	$i\sqrt{3}$	$-i\sqrt{3}$	0	a
H_3	2	-1	0	0	$-i\sqrt{3}$	$i\sqrt{3}$	0	a

set of primitive translations S_3 and S_3^{-1} do *not* belong to the same class and their characters have opposite sign.

The factor group G_k/T_k for L contains sixteen elements which form ten classes. There are 2 two-dimensional representations appropriate for L and 8 one-dimensional representations appropriate for M which are identical with those given in Table 8c. In Table 10c we present the characters of the two representations appropriate for L. Omitting

TABLE 10c

Characters for representations of the factor group G_k/T_k which are appropriate for L. The non-primitive translations τ are indicated. The phase factors ± 1 arising from 'even' or 'odd' primitive translations are omitted.

L	E	C_2 τ	$C_{2'}$	$C_{2''}$ τ	I	σ_h τ	σ_d	σ_v τ	Time inversion
L_1	2	0	0	0	0	0	2	0	a
L_2	2	0	0	0	0	0	-2	0	a

the phase factors ± 1 which arise from 'even' and 'odd' primitive translations leaves eight distinct classes.

The factor group G_k/T_k for R contains eight elements which form eight classes. The abstract group is therefore isomorphic with D_{2h}, the point group of M. There are 4 one-dimensional representations appropriate for R and also 4 one-dimensional representations appropriate for Σ which are identical with those given in Table 8e. In Table 10d we present the characters of the four representations appropriate for R. Omitting

TABLE 10d

Characters for representations of the factor group G_k/T_k which are appropriate for R. The non-primitive translations τ are indicated. The phase factors ± 1 arising from 'even' or 'odd' primitive translations are omitted. Time reversal symmetry causes R_1 to be degenerate with R_3 and R_2 to be degenerate with R_4.

R	E	$C_{2''}$ τ	σ_d	σ_h τ	Time inversion
R_1	1	1	1	1	b
R_2	1	1	-1	-1	b
R_3	1	-1	1	-1	b
R_4	1	-1	-1	1	b

the phase factors ± 1 which arise from 'even' or 'odd' primitive translations leaves four distinct classes. As we shall discuss later, time reversal symmetry causes extra degeneracies in the case of R. Thus, R_1 is degenerate with R_3 and R_2 with R_4.

The last factor group G_k/T_k we discuss in this framework is related to the plane G which is the $(k_z = \pi/c)$ top surface of the Brillouin zone. This factor group contains four elements which form four classes. There are 2 one-dimensional representations appropriate for G and 2 other one-dimensional representations appropriate for N which are identical with those given in Table 8f. In Table 10e we present the characters of the two representations appropriate for G. Omitting the phase factors ± 1 which arise from 'even' or 'odd' primitive translations leaves two distinct classes. As in the case of R time reversal symmetry causes an extra degeneracy, i.e., G_1 is degenerate with G_2.

TABLE 10e

Characters for representations of the factor group G_k/T_k which are appropriate for G. The non-primitive translation τ is indicated. The phase factors ± 1 arising from 'even' or 'odd' primitive translations are omitted. Time reversal symmetry causes G_1 to be degenerate with G_2.

G	E	σ_h τ	Time inversion
G_1	1	1	b
G_2	1	-1	b

With this last fact we find the general result that all representations for all **k**-points on the top-face of the Brillouin zone ($k_z = \pi/c$) i.e., for A, H, L, R, S and G are at least two-fold degenerate. This fact, often referred to as 'sticking together' of bands on the top-face of the Brillouin zone is related to the existence of the *two-fold* screw-axis perpendicular to the top-face [5].

TABLE 11

Compatibility relations between the various representations for Bloch states in a crystal with D_{6h} space group symmetry.

Γ_1^+	Γ_1^-	Γ_2^+	Γ_2^-	Γ_3^+	Γ_3^-	Γ_4^+	Γ_4^-	Γ_5^+	Γ_5^-	Γ_6^+	Γ_6^-
Δ_1	Δ_3	Δ_3	Δ_1	Δ_2	Δ_4	Δ_4	Δ_2	Δ_5	Δ_5	Δ_6	Δ_6
Σ_1	Σ_2	Σ_4	Σ_3	Σ_3	Σ_4	Σ_2	Σ_1	Σ_1, Σ_4	Σ_2, Σ_3	Σ_2, Σ_3	Σ_1, Σ_4
T_1	T_2	T_4	T_3	T_2	T_1	T_3	T_4	T_1, T_4	T_2, T_3	T_2, T_3	T_1, T_4
K_1	K_2	K_3	K_4	K_5	K_6			A_1	A_2	A_3	
T_1	T_2	T_4	T_3	T_1, T_4	T_2, T_3			Δ_1, Δ_2	Δ_3, Δ_4	Δ_5, Δ_6	
P_1	P_2	P_2	P_1	P_3	P_3						
M_1^+	M_1^-	M_2^+	M_2^-	M_3^+	M_3^-	M_4^+	M_4^-	H_1	H_2	H_3	
T_1	T_2	T_3	T_4	T_2	T_1	T_4	T_3	S_1	S_1	S_1	
Σ_1	Σ_2	Σ_2	Σ_1	Σ_3	Σ_4	Σ_4	Σ_3	P_1, P_2	P_3	P_3	
U_1	U_4	U_3	U_2	U_2	U_3	U_4	U_1				
L_1	L_2										
U_1, U_2	U_3, U_4										
R_1, R_2	R_3, R_4										

Knowing the character tables for the different factor groups $G_\mathbf{k}/T_\mathbf{k}$ describing the symmetry behavior of a given Bloch function $\psi_\mathbf{k}(\mathbf{r})$, the derivation of compatibility relations between neighboring $\mathbf{k}$-points of different degree of symmetry (Table 11) and the derivation of optical dipole selection rules (Table 12) is straightforward and needs not to be discussed in further detail. With this example we end the discussion on non-symmorphic space groups. We shall return to the example of D_{6h}^4 when we discuss the effect of spin-orbit coupling and of time reversal symmetry on the energy levels in a crystal.

TABLE 12

Single group selection rules for optical dipole transitions in a crystal with D_{6h}^4 space group symmetry.

	$\Gamma_1^\pm$	$\Gamma_2^\pm$	$\Gamma_3^\pm$	$\Gamma_4^\pm$	$\Gamma_5^\pm$	$\Gamma_6^\pm$
x, y	$\Gamma_6^\mp$	$\Gamma_6^\mp$	$\Gamma_5^\mp$	$\Gamma_5^\mp$	$\Gamma_3^\mp, \Gamma_4^\mp, \Gamma_6^\mp$	$\Gamma_1^\mp, \Gamma_2^\mp, \Gamma_5^\mp$
z	$\Gamma_2^\mp$	$\Gamma_1^\mp$	$\Gamma_4^\mp$	$\Gamma_3^\mp$	$\Gamma_5^\mp$	$\Gamma_6^\mp$

	A_1	A_2	A_3
x, y	A_3	A_3	A_1, A_2, A_3
z	A_1	A_2	A_3

	L_1	L_2
x	L_2	L_1
y, z	L_1	L_2

	$M_1^\pm$	$M_2^\pm$	$M_3^\pm$	$M_4^\pm$
x	$M_3^\mp$	$M_4^\mp$	$M_1^\mp$	$M_2^\mp$
y	$M_2^\mp$	$M_1^\mp$	$M_4^\mp$	$M_3^\mp$
z	$M_4^\mp$	$M_3^\mp$	$M_2^\mp$	$M_1^\mp$

	H_1	H_2	H_3
x, y	H_2, H_3	H_3, H_1	H_1, H_2
z	H_1	H_3	H_2

	K_1	K_2	K_3	K_4	K_5	K_6
x, y	K_5	K_6	K_5	K_6	K_1, K_3, K_5	K_2, K_4, K_6
z	K_4	K_3	K_2	K_1	K_6	K_5

1.4. DOUBLE GROUPS AND TIME REVERSAL SYMMETRY

With the inclusion of the electron spin several extensions of the crystal space group theory presented so far have to be made.

In addition to operating with $R = \{\alpha \mid \mathbf{t} + \tau\}$ in coordinate space, R has also to act on spin space. Its operation in spin space is described by the two-dimensional unitary matrix [1]

$$D_{1/2}(\alpha, \hat{n}) = \begin{pmatrix} \cos(\theta/2) \exp\left(-i\dfrac{\varphi + \psi}{2}\right) & -\sin(\theta/2) \exp\left(i\dfrac{\varphi - \psi}{2}\right) \\[2mm] \sin(\theta/2) \exp\left(-i\dfrac{\varphi - \psi}{2}\right) & \cos(\theta/2) \exp\left(i\dfrac{\varphi + \psi}{2}\right) \end{pmatrix}$$

(A26)

with the determinant $+1$. The angles θ, φ, ψ denote the usual Eulerian angles of a rotation by an angle α about a given axis along $\hat{n}$. The $D_{1/2}(\alpha, \hat{n})$ matrices may also be written involving α and $\hat{n}$ rather than θ, φ and ψ:

$$D_{1/2}(\alpha, \hat{n}) = \cos(\alpha/2)\, E - i \sin(\alpha/2)\boldsymbol{\sigma} \cdot \hat{n}, \tag{A27}$$

where $E = \begin{pmatrix} 1 & 0 \\ 0 & 1 \end{pmatrix}$ is the unit matrix and $\sigma_x = \begin{pmatrix} 0 & 1 \\ 1 & 0 \end{pmatrix}$, $\sigma_y = \begin{pmatrix} 0 & -i \\ i & 0 \end{pmatrix}$ and $\sigma_z = \begin{pmatrix} 1 & 0 \\ 0 & -1 \end{pmatrix}$ are the usual Pauli spin matrices. This form shows that the characters $X_{1/2}(\alpha) = 2 \cos(\alpha/2)$ depend only on the rotation angle α regardless of the direction of the axis of rotation. The translational part $\mathbf{t} + \tau$ has no influence in spin space. One striking property of the matrices $D_{1/2}(\alpha, \hat{n})$ is that rotations by an angle α and by an angle $\alpha + 2\pi$ about the same axis are *not* equivalent but lead to results with opposite sign. In other words spinors are transformed into their negatives by a rotation of 2π. Only rotations by multiples of 4π restore identity. Therefore, in representing the group G of operations $\{R\}$ the matrices $D_{1/2}(\alpha, \hat{n})$ are 'double valued'. In order to obtain single valued representations in the usual sense of the group G can be enlarged into $\mathbf{D}$ by including elements $\mathbf{R}$ containing $\{\varepsilon \mid 0\}$ the *non*-identical rotation by 2π. We then have with the new elements

$$\mathbf{R} = \{\varepsilon \mid 0\} R = \{\varepsilon \mid 0\}\, \{\alpha \mid \mathbf{t} + \tau\} = \{\alpha \mid \mathbf{t} + \tau\} \tag{A28}$$

enlarged the group G into the '*double group*' $\mathbf{D}$. The double group $\mathbf{D}$ can now formally be treated to obtain its irreducible representations as any group we considered so far.

The following rules are useful for deriving double group representations from already known single group representations.

(i) Every irreducible representation of the single group G is automatically an irreducible representation of the double group $\mathbf{D}$.

(ii) The difference between the number of classes of $\mathbf{D}$ and G is equal to the number of additional representations of $\mathbf{D}$.

(iii) Generally the elements $\{\alpha \mid \mathbf{t} + \tau\}$ and $\{\alpha \mid \mathbf{t} + \tau\}$ belong to different classes in the double group. The only exception to this rule is

(iv) If R is a *two*-fold rotation of G and G also contains another *two*-fold axis perpendicular to the axis of R, the elements R and $\mathbf{R}$ form a single class in the double group $\mathbf{D}$.

(v) For the 'old' representations of G the characters are given by $X(\mathbf{R}) = X(R)$ whereas for the additional representations one has $X(\mathbf{R}) = -X(R)$. This leads in the special case of (iv) to the result that $X(C_2) = X(\mathbf{C}_2) = 0$.

(vi) The additional irreducible representations Γ of $\mathbf{G}$ can be obtained from the 'old' representations Γ of G by forming the direct products $\Gamma \cdot D_{1/2}$ and decomposing it into irreducible representations according to

$$\Gamma \cdot D_{1/2} = \sum_j C_j \Gamma^{(j)}. \tag{A29}$$

The physical significance of Equation (A29) is evident: if the sum in (A29) contains more than one term, then the inclusion of the electron spin lifts degeneracies which were existing in the spin-free case.

As an example for these rules we shall present in Table 13a the complete double group character table for $\mathbf{D}_{6h}$ which is the group for e.g., the Γ-point of β-GaSe (see also discussion in Section 2.4). The inclusion of $\mathbf{E} = \{\varepsilon \mid 0\}$ increases the number of operation from twenty-four to forty-eight. The number of classes (and thus of representations) is increased from twelve to eighteen. The six extra representations are all two-dimensional. Point (iv) of the preceding discussion becomes valid and the operations C_2, $3C_2'$, and $3C_{2''}$ form classes with their 'partner' elements $\mathbf{C}_2$, $3\mathbf{C}_{2'}$ and $3\mathbf{C}_{2''}$. Accordingly, their characters for the additional representations are zero. The two-dimensional representation $D_{1/2}$ transforming like electron spinors is labelled Γ_7^+. The $+$ sign indicates the general rule which can be seen from Equation (A26) that the $\frac{1}{2}$ spinors have positive parity under space inversion. Considering any of the twelve single group representations the inclusion of the electron spin leads to the following relationships: the representations Γ_1^+, Γ_2^+, Γ_1^- and Γ_2^- become with the inclusion of spin all Γ_7^+ or Γ_7^-; the representations Γ_3^+, Γ_4^+, Γ_3^- and Γ_4^- become all Γ_8^+ or Γ_8^-. The two-dimensional representations Γ_5^+ and Γ_5^- split into the representations $\Gamma_8^+ + \Gamma_9^+$ and $\Gamma_8^- + \Gamma_9^-$ while the two-dimensional representations Γ_6^+ and Γ_6^- split into the representations $\Gamma_7^+ + \Gamma_9^+$ and $\Gamma_7^- + \Gamma_9^-$. The inclusion of the electron spin therefore only splits the two-fold orbit-degenerate levels Γ_5^+, Γ_5^-, Γ_6^+ and Γ_6^-. Optical selection rules and compatibility relations can be obtained for the double group representations in the same straightforward manner as for the single group representations. For the remaining small groups of wavevector $\mathbf{k}$ we only indicate in Tables 13b to 13m the extra representations. The characters for the 'partner' operations are the negatives of those for the 'standard' operations which are given only.

The idea of time reversal symmetry is related to the spatial inversion through the relativistic equivalence of space and time. This equivalence suggests to include the time reversal operation T into the symmetry operations which leave the system invariant. The operation T takes t into $-t$ and means that the system runs back in its evolution, i.e., if T is applied to any state ψ of the system, it creates the time-reversal conjugate state of ψ which would have its evolution back in time.

The requirements the time-reversal operator T has to meet in a quantum-mechanical system are the following:

(i) it has to leave the positions of all particles invariant i.e., $T\mathbf{r}_i T^{-1} = \mathbf{r}_i$;

(ii) it has to reverse the momentum of all particles i.e., $T\mathbf{p}_i T^{-1} = -\mathbf{p}_i$; and

TABLE 13a

Complete character table for the double group $\mathbf{D}_{6h}$.

D_{6h}	E	E	C_2 $\bar{C}_2$	$2C_3$	$2\bar{C}_3$	$2C_6$	$2\bar{C}_6$	$3C_2'$ $3\bar{C}_2'$	$3C_2''$ $3\bar{C}_2''$	I	$\bar{I}$	σ_h $\bar{\sigma}_h$	$2S_6$	$2\bar{S}_6$	$2S_3$	$2\bar{S}_3$	$3\sigma_d$ $3\bar{\sigma}_d$	$3\sigma_v$ $3\bar{\sigma}_v$	Time inversion
Γ_1^+	1	1	1	1	1	1	1	1	1	1	1	1	1	1	1	1	1	1	a
Γ_2^+	1	1	1	1	1	1	1	-1	-1	1	1	1	1	1	1	1	-1	-1	a
Γ_3^+	1	1	-1	1	1	-1	-1	1	-1	1	1	-1	1	1	-1	-1	1	-1	a
Γ_4^+	1	1	-1	1	1	-1	-1	-1	1	1	1	-1	1	1	-1	-1	-1	1	a
Γ_5^+	2	2	2	-1	-1	-1	-1	0	0	2	2	2	-1	-1	-1	-1	0	0	a
Γ_6^+	2	2	-2	-1	-1	1	1	0	0	2	2	-2	-1	-1	1	1	0	0	a
Γ_1^-	1	1	1	1	1	1	1	1	1	-1	-1	-1	-1	-1	-1	-1	-1	-1	a
Γ_2^-	1	1	1	1	1	1	1	-1	-1	-1	-1	-1	-1	-1	-1	-1	1	1	a
Γ_3^-	1	1	-1	1	1	-1	-1	1	-1	-1	-1	1	-1	-1	1	1	-1	1	a
Γ_4^-	1	1	-1	1	1	-1	-1	-1	1	-1	-1	1	-1	-1	1	1	1	-1	a
Γ_5^-	2	2	2	-1	-1	-1	-1	0	0	-2	-2	-2	1	1	1	1	0	0	a
Γ_6^-	2	2	-2	-1	-1	1	1	0	0	-2	-2	2	1	1	-1	-1	0	0	a
Γ_7^+	2	-2	0	1	-1	$\sqrt{3}$	$-\sqrt{3}$	0	0	2	-2	0	1	-1	$\sqrt{3}$	$-\sqrt{3}$	0	0	c
Γ_8^+	2	-2	0	1	-1	$-\sqrt{3}$	$\sqrt{3}$	0	0	2	-2	0	1	-1	$-\sqrt{3}$	$\sqrt{3}$	0	0	c
Γ_9^+	2	-2	0	-2	2	0	0	0	0	2	-2	0	-2	2	0	0	0	0	c
Γ_7^-	2	-2	0	1	-1	$\sqrt{3}$	$-\sqrt{3}$	0	0	-2	2	0	-1	1	$-\sqrt{3}$	$\sqrt{3}$	0	0	c
Γ_8^-	2	-2	0	1	-1	$-\sqrt{3}$	$\sqrt{3}$	0	0	-2	2	0	-1	1	$\sqrt{3}$	$-\sqrt{3}$	0	0	c
Γ_9^-	2	-2	0	-2	2	0	0	0	0	-2	2	0	2	-2	0	0	0	0	c

TABLE 13b

Character table for the extra representations of the double groups C_{6v} (Δ) and D_{3h} (K). The table corresponds to Table 8b. The characters for the 'partner' operations are the negatives of those for the 'standard' operations.

Δ	E	C_2 τ	$2C_3$	$2C_6$ τ	$3\sigma_d$	$3\sigma_v$ τ	
K	E	σ_h τ	$2C_3$	$2S_3$ τ	$3C_{2'}$	$3\sigma_v$ τ	Time inversion
Δ_7	2	0	1	$\sqrt{3}$	0	0	c
Δ_8	2	0	1	$-\sqrt{3}$	0	0	c
Δ_9	2	0	-2	0	0	0	c

TABLE 13c

Character table for the extra representations of the double group D_{2h} (M). The table corresponds to Table 8c.

M	E	C_2 τ	$C_{2'}$	$C_{2''}$ τ	I	σ_h τ	σ_d	σ_v τ	Time inversion
M_5^+	2	0	0	0	2	0	0	0	c
M_5^-	2	0	0	0	-2	0	0	0	c

TABLE 13d

Character table for the extra representations of the double group C_{3v} (P). The table corresponds to Table 8d. Time reversal symmetry causes P_5 to be degenerate with P_6.

P	E	$2C_3$	$3\sigma_v$ τ	Time inversion
P_4	2	1	0	c
P_5	1	-1	i	b
P_6	1	-1	$-i$	b

TABLE 13e

Character table for the extra representations of the double group C_{2v} (Σ, T, U). The table corresponds to Table 8e.

Σ	E	σ_d	$C_{2''}$ τ	σ_h τ	
T	E	σ_v τ	$C_{2'}$	σ_h τ	
U	E	σ_v τ	σ_d	C_2 τ	Time inversion
Σ_5	2	0	0	0	c

TABLE 13f

Character table for the extra representations of the double group C_s (F, B, N). The table corresponds to Table 8f. Time reversal symmetry causes F_3 to be degenerate with F_4.

F	E	σ_d	
B	E	σ_v τ	
N	E	σ_h τ	Time inversion
F_3	1	i	b
F_4	1	$-i$	b

TABLE 13g

Character table for the extra representations of the double factor group G_k/T_k of the line S. The table corresponds to Table 9. Time reversal symmetry causes S_2 to be degenerate with S_5 and S_3 to be degenerate with S_4.

S	E	$C_{2'}$	σ_h τ	σ_v	Time inversion
S_2	1	i	i	-1	b
S_3	1	$-i$	i	1	b
S_4	1	i	$-i$	1	b
S_5	1	$-i$	$-i$	-1	b

TABLE 13h

Character table for the extra representations of the double factor group G_k/T_k for the point A. The table corresponds to Table 10a. Time reversal symmetry causes A_4 to be degenerate with A_5.

A	E	C_2 τ	$2C_3$	$2C_6$ τ	$3C_{2'}$	$3C_{2''}$ τ	I	σ_h τ	$2S_6$	$2S_3$ τ	$3\sigma_d$	$3\sigma_v$ τ	Time inversion
A_4	2	0	-2	0	$2i$	0	0	0	0	0	0	0	b
A_5	2	0	-2	0	$-2i$	0	0	0	0	0	0	0	b
A_6	4	0	2	0	0	0	0	0	0	0	0	0	c

TABLE 13i

Character table for the extra representations of the double factor group G_k/T_k for the point H. The table corresponds to Table 10b. Time reversal symmetry causes H_4 to be degenerate with H_7 and H_5 to be degenerate with H_6.

H	E	$2C_3$	$3C_{2'}$	σ_h τ	S_3 τ	S_3^{-1} τ	$3\sigma_v$ τ	Time inversion
H_4	1	-1	-1	i	$-i$	i	i	b
H_5	1	-1	1	i	$-i$	i	$-i$	b
H_6	1	-1	1	$-i$	i	$-i$	i	b
H_7	1	-1	-1	$-i$	i	$-i$	$-i$	b
H_8	2	1	0	$2i$	i	$-i$	0	c
H_9	2	1	0	$-2i$	$-i$	i	0	c

TABLE 13k

Character table for the extra representations of the double factor group G_k/T_k for the point L. The table corresponds to Table 10c. Time reversal symmetry causes L_3 to be degenerate with L_4.

L	E	C_2 τ	$C_{2'}$	$C_{2''}$ τ	I	σ_h τ	σ_d	σ_v τ	Time inversion
L_3	2	0	$2i$	0	0	0	0	0	b
L_4	2	0	$-2i$	0	0	0	0	0	b

TABLE 13l

Character table for the extra representations of the double factor group G_k/T_k for the line R. The table corresponds to Table 10d. Time reversal symmetry causes R_5 to be degenerate with itself.

R	E	$C_{2''}$ τ	σ_d	σ_h τ	Time inversion
R_5	2	0	0	0	a

TABLE 13m

Character table for the extra representations of the double factor group $G_{\mathbf{k}}/T_{\mathbf{k}}$ for the plane G. The table corresponds to Table 10e. Time reversal symmetry causes G_3 and G_4 each to be degenerate with itself.

G	E	σ_h τ	Time inversion
G_3	1	i	a
G_4	1	$-i$	a

(iii) it has to reverse the spin-components like ordinary angular momenta, i.e., $T\sigma_i T^{-1} = -\sigma_i$.

These considerations imply that any magnetic field which is present in the Hamiltonian of the system must be reversed in the time reversal operation. This is automatically done if the field is produced by currents of the system's particles, since the reversal of their momenta reverses the field. However, external fields must explicitly be reversed.

With these requirements we find that T is an anti-linear and anti-unitary operator which can be factorized into a unitary operator U times complex conjugation K [1]:

$$T = U \cdot K. \tag{A30}$$

For the 'spin-less' case we then simply obtain $T = K$ and $T^2 = E$, such that the time-reversed conjugate state of ψ is given by $T\psi(\mathbf{r}) = \psi^*(\mathbf{r})$.

If we consider the electron spin, T can be written as $T = i\sigma_y K$ with $\sigma_y = \begin{pmatrix} 0 & -i \\ i & 0 \end{pmatrix}$ representing the y-component of the usual Pauli spin matrix [1]. The time-reversed conjugate state of a spinor $\begin{pmatrix} \psi_\alpha \\ \psi_\beta \end{pmatrix}$ is then given by

$$T\begin{pmatrix} \psi_\alpha \\ \psi_\beta \end{pmatrix} = \begin{pmatrix} \psi_\beta^* \\ -\psi_\alpha^* \end{pmatrix}. \tag{A31}$$

It follows from this relationship that $T^2 = E$ for any integral spin particle or for an even number of half-integral spin particles and that $T^2 = -E$ for odd number of half-integral spin particles.

The last result immediately leads to *Kramer's theorem* stating that all energy levels of a system containing an odd number of electrons must at least be two-fold degenerate independent of its spatial symmetry, provided that no external magnetic fields are present which would remove time-reversal symmetry. In a periodic crystal this theorem is fulfilled by the degeneracy of the states $\psi_{n\mathbf{k}}$ and $\psi_{n-\mathbf{k}}$ i.e., by $E_n(\mathbf{k}) = E_n(-\mathbf{k})$ regardless of spatial symmetry.

Since T is a symmetry operator of the Hamiltonian, we might expect the existence of additional degeneracies among the eigenfunctions of the system. This follows from $T\psi_{n\mathbf{k}}$ having the same eigenvalue as $\psi_{n\mathbf{k}}$; an additional degeneracy exists then if $\psi_{n\mathbf{k}}$ is linearly independent of $\psi_{n\mathbf{k}}$.

The 'Frobenius-Schur' test which has been modified for the application to space groups by Herring [15] is a convenient way to find out about the linear independence of ψ_n and $T\psi_n$. The test is based on the consideration of the representation matrices Γ and its complex conjugate Γ^* of ψ_n and ψ_n^*, respectively. Three possible cases may arise according to Wigner [1] and determine eventual additional degeneracies:

(a) Γ and Γ^* are equivalent and can be chosen to be real and identical;

(b) Γ and Γ^* are inequivalent;

(c) Γ and Γ^* are equivalent but *cannot* be chosen to be real and identical.

The results further depend on whether we have integral or half-integral spin. The test of Frobenius-Schur (or Herring) now consists in finding out which of the above three cases is correct for the representation under consideration. All it needs to perform the test are the character table and the diagonal elements of the group-multiplication table. The test requires to find the character of the *square* of each element R in the group G and to sum these characters over all R in G. If the sum is h, the order of G, we have case (a), if it is zero, we have case (b) and if it is $-h$, we have case (c).

The Frobenius-Schur test is impractical in this form if it should be applied to space groups which are virtually infinite in order.

Herring has shown that it is sufficient to limit the sum to certain elements R' whose squares R'^2 are in the factor group $G_{\mathbf{k}}/T_{\mathbf{k}}$ of the group $G_{\mathbf{k}}$ of the wavevector $\mathbf{k}$. The elements R' which meet this requirement are those which take $\mathbf{k}$ into $-\mathbf{k}$. If there are g such elements, Herring's test yields:

$$\sum_{R'} X(R'^2) = \begin{cases} g & \text{case (a)} \\ 0 & \text{case (b)} \\ -g & \text{case (c)} \end{cases} \tag{A32}$$

with $R'\mathbf{k} = -\mathbf{k}$. It is worth noting that, in performing the Frobenius-Schur (or Herring) test, the sum over elements can be reduced in the usual way to a sum over classes.

In Table 14 we show how these cases (a), (b) and (c) determine the existence of

TABLE 14

The dependence of additional degeneracies on the test cases (a), (b) and (c) of Frobenius-Schur and Herring.

case	Frobenius-Schur $\sum_R X(R^2)$	Herring $\sum_{R'} X(R'^2)$	Relation between Γ and Γ^*	Degeneracies for integral spin $T^2 = 1$	Degeneracies for half-integral spin $T^2 = -1$
(a)	h	g	Γ, Γ^* can be made real and equal	none	doubled
(b)	0	0	Γ, Γ^* are inequivalent	doubled	doubled
(c)	$-h$	$-g$	Γ, Γ^* are equivalent but cannot be made real and equal	doubled	none

additional degeneracies. From this table extra degeneracies can easily be found. In the cases (a) for half-integral spin systems and (c) for integral spin systems the representations Γ always appear pair-wise degenerate with themselves. In the case (b) for both spin systems the degeneracies are doubled and Γ always appears together with Γ^*. A more detailed investigation of case (b) shows that two wavefunctions of the *same* wavevector $\mathbf{k}$ are degenerate (i.e., bands stick together at this $\mathbf{k}$) if $\mathbf{k}$ is either equivalent to $-\mathbf{k}$ or belongs to the same star. If $\mathbf{k}$ and $-\mathbf{k}$ do not belong to the same star only the wavefunctions of $\mathbf{k}$ and $-\mathbf{k}$ are degenerate which is a confirmation of Kramer's theorem.

The 'sticking together' of bands usually happens for some special symmetry $\mathbf{k}$-points only. Two very general examples, however, can be given:

(i) Any general $\mathbf{k}$-point of a crystal with inversion symmetry is taken into $-\mathbf{k}$ by this inversion $R' = I$. Since $R'^2 = E$, case (a) of Herring's test holds. Thus, in the case of half-integral spin systems, at any $\mathbf{k}$-point in the Brillouin zone of a crystal with inversion symmetry, the wavefunctions are at least two-fold degenerate.

(ii) At any $\mathbf{k}$-point in a Brillouin zone surface which is perpendicular to a two-fold screw-axis C_2, the degeneracy is doubled. If there is *no* inversion present, $R' = C_2$ and Herring's test gives case (a) with spin and case (c) without spin (i.e., bands 'stick together' in the surface). If inversion is present, $R' = C_2, I$ and Herring's test gives case (b). Since in this case $\mathbf{k}$ and $-\mathbf{k}$ belong to the same star, bands also 'stick together' on the whole surface.

Let us now demonstrate these results on the example of the different groups of the wavevector $G_{\mathbf{k}}$ or their factor groups $G_{\mathbf{k}}/T_{\mathbf{k}}$ of the space group D_{6h}^4. Without spin all representations of all discussed small *single* groups belong to case (a) except the representations R_1, R_2, R_3, R_4 of the line R and the representations G_1, G_2 of the surface G, which belong to case (b). Thus R_1 is degenerate with R_3, R_2 with R_4 and G_1 with G_2. Since the other points of higher symmetry A, L, H and S in the surface plane of the Brillouin zone contain only two- and higher dimensional representations, bands 'stick together' at any $\mathbf{k}$-point in the surface plane of the hexagonal Brillouin zone.

With spin several extra degeneracies occur according to case (a) for R_5, G_3 and G_4 and according to case (b) for the pairs (P_5, P_6), (F_3, F_4), (S_2, S_5), (S_3, S_4), (A_4, A_5), (H_4, H_7), (H_5, H_6) and (L_3, L_4). The different cases according to Herring's test are marked in the Tables 13a to 13m. With this example we end the discussion on double groups and time reversal symmetry.

2. Symmetrized Basis Functions

2.1. USE OF A SYMMETRIZED BASIS

To solve the Schrödinger equation of a periodic crystal we usually expand the one electron wavefunction in a set of appropriate (Bloch) basis functions $\psi_{\mathbf{k}}(\mathbf{r}) = \sum_{\nu} c_{\nu} f_{\nu}(\mathbf{r}, \mathbf{k})$. The variation of the total energy then leads to the eigenvalue matrix equation

$$\langle f_{\nu}| H - E(\mathbf{k})|f_{\mu}\rangle = 0. \tag{A33}$$

The number of basis functions necessary for convergence and thus the order of the

Hamiltonian matrix to be diagonalized can increase for complex crystal structures such that the handling becomes very costly even on modern computers.

If, however, we restrict ourselves to **k**-vectors which lie in symmetry planes, along symmetry lines or at symmetry points of the Brillouin zone, the group of the wavevector $G_{\mathbf{k}}$ contains symmetry operations other than pure translations. In these cases it is possible to factorize the Hamiltonian matrix into non-interacting submatrices by use of *symmetrized* basis functions. The advantages of the factorization are mainly two-fold:

(i) The computation time to diagonalize a matrix varies with a power of 2 to 3 of the order of the matrix. It is therefore favorable to diagonalize several smaller matrices rather than one large matrix. Also, one might be limited in available computer storage space which would make the diagonalization of large matrices impossible.

(ii) The symmetrization and subsequent separate diagonalization of submatrices allow a direct identification of the energy eigenvalues in terms of their symmetry classification. Without symmetrization this can be achieved only by examining the corresponding eigenfunctions explicitly. The knowledge of the symmetries of the various eigenvalues allows e.g., to establish directly transition selection rules.

The choice of the kind of basis function depends on the individual problem and on the method of calculation which is used. Most of the methods (PW, APW, OPW, Pseudopotential) use plane waves or modified plane waves as basis functions which already possess the correct translational symmetry. The following discussion therefore describes the process of symmetrizing plane waves. Several discussions on this matter already exist in literature, some of which are J. C. Slater's book [14], a review article on the APW method by L. F. Mattheiss, J. H. Wood, and A. C. Switendick [16] and the thesis work of A. W. Luehrmann [17].

2.2. SYMMETRY OF PLANE WAVES

The one-electron wavefunction for a given wavevector **k** can be expanded in plane waves

$$\psi_{n\mathbf{k}}(\mathbf{r}) = \sum_{\mathbf{G}} a^n(\mathbf{G}, \mathbf{k}) \, e^{i(\mathbf{k}+\mathbf{G})\cdot\mathbf{r}}, \tag{A34}$$

where the **G**'s are reciprocal lattice vectors to warrant the Bloch form of $\psi_{n\mathbf{k}}(\mathbf{r})$. As discussed in the preceding section $\psi_{n\mathbf{k}}(\mathbf{r})$ transforms like one of the irreducible representations of the group $G_{\mathbf{k}}$. The space group $G_{\mathbf{k}}$ contains all operations R which leave **k** invariant (modulo a reciprocal lattice vector **G**). The effect of operating with $R = \{\alpha \mid \mathbf{t} + \boldsymbol{\tau}\}$ on a function $f(\mathbf{r})$ can be defined as

$$Rf(\mathbf{r}) = f(R^{-1}\mathbf{r}) = f(\alpha^{-1}\mathbf{r} - \alpha^{-1}(\mathbf{t} + \boldsymbol{\tau})) \tag{A35}$$

here we used the definition of the inverse element

$$R^{-1} = \{\alpha^{-1} \mid -\alpha^{-1}(\mathbf{t} + \boldsymbol{\tau})\}. \tag{A36}$$

In the case where $f(\mathbf{r})$ is a plane wave $e^{i(\mathbf{k}+\mathbf{G})\cdot\mathbf{r}}$ we have

$$\begin{aligned} Re^{i(\mathbf{k}+\mathbf{G})\cdot\mathbf{r}} &= e^{i(\mathbf{k}+\mathbf{G})\cdot(R^{-1}\mathbf{r})} \\ &= e^{iR(\mathbf{k}+\mathbf{G})\cdot\mathbf{r}} \\ &= e^{i\alpha(\mathbf{k}+\mathbf{G})\cdot\mathbf{r}} \, e^{-i\alpha\mathbf{k}\cdot\mathbf{t}} \, e^{-i\alpha\mathbf{k}\cdot\boldsymbol{\tau}} \, e^{-i\alpha\mathbf{G}\cdot\boldsymbol{\tau}} \end{aligned} \tag{A37}$$

The operation results in a new plane wave with wavevector $\alpha(\mathbf{k} + \mathbf{G})$ multiplied by phase factors arising from primitive and non-primitive translations.

The primitive translations $\mathbf{t}$ enter expression (A37) only through the phase factor $e^{-i\alpha\mathbf{k}\cdot\mathbf{t}} = e^{-i\mathbf{k}\cdot\mathbf{t}}$. Moroever, the complex conjugate of this factor is contained in the matrix elements $\Gamma_{\alpha\beta}^{(i)}(R)^*$ which will be used for the construction of symmetrized plane waves and multiplies expression (A37). Thus the results become independent of primitive translations $\mathbf{t}$. We like to add that sometimes in the literature (e.g., in Slater's book) the operation of R on a function is differently defined

$$Rf(\mathbf{r}) = f(R\mathbf{r}). \tag{A38}$$

This change in definition does, if consistently carried through, not influence the results of the symmetrization procedure. For a discussion of the different definitions see an article by S. Altmann [18].

To illustrate the advantages of the factorization let us consider an example. Suppose we want to calculate the electronic states at $\mathbf{k} = 0$ of a crystal with D_{6h}^4 symmetry, like β-GaSe. Suppose we include the first twenty neighboring shells of reciprocal lattice vectors $\mathbf{G}_i$ which yields a total of 137 unsymmetrized basis functions. The factorization according to the twelve irreducible representations $\Gamma_1^{\pm}$ to $\Gamma_6^{\pm}$ of D_{6h} (which is the group $G_\mathbf{k}$ at $\mathbf{k} = 0$) results in the following eleven submatrices whose dimensions are given in parentheses:

$$\Gamma_1^+ \ (14), \ \Gamma_2^+ \ (2), \ \Gamma_3^+ \ (13), \ \Gamma_5^+ \ (11), \ \Gamma_6^+ \ (9), \ \Gamma_1^- \ (2), \ \Gamma_2^- \ (11), \ \Gamma_3^- \ (1),$$
$$\Gamma_4^- \ (14), \ \Gamma_5^- \ (9), \ \Gamma_6^- \ (11).$$

The largest matrix is of order 14. The representation Γ_4^+ does not yet appear among the first 137 reciprocal lattice vectors which is partly due to the extreme c/a ration of 4.2 in β-GaSe. At $\mathbf{k}$-points of lower symmetry the reductions in matrix size due to symmetrization are less pronounced but still advantageous.

Applying all operations R of $G_\mathbf{k}$ (or more conveniently of the factor group $G_\mathbf{k}/T_\mathbf{k}$) on a 'prototype' plane wave $e^{i(\mathbf{k}+\mathbf{G})\cdot\mathbf{r}}$ we create new plane waves whose wavevectors have all the same length and which form a so-called 'star'. The number of distinct plane waves in a star can vary from 1 (all operations lead back to the prototype) to h_0 (each operation creates a new distinct plane wave), where h_0 is the order of the point groups $G_\mathbf{k}^0$ associated with $G_\mathbf{k}$. The plane waves of a star form the basis for a (generally reducible) representation D whose matrix elements are all zero except one element per column and per row which is a phase factor $e^{-i\alpha(\mathbf{k}+\mathbf{G})\cdot\mathbf{r}}$. The character of this representation for a given operation $R = \{\alpha \mid \tau\}$ is

$$X_D(R) = \sum_{\substack{(\mathbf{k}+\mathbf{G}) \text{ with} \\ \alpha(\mathbf{k}+\mathbf{G}) = (\mathbf{k}+\mathbf{G})}} e^{-i(\mathbf{k}+\mathbf{G})\cdot\tau} \tag{A39}$$

The largest possible star of h_0 distinct plane waves forms the 'regular representation' its characters being given by

$$X_{\text{reg}}(R) = h_0 \quad for \quad R = E$$
$$= 0 \quad \text{otherwise.} \tag{A40}$$

Each of these representations can be decomposed into irreducible representations of G_k according to Equation (A8). The special property of the regular representation is that each irreducible representation of G_k appears in the regular representation a number of times equal to the dimension of that representation. In other words, the invariant subspace V, spanned by h_0-dimensional star may be reduced into a set of irreducible subspaces V_α^j, where j labels the irreducible representations and where α ranges from 1 to l_j, the dimension of the jth irreducible representation. This may be written, if n_r is the number of irreducible representations as

$$V = \sum_{j=1}^{n_r} \sum_{\alpha=1}^{l_j} V_\alpha^j. \tag{A41}$$

The reduction of a subspace V which is spanned by any given star into irreducible subspaces is the essence of this discussion, i.e., we would like to find the correct linear combinations of plane waves of a star which transform according to the irreducible representations of G_k and thus generate a representation in block form.

2.3. GROUP PROJECTION OPERATORS

Before we give a recipe how to generate symmetrized plane waves, we like to recall the definition of a symmetrized function $\psi_\alpha^j(\mathbf{r}, \mathbf{k})$ which transforms according to one of the irreducible representations:

$$R\psi_\alpha^j(\mathbf{r}, \mathbf{k}) = \sum_{\beta=1}^{l_j} \Gamma^{(j)}(R)_{\beta\alpha} \psi_\beta^j(\mathbf{r}, \mathbf{k}) \tag{A42}$$

where, R is any symmetry operation of the group G_k, l_j is the dimension of the jth representation, and $\Gamma^{(j)}(R)_{\beta\alpha}$ is the $\beta\alpha$th element of the matrix representing the operation R. There are thus l_j functions which transform among themselves according to Equation (A42). These functions are degenerate, they are 'partners' in a basis for the jth irreducible representation. $\psi_\alpha^j(\mathbf{r}, \mathbf{k})$ is called the αth partner in the basis, and is said to transform according to the αth column of the matrices $\Gamma(R)$. Thus each partner transforms according to a *different* column of $\Gamma(R)$.

The objective now is to create these symmetrized functions $\psi_\alpha^j(\mathbf{r}, \mathbf{k})$ in terms of unsymmetrized basis functions (e.g., plane waves). This can be done by means of the 'full projection operator' [14]

$$\rho_{\beta\alpha}^j = \frac{l_j}{h} \sum_R [\Gamma^{(j)}(R)_{\alpha\beta}]^* R, \tag{A43}$$

where the sum runs over all operations R of the group G_k or the factor group G_k/T_k with dimension h. This 'full projection operator' is not, strictly speaking, a projection operator. Rather than being idempotent it satisfies the relation $\rho_{\alpha\beta}^j \rho_{\beta\alpha}^j = \rho_{\alpha\alpha}^j$. When applied to an arbitrary function $f(\mathbf{r})$, this operator yields a function $f_{\alpha\beta}^j(\mathbf{r})$ which has the symmetry properties defined by Equation (A42),

$$Rf_{\beta\alpha}^j(\mathbf{r}) = \sum_{\gamma=1}^{l_j} \Gamma^{(j)}(R)_{\gamma\alpha} f_{\beta\gamma}^j(\mathbf{r}). \tag{A44}$$

Thus, $f^j_{\beta\alpha}(\mathbf{r})$ which has been created by Equation (A43) from $f(\mathbf{r})$ transforms according to the αth column of the jth irreducible representation. Hence the l_j functions $f^j_{\beta 1}$, $f^j_{\beta 2} \ldots f^j_{\beta l_j}$ are partners in a basis for the jth irreducible representation. Note that β is *not* a symmetry index. Since β can according to Equation (A42) also vary from 1 to l_j, it is possible to create l_j different sets of partner functions from $f(\mathbf{r})$. Each set obeys Equation (A42) and thus forms a basis for the jth irreducible representation. It depends on the intrinsic symmetry properties of the considered function space, whether these sets of partner functions are zero, linearly dependent or linearly independent. The power of the 'full projection operator' $\rho^j_{\beta\alpha}$ is, that, if the protoptype function $f(\mathbf{r})$ lies in any way in the irreducible subspace V^j it will generate at least one complete set of basis functions for the subspace.

The usefulness of symmetrized basis functions is expressed in the following theorem:

We consider matrix elements of an operator O, which commutes with all operations of $G_\mathbf{k}$ (e.g., the unity operator or the Hamiltonian). The matrix elements are taken between symmetrized basis functions $f^j_{\alpha\beta}$ and $g^j_{\gamma\delta}$ obtained by the 'full projection operator'. The theorem says that

$$\langle f^j_{\alpha\beta} | \, O \, | g^k_{\gamma\delta} \rangle = \delta_{jk} \delta_{\beta\delta} \langle f | \, O \, | g^k_{\gamma\alpha} \rangle \tag{A45}$$

i.e., there are *no* matrix elements of O between functions which transform according to different irreducible representations and *no* matrix elements of O between functions transforming according to different columns of the same representation. The theorem further assures that matrix elements between two functions transforming according to the same column of the same representation are independent of that column index. In other words, in the case of multi-dimensional representations it is sufficient to calculate matrix elements for one partner (e.g., $f^j_{\beta 1}$) only. The elements involving the remaining partners $f^j_{\beta\alpha}$, $\alpha = 2 \ldots l_j$ lead to identical eigenvalues and just reflect the degeneracy of this representation. The functions $f^j_{\beta\alpha}$ can be obtained from $f^j_{\beta 1}$ by applying the 'shift operators' $\rho^j_{1\alpha}$ according to

$$\rho^j_{1\alpha} f^j_{\beta 1}(\mathbf{r}) = f^j_{\beta\alpha}(\mathbf{r}). \tag{A46}$$

The matrix element theorem thus assures the factorization of the Hamiltonian matrix into non-interacting sub-matrices according to the different irreducible representations. It furthermore puts into 'sub'-blocks those matrices which correspond to multi-dimensional representations.

A symmetrized function usually is a rather long (up to 48 terms) linear combination of unsymmetrized functions. Therefore the calculation of a single matrix element between symmetrized functions might appear to require the separate calculation of an enormous number (up to $48^2 = 2304$ terms) of unsymmetrized elements. The matrix element theorem also assures this number to be drastically reduced to the number of unsymmetrized functions (up to 48 terms) contained in one of the symmetrized functions. Furthermore, as Mattheiss *et al.* [16] have shown by choosing appropriate prototypes $g(\mathbf{r})$ the symmetrized functions $g^k_{\gamma\beta}(\mathbf{r})$ may all be written in the form $g^k_{11}(\mathbf{r})$. This can simplify the application of the symmetrization procedure, since only the $\Gamma^{(j)}_{11}$ matrix elements of the irreducible representations are then required. If however, all degenerate wavefunctions are desired, the remaining matrix elements $\Gamma^{(j)}_{\alpha\beta}$ are necessary. Tables of these matrices for all 230 space groups and the associated small groups G_k (and factor groups G_k/T_k) can be found in e.g., the books of Slater [14] and Kovalev [19].

When spin is considered, the size of the unsymmetrized basis is doubled by forming the direct product of the coordinate space with the two-dimensional spin space. Only a few modifications have to be made to the previously discussed procedure for symmetrizing plane waves. The 'full projection' operator is defined as before in Equation (A43) but with the understanding that

$$R(f(\mathbf{r})u_s) = \sum_{s'} f(R^{-1}\mathbf{r})u_{s'}D_{s's}^{1/2}(R), \tag{A47}$$

where $u_{s'}$ with $s' = 1$, 2 are the spin $\frac{1}{2}$ functions and where $D_{s's}^{1/2}(R)$ are the two-dimensional representation matrices transforming spinors as defined in Section 1.4. The sum over R in the projection operator (Equation (A43)) may be taken only over the h_0 elements of the single group rather than over the $2h_0$ elements of the double group. One additional subscript has to be given to functions extracted by the projection operator from spinors:

$$f_{\alpha\beta,s}^j = \rho_{\alpha\beta}^j(f(\mathbf{r})u_s). \tag{A48}$$

In this case we shall always assume that j labels one of the *double* valued representations.

There are basically two different ways to proceed in constructing symmetrized spinors.

(1) We can perform the single valued coordinate space symmetrization first, then multiply each symmetrized function $f_{\alpha\beta}^j$ by the two spin functions u_s and then operate with the full projection operator on these functions $f_{\alpha\beta}^j u_s$ to extract spinors which are symmetrized with respect to the double valued representations. This method has the advantage of being a two-step process. Once the coordinate space symmetrization has been performed the specific symmetry coefficients in the $f_{\alpha\beta}^j$ remain unaltered.

(2) On the other hand we can perform the spin symmetrization by starting with completely unsymmetrized spinors $\{f(\mathbf{r}), u_s\}$ which seems to be computationally simpler.

In the following section we shall give a detailed example of the symmetrization procedure applied to plane waves of a crystal with D_{6h}^4 symmetry.

2.4. EXAMPLE FOR SYMMETRIZED PLANE WAVES: NON-SYMMORPHIC SPACE GROUP D_{6h}^4

Let us consider a crystal with D_{6h}^4 symmetry, like e.g., β-GaSe. To remove most of the redundancy which occurs in considering a crystal-structure with a large c/a ratio ($c/a \sim 4.2$ for β-GaSe) we shall, however, assume in this example an ideal c/a ratio of $\sqrt{8/3}$.

In describing the hexagonal lattice it is often convenient to define the redundant but symmetric set of four unit vectors $\mathbf{u}_1$, $\mathbf{u}_2$, $\mathbf{u}_3$ and $\mathbf{u}_z$ with $\mathbf{u}_3 = -(\mathbf{u}_1 + \mathbf{u}_2)$. The set of primitive basis vectors $\mathbf{a}_1 = a\mathbf{u}_1$, $\mathbf{a}_2 = a\mathbf{u}_2$ and $\mathbf{a}_z = c\mathbf{u}_z$ spanning the hexagonal lattice is then increased by a fourth vector $\mathbf{a}_3 = a\mathbf{u}_3$. In this hexagonal system an arbitrary vector $\mathbf{t}$ is represented by the quartet of numbers (t_1, t_2, t_3, t_z) where each $t_i = \mathbf{t} \cdot \mathbf{u}_i$, i.e., by each of its four orthogonal projections onto the unit vectors. The reciprocal lattice vectors for a hexagonal lattice with $\mathbf{a}_1$, $\mathbf{a}_2$ and $\mathbf{a}_z$ as primitive vectors are $\mathbf{b}_1$, $\mathbf{b}_2$ and $\mathbf{b}_z$ which can be written in the four component system:

$$\mathbf{b}_1 = \frac{2\pi}{a} (1, 0, -1, 0)$$

$$\mathbf{b}_2 = \frac{2\pi}{a} (0, 1, -1, 0) \qquad\qquad (A49)$$

$$\mathbf{b}_z = \frac{2\pi}{c} (0, 0, 0, 1).$$

Thus an arbitrary reciprocal lattice vector may either be labelled by the coefficients of $\mathbf{b}_i$, i.e., $\mathbf{G} = (m_1, m_2, m_z)$ or by the quartet of coefficients of $\mathbf{u}_i$, i.e., $\mathbf{G} = (A, B, C, D)$ with $A = m_1$, $B = m_2$, $C = -(m_1 + m_2)$ and $D = m_z$. This second way of labelling reciprocal vectors of a hexagonal lattice is convenient in enumerating the various vectors of a star, since all rotations and reflections of the hexagonal group merely permute or change sign of the quartet of numbers.

In Table 15 we present the first twenty-four neighboring shells (stars) of reciprocal

TABLE 15

First 24 stars of reciprocal vectors of a hexagonal lattice with ideal c/a ratio corresponding to 233 G-vectors. Each star is identified by one prototype lattice vector whose coefficients (m_1, m_2, m_z) in terms of the $\mathbf{b}_i$ unit vectors are given. Also indicated is the classification in terms of the quartet $(ABCD)$ coefficients of $\mathbf{u}_i$. The number of G-vectors (PW) per star and their length in units of $2\pi/a$ are also listed. The last column (D_R) labels the representation which is induced by each star.

Star	Prototype m_1, m_2, m_z	Type	PW	G^2	D_R
1	000	$OOOO$	1	0	D_0
2	001	$OOOD$	2	0.61	D_1
3	100	$AO\bar{A}O$	6	1.15	D_2
4	002	$OOOD$	2	1.22	$D_{1'}$
5	101	$AO\bar{A}D$	12	1.31	D_4
6	102	$AO\bar{A}D$	12	1.68	D_4
7	003	$OOOD$	2	1.84	D_1
8	110	$AA\bar{B}O$	6	2.00	D_3
9	111	$AA\bar{B}D$	12	2.09	D_5
10	103	$AO\bar{A}D$	12	2.17	D_4
11	200	$AO\bar{A}O$	6	2.31	D_2
12	112	$AA\bar{B}D$	12	2.35	$D_{5'}$
13	201	$AO\bar{A}D$	12	2.39	D_4
14	004	$OOOD$	2	2.45	$D_{1'}$
15	202	$AO\bar{A}D$	12	2.61	D_4
16	104	$AO\bar{A}D$	12	2.71	D_4
17	113	$AA\bar{B}D$	12	2.72	D_5
18	203	$AO\bar{A}D$	12	2.95	D_4
19	210	$BA\bar{C}O$	12	3.06	D_6
20	005	$OOOD$	2	3.06	D_1
21	211	$BA\bar{C}D$	24	3.12	D_7
22	114	$AA\bar{B}D$	12	3.16	$D_{5'}$
23	105	$AO\bar{A}D$	12	3.27	D_4
24	212	$BA\bar{C}D$	24	3.29	D_7

lattice vectors of a hexagonal lattice with ideal c/a ratio. Each star is identified by its prototype lattice vector whose coefficients (m_1, m_2, m_z) in terms of the $\mathbf{b}_i$ unit vectors are given. These numbers are converted into the quartet notation and listed in the third column as general symbols $ABCD$. The fourth column contains the number of $\mathbf{G}$-vectors per star. This number varies from one to twenty-four according to the twenty-four operations of D_{6h}. The lengths of the $\mathbf{G}$-vectors are given in column 5 in units of $2\pi/a$. The last column contains symbols, labelling the different representations which are generated by the $\mathbf{G}$-vectors of each star.

Suppose we want to calculate the electronic states at $\mathbf{k} = 0$. The inclusion of the first twenty-four neighboring shells of $\mathbf{G}$-vectors would yield a total of 233 unsymmetrized plane waves. To symmetrize these plane waves we first inspect the various representations D_R which are generated by the stars. According to their characters for the operations of D_{6h} which can be calculated according to Equation (A3) we can distinguish between ten different representations. They are listed in Table 16a. The ten different

TABLE 16a

The representations D_0 to D_7 characterizing the ten different types of stars of D_{6h}^4 at $\mathbf{k} = 0(\Gamma)$ and their characters for the twelve classes of D_{6h}.

D	Prototype m_1, m_2, m_z	PW	E	C_2 τ	$2C_3$	$2C_6$ τ	$3C_2{}'$	$3C_2{}''$ τ	I	σ_h τ	$2S_6$	$2S_3$ τ	$3\sigma_d$	σ_v τ
D_0	0, 0, 0	1	1	1	1	1	1	1	1	1	1	1	1	1
D_1	0, 0, $(2m_z - 1)$	2	2	-2	2	-2	0	0	0	0	0	0	2	-2
$D_{1'}$	0, 0, $2m_z$	2	2	2	2	2	0	0	0	0	0	0	2	2
D_2	$m_1, 0, 0$	6	6	0	0	0	0	2	0	6	0	0	2	0
D_3	$m_1, m_1, 0$	6	6	0	0	0	2	0	0	6	0	0	0	2
D_4	$m_1, 0, m_z$	12	12	0	0	0	0	0	0	0	0	0	4	0
D_5	$m_1, m_1,$ $(2m_z - 1)$	12	12	0	0	0	0	0	0	0	0	0	0	-4
$D_{5'}$	$m_1, m_1, 2m_z$	12	12	0	0	0	0	0	0	0	0	0	0	4
D_6	$m_1, m_2, 0$	12	12	0	0	0	0	0	0	12	0	0	0	0
D_7	m_1, m_2, m_z	24	24	0	0	0	0	0	0	0	0	0	0	0

types of stars are characterized by the coefficients (m_1, m_2, m_z) of their prototype vectors. The $\mathbf{G}$-vector $(0, 0, 0)$ forms one star with a representation D_0. Because of the existence of non-primitive translations τ, stars of the type $(0, 0, m_z)$ generate the representations D_1 or $D_{1'}$ depending on whether m_z is odd or even. Stars in the $\mathbf{a}_1, \mathbf{a}_2$-plane of the type $(m_1, 0, 0)$, $(m_1, m_1, 0)$ and $(m_1, m_2, 0)$ generate the representations D_2, D_3 and D_6 respectively. Stars of the type $(m_1, 0, m_z)$ generate D_4, independent of m_z. In contrast to that, stars of the type (m_1, m_1, m_z) generate D_5 or $D_{5'}$ again depending on whether m_z is odd or even. Finally, the regular representation D_7 is generated by stars of the type (m_1, m_2, m_z).

The important point is how these ten different representations are reduced into the irreducible representations of D_{6h}. This can be done according to Equation (A8) knowing the characters of D. The results of how many times a given irreducible representation of D_{6h} appears in a given representation D_R are indicated in Table 16b. We see e.g., that

TABLE 16b

The representations D_0 to D_7 characterizing the ten different types of stars of D_{6h}^4 at $\mathbf{k} = 0(\Gamma)$ and their reduction into irreducible representations of D_{6h}^4 (Γ).

D_R	Irreducible representations of D_{6h}^4 at $\mathbf{k} = 0(\Gamma)$											
	Γ_1^+	Γ_2^+	Γ_3^+	Γ_4^+	Γ_5^+	Γ_6^+	Γ_1^-	Γ_2^-	Γ_3^-	Γ_4^-	Γ_5^-	Γ_6^-
D_0	1											
D_1			1							1		
$D_{1'}$	1							1				
D_2	1				1					1		1
D_3	1				1				1			1
D_4	1		1		1	1	1			1	1	1
D_5		1	1		1	1	1			1	1	1
$D_{5'}$	1			1	1	1		1	1		1	1
D_6	1	1			2				1	1		2
D_7	1	1	1	1	2	2	1	1	1	1	2	2

the identical representation Γ_1^+ appears in any type of star, except stars like $(0, 0, m_z)$ and (m_1, m_1, m_z) with m_z odd. We also see, that in stars of the type $(m_1, m_2, 0)$ and (m_1, m_2, m_z) the two-dimensional representations appear twice; i.e., two linearly independent sets of symmetrized combinations can be created by the projection operators $\rho_{\alpha\beta}^j$ from stars of this type. To examine this situation in more detail let us focus on the two-dimensional representation Γ_5^+. Generally, with the operators ρ_{11} and ρ_{12} we generate one set of basis functions and with ρ_{21} and ρ_{22} we generate a second set. If applied to stars of type D_0, D_1 or $D_{1'}$ these sets all vanish. If applied to stars of type D_2, D_3, D_4, D_5 or $D_{5'}$ the two sets of symmetrized basis functions are linearly dependent. However, if applied to stars of type D_6 or D_7 the two sets are linearly independent. For actual calculations, of course, only one partner of each set needs to be included, so only ρ_{11} and ρ_{21} need to be applied.

On the basis of the reduction of the various representations D_R into irreducible representations of D_{6h} we can now transform into blocks the representations generated by the 233 simple plane waves. The size of each block, i.e., the number of symmetrized plane waves for each irreducible representation are indicated in Table 17. We note that, instead of diagonalizing a matrix of order 233 using simple plane waves, we have now to diagonalize twelve different submatrices with dimensions ranging from four to nineteen, which results in a considerable reduction of computing time.

TABLE 17

Number of symmetrized plane waves obtained from the first 24 stars or 233 G-vectors of D_{6h}^4 at Γ (ideal c/a ratio).

Γ_1^+	Γ_2^+	Γ_3^+	Γ_4^+	Γ_5^+	Γ_6^+	Γ_1^-	Γ_2^-	Γ_3^-	Γ_4^-	Γ_5^-	Γ_6^-
19	5	15	4	18	14	4	14	6	18	14	18

To create the various symmetrized plane waves by the projection operator method we need, of course, to know some elements of the representation matrices $\Gamma^{(j)}(R)_{\alpha\beta}$. For the present example of D_{6h}^4 ($\mathbf{k} = 0$) the representation matrices are e.g. listed in Slater's book [14]. The final symmetry coefficients multiplying the simple plane waves are given in the Tables 18a to 18f. Separately for each type of star (labelled by the

TABLE 18a

Symmetry coefficients for the (trivial) D_0 representation.

D_0	$OOOO$
Γ_1^+	1

TABLE 18b

Symmetry coefficients for stars of the D_1 and $D_{1'}$ type. Only one G-vector is listed, the second is obtained by inversion. Its symmetry coefficient is the same for even representations and of opposite sign for odd representations.

D_1	$D_{1'}$	$OOOD$
Γ_3^+, Γ_4^-	Γ_1^+, Γ_2^-	1

TABLE 18c

Symmetry coefficients for stars of the D_2 and D_3 types. Only three of the six G-vectors are listed, the remaining three are obtained by inversion, their symmetry coefficients are obtained as described for Table 18b. ω stands for $e^{i2\pi/3}$.

D_2		$AO\bar{A}O$	$A\bar{A}OO$	$OA\bar{A}O$
	D_3	$A\bar{A}BO$	$B\bar{A}\bar{A}O$	$\bar{A}B\bar{A}O$
Γ_1^+	Γ_1^+	1	1	1
Γ_5^+	Γ_5^+	1	ω^2	ω
Γ_4^-	Γ_3^-	1	-1	-1
Γ_6^-	Γ_6^-	1	$-\omega$	$-\omega^2$

TABLE 18d

Symmetry coefficients for stars of the D_4, D_5 and $D_{5'}$ type. See caption of Table 18b for explanations.

D_4			$AO\bar{A}D$	$A\bar{A}OD$	$OA\bar{A}D$	$O\bar{A}AD$	$\bar{A}AOD$	$\bar{A}OAD$
	D_5	$D_{5'}$	$AA\bar{B}D$	$B\bar{A}\bar{A}D$	$\bar{A}B\bar{A}D$	$A\bar{B}AD$	$\bar{B}AAD$	$\bar{A}\bar{A}BD$
Γ_1^+, Γ_2^-	Γ_2^+, Γ_1^-	Γ_1^+, Γ_2^-	1	$(-1)^D$	$(-1)^D$	1	1	$(-1)^D$
Γ_3^+, Γ_4^-	Γ_3^+, Γ_4^-	Γ_4^+, Γ_3^-	1	$-(-1)^D$	$-(-1)^D$	1	1	$-(-1)^D$
Γ_5^+, Γ_5^-	Γ_5^+, Γ_5^-	Γ_5^+, Γ_5^-	1	$(-1)^D\omega^2$	$(-1)^D\omega$	ω	ω^2	$(-1)^D$
Γ_6^+, Γ_6^-	Γ_6^+, Γ_6^-	Γ_6^+, Γ_6^-	1	$-(-1)^D\omega$	$-(-1)^D\omega^2$	ω^2	ω	$-(-1)^D$

TABLE 18e

Symmetry coefficients for stars of the D_6 type. See caption of Table 18b for explanations.

D_6	$\bar{C}ABO$	$\bar{B}C\bar{A}O$	$A\bar{C}BO$	$BA\bar{C}O$	$\bar{C}BAO$	$\bar{A}BCO$
Γ_1^+	1	1	1	1	1	1
Γ_2^+	1	1	-1	-1	-1	1
$\Gamma_5^+(1)$	1	ω^2	0	0	0	ω
$\Gamma_5^+(2)$	0	0	ω	1	ω^2	0
Γ_3^-	1	-1	1	1	1	-1
Γ_4^-	1	-1	-1	-1	-1	-1
$\Gamma_6^-(1)$	1	$-\omega$	0	0	0	$-\omega^2$
$\Gamma_6^-(2)$	0	0	$-\omega^2$	-1	$-\omega$	0

representations D_0 to D_7) the coefficients are given for the various irreducible representations. Let us, e.g., examine star 11 with the prototype (200), which is of the type $(AO\bar{A}O)$ and contains six G-vectors and transforms like D_2. From Table 18c we find that a symmetrized function $f_{11}^{\Gamma_4^-}(\mathbf{r})$ which transforms like Γ_4^- has the following general form:

$$f_{11}^{\Gamma_4^-} = 1 \cdot (AO\bar{A}O) - 1 \cdot (A\bar{A}OO) - 1 \cdot (OA\bar{A}O) - 1 \cdot (\bar{A}OAO) +$$
$$+ 1 \cdot (\bar{A}AOO) + 1 \cdot (O\bar{A}AO) \tag{A50}$$

only the first three G-vectors and coefficients are actually listed in Table 18c. The remaining three coefficients are obtained by applying inversion to these G-vectors and using the first three coefficients with opposite sign (Γ_4^- is an odd representation). For star 11 the symmetrized function then becomes

$$f_{11}^{\Gamma_4^-} = (200) - (2\bar{2}0) - (020) - (\bar{2}00) + (\bar{2}20) + (0\bar{2}0). \tag{A51}$$

In cases of stars with non-vanishing third G-component $m_z = D$, the symmetry coefficients depend explicitly on D (with the exception of the representations D_1 and $D_{1'}$). This dependence is a consequence of the non-primitive translation $\tau = \frac{1}{2}\mathbf{a}_z$ which is associated with some of the operations. Let us finally examine a case in which two linearly independent symmetrized functions can be generated e.g., the case of the two-dimensional representation Γ_5^+. We choose star 19 with the (210) prototype and twelve G-vectors. This star is of the type $(BA\bar{C}O)$ (or $(ABCO)$) and transforms like D_6. Table 18e clearly exhibits the linear independence of the two functions transforming like $\Gamma_5^+(1)$ and $\Gamma_5^+(2)$ since they involve different G-vectors. So we have

$$f_{11}^{\Gamma_5^+} = 1 \cdot (\bar{3}10) + \omega^2 \cdot (\bar{2}30) + \omega \cdot (\bar{1}20) + 1 \cdot (3\bar{1}0) + \omega^2 \cdot (2\bar{3}0) + \omega \cdot (120)$$

and $$\tag{A52}$$

$$f_{21}^{\Gamma_5^+} = 1 \cdot (210) + \omega^2 \cdot (\bar{3}20) + \omega \cdot (1\bar{3}0) + 1 \cdot (\bar{2}\bar{1}0) + \omega^2 \cdot (3\bar{2}0) + \omega \cdot (\bar{1}30)$$

with $\omega = e^{i2\pi/3}$. Each of the functions contains only six of the twelve G-vectors of the star. This clear separation is a consequence of the choice of complex representation matrices in Slater's book [14] which we used here. Equivalent, real representations could be created by similarity transformations. However, the two derived symmetrized functions then would each include all twelve G-vectors. It can readily be shown from

TABLE 18f

Symmetry coefficients for stars of the D_7 type. See caption of Table 18b for explanations.

D_7	$\bar{C}ABD$	$\bar{B}CAD$	$A\bar{C}BD$	$BA\bar{C}D$	$\bar{B}ACD$	$A\bar{C}\bar{B}D$	$B\bar{C}AD$	$CA\bar{B}D$	$\bar{C}BAD$	$A\bar{B}CD$	$AB\bar{C}D$	$\bar{B}\bar{B}AD$
Γ_1^+, Γ_2^-	1	$(-1)^p$	$(-1)^p$	$(-1)^p$	1	1	1	$(-1)^p$	$(-1)^p$	$(-1)^p$	1	1
Γ_2^+, Γ_1^-	1	$(-1)^p$	$-(-1)^p$	$-(-1)^p$	-1	-1	1	$(-1)^p$	$-(-1)^p$	$(-1)^p$	1	-1
Γ_3^+, Γ_4^-	1	$-(-1)^p$	$-(-1)^p$	$-(-1)^p$	1	1	1	$-(-1)^p$	$-(-1)^p$	$-(-1)^p$	1	1
Γ_4^+, Γ_3^-	1	$-(-1)^p$	$(-1)^p$	$(-1)^p$	-1	-1	1	$-(-1)^p$	$(-1)^p$	$-(-1)^p$	1	-1
$\Gamma_5^+, \Gamma_5^-(1)$	1	$(-1)^p\omega^2$	0	0	0	0	ω^2	$(-1)^p$	0	$(-1)^p\omega$	ω	0
$\Gamma_5^+, \Gamma_5^-(2)$	0	0	$(-1)^p\omega$	$(-1)^p$	1	ω	0	0	$(-1)^p\omega^2$	0	0	ω^2
$\Gamma_6^+, \Gamma_6^-(1)$	1	$-(-1)^p\omega$	0	0	0	0	ω	$(-1)^p$	0	$-(-1)^p\omega^2$	ω^2	0
$\Gamma_6^+, \Gamma_6^-(2)$	0	0	$(-1)^p\omega^2$	$(-1)^p$	1	ω^2	0	0	$-(-1)^p\omega$	0	0	ω

Equations (A52) that, while the *first* function is obtained from the (210) prototype by applying the ρ_{11} operator, the *second* function may also be obtained by applying the ρ_{11} operator rather than the ρ_{21} operator but to a different prototype i.e. ($\overline{3}$10).

With this detailed example we terminate the discussion on symmetrized basis functions. The methods to generate them are straightforward and systematic when using projection operators and may well be done on a computer [20].

3. Symmetry Classification of Layer Compounds

3.1. STRUCTURAL ASPECTS

The materials discussed here are layer-like both in atomic arrangement and physical properties. It is clear that the definition 'layer-like' is very flexible and several classes of materials like e.g. the semimetals As, Sb and Bi, which shall not be included in the present discussion, may sometimes be called layer-like. Layer-compounds are generally only weakly ionic in character which allows anion-anion contact in the direction perpendicular to the layers. To the degree that they may be considered as ionic, they can conveniently be viewed as more or less perfect packings of large anions, with the smaller cations lying in interstices of these packings. The structure of layer compounds may also be described as a stacking of sandwiches that are several atomic-sheets thick. For most of the materials we shall discuss here, the sandwiches contain three atomic layers. The top and bottom sheets of the sandwiches are closed packed anions, while the middle sheet is cationic (see Figure 6). Some exceptions which shall also be considered are:

(i) The hexagonal network crystals graphite and boron nitride, in which the sandwiches consist of only one sheet (see Figure 6d),

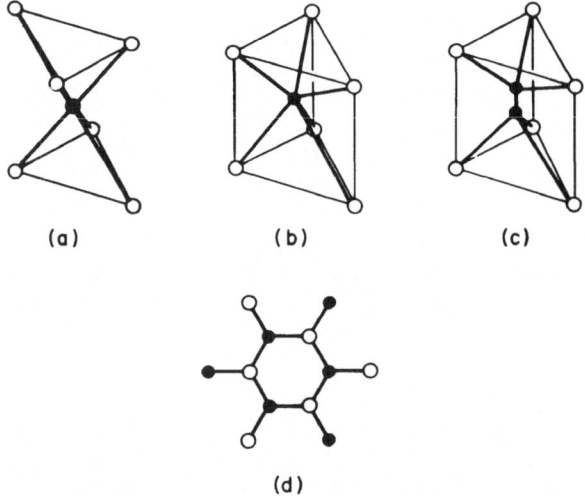

Fig. 6. Schematic atomic arrangement within one layer: (a) octahedral coordination (PbI$_2$); (b) trigonal prismatic coordination (MoS$_2$); (c) trigonal prismatic coordination with repeated cation sheet (GaSe); (d) planar (or slightly buckled) hexagonal network (boron nitride, graphite).

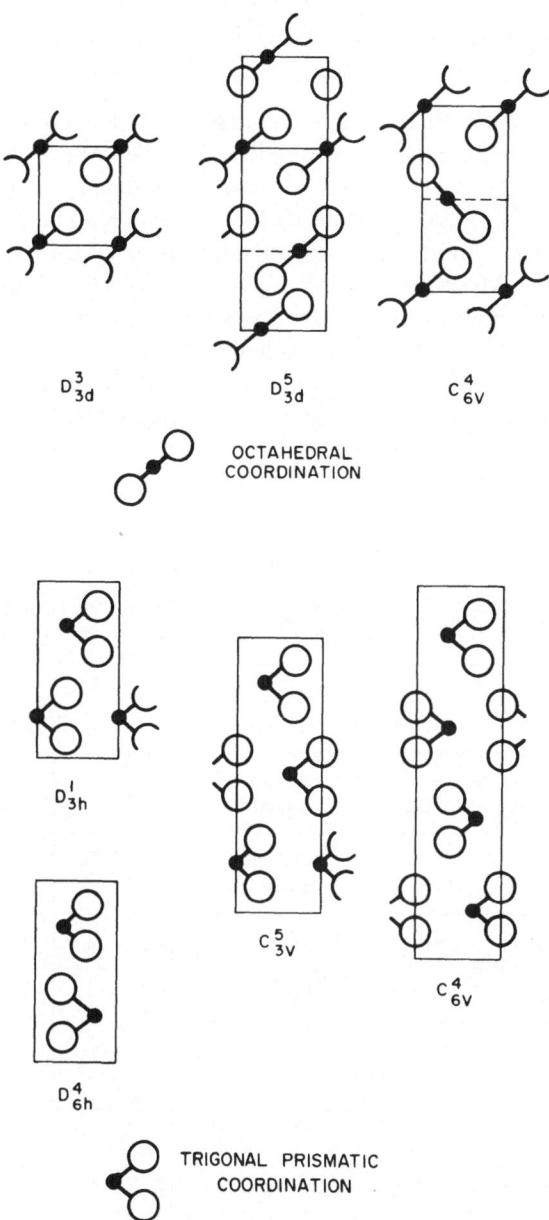

D_{3d}^3 D_{3d}^5 C_{6V}^4

OCTAHEDRAL
COORDINATION

D_{3h}^1

C_{3V}^5

C_{6V}^4

D_{6h}^4

TRIGONAL PRISMATIC
COORDINATION

Fig. 7.   Classification of some layered compounds according to type of coordination of cations and
type of stacking sequence. The figure corresponds to Table 19.

(ii) The compounds GaSe and GaS, in which the middle cation sheet is doubled, resulting in four sheets per sandwich (see Figure 6c), and

(iii) The compound Bi_2Te_3, in which five alternating anion cation sheets form one sandwich.

We shall, however, not discuss here the arsenic chalcogenides As_2S_3 and As_2Se_3 which are separately treated in a review article by Zallen and Blossey [21].

It is convenient to group the layer structures according to octahedral (AbC) or trigonal prismatic (AbA) coordination of the cations (b). The sandwiches stack in a variety of ways giving either pure octahedral, pure trigonal prismatic or mixed co-ordination polytypes. Moreover, within a given type of coordination, stacking of the layers varies like hexagonal or cubic or mixed close packing of the anions. Thus, an enormous variety of polytypes exists and their list is constantly enlarged by new crystallographic research work. It is known, e.g., that PbI_2 exhibits about three dozen polytypes corresponding to different stacking sequences [22]. The most simple stacking sequences together with their symmetries (Schoenfliess notation) are compiled in Table 19 and illustrated in Figure 7.

TABLE 19

Classification of layered compounds according to type of coordination of cations and type of stacking of layers. Only the most simple stacking sequences are listed.

Coordination of cations	Octahedral			Trigonal Prismatic		
close packing of anions	hex	cubic	simple mixed	hex	cubic	simple mixed
type of structure	CdI_2	$CdCl_2$	complex CdI_2	ε, β-GaSe MoS_2	γ-GaSe	δ-GaSe
space groups	D_{3d}^3	D_{3d}^5	C_{6v}^4	D_{6h}^4, D_{3h}^1	C_{3v}^5	C_{6v}^4
symmetry	hex	rhomb	hex	hex	rhomb	hex
layers per hex unit cell	1	3 (1 rhomb)	2	2	3 (1 rhomb)	4

Additions to Table 19 which we like to mention are:

(i) The BiI_3 structures. In this structure the cations are octahedrally coordinated and the anions are hexagonally close packed, but only $\frac{2}{3}$ of the cation sites are occupied. This increases the unit cell in the plane of the layers to two molecules per layer, whereas all structures mentioned in Table 19 contain only one molecule per layer. Due to the existence of voids and occupied cation sites, two different stacking polytypes exist, both corresponding to a hexagonal close-packing sequence of the anions. If the cations fall on top of each other, the hexagonal space group is D_{3d}^1 with one layer per unit cell; if the cations alternate with voids, the rhombohedral space group is C_{3i}^2 with one layer per unit cell. The corresponding hexagonal unit cell extends over three layers.

(ii) The hexagonal networks of graphite. Depending on hexagonal (graphite I) or cubic close-packing (graphite II), the space groups are the hexagonal group C_{6v}^4 with two layers per unit cell and the rhombohedral group C_{3v}^5 with one layer per unit cell, respectively. As for all other rhombohedral cases already mentioned, the corresponding hexagonal unit cell extends over three layers. The symmetry of the graphite structures resembles strongly those of the GaSe structures of Table 19. They differ from them by the absence of inversion symmetry which is present in MoS_2 or β-GaSe. This lack of inversion symmetry is due to some slight 'buckling' of the six fold C-rings in graphite. In many discussions of graphite the 'buckling' is neglected and the full hexagonal space group D_{6h}^4 is assumed (see Section 3.3). In boron nitride, in which the carbon atoms are alternatively replaced by boron and nitrogen, this 'buckling' is absent and inversion symmetry is restored.

(iii) The Bi_2Te_3 structure. In this structure the cations are octahedrally coordinated, and in addition to the CdI_2 arrangement another two atomic sheets of cations and anions are added. One layer of Bi_2Te_3 thus extends over *five* atomic sheets. The arrangement of the three anion sheets is cubic close-packed. The simple rhombohedral unit cell contains one layer and the atoms are placed in positions of D_{3d}^5.

It is worth mentioning that polytypes of layered compounds often are designated in terms of the number of *anion atomic sheets* per unit cell. Thus, the simple CdI_2 structure (group D_{3d}^3) would be 2H (H stands for hexagonal) or the rhombohedral modification of PbI_2 (group D_{3d}^5) would be 6R and so on. Unfortunately, this original definition has recently been altered by many authors and the designation is instead given in terms of the number of *layers* per unit cell. The 2H CdI_2 structure then becomes 1T (T stands for trigonal) and the 6R PbI_2 structure becomes 3R. Since this change of definition may lead to some confusion we shall use the Schoenfliess notation for the space groups throughout this article and indicate the number of layers to define the polytypes.

3.2. GROUP ANALYSIS OF SOME LAYER COMPOUNDS

Due to their sandwich structure most of the physical properties of layer compounds are predominantly determined by the atomic arrangement and interactions within one isolated layer. The interlayer interaction can then be regarded as a more or less strong perturbation on the *one*-layer system. This approach is in particular true for mechanical properties; the ratio of force constants between intralayer and interlayer interactions usually varies between 10 to 100. These facts have been emphasized by Zallen and Blossey [21]. As far as the electronic structure of individual electronic states is concerned, rather than an integral electronic screening charge, the isolated layer approach often leads to errors of the order of electron volts. In other words, some electronic states (e.g., the states close to gap in GaSe) may overlap strongly between layers and may lead to interlayer interactions of one eV. In spite of this quantitative difference, even for the electronic structure many effects and basic selection rules can be explained by simply considering isolated layers. It is therefore of interest, first to specify the symmetries of isolated layers. Thus, depending on the coordination of the cations, the following point groups describe the symmetry of isolated layers:

(i) D_{3d} for compounds with octahedral coordination of the cations like 1T-TaS$_2$ and for 'buckled' hexagonal networks like graphite (planar graphite has D_{6h} symmetry), and

(ii) D_{3h} for compounds with trigonal prismatic coordination of the cations like MoS$_2$ and GaSe and for planar hexagonal compound networks like boron nitride.

Also of interest are atomic site symmetries. Thus, in the octahedral materials the cation sees D_{3d} symmetry which allows for s-d_{z^2} mixing. The cations of trigonal prismatic crystals with doubled cation sheets like GaSe see C_{3v} symmetry which allows for s-p_z mixing. In all cases the anions see C_{3v} symmetry. For the hexagonal network materials, the atoms in 'buckled' graphite see C_{3v} symmetry, whereas the atoms in planar graphite and in boron nitride see D_{3h} symmetry.

For the quantitative computation of the electronic structure of layer compounds, the full three-dimensional space groups have to be considered. In the following we thus list the various three-dimensional symmetry groups of the most common layer-compounds, whose electronic properties shall be discussed in this article. The structures are described by nine different space groups, two of which (D_{6h}^4 and C_{6v}^4) are nonesymmorphic. In Table 20 we list eight prototypes for non-transition metal layer-

TABLE 20

Symmetry space groups of the most simple polytypes of some layered materials. The number of layers per unit cell is indicated in parentheses following the symmetry symbols. For the rhombohedral groups alternative three-layer hexagonal unit cells may be chosen. The symbols T, O, H refer to trigonal prismatic, octahedral or hexagonal network like coordination.

GaSe	T	$D_{6h}^4(2)$, $D_{3h}^1(2)$, $C_{3v}^5(3, 1)$, $C_{6v}^4(4)$
GaS	T	$D_{6h}^4(2)$
PbI$_2$	O	$D_{3d}^3(1)$, $D_{3d}^5(3, 1)$
SnS$_2$, Se$_2$	O	$D_{3d}^3(1)$
BiI$_3$	O	$D_{3d}^1(1)$, $C_{3i}^2(3, 1)$
C	H	$C_{6v}^4(2)$, $D_{6h}^4(2)$, $C_{3v}^5(3, 1)$
BN	H	$D_{6h}^4(2)$, $C_{3v}^5(3, 1)$
Bi$_2$Te$_3$	O	$D_{3d}^5(3, 1)$
MoS$_2$, Se$_2$	O	$D_{6h}^4(2)$, $C_{3v}^5(3, 1)$
TaS$_2$, Se$_2$	O, T	$D_{3d}^3(1)$, $D_{6h}^4(2)$
HfS$_2$, Se$_2$	O	$D_{3d}^3(1)$
ZrS$_2$, Se$_2$	O	$D_{3d}^3(1)$
NbS$_2$	O	$C_{3v}^5(3, 1)$
NbSe$_2$	O	$D_{6h}^4(2)$
TiS$_2$, Se$_2$	O	$D_{3d}^3(1)$

compounds and seven prototypes for transition metal layer-compounds. The various space groups for the most simple polytypes are given. Also indicated is the type of coordination (T = trigonal prismatic, O = octahedral, H = simple hexagonal network) in which the cations are found. The number of layers per unit cell is shown in parentheses following the space group symbols.

GaSe which exists in the trigonal prismatic form has been observed in four distinct polytypes which we shall all mention here because they represent the four most simple stacking sequences of trigonal prismatic layers [23]. The β- and ε-types with the space groups D_{6h}^4 and D_{3h}^1 respectively correspond to the hexagonal close-packing of anions; the γ-type (space group C_{3v}^5) corresponds to the cubic stacking sequence and the recently reported δ-type (space group C_{6v}^4) corresponds to the simple regular mixture of hexagonal and cubic stacking orders. Natural disordered mixings of hexagonal and cubic sequences are known as mixtures of ε- and γ-types GaSe. The more ionic GaS is invariably found in the β-modification with space group D_{6h}^4.

As already stated, PbI_2 has been found in a large variety of polytypes. The most simple polytype is the original CdI_2 grouping in which many octahedrally coordinated layer compounds are found (space group D_{3d}^3). While this polytype is characterized by a hexagonal close packing of the anions, the $CdCl_2$ arrangement in which PbI_2 is also found exhibits cubic close packing of the anions. The corresponding rhombohedral space group is D_{3d}^5.

SnS_2, $SnSe_2$, HfS_2, $HfSe_2$, ZrS_2, $ZrSe_2$, TiS_2, $TiSe_2$ and many other layer compounds (see tables in Ref. [24]) are only reported with D_{3d}^3 symmetry.

Interesting exceptions are TaS_2 and $TaSe_2$ which exists octahedrally coordinated with D_{3d}^3 symmetry and also trigonally prismatic coordinated, like MoS_2 and $MoSe_2$ with D_{6h}^4 symmetry.

TaS_2 or $TaSe_2$ thus represents an ideal material for studying the dependence of various physical properties on the type of coordination of the cations [25].

The compound NbS_2 has been reported to be like rhombohedral MoS_2 with space group C_{3v}^5.

The hexagonal network materials graphite and boron nitride have already been described in this section. Extensive structural details and crystallographic references for all mentioned layer compounds may e.g. be found in Ref. [24].

TABLE 21

Eight space groups (two non-symmorphic, D_{6h}^4, C_{6v}^4) in which most of the simple polytype layer compounds crystallize.

D_{6h}^4	C_{6v}^4	D_{3h}^1	D_{3d}^1	D_{3d}^3	D_{3d}^5	C_{3v}^5	C_{3i}^2
β-GaSe	δ-GaSe	ε-GaSe	BiI_3	CdI_2	$CdCl_2$	γ-GaSe	AsI_3
GaS	TaS_2			PbI_2	PbI_2	MoS_2	SbI_3
MoS_2, Se_2	CdI_2					NbS_2	BiI_3
TaS_2, Se_2	PbI_2			SnS_2, Se_2	Bi_2Te_3	C	
$NbSe_2$	C			HfS_2, Se_2		BN	
WS_2, Se_2				TaS_2			
C				TiS_2, Se_2			
BN				ZrS_2, Se_2			

Table 21 shows a reversed compilation of layered compounds grouped according to their space groups. The eight space groups indicated in Table 21 shall shortly be discussed in the following section.

3.3. DESCRIPTION OF EIGHT SPACE GROUPS OF LAYER COMPOUNDS

In this section we discuss the following eight space groups:

(i) hexagonal groups

$$D_{6h}^4, \; C_{6v}^4, \; D_{3d}^1, \; D_{3d}^3, \; D_{3h}^1,$$

(ii) rhombohedral groups

$$D_{3d}^5, \; C_{3v}^5, \; C_{3i}^2.$$

The two hexagonal space groups D_{6h}^4 and C_{6v}^4 are non-symmorphic, the remaining groups are symmorphic.

For convenience, the rhombohedral groups are also expressed in terms of a hexagonal coordinate system. In other words, each of the rhombohedral unit cells containing one layer can be associated with a hexagonal unit cell extending over three layers. The group operations are defined in a hexagonal lattice spanned by two unit vectors $\mathbf{a}_1$ and $\mathbf{a}_2$ with a 120° angle between them and a third unit vector $\mathbf{a}_z$ perpendicular to the basal plane defined by $\mathbf{a}_1$ and $\mathbf{a}_2$. To enumerate the group operations it is convenient to define a fourth redundant vector $\mathbf{a}_3 = -(\mathbf{a}_1 + \mathbf{a}_2)$ in the basal plane as displayed in Figure 5.

In Table 22 we list the point group operations of D_{6h}. The associated non-primitive

TABLE 22

Point group operations of D_{6h}. $\mathbf{a}_1$, $\mathbf{a}_2$, $\mathbf{a}_z$ are the basis vectors of the hexagonal unit cell. $\mathbf{a}_3 = -(\mathbf{a}_1 + \mathbf{a}_2)$ as shown in Figure 5. The associated non-primitive translations τ for the space group elements of D_{6h}^4 are indicated.

Number	Symbol	Non-primitive translation	Operation
1	E	–	identity
2, 3	$2C_6$	τ	$\pm \; \pi/3$ rotation about $\mathbf{a}_z$
4, 5	$2C_3$	–	$\pm \; 2\pi/3$ rotation about $\mathbf{a}_z$
6	C_2	τ	π rotation about $\mathbf{a}_z$
7, 8, 9	$3C_2{}'$	–	π rotation about $\mathbf{a}_1, \mathbf{a}_2, \mathbf{a}_3$
10, 11, 12	$3C_2{}''$	τ	π rotation about axes $\perp \mathbf{a}_1, \mathbf{a}_2, \mathbf{a}_3$ and $\perp \mathbf{a}_z$
13	I	–	inversion
14, 15	$2S_3$	τ	$\pm \; 2\pi/3$ rotation about $\mathbf{a}_z$ followed by a mirror plane $\perp \mathbf{a}_z$
16, 17	$2S_6$	–	$\pm \; \pi/3$ rotation about $\mathbf{a}_z$ followed by a mirror plane $\perp \mathbf{a}_z$
18	σ_h	τ	mirror plane $\perp \mathbf{a}_z$
19, 20, 21	$3\sigma_d$	–	mirror plane $\perp \mathbf{a}_1, \mathbf{a}_2, \mathbf{a}_3$ and $\| \mathbf{a}_z$
22, 23, 24	$3\sigma_v$	τ	mirror plane $\| \mathbf{a}_1, \mathbf{a}_2, \mathbf{a}_3$ and $\| \mathbf{a}_z$

translations $\tau = \frac{1}{2}\mathbf{a}_z$ for the space group elements of D_{6h}^4 are indicated. D_{6h} is the group of full hexagonal symmetry and the point groups of all the other space groups listed above are subgroups of D_{6h}. Thus, in Table 23 the various elements of D_{6h} are listed as they appear in the different point groups.

The non-symmorphic space group C_{6v}^4 is formally obtained from D_{6h}^4 by retaining only those twelve elements which do not contain inversion. Thus, the point group C_{6v} and the factor groups for $\mathbf{k}$-vectors inside the Brillouin zone are easily found. Because of *non*-symmorphicity the factor groups for $\mathbf{k}$-vectors on the surface of the Brillouin zone have to be constructed according to the prescriptions given in Section 1.3.3.

TABLE 23

Listing of six point groups and their elements which are connected to the discussed space groups of layer compounds. The labelling of the elements is with reference to Table 22.

Element	1	2, 3	4, 5	6	7, 8, 9	10, 11, 12	13	14, 15	16, 17	18	19, 20, 21	22, 23, 24
point group												
D_{6h}	×	×	×	×	×	×	×	×	×	×	×	×
C_{6v}	×	×	×	×							×	×
D_{3d}	×	×			×		×		×		×	
D_{3h}	×	×			×			×		×	×	
C_{3v}	×	×									×	
C_{3i}	×	×					×		×			

Specific care has to be taken in setting up the different space groups based on the D_{3d} point group. Formally D_{3d} contains identity, a three-fold axis, 3 two-fold axes perpendicular to the three-fold axis and making 120° angles with respect to each other as well as three reflection planes containing the three-fold axis but each being perpendicular to one of the two-fold axes. The ambiguity consists in the choice of the orientation of the two-fold axes and the reflection planes with respect to the crystal lattice vectors. In other words, the 3 two-fold axes can either be $C_{2'}$ (operations 7, 8, 9 in Table 22) as it is the case in D_{3d}^3 (PbI$_2$) or $C_{2''}$ (operations 10, 11, 12 in Table 22) as it is the case in D_{3d}^1 (BiI$_3$).

As mentioned earlier in Section 1.3.2, differences between the two cases show up, when considering the small groups $G_\mathbf{k}$ for certain $\mathbf{k}$-vectors.

With this discussion of some space groups met in layered crystals we like to conclude the chapter about symmetry considerations. Purposely ample space has been given to the discussion of examples which should help anyone who wants to undertake an actual calculation of the electronic structure of a layered compound. Additional information can be found, e.g., in the very useful table work of Refs. [14] or [19].

References

[1] E. P. Wigner: *Group Theory and Its Application to Quantum Mechanics*, Academic Press, New York 1959.
[2] M. Tinkham: *Group Theory and Quantum Mechanics*, McGraw Hill, New York 1964.
[3] L. M. Falicov: *Group Theory and Its Physical Applications*, University Press, Chicago 1972.
[4] M. I. Petrashen and E. D. Trifonov: *Applications of Group Theory in Quantum Mechanics*, Cambridge 1969.
[5] V. Heine: *Group Theory in Quantum Mechanics*, Pergamon Press, Oxford 1966.
[6] G. F. Koster: *Space Groups and Their Representations*, Academic Press, New York 1957.
[7] M. Lax: *Symmetry Principles in Solid State and Molecular Physics*, Wiley, New York 1974.
[8] F. Bassani and G. Pastori-Parravicini: *Electronic States and Optical Transitions in Solids*, Pergamon Press, Oxford 1975.
[9] G. F. Koster, J. O. Dimmock, R. G. Wheeler, and H. Statz: *Properties of the Thirty-Two Point Groups*, MIT Press, Cambridge 1969.
[10] C. Herring: *J. Franklin Inst.* **233** (1942), 525.
[11] R. J. Elliott and R. Loudon: *J. Phys. Chem. Solids*, **15** (1960), 146.

[12] J. Zak: *J. Math. Phys.* **1** (1960), 165.

[13] H. W. Streitwolf: *Deut. Akad. Wiss. Berlin*, Veröff. Phys.-Techn. Inst., 1962.

[14] J. C. Slater: *Quantum Theory of Molecules and Solids*, McGraw Hill, New York 1965, Vol. 2.

[15] C. Herring: *Phys. Rev.* **52** (1937), 361.

[16] L. F. Mattheiss, J. H. Wood and A. C. Switendick: in *Methods in Computational Physics*, ed. by
      B. Adler, S. Fernbach, M. Rotenberg, Academic Press, New York 1968, Vol. 8.

[17] A. W. Luehrmann: PhD Thesis, Chicago 1968, unpublished.

[18] S. Altmann: *Rev. Mod. Phys.* **37** (1965), 33.

[19] C. J. Bradley and A. P. Cracknell: *The Mathematical Theory of Symmetry in Solids*, Clarendon
      Press, Oxford 1972; and O. V. Kovalev: *Irreducible Representations of the Space Groups*, Gordon
      Breach, New York 1968.

[20] B. Renaud and M. Schlüter: *Helv. Phys. Acta*, **45** (1972), 66.

[21] R. Zallen and D. F. Blossey: review in the same series.

[22] J. I. Hanoka, K. Vedam and H. K. Henisch: *J. Phys. Chem. Solids, Supplement* (1967), 369.

[23] A. Kuhn, R. Chevalier and A. Rimsky: *Acta Cryst.* **B32** (1976), 1975.

[24] R. W. G. Wyckoff: *Crystal Structures*, Wiley, New York 1963, Vols. 1, 2.

[25] J. A. Wilson, F. J. Di Salvo and S. Mahajan: *Adv. Phys.* **24** (1975), 117.

CHAPTER B

METHODS USED FOR CALCULATING BAND
STRUCTURES OF LAYERED MATERIALS

C. Y. FONG

Dept. of Physics, University of California, Davis, Calif. 95616, U.S.A.

In this chapter we shall review the main aspects of various methods which have been applied to calculate the band structures of layer compounds. Our objective is to have the discussions self-contained such that the amount of information provided here will be in sufficient detail to analyze results. The methods will be discussed in the chronological order of appearance of layer compound calculations in the literature. Moreover, this list of methods of band calculations is by no means complete; only common methods as they have been applied to layer compounds are described in detail.

1. The Tight-Binding Method (LCAO)

The tight-binding method was first suggested by Bloch [1] in 1928. The central idea of the method is to solve the one electron Schrödinger equation in solids by using a set of basis functions which is a linear combination of atomic orbitals (LCAO method). The method has had a good deal of success in studying the energy band structure of insulators, especially concerning their *qualitative* features, like e.g. the various symmetry properties of electronic states in those crystals. Since the atomic orbitals are centered at different sites, the basis functions are generally not orthogonal to each other. One encounters the evaluation of multi-center integrals and an inconvenient form of eigenvalue equation due to the non-orthogonality of the basis functions. The complicated calculational procedures involved in the application of this method thus often precluded it from general usage. Various schemes have been developed to improve the method [2]. In this article, we shall restrict ourselves to the discussion of two improved schemes which have been applied to study the electronic properties of layered compounds. In addition to the *ab initio* approach using LCAO type basis functions [2], a modified semi-empirical scheme as suggested by Bromley and Murray [3] shall be discussed. The main feature of this semi-empirical approach is the approximation of the multi-center integrals by semi-empirically adjusted two-center integrals. A second empirical scheme is the Slater and Koster interpolation scheme [4] which was used by Mattheiss [5]. In this scheme, one also considers only two-center integrals which, however, are determined entirely empirically by fitting the results to first principle band calculations. Although the details of the various schemes are different, the choice of the appropriate atomic orbitals is an essential step common to any modification of the tight-binding method. We shall make a remark about this step before discussing the details of each scheme. The outline of our discussions is: In Section 1.1 the original formulation of the tight-binding method will be discussed. In particular we shall concentrate on the derivation of the eigenvalue equation, the choice of appropriate atomic orbitals, and the complications arising in the computational procedures. The Bromley and Murray scheme will be presented in Section 1.2. Details in this section will include the description of matrix elements, the evaluation of two-center integrals and their semi-empirical treatment in contrast to *ab initio* type approaches. Finally, in Section 1.3, we shall explain the Slater and Koster interpolation scheme. A discussion of the transformation of atomic orbitals to an orthogonal set and the empirical determinations of two equivalent forms of two center integrals will be included.

1.1. ORIGINAL FORMULATION OF THE TIGHT-BINDING METHOD

In order to appreciate the various modifications to the tight-binding method which have been applied to calculate the electronic energy band structures of layered compounds, we shall start with a discussion of the original formulation of the method and point out its characteristic difficulties.

The Schrödinger equation for one electron in a solid is given by

$$H\psi_{n\mathbf{k}}(\mathbf{r}) = \left\{ -\frac{\hbar^2}{2m} \nabla^2 + V(\mathbf{r}) \right\} \psi_{n\mathbf{k}}(\mathbf{r}) = E_{n\mathbf{k}}\psi_{n\mathbf{k}}(\mathbf{r}). \tag{B1}$$

The first term in the Hamiltonian is the kinetic energy and the second term is the crystal potential seen by the electron. $E_{n\mathbf{k}}$ is the energy of the electron in the nth band with crystal momentum, $\mathbf{k}$. The vector $\mathbf{k}$ is restricted to lie within the first Brillouin zone (BZ). $\psi_{n\mathbf{k}}$ describes the state of one electron in the solid.

The potential used in this method may be expressed as a linear combination of atomic potentials, $V_a(\mathbf{r})$, centered at the atomic sites of the lattice. That is

$$V(\mathbf{r}) = \sum_l \sum_\alpha V_a(\mathbf{r} - \mathbf{R}_l - \tau_\alpha), \tag{B2}$$

where $\mathbf{R}_l$ is a lattice vector and τ_α is the position vector of the αth atom within the lth unit cell, measured with respect to a properly chosen coordinate system within the unit cell. To solve Equation (B1) with the potential given by Equation (B2), the wavefunction, $\psi_{n\mathbf{k}}$, is expanded in terms of an LCAO set of basis functions, $\{\Phi_{j\mathbf{k}}\}$.

$$\psi_{n\mathbf{k}}(\mathbf{r}) = \sum_j A_{nj}\Phi_{j\mathbf{k}}(\mathbf{r}) \tag{B3}$$

with the Bloch functions

$$\Phi_{j\mathbf{k}}(\mathbf{r}) = \frac{1}{\sqrt{N}} \sum_{l,\alpha} e^{i\mathbf{k}\cdot\mathbf{R}_l} X_j(\mathbf{r} - \mathbf{R}_l - \tau_\alpha), \tag{B4}$$

where $X_j(\mathbf{r})$ is the jth atomic orbital of atom α and N is the number of lattice points per unit volume. Note that the Φ's are not orthogonal to each other because of the spatial extension of the atomic orbitals, X, in the crystal. $\{A_{nj}\}$ is the set of expansion coefficients to be determined by solving the eigenvalue problem. Substituting Equation (B3) into Equation (B1) and taking the inner product with $\Phi_{j'\mathbf{k}}(\mathbf{r})$, we obtain

$$\sum_j A_{nj}\langle \Phi_{j'\mathbf{k}}| H |\Phi_{j\mathbf{k}}\rangle = E_{n\mathbf{k}} \sum_j A_{nj}\langle \Phi_{j'\mathbf{k}} | \Phi_{j\mathbf{k}}\rangle. \tag{B5}$$

By defining $H_{j'j} = \langle \Phi_{j'\mathbf{k}}| H |\Phi_{j\mathbf{k}}\rangle$ and $S_{j'j} = \langle \Phi_{j'\mathbf{k}} | \Phi_{j\mathbf{k}}\rangle$. This equation can be written as

$$\sum_j A_{nj}(H_{j'j} - E_{n\mathbf{k}}S_{j'j}) = 0. \tag{B6}$$

Thus, for a set of nontrivial A_{nj}'s, we solve the determinatal equation,

$$\det(H_{j'j} - E_{n\mathbf{k}}S_{j'j}) = 0. \tag{B7}$$

This is the original form of the tight-binding method.

An indispensable step to calculate the matrix elements, $H_{j'j}$ and $S_{j'j}$ in Equation (B7) is the selection of appropriate atomic orbitals, $X_j(\mathbf{r})$ (Equation (B4)). Since the electronic properties of solids are determined by the outer shell electrons, the valence electrons of the constituent atoms, the choice depends explicitly on the compound of interest. The general principle for the selection of atomic orbitals is simplicity. In the case of layer-compounds it is essential to include in the LCAO basis the outer shell s- and p-orbitals (for transition metals the partially filled d-orbitals are necessary instead of the p-orbitals) of the metallic ions and the outer shell s- and p-orbitals of the non-metallic ions. These electrons play the dominant roles in determining the electronic states near the Fermi energy and the lower energy valence bands. If one is interested in the unfilled conduction bands, higher energy atomic orbitals should be included. For most the compounds reviewed by this article, our main interest is focused on the valence states and the states near the Fermi energy. Therefore, in most cases, the higher energy orbitals, such as the metallic p-like orbitals, for transition metals, are neglected. As we shall see later, Bromley and Murray [3] use s, d and p orbitals as mentioned earlier, while Mattheiss [5] uses only the metallic d-orbitals to study the electronic states near the Fermi energy.

The general usage of the original form of the tight-binding method, however, has been limited because of two complications which arise in solving Equation (B7):

(i) The multi-center integrals.

In evaluating the matrix elements such as $H_{j'j}$ and $S_{j'j}$ in Equation (B7), we encounter multi-center integrals. Let us consider, for example, the expression $H_{j'j}$:

$$H_{j'j} = \frac{1}{N}\sum_{l',l} e^{i\mathbf{k}\cdot(\mathbf{R}_l - \mathbf{R}_{l'})} \times$$
$$\times \langle X_{j'}(\mathbf{r} - \mathbf{R}_{l'} - \tau_\alpha)| - \frac{\hbar^2}{2m}\nabla^2 + V(\mathbf{r}) |X_j(\mathbf{r} - \mathbf{R}_l - \tau_\beta)\rangle. \tag{B8}$$

For crystals with more than one atom per unit cell the matrix elements of the potential energy term become quite involved. Even with the potential energy given by Equation (B2) and the approximation of taking into account only nearest neighbor interactions, the following two and three-center integrals contribute to the matrix elements:

$$\langle X_{j'}(\mathbf{r} - \tau_\alpha)| V_a(\mathbf{r} - \tau_\beta) |X_j(\mathbf{r} - \tau_\beta)\rangle \tag{B9}$$

and

$$\langle X_{j'}(\mathbf{r} - \tau_\alpha)| V_a(\mathbf{r} - \tau_\gamma) |X_j(\mathbf{r} - \tau_\beta)\rangle. \tag{B10}$$

The procedures for accurately evaluating these integrals can be extremely tedious.

(ii) The Secular Form of the Eigenvalue Equation.

Equation (B7) does not have the conventional form of the secular equation where the eigenvalue, E_{nk}, appears isolated in the matrix diagonal. In Equation (B7) E_{nk} appears in the diagonal terms as well as in the off-diagonal terms, because the Φ_{jk}'s are *non-orthogonal* and thus the overlap matrix, S, has nonvanishing off-diagonal matrix elements $S_{j'j}$. The procedures to find the eigenvalues from the unconventional form such as Equation (B7) by finding the zero's of a polynomial in E_{nk} will be discussed in a later part of this chapter. This approach, however, is generally very time consuming.

To circumvent the first difficulty (i), the simplest approach is to keep only the least complicated multi-center integrals – the two-center integrals and to neglect all three center integrals. This is actually the case for all LCAO calculations on layered compounds discussed in this article. However, this approximation generally yields poor quantitative results and further improvements are necessary.

One of the ways to resolve the second difficulty (ii), was suggested by Bromley and Murray [3]. They used the Cholesky decomposition method to find a transformation matrix which transforms Equation (B7) into the conventional secular equation.

1.2. THE BROMLEY-MURRAY SCHEME

Both the Bromley-Murray scheme and the Slater-Koster interpolation scheme though quite different in detail surmount the two main difficulties discussed in Section 1.1. In the present scheme the calculation of the matrix elements is simplified in a semi-empirical fashion. By keeping two-center integrals and by assuming nearest-neighbor interactions only, the matrix elements are first evaluated from atomic orbitals and atomic potentials. These matrix elements are then adjusted empirically.

We shall begin with a detailed description of the matrix elements and several ways of how to evaluate them numerically. Then, the semi-empirical treatment of the matrix elements will be discussed, and finally, we shall show how Equation (B7) may be transformed into the conventional form of a secular equation.

To calculate the matrix elements of Equation (B7), one has to know the wavefunctions Φ_{jk} of Equation (B4) and the atomic potentials of Equation (B2). We have mentioned earlier the importance of the choice of atomic orbitals used in LCAO calculations.

The explicit form of the atomic potential on the other hand depends on the choice of atomic configuration. In general, the non-metallic ions in the transition metal and non-transition metal dichalcogenides have only one neutral configuration, i.e., $(ns)^2(np)^4$, where n is the principal quantum number of the outer shell. However, transition metal ions may have several possible neutral configurations. The choice of a configuration may be made on physical grounds. For example, Mo atoms are known to have two possible neutral configurations, i.e., $4d^5 5s^1$ or $4d^4 5s^2$. Experimental evidence seems to indicate that Mo layer-compounds are only slightly ionic. Therefore, the configuration $4d^4 5s^2$ will be less appropriate to use than $4d^5 5s^1$ because the two s-like electrons may more easily be ionized to form a more ionic compound. With the proper choice of a configuration, atomic potentials as tabulated by Herman and Skillman [6] may be used to evaluate the matrix elements. Modification to the Herman-Skillman atomic potentials are sometimes necessary and shall be discussed after enumerating the various two-center integrals in the matrix elements $H_{j'j}$ and $S_{j'j}$.

As mentioned before, an important approximation in evaluating the matrix elements in most of the LCAO approaches is to retain only the nearest neighbor two-center integrals. Few approaches like e.g. the discrete variational scheme developed by Ellis and Painter [2], do not make this approximation. Matrix elements are integrated numerically as they appear in the original LCAO formulation. To discuss the integrals, let us first examine the expression of $H_{j'j}$ in Equation (B8). By using Equation (B2) and

$$\left(-\frac{\hbar^2}{2m} \nabla^2 + V_a(\mathbf{r})\right) |X_j(\mathbf{r})\rangle = E_j^{(0)} |X_j(\mathbf{r})\rangle \tag{B11}$$

we obtain

$$H_{j'j} = E_j^{(0)} S_{j'j} + \sum_l e^{-i\mathbf{k}\cdot\mathbf{R}_l} \langle X_{j'}(\mathbf{r} - \mathbf{R}_l - \boldsymbol{\tau}_\alpha)| \times$$
$$\times V_a(\mathbf{r} - \boldsymbol{\tau}_\beta) |X_j(\mathbf{r} - \boldsymbol{\tau}_\beta)\rangle, \qquad \alpha \neq \beta$$
$$+ \sum_{l,\gamma} \langle X_{j'}(\mathbf{r} - \mathbf{R}_l - \boldsymbol{\tau}_\alpha)| V_a(\mathbf{r} - \boldsymbol{\tau}_\gamma) |X_j(\mathbf{r} - \mathbf{R}_l - \boldsymbol{\tau}_\beta)\rangle, \qquad \alpha = \beta. \tag{B12}$$

In the nearest neighbor approximation l is restricted to lattice vectors pointing to unit cells such that the αth atom in that cell is the nearest neighbor of the βth atom in the reference unit cell. $S_{j'j}$ is given as follows:

$$S_{j'j} = \sum_l e^{-i\mathbf{k}\cdot\mathbf{R}_l} \langle X_{j'}(\mathbf{r} - \mathbf{R}_l - \boldsymbol{\tau}_\alpha) | X_j(\mathbf{r} - \boldsymbol{\tau}_\beta)\rangle. \tag{B13}$$

From Equations (B12) and (B13), three different forms of two-center integrals emerge:

$$\langle X_{j'}(\mathbf{r} - \mathbf{R}_l - \boldsymbol{\tau}_\alpha)| V_a(\mathbf{r} - \boldsymbol{\tau}_\beta) |X_j(\mathbf{r} - \boldsymbol{\tau}_\beta)\rangle \tag{B14}$$

$$\langle X_{j'}(\mathbf{r} - \mathbf{R}_l - \boldsymbol{\tau}_\alpha)| V_a(\mathbf{r} - \boldsymbol{\tau}_\beta) |X_j(\mathbf{r} - \mathbf{R}_l - \boldsymbol{\tau}_\alpha)\rangle \tag{B15}$$

and

$$\langle X_{j'}(\mathbf{r} - \mathbf{R}_l - \boldsymbol{\tau}_\alpha) | X_j(\mathbf{r} - \boldsymbol{\tau}_\beta)\rangle. \tag{B16}$$

A special case of Equation (B14) has been presented as Equation (B9) and is called 'bond potential integral'. The other two integrals (Equations (B15) and (B16)) are termed 'crystal field integral' and 'bond overlap integral' respectively.

As mentioned earlier, the atomic potential (in particular the exchange potential) used in one of these integrals, (Equation (B15)) should be modified from the standard Herman-Skillman form. In principle, this exchange potential is non-local. In [6] it was approximated by Slater's local expression ($\alpha = 1$)

$$V_{ex}(\mathbf{r}) = -6[\frac{3}{8\pi} \rho(\mathbf{r})]^{1/3}, \tag{B17}$$

where $\rho(\mathbf{r})$ is the total charge density. It was pointed out by Robinson et al. [7] that Equation (B17) often overestimates the contribution in the low density region due to the fact that screening effects have not been taken into account properly. In particular one would expect that crystal field integrals (Equation (B15)) are not accurately determined by this form because wavefunction in this integral are centered at sites different from the potential. To account for the proper screening effect, Bromley and Murray adopted the modification suggested in [7] which makes use of the following form of the exchange potential in Equation (B15):

$$V_{ex}^s(\mathbf{r}) = V_{ex}(\mathbf{r}) F(k_s/k_F) \quad \text{for} \quad |\mathbf{r}| \geq R_c \tag{B18}$$
$$= 0 \quad \text{otherwise}$$

and

$$F(\chi) = 1 - \tfrac{4}{3}\chi \tan^{-1}\left(\frac{2}{\chi}\right) + \frac{\chi^2}{2}\ln\left(1 + \frac{4}{\chi^2}\right)$$
$$- \frac{\chi^2}{6}\left[1 - \frac{\chi^2}{4}\ln\left(1 + \frac{4}{\chi^2}\right)\right], \tag{B19}$$

where $V_{ex}(\mathbf{r})$ is given in Equation (B17), $1/k_s$ is the Thomas-Fermi screening length and k_F is the Fermi momentum. χ is k_s/k_F. R_c is the radius of the highest energy core states.

The atomic orbitals $X_j(\mathbf{r})$ used in the wavefunction expansion may be taken from various Hartree-Fock calculations. While Herman and Skillman [6] tabulate the values of $rR_n(r)$ as defined by

$$X_j(\mathbf{r}) = R_n(r)Y_{l,m}(\theta, \varphi) \tag{B20}$$

on a mesh of r-points, expansions in terms of Slater functions (exponentials) are given e.g. by Clementi or Watson and Freeman [8]. A normalized Slater function may be written as

$$X_j^s(\mathbf{r}) = R_n^s(Z, r)Y_{l,m}(\theta, \varphi) \tag{B21}$$

with

$$R_n^s(Z, r) = [(2n)!]^{-1/2} (2Z)^{n+1/2} r^{n-1}e^{-Zr}. \tag{B22}$$

Up to about ten Slater functions are typically used to approximate s- or p-type orbitals.

Having decided upon the type of atomic orbitals and the atomic potential, the next step is to evaluate the two-center integrals in Equations (B14), (B15) and (B16). In order to facilitate the following discussion, we let

$$\mathbf{r}_A = \mathbf{r} - \mathbf{R}_l - \boldsymbol{\tau}_\alpha \tag{B23}$$

and

$$\mathbf{r}_B = \mathbf{r} - \boldsymbol{\tau}_\beta. \tag{B24}$$

A spheroidal polar coordinate system may be used for direct calculation. The relations between the spherical coordinate systems centered at the atomic sites and the new coordinate system are given by

$$\xi = \frac{1}{R}(r_A + r_B),$$

$$\eta = \frac{1}{R}(r_A - r_B), \tag{B25}$$

$$\phi = \phi_A = \phi_B,$$

where ϕ_A, ϕ_B, $\mathbf{r}_A$, $\mathbf{r}_B$ are shown in Figure 1 and R is the separation between the two nearest neighbors. From these definitions, η ranges from -1 to 1 whereas ξ varies from 1 to ∞.

To carry out the integrations over the spheroidal polar coordinates, the wavefunctions and potentials may be interpolated on a linear radial mesh centered at each atom. Simpson's rule may be used to perform the integrations in η and ξ. In practice, the

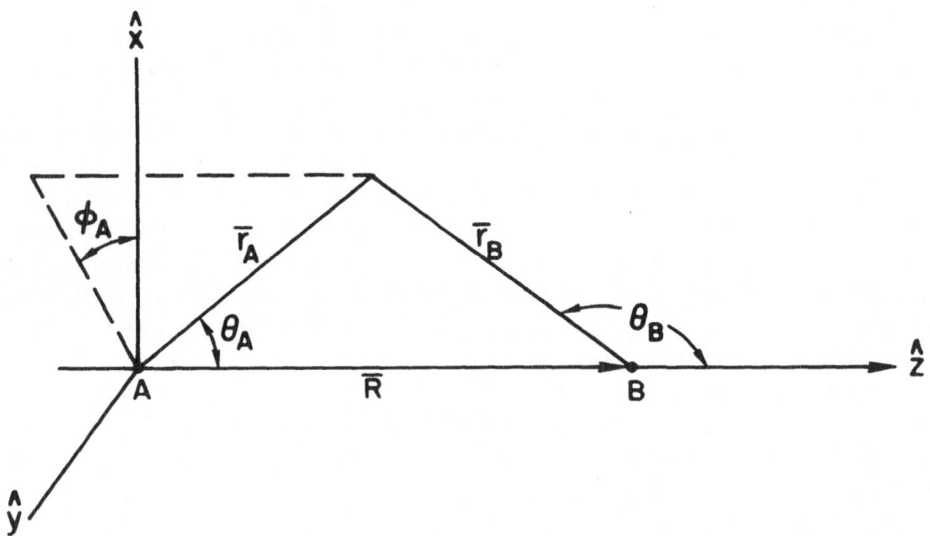

Fig. 1.   Spherical coordinates centered at sites A and B used to calculate two-center itegrals in the tight binding scheme.

upper limit of ξ is chosen to be 3, because the integrals converge very fast. This limit amounts to include contributions up to about one bond length on either side of each site.

Since the numerical integration of two-center integrals involving Slater functions is relatively extended, modified expansions are sometimes chosen. Each Slater function may e.g. be expanded in terms of gaussians as proposed by Steward [9].

$$X_j^s(\mathbf{r}) = \sum_i d_i g_{i,n,l,m}(\mathbf{r}) \tag{B26}$$

with

$$g_{i,n,l,m}(\mathbf{r}) = 2^{n+\frac{1}{2}}[(2n-1)!]^{-1/2}\left(\frac{2}{\pi}\right)^{1/4}\alpha_i^{(2n+1)/4}r^{n-1}$$
$$\times\, e^{-\alpha_i r^2}\, Y_{lm}(\theta,\,\varphi). \tag{B27}$$

The expansion coefficients are given by Steward [9] for $Z = 1$. The same coefficients, however, may be used for general Z-values if α_i in Equation (B27) is replaced by $\alpha_i \cdot Z^2$. Using a gaussian expansion, all two-center integrals can be solved analytically.

Sometimes symmetry properties of the crystal can be used to simplify the LCAO secular equation. For example, in approximating 2H–MoS$_2$ by a single layer, the reflection plane, σ_h, which passes through the metallic ions perpendicular to the c-axis is a useful symmetry operation. Wavefunctions associated with the chalcogenides can be combined into even and odd functions under σ_h, such as e.g., the s orbitals of the S-atoms in MoS$_2$.

$$\Phi_{\pm sk} = \Phi_{sk}(\mathbf{r} - \tau_{s1}) \pm \Phi_{sk}(\mathbf{r} - \tau_{s2}), \tag{B28}$$

where τ_{s1} and τ_{s2} are the position vectors of the two S-atoms in the unit cell measured with respect to the origin at the metallic atom. Hamiltonian matrix elements between different $\Phi_{\pm}$ states then vanish due to symmetry for any $\mathbf{k}$-vector in the two-dimensional

BZ. Such simplifications by symmetrization of the wavefunctions according to the irre-ducible representations of the small group G_k have been discussed extensively in Chapter A.

The matrix elements obtained by the method described above are often not suffi-ciently precise to give accurate band structures of crystals. One way to improve the result is to introduce a set of empirical scaling parameters multiplying these matrix elements. These parameters are usually determined by fitting the band structure to experimental data (especially optical data). For example, one may associate a reduction parameter for each type of two-center integral given in Equations (B14), (B15) and (B16). An alternative way is followed by Bromley and Murray [3] and by Bassani and Parravicini [10]. In the single layer approximation of the CdI$_2$ type layered compounds σ_h is a symmetry operation for all k-points in the two dimensional BZ. As mentioned above, the wavefunctions Φ_{jk} in Equation (B4) can be classified according to even (σ) or odd (π) behavior under this symmetry transformation and the corresponding inte-grals for Equations (B14) and (B16) are called σ bond and π bond integrals. Bromley and Murray associated a set of (three) parameters (X, Y, Z) with each of the crystal field integrals, the σ-bond integrals and the π-bond integrals respectively. These para-meters are then determined empirically by inspecting optical data. Structures in the optical spectrum of a solid originate from transitions taking place in regions inside the BZ of either large joint density of states or large transition matrix elements. Regions of large joint density of states are often associated with high symmetry k-points and are called critical points or van Hove singularities [11]. Observed optical structures may be assigned to such critical points in the BZ. Sometimes, however, symmetry induced vanishing of transition matrix elements can compensate the effect of large joint density of states contributions. It is therefore essential to determine the selection rules for various optical transitions by use of group theory (see Chapter A). Following these procedures a preliminary band structure calculation may be used to assign particular interband transitions to a few critical points. The empirical reduction parameters are then determined by quantitatively fitting these particular transition energies to experi-ment. Then the complete band structure may be calculated.

In the following we shall briefly describe a procedure of evaluating the tight binding secular equation (Equation (B7)). It is desirable to transform Equation (B7) into the conventional form of an eigenvalue problem. As we mentioned earlier, the non-vanishing overlap matrix S causes the particular form of Equation (B7). Bromley and Murray noted that the matrix S in Equation (B7) can be forced to be a real symmetric and positive definite matrix by properly choosing the phase factors of all wavefunctions. As shown by Cholesky [12], this real symmetric and positive definite matrix may be decomposed into a product of a matrix with vanishing matrix-elements above the principal diagonal and its transpose. Let this matrix be L, then S can be rewritten as

$$S = L\tilde{L}. \tag{B29}$$

Considering the following eigenvalue equation:

$$H\psi_{nk} = E_{nk}S\psi_{nk} \tag{B30}$$

we obtain by multiplying by L^{-1} on both sides and by using Equation (B29),

$$(L^{-1}H\tilde{L}^{-1})\tilde{L}\psi_{nk} = E_{nk}\tilde{L}\psi_{nk}. \tag{B31}$$

This is the conventional form of an eigenvalue equation for an operator H' and its eigenfunction $\psi'_{n\mathbf{k}}$, where

$$H' = L^{-1}H\tilde{L}^{-1},$$

$$\psi'_{n\mathbf{k}} = \tilde{L}\psi_{n\mathbf{k}}. \tag{B32}$$

1.3. THE SLATER-KOSTER INTERPOLATION SCHEME

Although the general usage of the tight-binding method is limited by the difficulties discussed in Section 1.1. and the quantitative results obtained by straightforward application of the method are often rather discouraging, the method does offer some attractive features:

(i) the band structure calculated from the LCAO method exhibits the correct symmetry properties – the degeneracies at various **k**-points in the BZ are consistent with the irreducible representations of the corresponding small groups, $G_{\mathbf{k}}$;

(ii) even at a general **k**-point in the BZ, the tight-binding method deals with only a small number of atomic orbitals chosen by the method discussed in Section 1.1, leading to a Hamiltonian matrix of small order which is easily solved compared to the large sized matrices used in most first-principles methods for calculating band structures. While the method of symmetrizing wavefunctions in first-principles calculations in order to reduce the Hamiltonian matrix into block form can only be used at high symmetry **k**-points in the *BZ*, an unsymmetrized tight-binding method may conveniently be used at general **k**-points.

Slater and Koster thus suggested the LCAO method to be used as an interpolation scheme. By treating the matrix elements $H_{j'j}$ or the two-center integrals as disposable parameters to fit the energies at a few high symmetry **k**-points, the tight binding method may then be used to calculate the complete band structure. The energies at the high symmetry **k**-points are usually calculated by first-principles methods using symmetrized wavefunctions.

In this interpolation scheme, a way of avoiding the form of Equation (B7) is to start with an orthogonal set of wavefunctions, $\{\varphi_j\}$, such that an equation of type (B31) will directly be obtained. Because of the different basis functions used, the matrix elements in the present case will differ from the ones given in Section 1.2.

The scheme starts with an orthonormal set of basis functions derived from atomic orbitals as suggested by Löwdin [13]. To derive these functions we define for convenience

$$X_j(\mathbf{r} - \mathbf{r}_\mu) \equiv X_j(\mathbf{r} - \mathbf{R}_l - \tau_\alpha) \tag{B33}$$

with X_j being normalized. j may be chosen as described in Section 1.1. The overlap matrix of these atomic orbitals has the form

$$\Delta_{jj'}(\mathbf{r}_\mu, \mathbf{r}_\nu) = \int X_j^*(\mathbf{r} - \mathbf{r}_\mu)X_{j'}(\mathbf{r} - \mathbf{r}_\nu)\,\mathrm{d}^3 r \quad \text{for} \quad \mu \neq \nu. \tag{B34}$$

The orthonormalized wavefunctions can now be written as

$$\varphi_j(\mathbf{r} - \mathbf{r}_\nu) = \sum_{j'} X_{j'}(\mathbf{r} - \mathbf{r}_\nu)(1 + \Delta)_{j'j}^{-1/2} \tag{B35}$$

and

$$(1 + \varDelta)_{j'j}^{-1/2} = \delta_{j'j} - \tfrac{1}{2}\varDelta_{j'j} + \tfrac{3}{8} \sum_p \varDelta_{j'p}\varDelta_{pj} - \cdots. \tag{B36}$$

It can be shown (Appendix of [4]) that the transformation (B35) preserves the symmetry of the atomic orbitals. Similar to Equation (B4), a Bloch state constructed from the functions φ_j is given by

$$\varPhi_{j\mathbf{k}}(\mathbf{r}) = \frac{1}{\sqrt{N}} \sum_\mu e^{i\mathbf{k}\cdot\mathbf{r}_\mu}\, \varphi_j(\mathbf{r} - \mathbf{r}_\mu). \tag{B37}$$

By forming the trial wavefunction as in Equation (B3) and following the procedures leading to Equation (B7), the secular equation can be simply expressed as

$$\det(H_{j'j} - \delta_{j'j}E_{n\mathbf{k}}) = 0, \tag{B38}$$

where $H_{j'j}$ is similar to Equation (B8) but with X_j replaced by φ_j. In Equation (B38), there are no off-diagonal overlap matrix elements because of the orthonormality of the $\{\varphi_j\}$'s, if $\varphi_{j'}$ and φ_j are centered at different sites. The diagonal matrix elements H_{jj} are equal for all orbitals with the same quantum number. We may denote them by $\varepsilon_{jj}(0)$, where 0 means the two orbitals j and j are centered at the same site. These energies may also be treated as disposable parameters.

As mentioned earlier, only two-center integrals are retained. They typically have the following form:

$$\langle\varphi_{j'}(\mathbf{r} - \mathbf{r}_\mu)|\, H(\mathbf{r} - \mathbf{r}_\nu)\, |\varphi_j(\mathbf{r} - \mathbf{r}_\nu)\rangle \quad \mu \neq \nu, \tag{B39}$$

where the $\mathbf{r}_\nu$ dependence of H occurs in the potential term. These two-center integrals in both the diagonal and the off-diagonal matrix elements may also be treated as disposable parameters. To reduce the total number of parameters, we shall present two typical types of simplifications.

(i) The crystal symmetry may relate integrals to each other, thus reducing the numbers of independent matrix elements in Equation (B39). Group theoretical rules presented in Chapter A allow to perform this symmetry reduction.

(ii) In practice, one mainly is interested in describing the states near the Fermi energy. Thus the number of atomic orbitals necessary to describe these energy bands can be reduced further. For example, the states near the Fermi energy in $TaSe_2$ are predominantly d-like, thus only d-orbitals are considered in the calculations of [5]. Finally, the number of interacting neighboring 'shells' of atoms may be reduced and within physical limits be determined by the ultimate number of free parameters one wishes to treat.

It follows from the preceding discussion that two main features of the Slater-Koster scheme differ from the Bromley-Murray scheme:

(i) All independent two-center integrals are explicitly treated as disposable empirical parameters whereas the Bromley-Murray scheme treats them semi-empirically;

(ii) Unlike the Bromley-Murray scheme, the two-center integrals are not restricted to nearest neighbors, since usually a larger number of free parameters can be fitted to the energies obtained by first-principles calculations.

To parametrize the two-center integrals, we shall present a set of symbols which characterize these integrals. There are two ways to express integrals such as given in Equation (B39) based on the fact that the set $\{\varphi_j\}$ has the symmetry of atomic orbitals:

(i) All orthonormalized wavefunctions are referred to a fixed rectangular coordinate system. The matrix elements are specified by the corresponding wavefunctions and the separation between the two centers. The set of $\{\varphi_j\}$ of one atom may contain the s, $p(p_x, p_y, p_z)$ and d(xy, xz, yz, $x^2 - y^2$, $3z^2 - r^2$) orbitals. In general, the vector $\mathbf{R} = \mathbf{r}_\mu - \mathbf{r}_\nu$ between the two atoms as shown in Figure 1 does not lie along the z-axis of the coordinate system. The cosines of the direction of $\mathbf{R}$ with respect to the fixed system may be denoted by (ξ, ζ, η). Then, the matrix element formed by the jth orbital of the center atom at $\mathbf{r}_\nu$ and the j'th orbital on the other atom separated by $\mathbf{R}$ may be named as $\varepsilon_{j'j}(\xi, \zeta, \eta)$. Combined with the $\varepsilon_{jj}(0)$'s, defined earlier, these parameters are fitted to an equal number of energies at high symmetry $\mathbf{k}$-points through Equation (B38). Then the complete band structure may be calculated.

(ii) All orthonormalized wavefunctions are expressed in terms of molecular orbitals. The integrals of Equation (B39) are similar to integrals appearing in problems of diatomic molecules. Therefore, the set of $\{\varphi_j\}$'s may be transformed such that the angular momenta of the resulting functions are quantized along the axis connecting the two atoms. For example, if φ_j and $\varphi_{j'}$ correspond to atomic p-orbitals, they may be used to form $p\sigma$, $p\pi_\pm$ molecular orbitals which are characterized by 0 and ± 1 angular momentum along $\mathbf{R}$. Similar combinations can be formed for s, d orbitals. These molecular orbitals with different quantum numbers are orthogonal to each other. If we take the $p\sigma$ orbital on the center atom and the $d\sigma$ orbital on its neighboring atom, the matrix element between the two orbitals may be termed $(pd\sigma)$. Unlike to case (i), integrals involving $p\pi_\pm$, $d\pi_\pm$ orbitals can change sign, if the atoms are interchanged.

A matrix element like the one given by Equation (B39) may then be expressed as linear combinations of terms similar to the one defined as $(pd\sigma)$. The expansion co-efficients are uniquely defined by deriving the molecular orbitals from the φ's. One may then parameterize these brackets to fit the energies as described at the end of (i). The two ways of expressing matrix elements are equivalent to each other. Their relations have been tabulated in [4].

The different tight-binding schemes discussed in this section have been applied to several transition metal and non-transition metal layer crystals. The various results shall be discussed in a later chapter.

2. The Augmented-Plane-Wave Method (APW)

Among the first-principles methods for calculating band structures of solids, the Augmented-plane-wave (APW) method has the most success in treating transition metals and their compounds. Aside from quantitative aspects, the results of this method often provide useful information about correct band orderings and effects of hybridization between states with different atomic symmetry properties. In this section, we shall consider the APW method in some detail. The basic concept of the method will be described in Section 2.1, Sections 2.2 and 2.3 will deal with the detailed construction of the potential and the method of calculating the energy eigenvalues, respectively. Finally, in Section 2.4 we shall give some comments on practical calculations.

2.1. BASIC CONCEPTS OF THE AUGMENTED-PLANE-WAVE METHOD

In 1937, Slater [14] proposed the Augmented-plane-wave (APW) method, which was based on the following physical picture. In a solid, as an electron approaches a nucleus, the potential energy becomes strong, negative and spherically symmetric similar to the potential of an isolated atom. In the interstitial regions, on the other hand, the potential energy of the electron varies only slowly and joins continuously with the strong potential near the nucleus. To develop an approximated, but reasonable potential for practical calculations, Slater assumed that the potential is spherically symmetric within a sphere of radius R_m about each nucleus and constant outside between the spheres. This model potential called muffin-tin-potential, has the following mathematical form:

$$V_m(\mathbf{r}) = \sum_{l,\alpha} V_\alpha(|\mathbf{r} - \mathbf{R}_l - \tau_\alpha|), \qquad |\mathbf{r} - \mathbf{R}_l - \tau_\alpha| \leq R_{m\alpha}$$

$$= V_0 \quad \text{otherwise,} \tag{B40}$$

where $\mathbf{R}_l$ and τ_α are similarly defined as in Equation (B2). $R_{m\alpha}$ the radius of the muffin-tin sphere of the αth atom is of the order of the core radius. V_α is the spherical potential, V_0 is a constant, in practice often set to be zero. Since the potential is separated into a spherically symmetric part and a constant part, it is convenient to solve the one-electron Schrödinger equation (Equation (B1)) in the two different regions separately and to match the solutions at $R_{m\alpha}$. Within the sphere, the wavefunction can be expanded in terms of spherical harmonics, whereas in the interstitial region of constant potential the wavefunction is expressed in terms of plane waves. The resultant wavefunction is called augmented plane wave (APW).

To present a mathematical description of an APW, we shall first consider the wavefunction within the muffin-tin sphere. Let $|\mathbf{r} - \tau_\alpha|$, θ, φ be the spherical coordinates at the αth atom, the wavefunction $\psi_{\mathbf{k}}(\mathbf{r})$ may be expressed in terms of spherical harmonics as:

$$\psi_{\mathbf{k}} = \sum_{l=0}^{\infty} \sum_{m=-l}^{l} A_{lm}(\mathbf{k}) Y_{lm}(\theta, \varphi) u_{\alpha l}(|\mathbf{r} - \mathbf{r}_\alpha|), \qquad |\mathbf{r} - \tau_\alpha| \leq R_{m\alpha}, \tag{B41}$$

where $u_{\alpha l}(|\mathbf{r}|)$ is the radial wavefunction which satisfies the following equation:

$$\frac{1}{r^2} \frac{d}{dr}\left(r^2 \frac{du_{\alpha l}}{dr}\right) + \left(\frac{l(l+1)}{r^2} + V_\alpha(r)\right)u_{\alpha l} = E u_{\alpha l}, \tag{B42}$$

where the factor $2m/\hbar^2$ is absorbed in the potential V_α and the energy E of the electron in the solid. Note that the solutions of Equation (B42) are not atomic-like because of different boundary conditions. The coefficients $A_{lm}(\mathbf{k})$ are determined by matching the wavefunction within the sphere to that in the interstitial region at the surface of the sphere ($r = R_{m\alpha}$).

To match the wavefunctions, we expand the plane-wave solution in the interstitial region also in terms of spherical harmonics:

$$e^{i\mathbf{k_G} \cdot \mathbf{r}} = e^{i\mathbf{k_G} \cdot \tau_\alpha} \sum_{l=0}^{\infty} \sum_{m=-l}^{l} 4\pi i^l j_l(k_G \cdot |\mathbf{r} - \tau_\alpha|) \, Y_{lm}^*(\theta_{\mathbf{k_G}}, \varphi_{\mathbf{k_G}}) \, Y_{lm}(\theta, \varphi), \tag{B43}$$

where $\mathbf{k_G} = \mathbf{k} + \mathbf{G}$, $\mathbf{G}$ is a reciprocal lattice vector, $k_G = |\mathbf{k_G}|$. $\theta_{\mathbf{k_G}}$ and $\varphi_{\mathbf{k_G}}$ are the polar and azimuthal angles of $\mathbf{k_G}$. By setting $|\mathbf{r} - \boldsymbol{\tau}_\alpha| = R_{m\alpha}$ in Equations (B41) and (B43) and by comparing the coefficients of $Y_{lm}(\theta, \varphi)$, we obtain

$$A_{lm}(\mathbf{k}) = 4\pi i^l \, e^{i\mathbf{k_G}\cdot\boldsymbol{\tau}_\alpha} \frac{j_l(k_G R_{m\alpha})}{u_{\alpha l}(R_{m\alpha}, E)} \, Y_{lm}^*(\theta_{\mathbf{k_G}}, \varphi_{\mathbf{k_G}}).$$ (B44)

Now, an APW can be written as

$$\psi_{\mathbf{k_G}}(\mathbf{r}) = e^{i\mathbf{k_G}\cdot\mathbf{r}} \quad \text{for} \quad r \geq R_{m\alpha}$$

$$= 4\pi e^{i\mathbf{k_G}\cdot\boldsymbol{\tau}_\alpha} \sum_{l=0}^{\infty} \sum_{m=-l}^{l} i^l \frac{j_l(k_G R_{m\alpha})}{u_{\alpha l}(R_{m\alpha}, E)} Y_{lm}^*(\theta_{\mathbf{k_G}}, \varphi_{\mathbf{k_G}}) \times$$

$$\times \, Y_{lm}(\theta, \varphi) u_{\alpha l}(|\mathbf{r} - \boldsymbol{\tau}_\alpha|) \quad \text{for} \quad r \leq R_{m\alpha}.$$ (B45)

This set of APW's for different $\mathbf{G}$'s is used as basis functions for ψ in Equation (B1). Note, that the energy E appears implicitly in the radial functions $u_{\alpha l}$.

2.2. CONSTRUCTION OF THE CRYSTAL POTENTIAL

Before discussing the construction of the crystal potential, two points are worth mentioning:

(i) At present, for most of the studies of the transition metal layer compounds spin-orbit interactions have been neglected. Therefore, in the following we shall neglect this interaction in constructing the crystal potential.

(ii) The most simple way to account for electron-electron interactions is to start with proper configurations of the constituent atoms from which the charge densities are calculated as described in Section 1.2. Then in principle the final form of the potential may be obtained self-consistently. Since this is a rather laborious process, most of the layer compound calculations reported in the literature are not carried through self-consistently [15].

As described in Section 2.1, the potential near the αth atomic site should be predominantly of atomic character. This potential is frequently derived from Hartree-Fock-Slater (HFS) self-consistent field calculations like those presented by Herman and Skillman [6]. The general form of the potential near αth atom can be written as

$$V_\alpha(\mathbf{r}) = V_{c\alpha}(\mathbf{r}) + V_{\text{ex},\alpha}(\mathbf{r}) \quad \text{for} \quad r \leq R_{m\alpha},$$ (B46)

where $V_{c\alpha}(\mathbf{r})$ is the Coulombic potential due to the atom α at the sphere center and to the neighboring atoms and where $V_{\text{ex},\alpha}(\mathbf{r})$ is the total exchange potential. The Coulombic potential (in Rydbergs) due to the center atom alone can be written as

$$V_{c\alpha 0}(\mathbf{r}) = -\frac{2Z}{r} + V_{\alpha 0}(\mathbf{r}),$$ (B47)

where Z is the atomic number, r is the distance from the nucleus measured in unit of the Bohr radius. The first term in Equation (B47) originates from the nucleus, while the second term is the direct electron-electron Coulomb interaction and satisfies Poisson's equation,

$$\nabla^2 V_{\alpha 0}(\mathbf{r}) = -8\pi \rho_{\alpha 0}(\mathbf{r})$$ (B48)

with

$$\rho_{\alpha 0}(\mathbf{r}) = \sum_{\substack{\text{occupied} \\ \text{states}}} |\psi_{nlm}(\mathbf{r})|^2 \tag{B49}$$

being the charge density of the αth atom. The wavefunctions ψ_{nlm} may be taken as HFS atomic wavefunctions given in Ref. [6].

The contributions to the Coulombic potential $V_{c\alpha}(\mathbf{r})$ due to nearby atoms (β) may be included by a method suggested by Löwdin [16]. The (lm)th component of $V_{c\beta 0}(\mathbf{r} - \tau_\beta)$, denoted by $F_{lm}(\mathbf{r}_\beta)$, with $\mathbf{r}_\beta = \mathbf{r} - \tau_\beta$, can be expanded into spherical harmonics about the α *site* as

$$F_{lm}(\mathbf{r}_\beta) = \sum_{l'm'} Y_{l'm'}(\theta_\alpha, \phi_\alpha) B_{l'm',lm}. \tag{B50}$$

The coefficients $B_{l'm', lm}$ are given by:

$$B_{l'm',lm} = \int \sin\theta_\alpha \, d\theta_\alpha \, d\phi_\alpha \, F_{lm}(\mathbf{r}_\beta) Y_{l'm'}^*(\theta_\alpha, \phi_\alpha). \tag{B51}$$

The relations between $(\theta_\alpha, \phi_\alpha)$ and $(r_\beta, \theta_\beta, \phi_\beta)$ can easily be derived from Figure 2.

$$\phi_\beta = \phi_\alpha, \tag{B52}$$

$$r_\beta \sin\theta_\beta = r_\alpha \sin\theta_\alpha, \tag{B53}$$

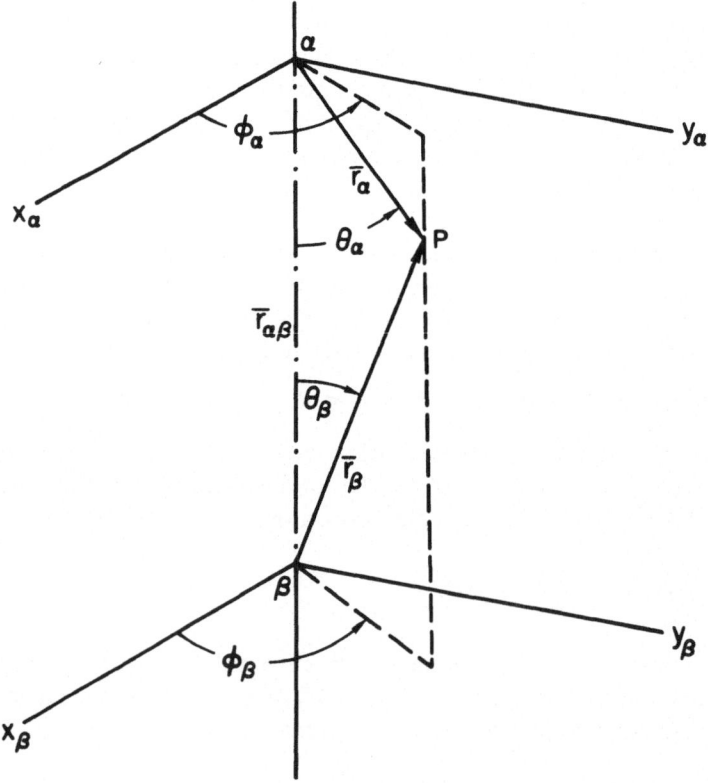

Fig. 2. Coordinate systems used for determining the crystal potential in APW calculations.

$$r_\beta \cos \theta_\beta + r_\alpha \cos \theta_\alpha = r_{\alpha\beta} \tag{B54}$$

and

$$r_\beta^2 = r_\alpha^2 + r_{\alpha\beta}^2 - 2r_\alpha r_{\alpha\beta} \cos \theta_\alpha, \tag{B55}$$

where $r_{\alpha\beta}$ is the distance between the α and β sites. From Equations (B54) and (B55) it is possible to express $\cos \theta_\beta$ as a function of r_β with fixed r_α. Furthermore, from Equation (B55), we get

$$r_\beta \, dr_\beta = r_\alpha r_{\alpha\beta} \sin \theta_\alpha \, d\theta_\alpha \tag{B56}$$

and Equation (B51) may be written as

$$B_{l'm',lm}(r_\alpha) = \frac{1}{r_\alpha r_{\alpha\beta}} \int_{|r_{\alpha\beta}-r_\alpha|}^{r_{\alpha\beta}+r_\alpha} r_\beta F_{lm}(\mathbf{r}_\beta) Y_{l'm'}^*(\theta_\alpha(r_\beta), \phi_\beta) \, dr_\beta \, d\phi_\beta, \tag{B57}$$

where $\theta_\alpha(r_\beta)$ indicates that θ_α is expressed as a function of r_β through the relations given in Equations (B53), (B54), and (B55). If $V_{c\beta 0}(\mathbf{r} - \boldsymbol{\tau}_\beta)$ is assumed to be spherically symmetric about the βth site, then

$$F_{00}(\mathbf{r}_\beta) = \frac{1}{\sqrt{4\pi}} f_0(r_\beta). \tag{B58}$$

The only nonvanishing B in Equation (B57) is

$$B_{00,00} = \frac{1}{2} \frac{1}{r_\alpha r_{\alpha\beta}} \int_{|r_{\alpha\beta}-r_\alpha|}^{r_{\alpha\beta}+r_\alpha} r_\beta f_0(r_\beta) \, dr_\beta \tag{B59}$$

assuming that $V_{c\alpha 0}(r - \boldsymbol{\tau}_\alpha)$ is also be spherically symmetric. In general, several $Y_{l'm'}(\theta_\alpha, \phi_\alpha) B_{l'm',lm}(r_\alpha)$ expressions contribute to $V_{c\beta 0}(\mathbf{r}_\alpha)$. The total Coulomb potential near the α-site with $r \leq R_{m\alpha}$ is given by

$$V_{c\alpha}(\mathbf{r}) = V_{c\alpha 0}(\mathbf{r}) + \sum_\beta V_{c\beta 0}(\mathbf{r}), \tag{B60}$$

where β is summed over an appropriate number of near neighbors determined by the crystal structure.

The exchange potential, $V_{\text{ex},\alpha}(\mathbf{r})$, may be approximated by Equation (B17). To include contributions of neighboring atoms, one first finds the total charge density at the point $\mathbf{r}$ by Löwdin's method in an analogous manner as used for the Coulomb potential in Equation (B60). Then, the total exchange potential within the muffin-tin sphere about αth atom is

$$V_{\text{ex},\alpha}(\mathbf{r}) = -6\left[\frac{3}{8\pi} \rho_\alpha(\mathbf{r})\right]^{1/3}. \tag{B61}$$

The total resulting potential, $V_\alpha(\mathbf{r})$, obtained this way is generally not spherically symmetric because of the inclusion of partially filled states and the anisotropic arrangement

of atoms in the crystal. The spherical portion of the muffin-tin potential may be determined by a weighted average of $V_\alpha(\mathbf{r})$ in Equation (B46) for three independent spatial
directions. Thus

$$V_\alpha(r) = \mathbf{V}_\alpha(\mathbf{r}) = V_{c\alpha}(r) + V_{ex,\alpha}(r). \tag{B62}$$

There are several ways to determine the constant potential V_0 between the muffin-tin
spheres by treating the non-linear charge density dependence of the exchange terms
differently. We shall mention two methods:

(i) Mattheiss in studying transition metal compounds [17] sets up a three-dimensional
grid distributed uniformly over the unit cell and evaluates $V(\mathbf{r}_i)$ at all grid points $\mathbf{r}_i$
in an analogous manner to calculate $V_\alpha(\mathbf{r})$ in Equation (B46) and then calculates the
average of all $V(\mathbf{r}_i)$.

(ii) The second method has been mentioned in [18]. The average Coulomb potential
is computed as in (i), whereas the contributions from the exchange interaction are
calculated by first averaging the charge density over the interstitial region and then
substituting the averaged density into an expression similar to Equation (B61).

After determining the value of V_0, the reference potential may be shifted to set the
constant portion of the muffin-tin potential equal to zero.

The muffin-tin potential is thus obtained by averaging the potential $V_\alpha(\mathbf{r})$, centered
at the αth atom within the muffin-tin sphere and the potential $V(\mathbf{r})$ in the interstitial
region separately. Consequently the averaging scheme introduces deviations and a
discontinuity at $r = R_{m\alpha}$ between the potential within the muffin-tin sphere and the
constant portion of the potential ($=0$) outside. It has been shown by DeCicco [21] that
the deviations within the muffin-tin sphere have negligible effect on the energy eigenvalues, whereas it is of more importance to account for deviations of $V(\mathbf{r})$ from zero in
the outside region. These deviations are also the primary reason for the discontinuity
of the muffin-tin potential at $r = R_{m\alpha}$. In the calculations of the transition metal layer
compounds corrections to the muffin-tin potential have been introduced outside the
muffin-tin spheres. To incorporate these corrections into the matrix elements between
APW's it is convenient to define the Fourier coefficients of the correcting potential as

$$V_\Delta(\mathbf{G}) = \frac{1}{N_p} \sum_{j=1}^{N_p} e^{-i\mathbf{G}\cdot\mathbf{r}_j} V_\Delta(\mathbf{r}_j) \tag{B63}$$

with

$$V_\Delta(\mathbf{r}_j) = V(\mathbf{r}_j),$$

where N_p is the total number of points of a grid in the interstitial region and $\mathbf{G}$ is a
reciprocal lattice vector. The resulting crystal potential thus obtained for the transition
metal layer compounds is

$$V(\mathbf{r}) = V_m(\mathbf{r}) + V_\Delta(\mathbf{r}), \tag{B64}$$

where $V_m(\mathbf{r})$ is the muffin-tin potential given in Equation (B40) and $V_\Delta(\mathbf{r})$ is given by

$$V_\Delta(\mathbf{r}) = V_\Delta(\mathbf{r}_j), \quad \text{for } \mathbf{r}_j \text{ in the interstitial region,}$$
$$= 0, \quad \text{otherwise.} \tag{B65}$$

2.3. CALCULATION OF THE ENERGY EIGENVALUES

Similar to the tight-binding method in finding the energy eigenvalues, a variational scheme is used to solve the one-electron Schrödinger equation. Details of the variational scheme of the two methods differ due to different properties of the basis functions. As we recall, the wavefunction given in Equation (B45) is required to be continuous at $r = R_{m\alpha}$. No requirement about the continuity of its derivative at the surface of the muffin-tin sphere has been imposed. Furthermore, strict continuity of the wavefunction would require an infinite sum over angular momenta within the muffin-tin sphere to match the plane wave portion for $r \geq R_{m\alpha}$. For finite sums the APW shows a slight discontinuity at $r = R_{m\alpha}$ in addition to the discontinuity in its derivative.

Several variational schemes for treating these discontinuous trial wavefunctions have been proposed [14, 20, 21]. The schemes suggested in [20] and [21] for obtaining an expression of the eigenvalues E start with a functional including discontinuity contributions at the muffin-tin spheres. With the use of modern computing facilities, one can extend the summation over angular momenta in Equation (B45) to render the effect of the discontinuity in the wavefunction itself negligible. In Slater's original formulation of the APW method [14] emphasis was put on the kinetic energy contribution arising from the discontinuity of the first derivative of the wavefunction at $r = R_{m\alpha}$. We shall limit ourselves to the discussion of this scheme.

We expand an arbitrary wavefunction at a given $\mathbf{k}$-point within the BZ into APW basis functions:

$$\psi_{\mathbf{k}}(\mathbf{r}) = \sum_{\mathbf{G}} A_{\mathbf{G}} \psi_{\mathbf{k}\mathbf{G}}(\mathbf{r}), \tag{B66}$$

where the APW's $\psi_{\mathbf{k}\mathbf{G}}$ are given by Equation (B45). By following the usual variational procedures, we obtain,

$$\sum_{\mathbf{G}'} (H - E)_{\mathbf{G},\mathbf{G}'} A_{\mathbf{G}'} = 0, \tag{B67}$$

where

$$(H - E)_{\mathbf{G},\mathbf{G}'} = \frac{1}{\Omega} \int_{\text{cell}} [(\nabla \psi_{\mathbf{k}\mathbf{G}}^*) \cdot (\nabla \psi_{\mathbf{k}\mathbf{G}'}) + (V(\mathbf{r}) - E)\psi_{\mathbf{k}\mathbf{G}}^* \psi_{\mathbf{k}\mathbf{G}'}] \, \mathrm{d}^3 r, \tag{B68}$$

where Ω is the volume of the unit cell. The number of $\mathbf{G}$'s in the summation of Equation (B67) is determined by the required accuracy of the eigenvalues and wavefunctions.

Comparing Equations (B68) and (B8) shows that different kinetic energy operators have been used, due to the existence of a surface integral taking into account the discontinuity in the derivative of the APW. We now present explicit expressions of APW matrix elements and explain how to determine the eigenvalues.

Corresponding to the different definitions of the wavefunction in the different regions of the unit cell the evaluation of matrix elements is divided into the following steps.

(i) The matrix element of the kinetic energy operator between two plane waves is

$$T_{\mathbf{G},\mathbf{G}'} = \frac{1}{\Omega} \int_I (\nabla e^{-i(\mathbf{k}+\mathbf{G})\cdot\mathbf{r}}) \cdot (\nabla e^{i(\mathbf{k}+\mathbf{G}')\cdot\mathbf{r}}) \, \mathrm{d}^3 r, \tag{B69}$$

where I means the interstitial region. By adding and subtracting the following term

$$\frac{1}{\Omega}\sum_\alpha e^{-i(\mathbf{G}-\mathbf{G}')\cdot\tau_\alpha}\int_\alpha (\nabla e^{-i(\mathbf{k}+\mathbf{G})\cdot\mathbf{r}})\cdot(\nabla e^{i(\mathbf{k}+\mathbf{G}')\cdot\mathbf{r}})\,\mathrm{d}^3 r, \tag{B70}$$

where the integration is inside the muffin-tin sphere at the αth atom, we obtain

$$T_{\mathbf{G},\mathbf{G}'} = (\mathbf{k}+\mathbf{G})\cdot(\mathbf{k}+\mathbf{G}')\delta(\mathbf{G}-\mathbf{G}') -$$

$$-\frac{4\pi}{\Omega}\sum_\alpha e^{-i(\mathbf{G}-\mathbf{G}')\cdot\tau_\alpha}(\mathbf{k}+\mathbf{G})\cdot(\mathbf{k}+\mathbf{G}')\times$$

$$\times\frac{R_{m\alpha}^2}{|\mathbf{G}-\mathbf{G}'|}j_1(|\mathbf{G}-\mathbf{G}'|\,R_{m\alpha}), \tag{B71}$$

where the first term is obtained by adding Equation (B70) to Equation (B69). The second contribution in Equation (B71) is just the subtracted term of Equation (B70). The factor $e^{-i\mathbf{G}\cdot\tau_\alpha}$ is the structure factor associated with the position of the αth in the unit cell.

(ii) The matrix element of $(H-E)_{\mathbf{G}\mathbf{G}'}$ with the muffin-tin potential has the form:

$$(H-E)_{\mathbf{G},\mathbf{G}'}^{m\alpha} = \frac{1}{\Omega}\int_{m\alpha} [(\nabla\psi_{\mathbf{k}\mathbf{G}}^*)\cdot(\nabla\psi_{\mathbf{k}\mathbf{G}'}) + (V_{m\alpha}(\mathbf{r})-E)\psi_{\mathbf{k}\mathbf{G}}^*\psi_{\mathbf{k}\mathbf{G}'}]\,\mathrm{d}^3 r, \tag{B72}$$

where $m\alpha$ indicates the contribution within the muffin-tin sphere at the αth site. The wavefunction in this expression is the one given in Equation (B45) with $r \le R_{m\alpha}$. By applying Green's theorem to the first term in Equation (B72) we obtain

$$(H-E)_{\mathbf{G},\mathbf{G}'}^{m\alpha} = \frac{1}{\Omega}\int_{m\alpha}\psi_{\mathbf{k}\mathbf{G}}^*[(-\nabla^2 + V_{m\alpha}(\mathbf{r})-E)\psi_{\mathbf{k}\mathbf{G}'}]\,\mathrm{d}^3 r +$$

$$+\frac{1}{\Omega}\int\psi_{\mathbf{k}\mathbf{G}}^*\left(\frac{\partial}{\partial n}\psi_{\mathbf{k}\mathbf{G}'}\right)\mathrm{d}S, \tag{B73}$$

where $\partial\psi/\partial n$ is the outward normal derivative of ψ at the surface of the muffin-tin sphere. The first term vanishes since the expression in the square bracket equals the one given in Equation (B42) with $V_\alpha(r)$ replaced by $V_{m\alpha}(\mathbf{r})$. The surface integral can easily be evaluated by using Equation (B45).

$$(H-E)_{\mathbf{G},\mathbf{G}'}^{m\alpha} = \frac{4\pi}{\Omega}e^{i(\mathbf{G}-\mathbf{G}')\cdot\tau_\alpha}R_{m\alpha}^2\sum_{l=0}^{\infty}j_l(|\mathbf{k}+\mathbf{G}|\,R_{m\alpha})j_l(|\mathbf{k}+\mathbf{G}'|\,R_{m\alpha})\times$$

$$\times(2l+1)P_l(\cos\theta_{\mathbf{G},\mathbf{G}'})\frac{u_{\alpha l}'(R_{m\alpha},E)}{u_{\alpha l}(R_{m\alpha},E)}, \tag{B74}$$

where the P_l's are Legendre polynomials resulting from the addition theorem of the $Y_{lm}(\theta_{\mathbf{k}\mathbf{G}},\varphi_{\mathbf{k}\mathbf{G}})$'s. $\theta_{\mathbf{G},\mathbf{G}'}$ is the angle between $\mathbf{k}+\mathbf{G}$ and $\mathbf{k}+\mathbf{G}'$ and $u_{\alpha l}'(R_{m\alpha})$ is

$$\left.\frac{\mathrm{d}u_{\alpha l}(r)}{\mathrm{d}r}\right|_{r=R_{m\alpha}}.$$

(iii) The matrix element $-E$ between the two plane waves can be derived by following the method described in (i)

$$(-E)^{\mathrm{I}}_{\mathbf{GG'}} = -E\delta(\mathbf{G} - \mathbf{G'}) + E\,\frac{4\pi}{\Omega}\sum_{\alpha} e^{-i(\mathbf{G}-\mathbf{G'})\cdot\boldsymbol{\tau}_{\alpha}}\,(\mathbf{k} + \mathbf{G})\cdot(\mathbf{k} + \mathbf{G'}) \times$$

$$\times \frac{R_{m\alpha}^2}{|\mathbf{G} - \mathbf{G'}|}\,j_1(|\mathbf{G} - \mathbf{G'}|\,R_{m\alpha}). \tag{B75}$$

(iv) The matrix element of the correction to the muffin-tin potential outside the muffin-tin spheres has the form:

$$(V_A)_{\mathbf{G},\mathbf{G'}} = \frac{1}{\Omega}\int_I e^{-i(\mathbf{k}+\mathbf{G})\cdot\mathbf{r}}\,V_A(\mathbf{r})\,e^{i(\mathbf{k}+\mathbf{G'})\cdot\mathbf{r}}\,\mathrm{d}^3 r, \tag{B76}$$

where $V_A(\mathbf{r})$ is given by Equation (B65). Since $V_A(\mathbf{r})$ vanishes within the muffin-tin spheres, we may extend the range of integration to the whole unit cell. Then, this matrix element is just the Fourier transform of $V_A(\mathbf{r})$. By using Equation (B63), we obtain

$$(V_A)_{\mathbf{G},\mathbf{G'}} = V_A(\mathbf{G} - \mathbf{G'}). \tag{B77}$$

Combining the results in Equations (B71), (B74), (B75), and (B77), the resulting APW matrix element can be written as

$$(H - E)_{\mathbf{G},\mathbf{G'}} = [(\mathbf{k} + \mathbf{G})\cdot(\mathbf{k} + \mathbf{G'}) - E]\,\delta(\mathbf{G} - \mathbf{G'})$$

$$+ \frac{4\pi}{\Omega}\sum_{\alpha} e^{-i(\mathbf{G}-\mathbf{G'})\cdot\boldsymbol{\tau}_{\alpha}}\,F_{\alpha\mathbf{GG'}} + V_A(\mathbf{G} - \mathbf{G'}),$$

where

$$F_{\alpha\mathbf{GG'}} = R_{m\alpha}^2\left\{-[(\mathbf{k} + \mathbf{G})\cdot(\mathbf{k} + \mathbf{G'}) - E]\frac{j_1(|\mathbf{G} - \mathbf{G'}|\,R_{m\alpha})}{|\mathbf{G} - \mathbf{G'}|} + \right.$$

$$+ \sum_{l=0}^{\infty}(2l + 1)P_l(\cos\theta_{\mathbf{G},\mathbf{G'}})j_l(|\mathbf{k} + \mathbf{G}|\,R_{m\alpha})j_l(|\mathbf{k} + \mathbf{G'}|\,R_{m\alpha}) \times$$

$$\left. \times\,\frac{u'_{\alpha l}(R_{m\alpha}, E)}{u_{\alpha l}(R_{m\alpha}, E)}\right\}. \tag{B78}$$

2.4. COMMENTS ON COMPUTATIONAL DETAILS

We like to make the following comments about Equation (B78):

(i) The structure factor $e^{-i(\mathbf{G}-\mathbf{G'})\cdot\boldsymbol{\tau}_{\alpha}}$ accounts for the relative phase of the αth atom in the unit cell to the APW matrix element. In most of the transition metal layer compounds the crystals possess inversion symmetry. This reduces the structure factor and thus the total APW matrix-elements to real quantities, for an appropriately chosen origin. The process of solving the eigenvalue equation is then reduced by about a factor of five.

(ii) The calculation of the logarithmic derivative $[u'_{\alpha l}(R_{m\alpha}, E)]/[u_{\alpha l}(R_{m\alpha}, E)]$ requires some extra computation. In the following, we shall discuss the Numerov scheme discussed in [18] to obtain $u_{\alpha l}(R_{m\alpha}, E)$ from a differential equation. The logarithmic derivative can then be calculated by numerical differentiation and interpolation of the following one-dimensional differential equation:

$$R''_{\alpha l} \equiv \frac{d^2 R_{\alpha l}}{dr^2} = \left[V_{m\alpha}(r) + \frac{l(l+1)}{r^2} - E \right] R_{\alpha l} \equiv Q_l(r) R_{\alpha l}, \tag{B79}$$

where $R_{\alpha l} = r u_{\alpha l}(r)$. In the Numerov scheme one starts by specifying the independent variable r on a discrete mesh and by substituting the solution $R_{\alpha l}(r)$ at its boundaries by exact asymtotic solutions. Then integration is performed by a discrete difference method. To do so, let $x = \log(r)$ and $X = (1/\sqrt{r}) R_{\alpha l}(r)$. Equation (B79) may then be written as

$$X'' = [e^{2x}(V_{m\alpha} - E) + (l + \tfrac{1}{2})^2] X. \tag{B80}$$

As $x \to -\infty$, the first term in Equation (B80) vanishes. Since $V_{m\alpha} \underset{r \to 0}{\to} -(2Z/r)$, the above condition is expressed as

$$e^x 8Z \ll 1 \tag{B81}$$

with $l = 0$. Let x_1 be the value of x which satisfies Equation (B81). The solution X near x_1 which is regular at $x \to -\infty$ is given by

$$X(x_i) = X(x_1) e^{(l + \frac{1}{2})(x_i - x_1)}. \tag{B82}$$

$R_{\alpha l}(r_{i+1})$ at a neighboring mesh point can be found by using a Taylor series expansion about r_i.

$$R_{\alpha l}(r_{i+1}) = R_{\alpha l}(r_i) + h R'_{\alpha l}(r_i) + \frac{h^2}{2!} R''_{\alpha l}(r_i) + \cdots, \tag{B83}$$

where $h = r_{i+1} - r_i$. A similar expression for $R_{\alpha l}(r_{i-1})$ can also be found.

$$R_{\alpha l}(r_{i-1}) = R_{\alpha l}(r_i) - h R'_{\alpha l}(r_i) + \frac{h^2}{2!} R''_{\alpha l}(r_i) + \cdots \tag{B84}$$

with $h = r_i - r_{i-1}$. By adding Equations (B83) to (B84), we obtain the following difference equation

$$R_{\alpha l}(r_{i+1}) - 2 R_{\alpha l}(r_i) + R_{\alpha l}(r_{i-1}) = h^2 R''_{\alpha l}(r_i) + \frac{h^4}{12} R^{iv}_{\alpha l}(r_i) + \cdots, \tag{B85}$$

where $R^{iv}_{\alpha l}(r_i)$ is the fourth order derivative of $R_{\alpha l}(r)$ evaluated at r_i. By neglecting the fourth and high order derivatives, Equation (B85) may be used as recursion relation to define the second derivative.

The following modification of the above method, called Numerov scheme, provides more accurate solutions $R_{\alpha l}(r)$ with reasonably small spacings h ($\sim 10^{-3} Z^{-1/3}$, where Z is the atomic number of the constituent atom).

By taking the second order derivative of Equation (B85) and neglecting the sixth order derivative, we get

$$R''_{\alpha l}(r_{i+1}) - 2 R''_{\alpha l}(r_i) + R''_{\alpha l}(r_{i-1}) = h^2 R^{iv}_{\alpha l}(r_i). \tag{B86}$$

Using Equation (B86), the difference equation (Equation (B85)) can be expressed in terms of the second order derivatives of $R_{\alpha l}$ alone. The recursion formula of the differential equation (Equation (B83)) then has the form:

$$R_{\alpha l}(r_{i+1}) = \left[1 - \frac{h^2}{12} Q_l(r_{i+1})\right]^{-1} \{[2R_{\alpha l}(r_i) - R_{\alpha l}(r_{i-1})] +$$

$$+ h^2[\tfrac{5}{6} Q_l(r_i)R_{\alpha l}(r_i) + \tfrac{1}{12} Q_l(r_{i-1})R_{\alpha l}(r_{i-1})]\}. \tag{B87}$$

From this equation the logarithmic derivates $[u'_{\alpha l}(R_{m\alpha},E)]/[u_{\alpha l}(R_{m\alpha},E)]$, may be calculated by recursive numerical interpolation starting with the boundary conditions (Equation (B82)) for small r-values.

Further simplifications may be made for the energy dependence of the logarithmic derivative. According to earlier APW-calculations of various compounds [18] the logarithmic derivatives for $l \geq 4$ depend linearly on the energy in a rather large range. Therefore, in practice, for $l \geq 4$, one may approximate

$$\frac{u'_{\alpha l}(R_{m\alpha}, E)}{u_{\alpha l}(R_{m\alpha}, E)} = \gamma^0_{\alpha l} + \gamma^1_{\alpha l} E, \tag{B88}$$

where $\gamma^0_{\alpha l}$ and $\gamma^1_{\alpha l}$ are constants determined by fitting to some calculated values of the logarithmic derivative. For $l < 4$, however, exact calculations of the energy dependence of the derivatives are necessary. After evaluating these terms and substituting them into Equation (B67), the secular equation for nontrivial solutions of A_G is given by

$$\det(H - E) = 0. \tag{B89}$$

Note that the eigenvalue E implicitly appears in the off-diagonal terms through the logarithmic derivatives. Therefore, standard diagonalization routines are not applicable. The customary scheme [18] to obtain the eigenvalues is to evaluate the determinant as a function of E over a range of interest and to locate the zeros by an iterative inverse interpolation method (like e.g., a 4-point Lagrangian method). The fastest method for evaluating determinants as arising in present case is the Gaussian elimination method with pivoting [18], which is contained in most computer libraries.

Finally, the equation of the eigenvectors has to be solved. For each eigenenergy E_c the linear homogeneous system:

$$\sum_{G'} (H - E_c)_{G,G'} A_{G'} = 0 \tag{B90}$$

has to be solved which may be done by standard procedures like Gaussian elimination.

The APW method described in detail has been applied to several 1T and 2H transition metal layer compounds for high symmetry **k**-points and lines in the Brillouin zone by Mattheiss [15]. It was found to be necessary to have about 160 and 280 simple unsymmetrized APW's for the two structures respectively in order to converge in energy to within 0.05 eV. At symmetry points or lines, a set of symmetrized APW basis functions was formed according to the procedures discussed in Chapter A, which reduced the determinant sizes to as much as 60×60. The results of these calculations shall be presented in a later chapter.

3. The Empirical Pseudopotential Method (EPM)

As we have seen in the two previous sections one of the main tasks of calculating the electronic energy band structure is to construct the one-electron crystal potential. Two difficulties are often met in constructing the potential;

(i) laborious computational procedures; and

(ii) the lack of accurate but computationally convenient forms of the exchange potential. This latter difficulty is usually the reason for limiting the accuracy of the calculated results to about 0.5 eV compared with experimental data.

The empirical pseudopotential method avoids the explicit construction of a one electron potential in solids from atomic potentials and wavefunctions. In its earlier applications, the potential of a valence electron was simulated by a weak effective local potential – a local pseudopotential. The Fourier coefficients of this pseudopotential are then treated as disposable parameters to be determined by fitting calculated results to experimental spectroscopic data. This method is very successful in empirically determining the band structures of group IV, III–V and II–VI semiconductors and insulators [22]. The important peaks in the linear optical spectra derived from these band structures generally agree with experimental data to within 0.1 eV. This method has also been applied to calculate the band structures of non-transition metal layer compounds.

The method has then been modified by augmenting it with a non-local pseudopotential [23], so that one is able to calculate band structures of noble and transition metals and their compounds.

Very recently pseudopotentials have been used to perform *ab-initio*-like self-consistent calculations for complicated systems like surfaces, molecules or lattice defects. In these calculations the potential fourier-transforms are *not* treated as disposable fitting parameters, but determined *self-consistently* by an iterative solution of Hartree-Fock-Slater like equations. Since this method has not been applied to layer-compounds yet, we shall not present a detailed discussion of it here, but refer the reader to the original papers [24] and [25].

In the following, we shall first present some general considerations about the empirical pseudopotential method. The specific characteristics of the local pseudopotential method including the treatment of spin-orbit effects shall then be presented, followed by a discussion of the non-local pseudopotential method. Finally, we shall comment on some fitting procedures.

3.1. GENERAL CONSIDERATIONS OF THE PSEUDOPOTENTIAL METHOD

The effect of a pseudopotential acting onto a valence electron in a solid is described to be equivalent to an effective potential which the electron experiences. There are several ways to formulate this effective potential under the restriction that the energy of the valence electron remains unchanged as compared to the exact crystal potential case. A comprehensive review of the various formulations has been given by Cohen and Heine [22]. Here, we shall discuss only the form of the pseudopotential suggested by Phillips and Kleinman [26] based on the orthogonalized plane wave (OPW) method [27] to motivate our discussions in Sections 3.2 and 3.3.

As for other band theories, we start with the one-electron Schrödinger equation as given in Equation (B1). If we assume that $V(\mathbf{r})$ is obtained by a self-consistent Hartree or Hartree-Fock approximation, the solutions of Equation (B1) in principle, provide a complete description of the energies and the corresponding wavefunctions for all possible one-electron states of the crystal ground state. These states range from the occupied inner core levels over the valence states to the unoccupied conduction states.

As we have discussed before, many interesting electronic properties characterizing a crystal are determined by its valence states. The wavefunctions characterizing these states are required to be orthogonal to those describing the core states. In the OPW method, the wavefunction of a valence state is thus expressed in terms of a linear combination of orthogonalized-plane-waves (OPW's) as typically given by

$$|\text{OPW}\rangle = |\mathbf{k}\rangle - \sum_c |c\rangle \langle c | \mathbf{k}\rangle, \tag{B91}$$

where $|\mathbf{k}\rangle$ is a plane wave with wave vector $\mathbf{k}$ normalized to the unit cell, that is $(1/\sqrt{\Omega})e^{i\mathbf{k}\cdot\mathbf{r}}$, Ω is the unit cell volume. $|c\rangle$ is the Bloch function of a core state. Typical expressions of $|c\rangle$ are given in Equation (B4) with j labeling the inner core orbital. The OPW in Equation (B91) is consistent with the physical picture of the crystal potential described before. In the interstitial region, where $\langle c | \mathbf{k}\rangle$ is small, the wave function varies slowly as does the potential. However, near the core region, the wavefunction resembles an atomic wavefunction as the potential approaches the atomic potential. The strong oscillatory behavior of the wavefunction for the valence states is thus characterized by the second term in Equation (B91). Phillips and Kleinman [26] extended the idea of the OPW and formulated a pseudopotential for the valence electron.

Let us assume that the wavefunction, $\psi_{n\mathbf{k}}$, is expressed as a linear combination of the OPW's as given in Equation (B91) and let us define the sum of plane waves, $\sum_\mathbf{G} A_{n\mathbf{G}} e^{i(\mathbf{k}+\mathbf{G})\cdot\mathbf{r}} \equiv \Phi_{n\mathbf{k}}(\mathbf{r})$, with expansion coefficients $A_{n\mathbf{G}}$, to be the *pseudo*wave function, which describes the *smooth* part of the exact wavefunction, $\psi_{n\mathbf{k}}$. A schematic comparison of $\psi_{n\mathbf{k}}(\mathbf{r})$ and $\Phi_{n\mathbf{k}}(\mathbf{r})$ is shown in Figure 3. The actual wavefunction can now be written as

$$\psi_{n\mathbf{k}}(\mathbf{r}) = \Phi_{n\mathbf{k}}(\mathbf{r}) - \sum_c |c\rangle \langle c | \Phi_{n\mathbf{k}}\rangle. \tag{B92}$$

It is easily shown that this wavefunction is orthogonal to all core states, i.e., that $\langle c | \psi_{n\mathbf{k}}\rangle = 0$. Substituting Equation (B92) into Equation (B1) and rearranging the terms, we obtain,

$$\left\{ -\frac{\hbar^2}{2m} \nabla^2 + V_p(\mathbf{r}) \right\} \Phi_{n\mathbf{k}}(\mathbf{r}) = E_{n\mathbf{k}} \Phi_{n\mathbf{k}}(\mathbf{r}) \tag{B93}$$

with

$$V_p(\mathbf{r}) = V(\mathbf{r}) + \sum_c (E_{n\mathbf{k}} - E_c)|c\rangle \langle c|, \tag{B94}$$

where E_c is the energy of a core state $|c\rangle$, and $V_p(\mathbf{r})$ is called the pseudopotential. The last two equations are the essence of the Phillips and Kleinmans's formulation of the pseudopotential method.

To elucidate the significance of Equations (B93) and (B94), we consider the following points:

(i) In Equation (B94) the energy, $E_{n\mathbf{k}}$, for the valence state is always greater than the one of the core states, E_c. Therefore, the second term is a repulsive potential.

$$V_R(\mathbf{r}) = \sum_c (E_{n\mathbf{k}} - E_c)|c\rangle \langle c|. \tag{B95}$$

82

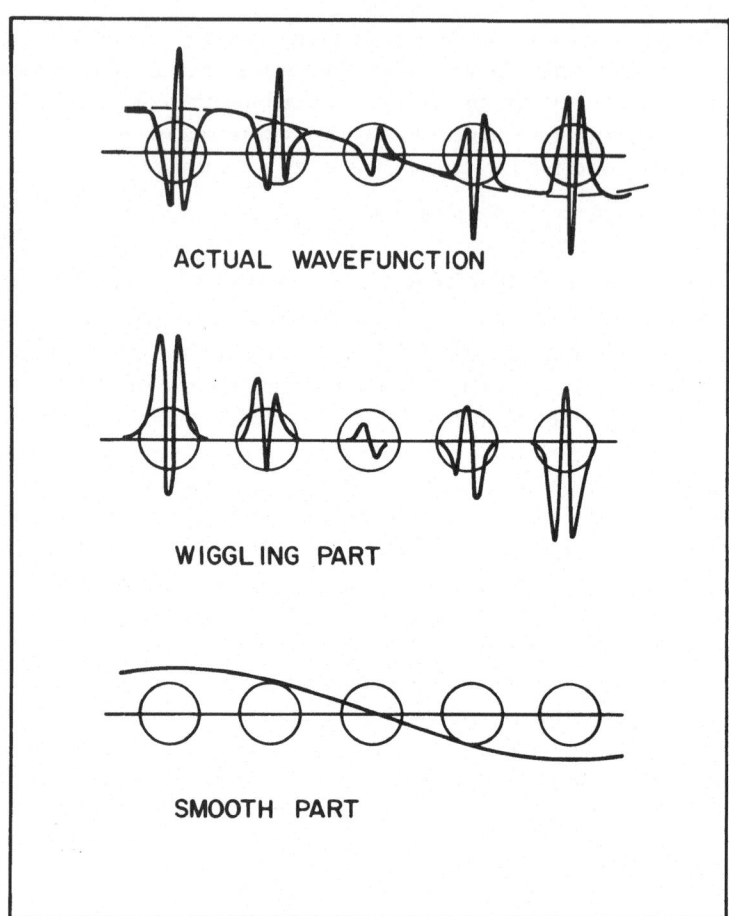

Fig. 3. Schematic comparison of the actual wavefunction $\psi_{n\mathbf{k}}$ and the smooth pseudowave-function $\Phi_{n\mathbf{k}}$.

Its effect is to cancel the attractive crystal potential, $V(\mathbf{r})$, in the core region where $|c\rangle$ is non-vanishing. The resultant pseudopotential, $V_p(\mathbf{r})$, is weak, but energy dependent and non-local.

(ii) The projection operator upon core states, $|c\rangle\langle c|$, in Equation (B94) or (B95) can be expressed in terms of spherical harmonics. Therefore, the operator projects out different angular momenta when it acts on an arbitrary function. We may write

$$V_R(\mathbf{r}) = V_s(\mathbf{r}) + V_p(\mathbf{r}) + V_d(\mathbf{r}) + \cdots, \tag{B96}$$

where s, p, d correspond to the different angular momenta, contained in the core states. This reflects the non-locality of the repulsive potential. A second source of non-locality, of course, is the (Hartree-Fock-type) self-consistent crystal potential $V(\mathbf{r})$ due to the presence of exchange terms.

(iii) Equation (B93) is the Schrödinger equation with a weak pseudopotential for the smooth pseudowave function, $\Phi_{n\mathbf{k}}$. It is important to note that the eigenvalues of this equation correspond to the exact energies of the valence states.

Therefore, what the transformation of Equation (B92) has accomplished is that one may obtain the energies of valence states alone by solving a Schrödinger equation with a weak potential. The corresponding pseudowave functions provide a proper description of the valence states away from the core regions as illustrated in Figure 3. Close to the cores the pseudowave function has to be orthogonalized to core states according to Equation (B91) to be correct.

To rigorously construct the pseudopotential from first principle, one encounters the same difficulties present in the OPW method. Therefore, in practice, the power of the pseudopotential method is in its empirical or semi-empirical approach. Several schemes have been developed in the past to empirically determine the pseudopotential. An extensive review of the schemes are given by Cohen and Heine [22]. In this article, we shall restrict ourselves to the discussions of the scheme as it has been applied to calculate the band structures of layer compounds. As mentioned briefly in the introductory remarks, this method involves both forms of potentials, the local and the non-local empirical pseudopotential.

3.2. THE LOCAL EMPIRICAL PSEUDOPOTENTIAL METHOD

The purpose of this discussion is to describe details of the local empirical pseudopotential method, the empirical parameters and the way how the energy eigenvalues are obtained. Details of fitting procedures will be commented on later in Section 3.4. For compounds composed of heavy elements, spin-orbit effects may be important. Methods to take the spin-orbit effect into account within the formulation of the empirical pseudopotential method will also be described.

The use of the local EPM is particularly suited for *non*-transition metal layer compounds, in which the valence states are derived from outer s and p orbitals alone.

For example, in GaSe, the valence electrons originate from a $(4s)^2(4p)^1$Ga configuration and a $(4s)^2(4p)^4$Se configuration. The pseudowavefunctions characterizing the valence states are expected to have predominantly s and p-like characters. Thus from Equation (B96) the contributions of V_s and V_p to $V_R(\mathbf{r})$ are essential when acting on the pseudo wavefunction. The cancellation of $V_R(\mathbf{r})$ and the attractive potential $V(\mathbf{r})$ in the core region is nearly perfect in this case as shown by Cohen and Heine [28]. Instead of reproducing their calculations, we make the following argument to support this conclusion: In Equation (B92) the oscillatory portion of the wave function $\psi_{n\mathbf{k}}$ due to the non-vanishing $\langle c \mid \Phi_{n\mathbf{k}} \rangle$ implies a large kinetic energy of the valence electrons in the core region. Although the crystal potential is strong and attractive in this region, the electron spends essentially no time in this region because of its large kinetic energy. It therefore, does not feel the strong attractive potential. The resulting effective pseudopotential is thus quite weak. It can also be shown [22], that except for first row elements the s-p non-locality is small and may be neglected in the repulsive potential and thus in the resulting pseudopotential. As we mentioned before, the crystal potential may also be a non-local due to the presence of exchange terms. In the early applications of this method, however, a completely local approximation $V_L(\mathbf{r})$ is made to the resulting total pseudopotential which seems to reproduce valence energies correctly within ± 0.1 eV.

To parametrize this local pseudopotential let us consider the one-electron Schrödinger equation:

$$\left\{ -\frac{\hbar^2}{2m} \nabla^2 + V_L(\mathbf{r}) \right\} \Phi_{n\mathbf{k}}(\mathbf{r}) = E_{n\mathbf{k}} \Phi_{n\mathbf{k}}(\mathbf{r}), \tag{B97}$$

where

$$V_L(\mathbf{r}) = V_L(\mathbf{r} + \mathbf{R}_l) \tag{B98}$$

is periodic. It may thus be expanded in a Fourier series as

$$V_L(\mathbf{r}) = \sum_{\mathbf{G}} V_L(\mathbf{G}) \, e^{i\mathbf{G}\cdot\mathbf{r}}, \tag{B99}$$

where $\mathbf{G}$ is a reciprocal lattice vector and $V_L(\mathbf{G})$ is the form-factor given by

$$V_L(\mathbf{G}) = \frac{1}{\Omega} \int_{\text{cell}} V_L(\mathbf{r}) \, e^{-i\mathbf{G}\cdot\mathbf{r}} \, d^3r, \tag{B100}$$

where Ω is the volume of the unit cell. In general, if there are more than one atom per unit cell one may write

$$V_L(\mathbf{r}) = \sum_{\alpha=1}^{n} V_{\alpha L}(\mathbf{r} - \boldsymbol{\tau}_\alpha) \tag{B101}$$

with n the total number of atoms per unit cell, and $V_{\alpha L}(\mathbf{r})$ the atomic pseudopotential of the αth atom. We then obtain the Fourier transform

$$V_L(\mathbf{G}) = \frac{1}{n} \sum_{\alpha=1}^{n} e^{-i\mathbf{G}\cdot\boldsymbol{\tau}_\alpha} \frac{n}{\Omega} \int_{\text{cell}} V_L(\mathbf{r}) \, e^{-i\mathbf{G}\cdot\mathbf{r}} \, d^3r. \tag{B102}$$

If among the n atoms in the unit cell, there are m different species of atoms, such as Ga and Se then Equation (B102) may be rewritten as

$$V_L(\mathbf{G}) = \sum_{\beta=1}^{m} S_\beta(\mathbf{G}) \frac{n}{\Omega} \int_{\text{cell}} V_{\beta L}(\mathbf{r}) \, e^{-i\mathbf{G}\cdot\mathbf{r}} \, d^3r \tag{B103}$$

and

$$S_\beta(\mathbf{G}) = \frac{1}{n} \sum_{\alpha_\beta=1}^{n_\beta} e^{-i\mathbf{G}\cdot\boldsymbol{\tau}_{\alpha\beta}}, \tag{B104}$$

where $S_\beta(\mathbf{G})$ is called the structure factor for β-type atoms. α_β numbers the α-atom belonging to βth species, $V_{\beta L}(\mathbf{r})$ and $\boldsymbol{\tau}_{\alpha_\beta}$ are the corresponding local atomic pseudopotentials and position vectors. n_β is the number of atoms of the βth species. Usually the atomic pseudopotentials are assumed to be spherically symmetric, then $V_{\beta L}(\mathbf{G})$ depends only on the magnitude of the $\mathbf{G}$-vector. Now, Equation (B109) can be written as

$$V_L(\mathbf{r}) = \sum_{\mathbf{G}} \left[\sum_{\beta=1}^{m} S_\beta(\mathbf{G}) V_{\beta L}(|\mathbf{G}|) \right] e^{i\mathbf{G}\cdot\mathbf{r}}. \tag{B105}$$

The empirical parameterization of the local pseudopotential is accomplished by treating the atomic form factors $V_{\beta L}(|\mathbf{G}|)$ as disposable parameters. In principle there is an infinite number of parameters, $V_{\beta L}(|\mathbf{G}|)$, for an infinite number of $\mathbf{G}$-vectors. Since, however, the pseudopotential for the valence electrons is very weak, $V_{\beta L}(|\mathbf{G}|) \to 0$, as $|\mathbf{G}| \to \infty$, and we may retain $V_{\beta L}(|\mathbf{G}|)$'s only up to a certain $|\mathbf{G}|_{max}$. Equation (B105) can be rewritten as

$$V_L(\mathbf{r}) = \sum_{\mathbf{G}}^{|\mathbf{G}|_{max}} V_L(\mathbf{G}) \, e^{i\mathbf{G}\cdot\mathbf{r}} \tag{B106}$$

and

$$V_L(\mathbf{G}) = \sum_{\beta=1}^{m} S_\beta(\mathbf{G}) V_{\beta L}(|\mathbf{G}|).$$

This finite set of $V_{\beta L}(|\mathbf{G}|)$'s is fitted to experimental data such as optical reflectivity and density of states derived from photoemission experiments.

If a specific atom plays similar roles in two different crystals, the atomic pseudopotential $V_{\beta L}(\mathbf{r})$ is not expected to change drastically from one crystal to the other. For example, in GaAs and GaSb, the Ga-atoms may get ionized similarly and the respective Ga-pseudopotentials should be quite similar in both cases. Thus, the set of atomic Ga-form factors can be scaled and be transferred from one compound to the other by accounting for the difference in the lattice constants. Finer adjustments are sometimes needed. The procedures of scaling will be discussed later. By applying the method to crystals with zincblende or rocksalt structure [22] $V_L(\mathbf{r})$ is often expressed in terms of symmetric and antisymmetric form factors. Since there are two atoms of different species in the unit cell of these crystals, the symmetric and the antisymmetric form factors are defined as

$$
\begin{aligned}
V_S(|\mathbf{G}|) &= \tfrac{1}{2}(V_{1L}(|\mathbf{G}|) + V_{2L}(|\mathbf{G}|)), \\
V_A(|\mathbf{G}|) &= \tfrac{1}{2}(V_{1L}(|\mathbf{G}|) - V_{2L}(|\mathbf{G}|)),
\end{aligned}
\tag{B107}
$$

where V_S and V_A are fitted to the experimental data. For crystals with more than two atoms per unit cell and for purposes of retaining the characters of the individual atomic pseudopotential form factors, the scheme of using V_S and V_A is not as convenient as the expression given in Equation (B105). Reasons for limiting the difference in atomic form factors for different crystals are not only to avoid misinterpretation of experimental data but also to construct consistent atomic pseudopotentials.

Eigenvalues are calculated by substituting Equation (B106) into Equation (B97)

$$\left\{ -\frac{\hbar^2}{2m} \nabla^2 + \sum_{\mathbf{G}}^{|\mathbf{G}|_{max}} V_L(\mathbf{G}) \, e^{i\mathbf{G}\cdot\mathbf{r}} \right\} \Phi_{n\mathbf{k}}(\mathbf{r}) = E_{n\mathbf{k}} \Phi_{n\mathbf{k}}(\mathbf{r})$$

with the pseudowave functions, $\Phi_{n\mathbf{k}}(\mathbf{r})$, expanded in plane waves,

$$\Phi_{n\mathbf{k}}(\mathbf{r}) = \sum_{\mathbf{G}} A_{n\mathbf{k}}(\mathbf{G}) \, e^{i(\mathbf{k}+\mathbf{G})\cdot\mathbf{r}} \tag{B109}$$

and the coefficients $A_{n\mathbf{k}}(\mathbf{G})$ determined by a variational method. Assuming $\Phi_{n\mathbf{k}}(\mathbf{r})$ to be normalized, we minimize the total energy functional $F = \langle \Phi_{n\mathbf{k}} | H_p - E_{n\mathbf{k}} | \Phi_{n\mathbf{k}} \rangle$ with respect to $A_{nk}(\mathbf{G})^*$. For non-trivial solutions of $A_{nk}(\mathbf{G})$, we get the usual form of the secular equation,

$$\det\left|\{|\mathbf{k} + \mathbf{G}|^2 - E_{nk}\}\, \delta(\mathbf{G} - \mathbf{G}') + V_L(\mathbf{G}' - \mathbf{G})\right| = 0. \tag{B110}$$

with $\hbar^2/2m$ included in E_{nk} and V.

Independent of the cutoff for the expansion of the local pseudopotential at $|\mathbf{G}|_{max}$, the series of Equation (B109) has to be cutoff at some energy $E_1 = |\mathbf{G}|^2$, which determines the size of the secular equation.

For most crystals, one is interested in the lowest ten to thirty bands only. These bands include the occupied valence bands and the low lying conduction bands, which are important to explain optical and photoemission data. Therefore, a legitimate criterion to determine the cutoff is to include sufficient terms in Equation (B109) so that energies and wave functions of the interesting bands converge within some given tolerance. For layer compounds, some few hundred terms may be needed to achieve this convergence. To reduce this enormous size of the secular equation, a perturbation scheme suggested first by Löwdin [29] and later modified by Brust [30] is often used. We shall outline the procedures of this scheme.

The contribution of plane waves to the sum in Equation (B109) is divided into two groups, specified by two control energies, E_1 and E_2. If $|\mathbf{k} + \mathbf{G}|^2 \leq E_1$ for a given plane wave it is explicitly included in Equation (B109). Plane waves with $E_1 < |\mathbf{k} + \mathbf{G}|^2 \leq E_2$ are treated as perturbation in second order, which corresponds to effectively folding down a large matrix into a smaller one. The resulting matrix elements of the small folddown matrix have the form:

(i) For the diagonal matrix elements, the $|\mathbf{k} + \mathbf{G}|^2$ term in Equation (B110) will be replaced by

$$|\mathbf{k} + \mathbf{G}|^2 \to |\mathbf{k} + \mathbf{G}|^2 + \sum_{\mathbf{G}''} \frac{V_L(\mathbf{G} - \mathbf{G}'') V_L(\mathbf{G}'' - \mathbf{G})^*}{|\mathbf{k} + \mathbf{G}|^2 - |\mathbf{k} + \mathbf{G}''|^2}, \tag{B111}$$

where $|\mathbf{k} + \mathbf{G}|^2 \leq E_1$ and $E_1 < |\mathbf{k} + \mathbf{G}''|^2 \leq E_2$.

(ii) For the off-diagonal matrix elements, the term $V_L(\mathbf{G}' - \mathbf{G})$ in Equation (B110) is modified in the following manner:

$$V_L(\mathbf{G}' - \mathbf{G}) \to V_L(\mathbf{G}' - \mathbf{G}) + \sum_{\mathbf{G}''} \frac{V_L(\mathbf{G} - \mathbf{G}'')^* \, V_L(\mathbf{G}' - \mathbf{G}'')}{E - |\mathbf{k} + \mathbf{G}''|^2}, \tag{B112}$$

where $|\mathbf{k} + \mathbf{G}|^2$, $|\mathbf{k} + \mathbf{G}''|^2$ are in the range specified in (i), and where $|\mathbf{k} + \mathbf{G}'|^2 \leq E_1$. E is an average energy of the bands under consideration whose choice is not very critical. The choice of E_1 and E_2 is guided by convergence criteria as discussed before.

For layer compounds, typical sizes of the secular equation range from 80×80 to 200×200, depending on the number of atoms per unit cell, with another 200 to 300 plane waves contributing via the perturbation scheme.

The clever choice of an initial set of atomic form factors is advised for economic reasons. To avoid misinterpretation of experimental data too, it is advisable to start with a set of atomic form factors which has been obtained by scaling the potentials used in other crystals. This is usually possible since most crystals with non-transition elements as constituents have been studied by the local empirical pseudopotential method. The following scheme for scaling potentials has been used by Fong and Cohen [31].

Let a_1, a_2, Ω_1, Ω_2 and n_1, n_2 be the lattice constants, the unit cell volumes and the number of atoms per unit cell for crystals 1 and 2 respectively. Let us furthermore, assume that the atomic form factors of the β-species in crystal 2 have already been determined. With the use of Equation (B103) the form factors of the β-species in crystal 1 may then easily be calculated. Since the $\mathbf{G}$-vectors in expression (B103) depend on the lattice constant, it is convenient to define a dimensionless, lattice constant independent quantity $\mathbf{g}$, such that

$$\mathbf{G} = \mathbf{g}\left(\frac{2\pi}{a}\right). \tag{B113}$$

With this definition the atomic form factors for crystal 1 $V_{\beta L,1}(|\mathbf{G}_1|)$ may be written as

$$V_{\beta L,1}(|\mathbf{G}_1|) = \left(\frac{n_1\Omega_2}{n_2\Omega_1}\right)\left(\frac{n_2}{\Omega_2}\right) \int\limits_{\text{cell}} V_{\beta L,2}(r)\, e^{i\mathbf{g}\cdot\mathbf{r}(2\pi/a_2)(a_2/a_1)}\, \mathrm{d}^3 r. \tag{B114}$$

The integral with the normalization factor n_2/Ω_2 is just $V_{\beta L,2}(|\mathbf{G}_2|(a_2/a_1))$. Therefore, we obtain

$$V_{\beta L,1}(|\mathbf{G}_1|) = \left(\frac{n_1\Omega_2}{n_2\Omega_1}\right) V_{\beta L,2}\left(|\mathbf{G}_2|\frac{a_2}{a_1}\right). \tag{B115}$$

In practice, the calculation of $V_{\beta L,1}$ according to Equation (B115) is best done graphically. The known $V_{\beta L,2}(|\mathbf{G}_2|)$ is plotted as a function of $|\mathbf{G}_2|^2$ in units of $(2\pi/a_2)^2$. A smooth curve may then be fitted to the set of discrete $\mathbf{G}_2$-vectors.

The corresponding form factors $V_{\beta L,2}(|\mathbf{G}_2|(a_2/a_1))$ can easily be obtained from the smooth curve at the abscissas $(a_2/a_1)^2 g^2$ and the final form factors $V_{\beta L,1}(|\mathbf{G}_1|)$ are obtained by multiplying $V_{\beta L,2}(|\mathbf{G}_2|(a_2/a_1))$ by the volume scaling factor $(n_1\Omega_2/n_2\Omega_1)$. An alternative scaling scheme [32] has been proposed, which consists in expressing all reciprocal lattice vectors in absolute units of inverse length, i.e. a_0^{-1}. After an initial set of atomic form factors has been obtained this way, fine adjustments can be made by finding the best fit of the results to experimental data.

For layer compounds, such as PbI_2 or BiI_3, composed of heavy elements, spin-orbit effects should not be neglected if quantitative results are desired. The treatment of spin-orbit effects within the pseudopotential formalism was first introduced by Weisz [33]. Basically one goes back to the original OPW formulation, introduces spin-orbit effects and retransforms to a pseudopotential.

The Hamiltonian characterizing the spin-orbit interaction is given by

$$H_{\text{s.o.}} = \frac{\hbar^2}{4m^2 c^2}\left[\nabla V(\mathbf{r}) \times \mathbf{p}\right]\cdot\boldsymbol{\sigma} \equiv \Lambda\cdot\boldsymbol{\sigma}, \tag{B116}$$

where $\nabla V(\mathbf{r})$ is the gradient of the real crystal potential, $\mathbf{p}$ is the momentum operator and $\boldsymbol{\sigma}$ are the Pauli spin matrices. Equation (B92) may be considered as a transformation between the real wave function and the pseudowave function, that is

$$\psi_{n\mathbf{k}}(\mathbf{r}) = (1 - P_c)\Phi_{n\mathbf{k}} \tag{B117}$$

with $P_c = I_s \sum_c |c\rangle \langle c|$ and I_s being the identity operator in spin space. Then, to account for spin-orbit effects in the pseudopotential framework the following term has to be considered:

$$H^P_{s.o.} = (1 - P_c) + H_{s.o.}(1 - P_c).$$ (B118)

The matrix elements of $H^P_{s.o.}$ between two plane wave components of the pseudowave function can formally be written as

$$\langle \mathbf{k}', s' | H^P_{s.o.} | \mathbf{k}, s \rangle = \sum_\beta S_\beta(\mathbf{k} - \mathbf{k}') \langle s' | \sigma | s \rangle \cdot \{ \Lambda_{pp} + \Lambda_{cp} + \Lambda_{pc} + \Lambda_{cc} \},$$

(B119)

where $\mathbf{k}$ and $\mathbf{k}'$ are the momenta of the plane waves, and $|s\rangle$, $|s'\rangle$ are electron spin eigenfunctions. Λ indicates matrix elements between states labelled by the subscripts p and c, which stand for plane waves and core states. It was shown by Weisz [33] that the dominant contribution of the four matrix elements in Equation (B119) is Λ_{cc}. Thus by retaining only this term, spin-orbit effects between pseudo wavefunctions are approximated by the spin-orbit interaction induced changes among the core states which are felt by the valence states via the orthogonality requirement. Since for non-transition metal layer compounds, the valence electrons are mostly of s and p character, and since the core s electrons do not show spin-orbit effects, the only contribution to the spin-orbit splitting of the valence states originates from the p core states. Evaluating the orbital part of Equation (B117) the matrix elements of Equation (B119) take the following form:

$$(H^P_{s.o.})_{\mathbf{k}'s'\mathbf{k}s} = -i \langle s' | \sigma | s \rangle \cdot (\hat{k}' \times \hat{k}) \sum_\beta S_\beta(\mathbf{k} - \mathbf{k}') \lambda_\beta(k', k),$$ (B120)

where $\hat{k}$ and $\hat{k}'$ are unit vectors along $\mathbf{k}$ and $\mathbf{k}'$. $\lambda_\beta(k', k)$ is the product of two radial integrals $B^*_{nl}(k')$ and $B_{nl}(k)$ multiplied by a constant which is treated as disposable parameter to characterize the strength of the core spin-orbit interaction. The radial integrals $B_{nl}(k)$, which originate from the orthogonality integrals $\langle c | e^{i\mathbf{k} \cdot \mathbf{r}} \rangle$, are given by

$$B_{nl}(k) = \int_0^\infty i^l [4\pi(2l + 1)]^{1/2} j_l(kr) R_{nl}(r) r^2 \, dr,$$ (B121)

where j_l is the spherical Bessel function of order l. $R_{nl}(r)$ is the radial function of an atomic core orbital with principal quantum number n and angular momentum quantum number l. These radial wave functions may be taken from the tabulated values of Herman and Skillman [6]. Note the 'i' and the complex spin matrix element $\langle s' | \sigma | s \rangle$ in Equation (B120) which cause the pseudopotential Hamiltonian matrix with spin-orbit interaction to be complex hermitian and twice the size as without spin-orbit interaction. This is of course reflected by the fact that in general spin-orbit interaction removes the two-fold spin-degeneracy of electron bands. If, however, the crystal under consideration possesses inversion symmetry, all energy bands remain even with spin-orbit interaction at least two-fold degenerate throughout the Brillouin zone, and the wave functions of the degenerate states are simply related to each other by the time reversal operation. The spin-orbit Hamiltonian matrix of double size, therefore, carries symmetry induced redundancy in itself and should conveniently be reduced. Even though this is mathematically possible [34], no practical scheme has yet been proposed.

3.3. THE NON-LOCAL EMPIRICAL PSEUDOPOTENTIAL METHOD

If the *local* empirical pseudopotential method is applied to crystals in which d-states play an important role, their energies are always found too high with respect to the s and p-state energies. This result is not surprising, since the *local* pseudopotential method assumes the cancellation effect to be nearly perfect for *all* valence and conduction states including d-states. Physically, however, the outermost d-shell electrons do not experience the same weak pseudopotential caused by complete cancellation. This is most obvious if the core does not contain d-states at all as it is the case in potassium where the unoccupied $3d$-atomic orbitals constitute the low lying conduction bands. The core states of potassium have only s and p-like character and the second term in Equation (B94) vanishes whenever the projection operator of the repulsive potential acts on a wave function of d-character. The potential for the d-states in this case is just the strong attractive real crystal potential. Similarly, for transition metal compounds, where the atomic configuration of the valence electrons for the metallic atom can be expressed as $(n-1)\,d^x ns^y$. While n is the principal quantum number of the outermost s-shell, the d electrons belong to shell $n-1$. Here x ranges from 1 to 9 and y may be either 1 or 2. (For example, Nb has a $4d^3 5s^2$ configuration.) The d-electrons of shell $(n-1)$ experience different cancellation effects than do the s or p electrons of shell n.

One way to treat the d-electrons within the pseudopotential formalism is to construct a non-local model potential specifically acting upon states with $l = 2$ to compensate the over cancellation effect of the local pseudopotential. The parameters characterizing this model potential and the remaining standard form factors of the local pseudopotential may then be determined empirically. The non-locality may mathematically be expressed by the use of an angular dependent projection operator in the model potential.

The first application of this method is due to Lee and Falicov [35] who used a spherically symmetric, attractive square well combined with a $l = 2$ projection operator to characterize the potential for the conduction d-states in potassium. The form of the potential is in this case

$$V_{NL}(\mathbf{r}) = \sum_l \mathbf{P}_2^+ U(|\mathbf{r} - \mathbf{R}_l|)\mathbf{P}_2 \tag{B122}$$

with

$$U(|\mathbf{r} - \mathbf{R}_l|) = -A_2, \quad \text{for} \quad |\mathbf{r} - \mathbf{R}_l| \le R_s,$$

$$= 0 \quad \text{otherwise,}$$

where $\mathbf{R}_l$ is a lattice vector, $\mathbf{P}_2$ and $\mathbf{P}_2^+$ are the $l = 2$ projection operator and its adjoint. R_s is the radius of the K^+ ion and A_2 is the depth of the square well which is treated as a disposable parameter. The total pseudopotential for potassium then has the following form:

$$V_p(\mathbf{r}) = V_L(\mathbf{r}) + V_{NL}(\mathbf{r}). \tag{B123}$$

To calculate the matrix elements of $V_{NL}(\mathbf{r})$ with plane waves, we expand the plane waves into spherical harmonics as given in Equation (B43). The different types of matrix elements are:

(i) The diagonal matrix elements:

$$\langle \mathbf{k} + \mathbf{G}| \; V_{NL}(\mathbf{r}) \, |\mathbf{k} + \mathbf{G}\rangle = \frac{10\pi}{\Omega} A_2 R_s^3 [j_2^2(|\mathbf{k} + \mathbf{G}| \, R_s) -$$

$$- j_1(|\mathbf{k} + \mathbf{G}| \, R_s) j_3(|\mathbf{k} + \mathbf{G}| \, R_s)], \qquad \text{(B124)}$$

where Ω is the volume of the unit cell and j_l are spherical Bessel functions.

(ii) The off-diagonal matrix element with $|\mathbf{k} + \mathbf{G}| = |\mathbf{k} + \mathbf{G}'|$ (elastic scattering):

$$\langle \mathbf{k} + \mathbf{G}'| \; V_{NL}(\mathbf{r}) \, |\mathbf{k} + \mathbf{G}\rangle = \frac{10\pi}{\Omega} A_2 R_s^3 P_2(\cos \theta_{\mathbf{G}',\mathbf{G}}) \times [j_2^2(|\mathbf{k} + \mathbf{G}| \, R_s) -$$

$$- j_1(|\mathbf{k} + \mathbf{G}| \, R_s) j_3(|\mathbf{k} + \mathbf{G}| \, R_s)], \qquad \text{(B125)}$$

where $P_2(\cos \theta_{\mathbf{G}',\mathbf{G}})$ is the Legendre polynomial of order 2 and where the angle $\theta_{\mathbf{G}',\mathbf{G}}$ is the taken between $\mathbf{k} + \mathbf{G}'$ and $\mathbf{k} + \mathbf{G}$.

(iii) The off-diagonal matrix element with $|\mathbf{k} + \mathbf{G}| \neq |\mathbf{k} + \mathbf{G}'|$ (inelastic scattering):

$$\langle \mathbf{k} + \mathbf{G}'| \; V_{NL}(\mathbf{r}) \, |\mathbf{k} + \mathbf{G}\rangle = 20\frac{\pi}{\Omega} A_2 P_2(\cos \theta_{\mathbf{G}',\mathbf{G}}) \frac{R_s^2}{|\mathbf{k} + \mathbf{G}|^2 - |\mathbf{k} + \mathbf{G}'|^2} \times$$

$$\times [|\mathbf{k} + \mathbf{G}| \, j_3(|\mathbf{k} + \mathbf{G}| \, R_s) j_2(|\mathbf{k} + \mathbf{G}'| \, R_s) -$$

$$- |\mathbf{k} + \mathbf{G}'| \, j_3(|\mathbf{k} + \mathbf{G}'| \, R_s) j_2(|\mathbf{k} + \mathbf{G}| \, R_s].$$

$$\text{(B126)}$$

For most transition metal compounds, the partially filled d-states become very important in determining the properties of valence and conduction bands. Since, however, the potential for the partially occupied d-states is stronger than that for the unoccupied conduction states in potassium, a straightforward application of $V_{NL}(\mathbf{r})$ as given in Equation (B122) would cause serious convergence problems. The general approach, however, is rather appealing because of its simplicity: $V_{NL}(\mathbf{r})$ determines the width of the d-bands and the relative energy between d-states and s and p-like states. At the same time, the local pseudopotential, $V_L(\mathbf{r})$, determines the energies of s and p bands as well as $s - d$ hybridization effects. To resolve the convergence problem, Fong and Cohen [23] introduced a damping factor into the off-diagonal matrix elements of $V_{NL}(\mathbf{r})$. The modified matrix elements which replace the elastic and inelastic scattering matrix elements of Equations (B125) and (B126) have the forms

$$\langle \mathbf{k} + \mathbf{G}'| \; V_{NL}(\mathbf{r}) \, |\mathbf{k} + \mathbf{G}\rangle \rightarrow \exp\left[-\alpha\left(\frac{|\mathbf{k} + \mathbf{G}'| - \kappa}{2k_F}\right)^2\right] \times$$

$$\times \langle \mathbf{k} + \mathbf{G}'| \; V_{NL}(\mathbf{r}) \, |\mathbf{k} + \mathbf{G}\rangle \exp\left[-\alpha\left(\frac{|\mathbf{k} + \mathbf{G}| - \kappa}{2k_F}\right)^2\right] \quad \text{(B127)}$$

here k_F is the Fermi momentum of the corresponding transition metal and α and κ are two additional empirical parameters. Since the diagonal matrix elements of $V_{NL}(\mathbf{r})$ are kept the same form as shown in Equation (B124), and since $V_{NL}(\mathbf{r})$ has not been included into the Löwdin-Brust perturbation scheme, the strict non-locality of V_{NL} is violated. This violation, however, has no quantitative consequences.

The method has been applied to NbSe$_2$ [36]. Because of the large c/a ratio (~ 4) of this crystal, the size of the secular equation has to be at least of the order of 190×190. An additional 300 plane waves contribute to the perturbation scheme containing the local pseudopotential to warrant energy convergence to within 0.2 eV.

Unlike the non-transition metal elements, only a few transition metals have been studied by the empirical pseudopotential method. Therefore, a certain difficulty in choosing the initial pseudopotentials for transition metal compounds exists. Existing *ab-initio* calculations may be used to determine the empirical pseudopotential by fitting the results to the existing band structures. The result may then be improved by further adjustments.

3.4. COMMENTS ON FITTING PROCEDURES

Before proceeding to comment on the fitting procedures, we shall first summarize the discussions of the previous sections about how to obtain the energies at a few **k** points in the Brillouin zone if one has a set of atomic pseudopotentials for the constituent atoms. The band energies $E_n(\mathbf{k})$ for a given **k**-point in the Brillouin zone are determined by first specifying the control energies E_1 and E_2, by setting up the Hamiltonian matrix and by diagonalizing it. We illustrate the process of a pseudopotential band structure calculation by the schematic block diagram shown in Figure 4. After determining the appropriate control energies E_1 and E_2, we repeat the processes at various **k**-points to obtain the complete band structure.

Four parameters usually define the non-local pseudopotential, $V_{NL}(\mathbf{r})$, which can easily be determined. The local pseudopotential, however, may cause complications. For physical reasons of considering most of the scattering power of the potential, $|\mathbf{G}|_{\max} \geq 2k_F$, where k_F is the Fermi momentum estimated from the density of the electrons in the crystal. The number of **G**-vectors with lengths smaller than the cutoff $|\mathbf{G}|_{\max}$ in Equation (B106) depends strongly on the crystal structure. Some of the layer compounds in hexagonal structures have large c/a ratios as for example, NbSe$_2$, with $c/a \approx 4$. With a reasonable cutoff value of $|\mathbf{G}|_{\max}$, the number of atomic form factors for each constituent atom is of the order of thirty. To manage such a large number of parameters, two schemes are often used. One is to group form factors; adjustments will then refer to these groups. This is suggested by the fact that structure factors may vanish at certain **G**-vectors and the corresponding form factors do not have any effect on the energies. A natural grouping of form factors is thus given by the intervals between vanishing structure factors.

Another scheme is to use a continuous interpolation formula as suggested by Animalu and Heine [37], and as used in [32].

$$V_{\beta L}(|\mathbf{G}|) = \frac{A_{\beta 1}(G^2 - A_{\beta 2})}{\exp(A_{\beta 3}(G^2 - A_{\beta 4})) + 1} \tag{B128}$$

which has only $A_{\beta i}$, $i = 1, 2, 3, 4$ as empirical parameters. The form of Equation (B128) is dictated by the general form of atomic Fourier transforms [22].

The fitting procedures involved in the empirical pseudopotential method are as follows:

(i) Find a starting set of pseudopotentials by scaling the atomic form factors from

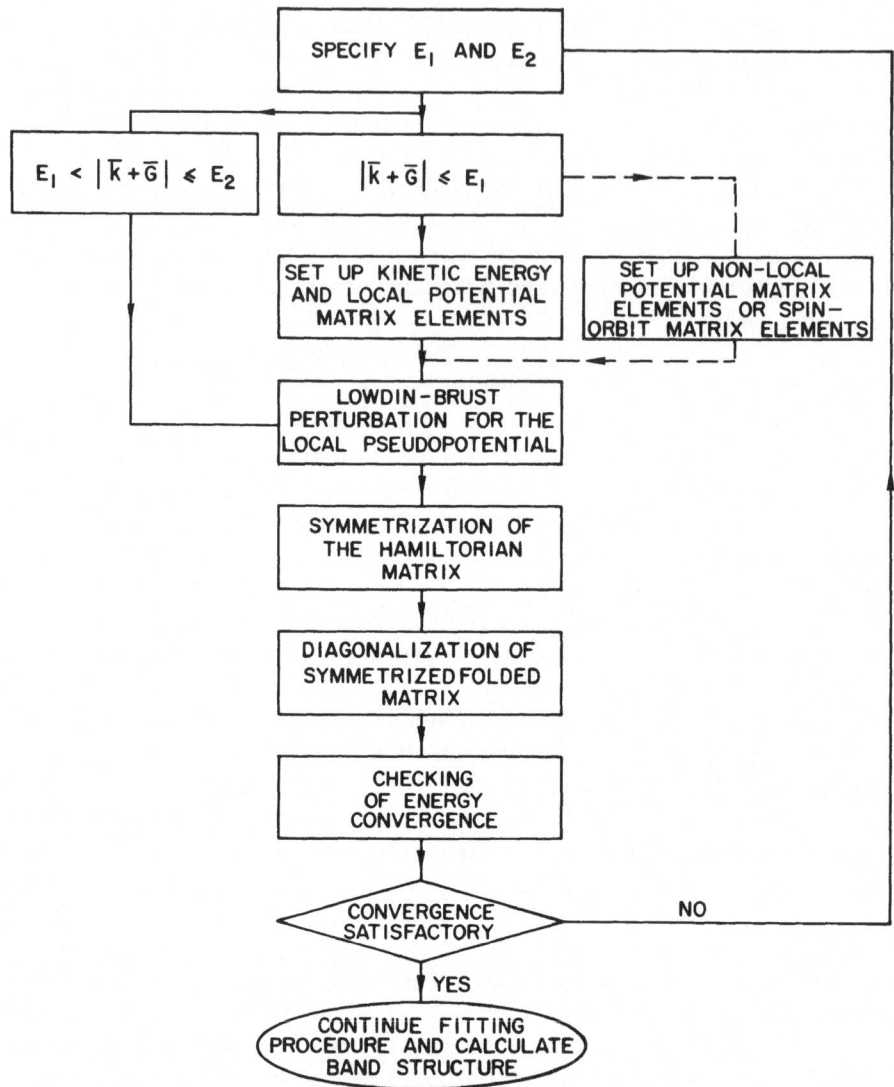

Fig. 4. Block-diagram for determining converged eigenvalues by the empirical pseudopotential method.

other crystals with known form factors. For the non-local pseudopotential, k, α can also be scaled, A_2 and R_s may be used directly from other crystals without scaling.

(ii) Check the convergence.

(iii) Study the experimental data as described in Section 1. From the study of the optical properties of a crystal we are able to assign certain structures to transitions between valence and the conduction bands close to high symmetry k-points. Moreover, the information obtained from examining the results of photoemission measurements provides energy separations within the valence bands.

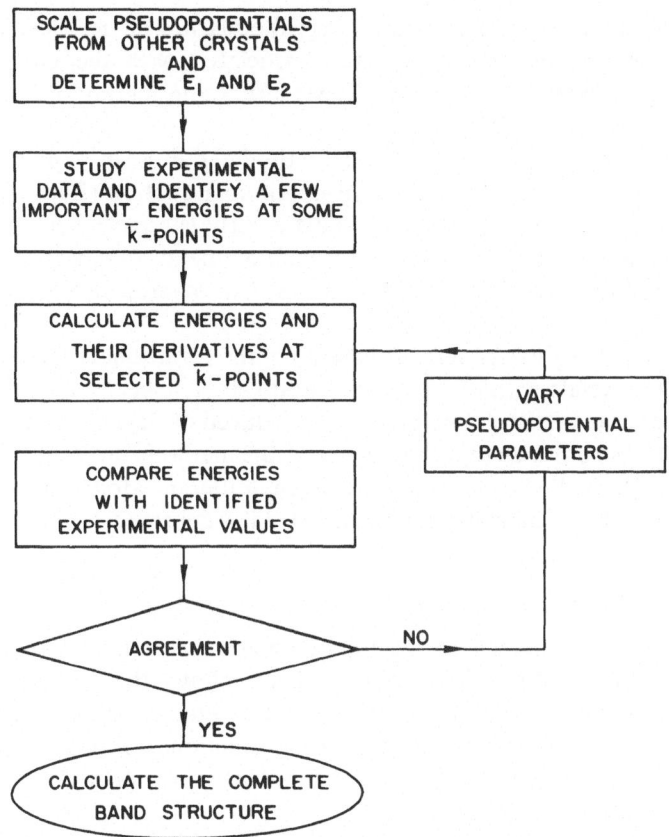

Fig. 5.   Block-diagram for determining the empirical pseudopotential parameters.

(iv) Calculate the derivatives of the energies at certain **k**-points with respect to the empirical parameters, either by perturbation theory or by repeating the complete calculations for slightly changed parameters. These derivatives may be used to estimate the total variation of the parameters necessary to fit the energies to the identified experimental values. Automatic numerical non-linear fitting procedures have been used for this purpose. Thorough inspection of the results after each step and individual adjustments, however, often lead faster to satisfactorily converged results. The procedures are schematically summarized in form of a block diagram in Figure 5.

The EPM has been applied to a large number of transition metal and non-transition metal layer compounds. The various results shall be presented in a later chapter.

4. The Green's Function Method (KKR)

The Green's function method has since a long time been recognized as a powerful tool in solving scattering problems. In 1947, Korringa [38] suggested the application of the Green's function method to solid state physics, i.e., to view the electronic states in solids as a result of multiple scattering on atomic potentials. However, the general application of the method has not been very successful until Kohn and Rostoker [39] proposed in 1954 a variational scheme to treat the integral equation of the one electron wave-function and to overcome the singularities associated with the Green's function. Honoring these fundamental papers, the Green's function method used in solid state physics sometimes is called KKR method.

The application of the Green's function method requires an explicit form of the potential energy. In practice, the muffin-tin potential Equation (B40) is often a convenience choice. This type of potential is particularly successful in studying simple metals, but has also been used for noble and transition metals [40]. Recently the muffin-tin potential approximation has also been applied to layered transition metal compounds [41].

In Section 4.1 we shall give a general discussion of the variational procedure suggested in [39] for simple crystals with one atom per unit cell. The generalization to complex crystals with more than one atom per unit cell introduced by Segall [42] will be discussed in Section 4.2. It will be shown that the essence of the method consists of characterizing the different phases of the waves scattered on the different atoms in the unit cell. A detailed discussion of the calculation of the involved structure factors will thus be given in Section 4.3.

4.1. VARIATIONAL PROCEDURE FOR SIMPLE STRUCTURES

The equation to be solved in the Green's function method is an integral equation for the electron wavefunction. Kohn and Rostoker [39] have shown that this integral equation can be derived from a functional of the wavefunction and that the eigenvalues can be obtained by applying a variational technique to the functional. To illustrate the essential features of this, we shall for clarity, restrict the discussions in this section to simple crystal with one atom per unit cell.

First an integral equation for the wavefunction will be derived by use of the conventional Green's function method. Then, a functional from which the integral equation can be easily obtained, will be constructed. By using a trial wavefunction and by applying the usual variational procedures to this functional, it is possible to derive a determinantal equation for the energy of an electron in a bound state in the solid. In particular, the matrix elements of the determinantal equation shall be evaluated for a muffin-tin potential and the treatment of the singularity of the Green's function shall be discussed.

To simplify the discussions, we use atomic units (i.e. Bohr radius, Rydberg) in all equations. The Schrödinger equation then reads:

$$(-\nabla^2 + V(\mathbf{r}))\psi_{\mathbf{k}}(\mathbf{r}) = E\psi_{\mathbf{k}}(\mathbf{r}). \tag{B129}$$

The corresponding Green's function satisfies the following equation:

$$(\nabla^2 + E)G_{\mathbf{k}}(\mathbf{r}, \mathbf{r}') = \delta(\mathbf{r} - \mathbf{r}') \tag{B130}$$

with the periodic boundary conditions,

$$G_{\mathbf{k}}(\mathbf{r} + \mathbf{R}_l, \mathbf{r}') = e^{i\mathbf{k}\cdot\mathbf{R}_l} G_{\mathbf{k}}(\mathbf{r}, \mathbf{r}'), \tag{B131}$$

where $\mathbf{R}_l$ is a lattice vector. The solution of Equations (B130) and (B131) can be written as

$$G_{\mathbf{k}}(\mathbf{r}, \mathbf{r}') = -\frac{1}{\Omega} \sum_{\mathbf{G}} \frac{e^{i(\mathbf{k}+\mathbf{G})\cdot(\mathbf{r}-\mathbf{r}')}}{(\mathbf{k}+\mathbf{G})^2 - E}, \tag{B132}$$

where Ω is the volume of the unit cell and $\mathbf{G}$ is a reciprocal lattice vector. By multiplying $G_{\mathbf{k}}$ to Equation (B129) and $\psi_{\mathbf{k}}^*(\mathbf{r})$ to Equation (B130) and by integrating over a unit cell, we obtain the integral equation for the bound states $\psi_{\mathbf{k}}(\mathbf{r})$:

$$\psi_{\mathbf{k}}(\mathbf{r}) = \int_{\Omega} G_{\mathbf{k}}(\mathbf{r}, \mathbf{r}')V(\mathbf{r}')\psi_{\mathbf{k}}(\mathbf{r}')\, \mathrm{d}^3 r'. \tag{B133}$$

The energy E is contained implicitly in the expression of $G_{\mathbf{k}}(\mathbf{r}, \mathbf{r}')$. The treatment of scattering states for energies above the ionization energy is not presented here. It is of less importance for standard bandstructure calculations. For its discussion the reader is referred to the literature.

Instead of directly calculating $\psi_{\mathbf{k}}(\mathbf{r})$ from Equation (B133), Kohn and Kostoker considered the following functional of $\psi_{\mathbf{k}}(\mathbf{r})$:

$$\Lambda = \int_{\Omega} \psi_{\mathbf{k}}^*(\mathbf{r})V(\mathbf{r})\psi_{\mathbf{k}}(\mathbf{r})\, \mathrm{d}^3 r - \int_{\Omega}\int_{\Omega} \psi_{\mathbf{k}}^*(\mathbf{r})V(\mathbf{r})G_{\mathbf{k}}(\mathbf{r}, \mathbf{r}')V(\mathbf{r}')\psi_{\mathbf{k}}(\mathbf{r}')\, \mathrm{d}^3 r\, \mathrm{d}^3 r'. \tag{B134}$$

By varying Λ with respect to $\psi_{\mathbf{k}}^*(\mathbf{r})$, Equation (B133) can be obtained. We may now apply the conventional variational method to the functional Λ to find the eigenvalues E. For this let $\psi_{\mathbf{k}}(\mathbf{r})$ be expressed as a linear combination of certain basis functions:

$$\psi_{\mathbf{k}}(\mathbf{r}) = \sum_p C_{p\mathbf{k}}\Phi_p(\mathbf{r}). \tag{B135}$$

Then, Equation (B134) can be rewritten as

$$\Lambda = \sum_{p,q} C_{p\mathbf{k}}^* \Lambda_{pq}^{\mathbf{k}} C_{q\mathbf{k}}, \tag{B136}$$

where

$$\Lambda_{pq}^{\mathbf{k}} = \int_{\Omega} \Phi_p^*(\mathbf{r})V(\mathbf{r})\Phi_q(\mathbf{r})\, \mathrm{d}^3 r - \int_{\Omega}\int_{\Omega} \Phi_p^*(\mathbf{r})V(\mathbf{r})G_{\mathbf{k}}(\mathbf{r}, \mathbf{r}')V(\mathbf{r}')\Phi_q(\mathbf{r}')\, \mathrm{d}^3 r\, \mathrm{d}^3 r'. \tag{B137}$$

By varying Λ with respect to $C_{p\mathbf{k}}^*$ one obtains a set of linear equations,

$$\sum_q \Lambda_{pq}^{\mathbf{k}} C_{q\mathbf{k}} = 0. \tag{B138}$$

For non-trivial solutions of $C_{q\mathbf{k}}$, the determinantal equation is given by

$$\det \Lambda_{pq}^{\mathbf{k}} = 0. \tag{B139}$$

The values of E which solve Equation (B139) are the eigenenergies of the electronic states at the point $\mathbf{k}$ in the BZ.

The formulation given above applies to a general potential, $V(\mathbf{r})$. If, however, the potentials centered at $\mathbf{r}$ and $\mathbf{r}'$ overlap, the evaluation of the second term in Equation (B137) can become quite involved. In practice, therefore, the Green's function method is mostly used in connection with non-overlapping muffin-tin potentials. The form of a muffin-tin potential has been given in Equation (B40) with V_0 equal to zero. With this choice of potential, the spherical portion of an APW can be used as a basis function, that is

$$\Phi_p(\mathbf{r}) = \Phi_{lm}(\mathbf{r}) = u_l(r)Y_{lm}(\theta, \varphi) \tag{B140}$$

with

$$u_l(0) = \text{finite and } u_l(R_m) = 1.$$

where R_m is the radius of the muffin-tin sphere.

To evaluate the matrix elements in Equation (B137), the singularity of the Green's function at $\mathbf{r} = \mathbf{r}'$ must be carefully treated. The following asymtotic procedure was suggested in [39]:

$$\Lambda_{pq}^{\mathbf{k}} = \lim_{\varepsilon \to 0} \Lambda_\varepsilon , \tag{B141}$$

where

$$\Lambda_\varepsilon = \int\limits_{r < R_m - 2\varepsilon} \mathrm{d}^3 r\, \Phi_p^*(\mathbf{r})V(\mathbf{r})\left\{\Phi_q(\mathbf{r}) - \int\limits_{r' < R_m - \varepsilon} \mathrm{d}^3 r'\, G_{\mathbf{k}}(\mathbf{r}, \mathbf{r}')V(\mathbf{r}')\Phi_q(\mathbf{r}')\right\}. \tag{B142}$$

Important details for evaluating this matrix element are in particular:

(i) The transformation of the volume integral in the curly bracket of Equation (B142) into a surface integral. By using Equation (B129) to express $V(\mathbf{r}')$ in terms of V^2 and E and by applying the divergence theorem and by using Equation (B130), it follows that

$$\int\limits_{r' < R_m - \varepsilon} \mathrm{d}^3 r'\, G_{\mathbf{k}}(\mathbf{r}, \mathbf{r}')V(\mathbf{r}')\Phi_q(\mathbf{r}') = \Phi_q(\mathbf{r}) +$$

$$+ \int\limits_{r' = R_m - \varepsilon} \left[G_{\mathbf{k}}(\mathbf{r}, \mathbf{r}')\frac{\partial \Phi_q(\mathbf{r}')}{\partial r'} - \Phi_q(\mathbf{r}')\frac{\partial}{\partial r'}\, G_{\mathbf{k}}(\mathbf{r}, \mathbf{r}')\right]\mathrm{d}S', \tag{B143}$$

where $\mathrm{d}S'$ is the surface element of the muffin-tin sphere with radius $R_m - \varepsilon$. With this form, Λ_ε is simplified as

$$\Lambda_\varepsilon = - \int\limits_{r < R_m - 2\varepsilon} \mathrm{d}^3 r\, \Phi_p^*(\mathbf{r})V(\mathbf{r}) \int\limits_{r' = R_m - \varepsilon} \times$$

$$\times \mathrm{d}S'\left[G_{\mathbf{k}}(\mathbf{r}, \mathbf{r}')\frac{\partial}{\partial r'}\, \Phi_q(\mathbf{r}') - \Phi_q(\mathbf{r}')\frac{\partial}{\partial r'}\, G_{\mathbf{k}}(\mathbf{r}, \mathbf{r}')\right]. \tag{B144}$$

(ii) The transformation of the remaining volume integral in Equation (B144) into a surface integral. Since r is restricted to be smaller than r', the same procedures as used for obtaining Equation (B143) may be used again, so that

$$\int\limits_{< R_m - 2\varepsilon} \mathrm{d}^3 r\, \Phi_p^*(\mathbf{r})V(\mathbf{r})G_{\mathbf{k}}(\mathbf{r}, \mathbf{r}') = \int\limits_{r = R_m - 2\varepsilon} \mathrm{d}S \times$$

$$\times \left[\frac{\partial \Phi_p^*(\mathbf{r})}{\partial r}\, G_{\mathbf{k}}(\mathbf{r}, \mathbf{r}') - \Phi_p^*(\mathbf{r})\frac{\partial}{\partial r}\, G_{\mathbf{k}}(\mathbf{r}, \mathbf{r}')\right], \tag{B145}$$

where dS is the surface element of the muffin-tin sphere with radius $R_m - 2\varepsilon$. Equation (B144) can now be rewritten by using Equation (B145)

$$\Lambda_\varepsilon = \int_{r = R_m - 2\varepsilon} dS \int_{r' = R_m - \varepsilon} dS' \left[\frac{\partial \Phi_p^*(\mathbf{r})}{\partial r} - \Phi_p^*(\mathbf{r}) \frac{\partial}{\partial r} \right] \times$$

$$\times \left[\Phi_q(\mathbf{r}') \frac{\partial}{\partial r'} G_\mathbf{k}(\mathbf{r}, \mathbf{r}') - G_\mathbf{k}(\mathbf{r}, \mathbf{r}') \frac{\partial}{\partial r'} \Phi_q(\mathbf{r}') \right]. \tag{B146}$$

(iii) A convenient expression of the Green's function. Because of the particular form of the basis functions $\Phi_p(\mathbf{r})$ given in Equation (B140) it is advantageous to express the Green's function also in terms of spherical harmonics. Within a muffin-tin sphere, the Green's function can be written as a particular solution which is singular at $\mathbf{r} = \mathbf{r}'$ and a general solution which satisfies the homogeneous form of Equation (B130). The former solution can be written as

$$G_\mathbf{k}^0(\mathbf{r}, \mathbf{r}') = \frac{1}{4\pi} \frac{\cos(k\,|\mathbf{r} - \mathbf{r}'|)}{|\mathbf{r} - \mathbf{r}'|},$$

$$= \begin{cases} k \sum_{lm} j_l(kr) n_l(kr') Y_{lm}(\theta, \varphi) Y_{lm}^*(\theta', \varphi') & \text{for} \quad r < r', \\ k \sum_{lm} j_l(kr') n_l(kr) Y_{lm}^*(\theta, \varphi) Y_{lm}(\theta', \varphi') & \text{for} \quad r > r', \end{cases} \tag{B147}$$

where $k^2 = E$, and j_l and n_l are the spherical Bessel and Neuman functions respectively. The latter solution is a linear combination of the product of j's:

$$G_\mathbf{k}^h(\mathbf{r}, \mathbf{r}') = \sum_{lm} \sum_{l'm'} A_{lm,l'm'} j_l(kr) j_{l'}(k'r') Y_{lm}(\theta, \varphi) Y_{l'm'}^*(\theta', \varphi'), \tag{B148}$$

where $A_{lm,l'm'}$ is called structure factor; its explicit form will be given later. By combining Equations (B147) and (B148), the expression of $G(\mathbf{r}, \mathbf{r}')$ for $r < r' < R_m$ has the following form:

$$G(\mathbf{r}, \mathbf{r}') = \sum_{lm} \sum_{l'm'} \{ A_{lm,l'm'} j_l(kr) j_{l'}(kr') +$$

$$+ k j_l(kr) n_{l'}(kr') \delta_{ll'} \delta_{mm'} \} Y_{lm}(\theta, \varphi) Y_{l'm'}^*(\theta', \varphi'). \tag{B149}$$

Now, it is straightforward to substitute Equation (B149) into Equation (B146) and to carry out the surface integrals. The final form of the matrix element with $\varepsilon \to 0$ is then given as

$$\Lambda_{lm,l'm'}^\mathbf{k} = R_m^4 [u_l u_{l'} (L_l j_l - j_l')]_{R_m} [(A_{lm,l'm'} j_{l'}' + k n_{l'} \delta_{ll'} \delta_{mm'}) -$$

$$- (A_{lm,l'm'} j_{l'} + k n_{l'} \delta_{ll'} \delta_{mm'}) L_{l'}]_{R_m}, \tag{B150}$$

where

$$L_l = \frac{1}{u_l} \frac{\partial}{\partial r} u_l \quad \text{and} \quad j_l' = \frac{\partial}{\partial r} j_l(kr).$$

The subscript R_m means all the functions in the brackets are evaluated at R_m. Since $u_l(R_m)$ and $u_{l'}(R_m)$ are chosen equal to 1, the first bracket in Equation (B150) can simply be written as $(L_l j_l - j'_l)_{R_m}$. Substituting Equation (B150) into Equation (B139) and dividing each row by $R_m^2 (L_l j_l - j'_l)_{R_m}$ and each column by $-R_m^2 (L_{l'} j_{l'} - j'_{l'})_{R_m}$, we obtain

$$\det \left| A_{lm,l'm'} + k \delta_{ll'} \delta_{mm'} \left(\frac{n_l L_l - n'_l}{L_{l'} j_{l'} - j'_{l'}} \right)_{R_m} \right| = 0. \tag{B151}$$

The **k**-dependence in Equation (B151) is in the expression of $A_{lm,l'm'}$, whereas $k^2 = E$ and the logarithmic derivatives L_l contain the energy. Therefore, Equation (B151) represents an $E = E(\mathbf{k})$ relation. The logarithmic derivative L_l may be calculated by numerically integrating Equation (B42). The details have been discussed in Section 2.

Since most layer compounds consist of two or more atoms per unit cell, the method discussed above has to be generalized in order to be applicable to these cases and it shall be presented in the following section.

4.2. GENERALIZATION TO COMPLEX CRYSTALS

To generalize the previous results to crystals with more than one atom per unit cell, we shall present here an approach first suggested by Segall [42]. In this approach, the potential is restricted to be of the non-overlapping muffin-tin form centered at each atomic site. The matrix elements of Λ_{pq}^k involving surface integrals centered at the same atom are identical to those given in Section 4.1. However, results for matrix elements with surface integrals centered at different atomic sites are different.

Instead of using Equation (B135) as trial wavefunction, the basis functions for the complex crystal case are given as linear combination of wavefunctions centered at each atomic site in the unit cell, that is

$$\psi_{\mathbf{k}}(\mathbf{r}) = \sum_{p,\tau} C_{\tau p \mathbf{k}} \Phi_p (\mathbf{r} - \tau), \tag{B152}$$

where $\Phi_p(\mathbf{r} - \tau)$ has the same form as given in Equation (B140) except that it is centered at τ and satisfies the following equation:

$$-\nabla^2 \Phi_p(\mathbf{r} - \tau) + V(|\mathbf{r} - \tau|) \Phi_p(\mathbf{r} - \tau) = E \Phi_p(\mathbf{r} - \tau) \tag{B153}$$

with $V(|\mathbf{r} - \tau|)$ the muffin-tin potential centered at τ. The summation over p runs up to a certain maximum angular momentum depending on the constituent atoms and the electronic states of interest.

The assumption of non-overlapping muffin-tin potentials simplifies considerably further calculations. By following all the steps from Equation (B136) to Equation (B145) for one atomic site and by using Equation (B153), the generalized form of Equation (B146) which includes terms of all atoms in the unit cell is given by

$$\Lambda_{p,q}^{\mathbf{k}} = \sum_{\alpha,\beta} \left\{ \int\limits_{R_{m\alpha} - 2\varepsilon} dS_\alpha \left[\frac{\partial \Phi_p^*(\mathbf{r} - \tau_\alpha)}{\partial r} - \Phi_p^*(\mathbf{r} - \tau_\alpha) \frac{\partial}{\partial r} \right] \right\} \times$$

$$\times \left\{ \int\limits_{R_{m\beta} - \varepsilon} dS_\beta \left[\Phi_q(\mathbf{r}' - \tau_\beta) \frac{\partial}{\partial r'} G_{\mathbf{k}}(\mathbf{r}, \mathbf{r}') - G_{\mathbf{k}}(\mathbf{r}, \mathbf{r}') \frac{\partial}{\partial r'} \Phi_q(\mathbf{r}' - \tau_\beta) \right] \right\}, \tag{B154}$$

where $R_{m\alpha}$ or $R_{m\beta}$ is the radius of the muffin-tin sphere of the αth or the βth atom in the unit cell, respectively.

In this complex crystal case, $\Lambda^k_{p,q}$ consists of two kinds of contributions:

(i) $\Lambda^k_{p,q}(\alpha, \alpha)$, the diagonal terms, which involve the product of surface integrals with $\alpha = \beta$. The explicit expression in terms of spherical Bessel functions and $u_l(r)$'s can be obtained in exactly the same way as given before in Equations (B147) to (B150).

(ii) $\Lambda^k_{p,q}(\alpha, \beta)$, the off-diagonal terms, which do not appear in the case of the simple crystal and which involve the product of surface integrals with $\alpha \neq \beta$:

$$\Lambda^k_{p,q}(\alpha, \beta) = \int_{R_{m\alpha}-2\varepsilon} \left\{ dS_\alpha \left[\frac{\partial \Phi^*_p(\mathbf{r} - \mathbf{\tau}_\alpha)}{\partial r} - \Phi^*_p(\mathbf{r} - \mathbf{\tau}_\alpha) \frac{\partial}{\partial r} \right] \int_{R_{m\beta}-\varepsilon} dS_\beta \times \right.$$
$$\left. \times \left[\Phi_q(\mathbf{r}' - \mathbf{\tau}_\beta) \frac{\partial}{\partial r'} G_k(\mathbf{r}, \mathbf{r}') - G_k(\mathbf{r}, \mathbf{r}') \frac{\partial}{\partial r'} \Phi_q(\mathbf{r}' - \mathbf{\tau}_\beta) \right] \right\}. \tag{B155}$$

This expression shall be transformed similarly to Equation (B150). In terms of spherical harmonics about two centers, $\mathbf{\tau}_\alpha$ and $\mathbf{\tau}_\beta$, the Green's function can be written as

$$G_k(\mathbf{r}, \mathbf{r}') = G^{\alpha,\beta}_k(\mathbf{r}_\alpha - \mathbf{r}_\beta + \mathbf{\tau}_\alpha - \mathbf{\tau}_\beta),$$
$$= \sum_{l,m} \sum_{l'm'} A^{\alpha,\beta}_{lm,l'm'} j_l(kr_\alpha) j_{l'}(kr_\beta) Y_{lm}(\theta_\alpha, \varphi_\alpha) Y^*_{l'm'}(\theta_\beta, \varphi_\beta), \tag{B156}$$

where $\mathbf{r}_\alpha = \mathbf{r} - \mathbf{\tau}_\alpha$ and $\mathbf{r}_\beta = \mathbf{r} - \mathbf{\tau}_\beta$, with $r_\alpha + r_\beta < |\mathbf{\tau}_\alpha - \mathbf{\tau}_\beta|$. This latter condition results from the non-overlapping nature of muffin-tin spheres. θ_α, φ_α are the polar angles of $\mathbf{r}_\alpha$. Substituting Equation (B156) into Equation (B155) and carrying out the angular integrations, we get

$$\Lambda^k_{p,q}(\alpha, \beta) = -R^2_{m\alpha} R^2_{m\beta} A^{\alpha,\beta}_{lm,l'm'}(L_l j_l - j'_l)_{R_{m\alpha}}(L_{l'} j_{l'} - j'_{l'})_{R_{m\beta}}. \tag{B157}$$

Combining Equation (B150) for the diagonal term and Equation (B157) for the off-diagonal term the determinantal equation is

$$\det \left| A^{\alpha\beta}_{lm,l'm'} + k\delta_{ll'}\delta_{mm'}\delta_{\alpha,\beta} \left(\frac{n_l L_l - n'_l}{L_{l'} j_{l'} - j'_{l'}} \right)_{R_{m\alpha}} \right| = 0. \tag{B158}$$

To find the energies at a k-point from Equation (B158) it is necessary to evaluate the structure factors $A^{\alpha,\alpha}_{lm,l'm'}$ and $A^{\alpha,\beta}_{lm,l'm'}$. Their expressions will be derived in the next section.

4.3. CALCULATION OF STRUCTURE FACTORS

Since the structure factor $A_{lm,l'm'}$ in Equation (B151) is the same as in the diagonal matrix element given in Equation (B158) it is sufficient to consider the generalized structure factors $A^{\alpha,\alpha}_{lm,l'm'}$ and $A^{\alpha,\beta}_{lm,l'm'}$. The derivation of these two terms differs due to the singularity of the Green's function which has to be considered in the case of $A^{\alpha,\alpha}_{lm,l'm'}$, whereas for $A^{\alpha,\beta}_{lm,l'm'}$ the Green's function will never be singular as shown in the last section. In the following, we shall first derive two expressions of $A^{\alpha,\alpha}_{lm,l'm'}$, one of which is useful to illustrate certain relations among the $A^{\alpha,\alpha}_{lm,l'm'}$'s and the other being adequate for carrying out numerical calculations. An expression of $A^{\alpha,\beta}_{lm,l'm'}$ will then be given.

Let us begin with the derivation of formal expressions for $A^{\alpha,\beta}_{lm,l'm'}$ from which certain relations among the diagonal structure factors with different $\{lm, l'm'\}$ can be deduced. The Green's functions given in Equations (B132) and (B147) can be considered as centered at the αth atom for both vectors $\mathbf{r}$ and $\mathbf{r}'$. The difference between these two functions is just the first term at right hand side of Equation (B149), therefore, we may write:

$$\sum_{lm} \sum_{l'm'} A^{\alpha,\alpha}_{lm,l'm'} j_l(kr) j_{l'}(kr) Y_{lm}(\theta, \varphi) Y^*_{l'm'}(\theta', \varphi') =$$

$$= -\frac{1}{\Omega} \sum_{\mathbf{G}} \frac{e^{i(\mathbf{k}+\mathbf{G})\cdot(\mathbf{r}-\mathbf{r}')}}{(\mathbf{k}+\mathbf{G})^2 - E} - \sum_{lm} \sum_{l'm'} k j_l(kr) n_{l'}(kr') \times$$

$$\times \delta_{ll'}\delta_{mm'} Y_{lm}(\theta, \varphi) Y^*_{l'm'}(\theta', \varphi') \quad \text{for} \quad r, r' < R_{m\alpha}. \tag{B159}$$

The plane waves $e^{i(\mathbf{k}+\mathbf{G})\cdot\mathbf{r}}$ and $e^{-i(\mathbf{k}+\mathbf{G})\cdot\mathbf{r}'}$ are expanded in spherical harmonics as given in Equation (B43). The expression of $A^{\alpha,\alpha}_{lm,l'm'}$ can now be obtained by identifying the coefficients of $Y_{lm}(\theta, \varphi)$ and $Y^*_{l'm'}(\theta', \varphi')$ in Equation (B159).

$$A^{\alpha,\alpha}_{lm,l'm'} = -\frac{(4\pi)^2}{\Omega} i^{l-l'} \frac{1}{j_l(kr)j_{l'}(kr')} \sum_{\mathbf{G}} \times$$

$$\times \frac{j_l(|\mathbf{k}+\mathbf{G}|\, r) j_{l'}(|\mathbf{k}+\mathbf{G}|\, r')}{(\mathbf{k}+\mathbf{G})^2 - E} Y^*_{lm}(\theta_{\mathbf{k}+\mathbf{G}}, \varphi_{\mathbf{k}+\mathbf{G}}) \times$$

$$\times Y_{l'm'}(\theta_{\mathbf{k}+\mathbf{G}}, \varphi_{\mathbf{k}+\mathbf{G}}) - k\delta_{ll'}\delta_{mm'} \times$$

$$\times \frac{n_{l'}(kr')}{j_l(kr')} \quad \text{for} \quad r < r' < R_{m\alpha}, \tag{B160}$$

where $\theta_{\mathbf{k}+\mathbf{G}}$ and $\varphi_{\mathbf{k}+\mathbf{G}}$ are the polar angles of $\mathbf{k}+\mathbf{G}$ with respect to a fixed coordinate system. From the definition of $A^{\alpha\alpha}_{lm,l'm'}$ given in Equation (B148) the expression in the right hand side of Equation (B160) is independent of the choice of r and r'. Furthermore from Equation (160) it is easily seen that

$$A^{\alpha,\alpha}_{lm,l'm'} = A^{\alpha,\alpha*}_{l'm',lm}.$$

This relation among the $A^{\alpha,\alpha}_{lm,l'm'}$'s originates from the hermiticity of $G_\mathbf{k}(\mathbf{r}, \mathbf{r}')$. Although it is possible by using the asymptotic form of $j_l(x)$ as $x \to \infty$ to show that the lattice sum in Equation (B160) is absolutely convergent, in practical cases this sum may take many terms to converge for finite $R_{m\alpha}$. In addition, as seen from Equation (B159) the $A^{\alpha,\alpha}_{lm,l'm'}$ are not linearly independent from each other. It is therefore inconvenient to use Equation (B160) for practical calculations. We shall thus derive an alternative expression for $A^{\alpha,\alpha}_{lm,l'm'}$ in terms of a small number of independent constants and a fast converging series. As seen from Equation (B132) the Green's function depends on the difference of $\mathbf{r}$ and $\mathbf{r}'$. Defining

$$\mathbf{R} = \mathbf{r} - \mathbf{r}' \tag{B162}$$

the Green's function can be written as

$$G_\mathbf{k}(\mathbf{R}) = -\frac{1}{\Omega} \sum_{\mathbf{G}} \frac{e^{i(\mathbf{k}+\mathbf{G})\cdot\mathbf{R}}}{(\mathbf{k}+\mathbf{G})^2 - E},$$

$$= -\frac{1}{4\pi}\frac{\cos kR}{R} + \sum_{L,M} D_{LM}j_L(kR)Y_{LM}(\Theta, \Phi) \quad \text{for} \quad R \le 2R_{ma},$$

(B163)

where the second equality is just the sum of the expressions of the particular (Equation (B147)) and the general solutions (Equation (B148)) in terms of the variable $\mathbf{R}$.

Θ and Φ are the polar angles of $\mathbf{R}$ with respect to a fixed coordinate system. By expanding the plane wave $e^{i(\mathbf{k}+\mathbf{G})\cdot\mathbf{R}}$ in terms of spherical harmonics (Equation (B43)), multiplying it by $Y^*_{LM}(\Theta, \Phi)$ and by integrating over the solid angle of $\mathbf{R}$ in Equation (B163), the quantities D_{LM} are given as

$$D_{LM} = -\frac{4\pi}{\Omega}i^L\frac{1}{j_L(kR)} \times \sum_{\mathbf{G}} j_L(|\mathbf{k} + \mathbf{G}| R)\frac{Y^*_{LM}(\Theta_{\mathbf{k}+\mathbf{G}}, \Phi_{\mathbf{k}+\mathbf{G}})}{(\mathbf{k} + \mathbf{G})^2 - E} +$$

$$+ \frac{k}{4\pi}\delta_{L0}\delta_{MO}\cot(kR),$$

(B164)

$D^*_{LM} = (-1)^L D_{L-M}$ from the property of spherical harmonics. The summation over $\mathbf{G}$ in Equation (B164) can be evaluated by Ewald's method [43]. Let us define

$$F(\eta) = \sum_{\mathbf{G}} j_L(|\mathbf{k} + \mathbf{G}| R)\frac{Y_{LM}(\Theta_{\mathbf{k}+\mathbf{G}}, \Phi_{\mathbf{k}+\mathbf{G}})}{(\mathbf{k} + \mathbf{G})^2 - E} \times e^{\{E-(\mathbf{k}+\mathbf{G})^2\}/\eta}$$

(B165)

then the sum in Equation (B164) is just $F(\infty)$. Ewald showed that $\lim_{\eta\to\infty}\{F(\eta) - F(\infty)\}$ coverges fast to zero. We may therefore choose a reasonably large value of η to evaluate $F(\eta)$ for approximating $F(\infty)$ without appreciable error. After calculating D_{LM} this way $A^{\alpha,\alpha}_{lm,l'm'}$ may be expressed in terms of D_{LM}'s. From Equations (B148) and (B164) we get

$$\sum_{l,m}\sum_{l'm'} A^{\alpha,\alpha}_{lm,l'm'}j_l(kr)j_{l'}(kr')Y_{lm}(\theta, \varphi)Y^*_{l'm'}(\theta', \varphi') = \sum_{L,M} D_{LM}j_L(kr)Y_{LM}(\Theta, \Phi)$$

(B166)

using

$$j_L(kR)Y_{LM}(\Theta, \Phi) = \frac{1}{4\pi i^L}\int d\Omega_{\mathbf{k}}\, e^{i\mathbf{k}\cdot\mathbf{R}}Y_{LM}(\theta_{\mathbf{k}}, \varphi_{\mathbf{k}})$$

$$= \frac{4\pi}{i^L}\sum_{lm}\sum_{l'm'} i^{(l-l')}C_{LM,lm,l'm'} \times$$

$$\times j_l(kr)j_{l'}(kr')Y_{lm}(\theta, \varphi)Y^*_{l'm'}(\theta', \varphi'),$$

(B167)

where $C_{Lm,lm,l'm'} = \int d\Omega_{\mathbf{k}} Y_{LM}(\theta_{\mathbf{k}}, \varphi_{\mathbf{k}})Y^*_{lm}(\theta_{\mathbf{k}}, \varphi_{\mathbf{k}})Y_{l'm'}(\Theta_{\mathbf{k}}, \Phi_{\mathbf{k}})$ is called Gaunt Coefficient, and where $\mathbf{k}$ is a vector with magnitude $k = \sqrt{E}$. By comparing the coefficients of $Y_{lm}(\theta, \varphi)Y_{l'm'}(\theta', \varphi')$ in Equations (B166) and (B167) the final result of $A^{\alpha,\alpha}_{lm,l'm'}$ is

$$A^{\alpha,\alpha}_{lm,l'm'} = 4\pi i^{(l-l')}\sum_L \frac{1}{i^L}D_{LM}C_{Lm-m',lm,l'm'},$$

(B168)

where the sum over L is from $|l - l'|$ to $|l + l'|$. The evaluation of $A^{\alpha,\alpha}_{lm,l'm'}$ can now be done very efficiently by a proper choice of η and by using a table of Gaunt coefficients.

In the case of $\alpha \neq \beta$, the Green's function is never singular due to the assumption of non-overlapping muffin-tin spheres. In terms of r_α, r_β, τ_α and τ_β defined in Equation (B156) the expression of $G_k(r, r')$ given in Equation (B132) can be rewritten as

$$G_k^{\alpha\beta}(\mathbf{r}_\alpha - \mathbf{r}_\beta + \tau_\alpha - \tau_\beta) = -\frac{1}{\Omega} \sum_G \frac{e^{i(\mathbf{k}+\mathbf{G})\cdot(\tau_\alpha - \tau_\beta)}}{(\mathbf{k}+\mathbf{G})^2 - E} e^{i(\mathbf{k}+\mathbf{G})\cdot(\mathbf{r}_\alpha - \mathbf{r}_\beta)}. \qquad (B169)$$

Let $\mathbf{R} = \mathbf{r}_\alpha - \mathbf{r}_\beta$, then $\mathbf{G}_k^{\alpha\beta}$ can be expanded in terms of spherical harmonics as functions of Θ and Φ:

$$G_k^{\alpha\beta}(\mathbf{R} + \tau_\alpha - \tau_\beta) = -\frac{1}{\Omega} \sum_G e^{i(\mathbf{k}+\mathbf{G})\cdot\mathbf{R}} \frac{e^{i(\mathbf{k}+\mathbf{G})\cdot(\tau_\alpha - \tau_\beta)}}{(\mathbf{k}+\mathbf{G})^2 - E} =$$

$$= \sum_{L,M} D_{LM}^{\alpha\beta} j_L(kR) Y_{LM}(\Theta, \Phi). \qquad (B170)$$

Now, by using Equation (B43) for $e^{i(\mathbf{k}+\mathbf{G})\cdot\mathbf{R}}$ one obtains after some algebra

$$D_{LM}^{\alpha\beta} = -4 \frac{\pi}{\Omega} \frac{i^L}{j_L(kR)} \sum_G \frac{e^{i(\mathbf{k}+\mathbf{G})\cdot(\tau_\alpha - \tau_\beta)}}{(\mathbf{k}+\mathbf{G})^2 - E} \times$$

$$\times j_L(|\mathbf{k}+\mathbf{G}|\, R) Y_{LM}^*(\theta_{\mathbf{k}+\mathbf{G}}, \varphi_{\mathbf{k}+\mathbf{G}}). \qquad (B171)$$

The summation $\mathbf{G}$ can be evaluated by Ewald's method as discussed before. Finally, by equating Equation (B156) to Equation (B170) and by following the steps given in Equations (B166) and (B167) the desired expression of the structure factor $A_{l'm,l'm'}^{\alpha,\beta}$ can be written as

$$A_{lm,l'm'}^{\alpha,\beta} = 4\pi i^{(l-l')} \sum_L \frac{1}{i^L} \times D_{L,m-m'}^{\alpha\beta} C_{L,m-m',lm,l'm'}, \qquad (B172)$$

where the Gaunt coefficients are defined as before.

In summary, we have discussed in detail the Green's function method for simple and complex crystals. The important physical quantities such as the logarithmic derivatives, $n_l L_l - n'_l, L_l j_l - j'_l$, and structure factors can be calculated by methods discussed in Section 2.2 and from Equations (B168) and (B172). The energies at a given $\mathbf{k}$-point can then be obtained by solving the determinantal equation given by Equation (B151) or by Equation (B158) depending on the structural complexity of the crystal under consideration.

The application of the Green's method to the layer compound TiS_2 has been carried out by Myron and Freeman [41]. The procedures of constructing the muffin-tin potential are the same as used in the APW method. States up to $l = 2$ and an Ewald parameter η of $(\pi/a)^2$ where a is the lattice constant have been used to calculate the energies along some symmetry directions. The resulting band structure shall be compared with results obtained by other methods in a later chapter.

5. The Method of Linear Combination of Muffin-Tin Orbitals (LCMTO)

The method to be discussed in this section is closely related to the APW and the Green's function methods. It uses the solutions of the Schrödinger equation for a muffin-tin

potential as basis and thus offers a convenient way to treat the non-muffin-tin corrections which cause serious limitations for the application of the Green's function
method.

One of the major difficulties met in the energy band structure calculations is the
choice of proper trial wavefunctions. To obtain good convergence the trial wavefunction should simultaneously show two characters:

(i) atomic like behavior near the nuclei, and

(ii) free electron like behavior in the interstitial region. As we have discussed in
Section 2.2, a solution for a distinct muffin-tin potential (Equation (B42)) does approximate the behavior of an electron in a solid. This solution, however, is normalized only
when E is strictly an eigenvalue of the muffin-tin potential problem, which restricts the
use of these solutions as trial wavefunctions. Recently, Andersen [44] has modified the
solutions of a distinct muffin-tin potential such that they are reasonable descriptions of
the electrons in a solid and also are normalizable for any arbitrary value of E. These new
wavefunctions are called muffin-tin orbitals (MTO's). A linear combination of such
MTO's centered at different atomic sites may be used as fast convergent basis function
set. We shall call this set linear combination of muffin-tin orbitals (LCMTO's).

To see the advantage of using LCMTO's as basis functions for band calculations, let
us point out some major limitations which prevent the Green's function method from
having more general usage:

(i) For mathematical simplicity applications of the Green's function method are
restricted to cases in which the potential is well approximated by non-overlapping
muffin-tin potentials. Hence, satisfactory results were obtained mainly for close packed
crystals whose potentials are nearly spherically symmetric. The accuracy of band
structures for crystals with either open structures or strongly covalent character using
the Green's function method has not been discussed thoroughly enough in the literature.

(ii) The chosen variational scheme of minimizing a particular functional instead of
the energy expectation value of the Hamiltonian raises ambiguities about how to treat
perturbations. To surmount these difficulties of the Green's function method, Andersen
and Kasowski [45] suggested the Hamiltonian to be minimized using a set of muffin-tin
orbitals. This way the non-muffin-tin corrections can be treated either as a perturbation
or directly as part of the matrix elements in the secular equation.

Since the LCMTO involves MTO's centered at different atomic sites similar multi-
center integrals as in the LCAO method are encountered in setting up the Hamiltonian
matrix. Special techniques will be discussed to express these multi-center integrals in
terms of one-center integrals.

The discussions of the method shall be as follows: In Section 5.1 the mathematical
definition of a MTO will be given, the Bloch wavefunction (LCMTO) will then be constructed from the MTO's. We shall in Section 5.2 describe the use of LCMTO's as basis
functions for solving the eigenvalue problem at any k-point in the Brillouin zone. Typical
Hamiltonian matrix elements with the full potential (the muffin-tin potential and its
non-spherical corrections) will be derived in detail, including the reduction of multi-
center integrals to one-center integrals. Finally in Section 5.3 a few comments about
practical calculations will be presented.

5.1. MATHEMATICAL FORMULATION OF MUFFIN-TIN ORBITALS

Let us consider the solution of the Schrödinger equation for distinct isolated muffin-tin potential centered at an atomic site:

$$\Phi_{lm}(\mathbf{r}, E) = i^l Y_{lm}(\hat{r}) u_l(r, E) \quad \text{for} \quad r \le R_m$$
$$= i^l Y_{lm}(\hat{r}) t_l(r, E) \quad \text{for} \quad r \ge R_m, \tag{B173}$$

where R_m is the muffin-tin radius. $Y_{lm}(\hat{r}) = Y_{lm}(\theta, \varphi)$ are the spherical harmonics which depend on the angular coordinates of $\mathbf{r}$. $u_l(r, E)$ is the regular radial solution inside the muffin-tin sphere which satisfies Equation (B42). The form of the solution outside the muffin-tin sphere and its asymptotic behavior at large r are given by:

$$t_l(r, E) = \sqrt{E} \left\{ n_l(\sqrt{E}\, r) - \cot \delta_l j_l(\sqrt{E}\, r) \right\}$$

$$\text{for} \quad r \to \infty, t_l \to \frac{1}{r} \quad \text{for} \quad E > 0, \tag{B174}$$

$$t_l \to \frac{1}{r} e^{r\sqrt{|E|}} \quad \text{for} \quad E < 0 \text{ in general,}$$

$$t_l \to \frac{1}{r} e^{-r\sqrt{|E|}} \quad \text{for } E \text{ being an eigenvalue.}$$

where δ_l is a phase shift. n_l and j_l are the spherical Neumann and Bessel functions. For $E > 0$, Φ_{lm} can be δ-function normalized, whereas for $E < 0$, it is normalizable only when E is an eigenvalue. However, in a solid, as we have shown in Section 2.2, E generally differs from the eigenvalue of an isolated muffin-tin potential. Hence, Φ_{lm} is not a convenient basis function to be used for a solid.

We define the muffin-tin orbital (MTO) $X_{lm}(\mathbf{r}, E)$ by adding the solution of the homogeneous wave equation to $\Phi_{lm}(\mathbf{r}, E)$, so that the resulting normalizable wave function centered at the origin is:

$$X_{lm}(\mathbf{r}, E) = \Phi_{lm}(\mathbf{r}, E) + C_l(E) J_{lm}(\mathbf{r}, E) \quad \text{for} \quad r \le R_m$$
$$= - S_l(E) K_{lm}(\mathbf{r}, E) \quad \text{for} \quad r > R_m, \tag{B175}$$

where

$$J_{lm}(\mathbf{r}, E) = \varphi_{lm}(\mathbf{r}, E) - \sum_c |c\rangle \langle c | \varphi_{lm}\rangle$$

with

$$\varphi_{lm}(\mathbf{r}, E) = \left(\frac{i}{k_0}\right)^l Y_{lm}(\hat{r}) j_l(k_0 r)$$

is an OPW-type wavefunction and where

$$K_{lm}(\mathbf{r}, E) = i^l Y_{lm}(\hat{r}) K_l(\mathbf{r}, E)$$

$$= k_0^{l+1} i^l Y_{lm}(\hat{r}) \begin{cases} i j_l(k_0 r) - n_l(k_0 r), & E \le 0 \\ - n_l(k_0 r), & E > 0 \end{cases}$$

is the usual asymptotic spherical expansion of a plane wave and where $k_0^2 = E - V_0$, with V_0 to be the constant part of the muffin-tin potential. The $|c\rangle$'s are the core states at the chosen atomic site. It is easily shown that by matching the logarithmic derivatives of the wavefunction at $r = R_m$, and by using the Wronskian of the spherical Bessel functions, the coefficients $S_l(E)$ can be expressed as

$$S_l(E) = R_m^2 \left\{ J_l \frac{\partial u_l}{\partial r} - u_l \frac{\partial}{\partial r} J_l \right\}_{r=R_m} \tag{B176}$$

The coefficients, $C_l(E)$, are then determined from the continuity of the wavefunctions at $r = R_m$.

$$C_l(E) = -R_m^2 \left\{ K_l \frac{\partial u_l}{\partial r} - u_l \frac{\partial}{\partial r} K_l \right\}_{r=R_m} \tag{B177}$$

The ratio of C_l/S_l can also be written in terms of the phase shift, δ_l, that is,

$$\frac{C_l}{S_l} = k_0^{2l+1}(\cot \delta_l - i) \quad \text{for} \quad E \le 0$$
$$= k_0^{2l+1} \cot \delta_l \quad \text{for} \quad E > 0. \tag{B178}$$

Unlike the usual definition of the normalization, Andersen and Kasowski [46] suggested the normalization of the MTO to be specified by

$$\langle J_{lm}(\mathbf{r})| \, k_0^2 + V_{MT} - E \, |\Phi_{lm} \rangle = S_l(E), \tag{B179}$$

where V_{MT} is the muffin-tin potential which will be defined in the next section. For convenience we drop the parameter E indicating the explicit energy dependence of the various functions. Because of the presence of J_{lm} and K_{lm} in Equation (B175), $X_{lm}(\mathbf{r})$ is no longer the exact solution of an isolated muffin-tin potential. However, it is composed of localized and itinerant portions which describe properly the behavior of an electron in a solid.

To apply the MTO to a solid, the long range nature of $K_{lm}(\mathbf{r})$ deserves more specific discussions. By making use of the addition theorem of $K_{lm}(\mathbf{r})$ [44], its expansion around $\mathbf{Q}' = (\mathbf{R}' + \boldsymbol{\tau}')$ due to a function centered at $\mathbf{Q}$ is given by

$$K_p(\mathbf{r} - \mathbf{Q}) = \sum_{s,t} 4\pi C_{pks} k_0^{l_p + l_s - l_t} K_t^*(\mathbf{Q} - \mathbf{Q}') J_s(\mathbf{r} - \mathbf{Q}') \tag{B180}$$

for

$$|\mathbf{r} - \mathbf{Q}'| < |\mathbf{Q} - \mathbf{Q}'|,$$

where p, s, t stand for different (lm) values. C_{pst} is the Gaunt coefficient, $\int Y_p(\hat{r}) Y_s^*(\hat{r}) Y_t(\hat{r}) \, d\Omega$. If the crystal is divided into atomic Wigner-Seitz (WS) cells, Equation (B180) holds for $|\mathbf{r} - \mathbf{Q}|$ within the WS cell centered at $\mathbf{Q}$. Then, $K_{lm}(\mathbf{r} - \mathbf{Q})$ in Equation (B175) may be written as

$$K_{lm}(\mathbf{r} - \mathbf{Q}) = K_p(\mathbf{r} - \mathbf{Q}) \quad \text{for} \quad b_{WS}^Q \ge |\mathbf{r} - \mathbf{Q}| > R_m$$
$$= \sum_{\mathbf{Q}'} (1 - \delta_{\mathbf{Q},\mathbf{Q}'}) \sum_{s,t} 4\pi C_{pst} k_0^{l_p + l_s - l_t} K_t^*(\mathbf{Q} - \mathbf{Q}') J_s(\mathbf{r} - \mathbf{Q}') \tag{B181}$$

for

$$b_{WS}^{Q'} > |\mathbf{r} - \mathbf{Q}'|,$$

where b_{WS}^Q is the distance between the boundary of the WS cell and its center $\mathbf{Q}$.

$J_s(\mathbf{r} - \mathbf{Q}')$ is defined in Equation (B175) with core states centered at $\mathbf{Q}'$. Equation (B181) thus defines the function K_{lm} with its center at $\mathbf{Q}$ for arbitrary $\mathbf{r}$ in terms of expansion around different centers $\mathbf{Q}'$.

The Bloch wavefunction for α-type sites constructed from the MTO's defined by Equation (B175) has the following form:

$$\psi_{lm,\alpha}(\mathbf{r},\,\mathbf{k}) = \frac{1}{\sqrt{N}} \sum_n e^{i\mathbf{k}\cdot\mathbf{R}_n}\, X_{lm}(\mathbf{r} - \boldsymbol{\tau}_\alpha - \mathbf{R}_n), \tag{B182}$$

X_{lm} is centered at $\mathbf{R}_n + \boldsymbol{\tau}_\alpha$ where $\boldsymbol{\tau}_\alpha$ is the position vector of an atom within the unit cell specified by $\mathbf{R}_n$. $\mathbf{k}$ is as usually restricted to the first Brillouin zone. By using Equation (B181) we are able to express Equation (B182) in terms of spherical Bessel functions only:

$$\begin{aligned}
\psi_{lm,\alpha}(\mathbf{r},\,\mathbf{k}) = {}& \frac{1}{\sqrt{N}} \sum_n e^{i\mathbf{k}\cdot\mathbf{R}_n}\{[\varPhi_{lm}(\mathbf{r} - \boldsymbol{\tau}_\alpha - \mathbf{R}_n) + \\
& + C_l J_{lm}(\mathbf{r} - \boldsymbol{\tau}_\alpha - \mathbf{R}_n)]\varTheta(R_m - |\mathbf{r} - \boldsymbol{\tau}_\alpha - \mathbf{R}_n|) - \\
& - S_l K_{lm}(\mathbf{r} - \boldsymbol{\tau}_\alpha - \mathbf{R}_n)[\varTheta(b_{WS}^{\boldsymbol{\tau}_\alpha + \mathbf{R}_n} - |\mathbf{r} - \boldsymbol{\tau}_\alpha - \mathbf{R}_n|) - \\
& - \varTheta(R_m - |\mathbf{r} - \boldsymbol{\tau}_\alpha - \mathbf{R}_n|)] - S_l \sum_{\beta,n'} e^{i\mathbf{k}\cdot\mathbf{R}_{n'}} \times \\
& \times (1 - \delta(\boldsymbol{\tau}_\alpha - \boldsymbol{\tau}_\beta + \mathbf{R}_{n'}))\sum_{s,t} 4\pi C_{lm,s,t} k_0^{l+l_s-l_t} \times \\
& \times K_t^*(\boldsymbol{\tau}_\alpha - \boldsymbol{\tau}_\beta + \mathbf{R}_{n'}) J_s(\mathbf{r} - \boldsymbol{\tau}_\beta - \mathbf{R}_n)\varTheta(b_{WS}^{\boldsymbol{\tau}_\beta + \mathbf{R}_n} - |\mathbf{r} - \boldsymbol{\tau}_\beta - \mathbf{R}_n|)\} \\
= {}& \frac{1}{\sqrt{N}} \sum_n e^{i\mathbf{k}\cdot\mathbf{R}_n} \{X_{lm}(\mathbf{r} - \boldsymbol{\tau}_\alpha - \mathbf{R}_n) - \\
& - S_l \sum_{\beta,s} B_{lm,s}^{\mathbf{k}}(\alpha,\,\beta) J_s(\mathbf{r} - \boldsymbol{\tau}_\beta - \mathbf{R}_n)\varTheta(b_{WS}^{\boldsymbol{\tau}_\beta + \mathbf{R}_n} - |\mathbf{r} - \boldsymbol{\tau}_\beta - \mathbf{R}_n|)\},
\end{aligned} \tag{B183}$$

where the Heavyside function is given by

$$\varTheta(x - y) = 1 \quad \text{for} \quad x > y$$

$$= 0 \quad \text{otherwise}$$

$$\begin{aligned}
X_{lm}(\mathbf{r} - \boldsymbol{\tau}_\alpha - \mathbf{R}_n) = {}& \{\varPhi_{lm}(\mathbf{r} - \boldsymbol{\tau}_\alpha - \mathbf{R}_n) + \\
& + C_l J_{lm}(\mathbf{r} - \boldsymbol{\tau}_\alpha - \mathbf{R}_n)\}\varTheta(R_m - |\mathbf{r} - \boldsymbol{\tau}_\alpha - \mathbf{R}_n|) - \\
& - S_l K_{lm}(\mathbf{r} - \boldsymbol{\tau}_\alpha - \mathbf{R}_n)[\varTheta(b_{WS}^{\boldsymbol{\tau}_\alpha + \mathbf{R}_n} - \\
& - |\mathbf{r} - \boldsymbol{\tau}_\alpha - \mathbf{R}_n|)\,\varTheta(R_m - |\mathbf{r} - \boldsymbol{\tau}_\alpha - \mathbf{R}_n|)
\end{aligned}$$

and

$$\begin{aligned}
B_{lm,s}^{\mathbf{k}}(\alpha,\,\beta) = {}& \sum_{n'} e^{i\mathbf{k}\cdot\mathbf{R}_{n'}}(1 - \delta(\boldsymbol{\tau}_\alpha - \boldsymbol{\tau}_\beta + \mathbf{R}_{n'})) \times \\
& \times \sum_t 4\pi C_{lm,s,t} k_0^{l+l_s-l_t} K_t^*(\boldsymbol{\tau}_\alpha - \boldsymbol{\tau}_\beta + \mathbf{R}_{n'}).
\end{aligned}$$

The functions X_{lm} are now restricted to within one WS cell, while the overlap of the MTO's into neighboring WS cells is defined by the expansion coefficients $B^{\mathbf{k}}_{lm,s}$.

Similar to the variational scheme used in the LCAO method, we treat the wavefunction in Equation (B183) as a basis function. The total trial wavefunction is then given by

$$\Psi(\mathbf{r}, E) = \sum_{lm} \sum_{\alpha} A_{lm,\alpha} \psi_{lm,\alpha}(\mathbf{r}, \mathbf{k}). \tag{B184}$$

The coefficients, $A_{lm,\alpha}$ are determined by minimizing the energy expectation value of the crystal Hamiltonian. The details of the variational procedure will be presented in the next section.

5.2. VARIATIONAL SCHEME USING A LINEAR COMBINATION OF MUFFIN-TIN ORBITALS

In this section, we shall explain in detail the use of LCMTO's for calculating the electronic band structure of a solid. A brief comment on the construction of the crystal potential will be given first followed by the derivation of LCMTO Hamiltonian matrix elements with special emphasis on the technique of reducing the multi-center integrals to one-center integrals.

Similar to the procedures used in the APW method, we divide the crystal into atomic WS cells. The potential at any point within the WS cell is given as sum of Coulomb and exchange potentials:

$$V(\mathbf{r}) = \sum_{n,\alpha} V_{\text{Coul}}(\mathbf{r} - \mathbf{\tau}_\alpha - \mathbf{R}_n) + \alpha \left\{ \sum_{n,\alpha} \rho(\mathbf{r} - \mathbf{\tau}_\alpha - \mathbf{R}_n) \right\}^{1/3}, \tag{B185}$$

where a local (Slater) form of the exchange potential has been used. The overlapping charge distribution in Equation (B185) may be calculated from the atomic wavefunctions of Herman and Skillman [6] with an assumed atomic configuration of the constituent atoms. $V(\mathbf{r})$ may then be expanded in spherical harmonics by numerical Monte Carlo or de Fantine methods:

$$V(\mathbf{r}) = V_0(r) + \sum_{\substack{l,m \\ l \neq 0}} V_l(r) Y_{lm}(\hat{r}) = V_{MT}(\mathbf{r}) + \Delta V(\mathbf{r}) \tag{B186}$$

and

$$V_l(r) = \int_{WS} V(\mathbf{r}) Y^*_{lm}(\hat{r}) \, d\Omega.$$

A factor of $1/\sqrt{4\pi}$ has been absorbed in $V_0(r)$. The muffin-tin potential around an atom is defined as

$$V_{MT}(\mathbf{r}) = V_0(r), \quad r \leq R_m$$

$$= V_0 \quad \text{otherwise.} \tag{B187}$$

where V_0 is the average potential for $R_m < r \leq b_{WS}$. The other term, $\Delta V(\mathbf{r})$ in Equation (B186) characterizes the deviations from the spherically symmetric potential inside the muffin-tin sphere and is the full potential in the interstitial region.

Having determined the crystal potential, we then use the wavefunction given in

Equation (B184) as trial wavefunction to minimize the Hamiltonian. The secular equation with fixed E or k_0 is obtained from

$$\frac{\partial}{\partial A_{p\alpha'}^*} \left\langle \sum_{p,\alpha'} A_{p\alpha'}\psi_{p,\alpha'} \right| - \nabla^2 +$$

$$+ \sum_{n,\alpha} V(\mathbf{r} - \boldsymbol{\tau}_\alpha - \mathbf{R}_n) - \varepsilon \left| \sum_{q,\alpha''} A_{q\alpha''}\psi_{q\alpha''} \right\rangle = 0, \tag{B188}$$

where ε is the usual Lagrange multiplier. In Equation (B188) we have expressed the Hamiltonian in the same form as given in the LCAO and the APW methods. A typical matrix element in the secular equation has the following form:

$$\langle \psi_{p,\alpha'}| - \nabla^2 + \sum_{n,\alpha} V(\mathbf{r} - \boldsymbol{\tau}_\alpha - \mathbf{R}_n) - \varepsilon |\psi_{q,\alpha''}\rangle$$

$$= \frac{1}{N} \sum_{n'n''} e^{i\mathbf{k}\cdot(\mathbf{R}_{n'} - \mathbf{R}_{n''})} I_{p\alpha'n';q\alpha''n''}. \tag{B189}$$

$I_{p\alpha'n';q\alpha''n''}$ in Equation (B189) involves one-, two- and three-center integrals. However, by using Equation (B183), we are able to express the multi-center integrals in terms of one-center integrals.

To simplify the notation let us omit the Θ-functions of Equation (B183) and the arguments α, β in the various functions. Then, $I_{p\alpha'n';q\alpha''n''}$ may be written as

$$I_{p\alpha'n';q\alpha''n''} = \left\langle X_p - S_{l_p} \sum_{\beta_1,s_1} B_{p,s_1}^{\mathbf{k}} J_{s_1} \right| \times$$

$$\times \left\{ -\nabla^2 + \sum_{n,\alpha} V(\mathbf{r} - \boldsymbol{\tau}_\alpha - \mathbf{R}_n) - \varepsilon \right\} \times$$

$$\times \left| X_q - S_{l_q} \sum_{\beta_2,s_2} B_{q,s_2}^{\mathbf{k}} J_{s_2} \right\rangle. \tag{B190}$$

Due to their definition given in Equation (B183) the spherical Bessel functions are non-vanishing only within their respective WS cells. Thus multi-center integrals in Equation (B190) are now reduced to one-center integrals. The different contributions to the matrix element may be separated into:

(i) $\langle X_p| - \nabla^2 + \sum_{n,\alpha} V(\mathbf{r} - \boldsymbol{\tau}_\alpha - \mathbf{R}_n) - \varepsilon |X_q\rangle$

$$= \langle \Phi_p| \, \Delta V' \, |X_q\rangle_{R_m} + C_{l_p}^* \langle J_p| \, V' \, |X_q\rangle_{R_m} +$$

$$+ S_{l_p}^* S_{l_q} \langle K_p| \, \Delta V' \, |K_q\rangle_{WS,R_m} \, \delta_{nn'} \delta_{nn''} \delta_{\alpha\alpha'} \delta_{\alpha\alpha''}, \tag{B191}$$

where $\Delta V' = E - \varepsilon + \Delta V$ for $r < R_m$ and $\Delta V' = k_0^2 - \varepsilon + \Delta V$ for $r > R_m$ and where $V'(\mathbf{r}) = k_0^2 - \varepsilon + V(\mathbf{r})$. The range of the first two matrix elements is restricted to within the muffin-tin spheres whereas the last term is restricted to the region outside the muffin-tin spheres and inside the WS cell.

(ii) $\left\langle \sum_{\beta_1,s_1} B_{ps_1}^{\mathbf{k}} J_{s_1} \right| - \nabla^2 + \sum_{n,\alpha} V(\mathbf{r} - \boldsymbol{\tau}_\alpha - \mathbf{R}_n) - \varepsilon \left| X_q \right\rangle$

$$= \left\{ \sum_{s_1, \beta_1}' B_{ps_1}^{\mathbf{k}*} \langle J_{s_1} | - \nabla^2 + V(\mathbf{r}) - \varepsilon \, | X_q \rangle + \right.$$

$$\left. + B_{pq}^{\mathbf{k}*} \langle J_q | - \nabla^2 + V(\mathbf{r}) - \varepsilon \, | X_q \rangle \right\} \delta_{\alpha\alpha''} \delta_{n'n} \delta_{n''} . \tag{B192}$$

The prime on the sum excludes the $s_1 = q$ term centered at $\beta_1 = \alpha''$. The second matrix element at the right-hand side which represents the diagonal term and can further be simplified. By adding and subtracting E and Equation (B179), we obtain

$$\langle J_q | - \nabla^2 + V(\mathbf{r}) - \varepsilon \, | X_q \rangle = \langle J_q | V' \, | X_q \rangle_{WS} - S_{l_q} . \tag{B193}$$

With this result the total expression of (ii) can be written as:

$$\left\langle \sum_{\beta_1, s_1} B_{ps_1}^{\mathbf{k}} J_{s_1} \middle| - \nabla^2 + \sum_{n, \alpha} V(\mathbf{r} - \boldsymbol{\tau}_\alpha - \mathbf{R}_n) - \varepsilon \, \middle| X_q \right\rangle$$

$$= \left\{ \sum_{\beta_1, s_1} B_{ps_1}^{\mathbf{k}*} \langle J_{s_1} | V' \, | X_q \rangle_{WS} - S_{l_q} B_{pq}^{\mathbf{k}*} \right\} \delta_{nn'} \delta_{nn''} \delta_{\beta_1, \alpha} \delta_{\alpha\alpha''} . \tag{B194}$$

(iii)

$$\left\langle \sum_{s_1, \beta_1} B_{ps_1}^{\mathbf{k}} J_{s_1} \middle| - \nabla^2 + \sum_{n, \alpha} V(\mathbf{r} - \boldsymbol{\tau}_\alpha - \mathbf{R}_n) - \varepsilon \middle| \sum_{s_2 \beta_2} B_{qs_2}^{\mathbf{k}} J_{s_2} \right\rangle$$

$$= \sum_{s_1, \beta_1} \sum_{s_2, \beta_2} B_{ps_1}^{\mathbf{k}*} \langle J_{s_1} | V' \, | J_{s_2} \rangle_{WS} \, B_{qs_2}^{\mathbf{k}} \delta_{nn'} \delta_{nn''} \delta_{\beta_1, \alpha} \delta_{\beta_2, \alpha} . \tag{B195}$$

All the integrations in the last four equations are to be carried out within the WS cells. The final expression of the matrix element of the Hamiltonian between the LCMTO's (Equation (B189)) can now be written down by combining results of Equations (B191), (B194) and (B195):

$$\langle \psi_{p, \alpha'} | - \nabla^2 + \sum_{n, \alpha} V(\mathbf{r} - \boldsymbol{\tau}_\alpha - \mathbf{R}_n) - \varepsilon \, | \psi_{q, \alpha''} \rangle$$

$$= \left\{ \langle \Phi_p | \Delta V' \, | X_q \rangle_{R_m} + C_{l_p}^* \langle J_p | V' \, | X_q \rangle_{R_m} + \right.$$

$$\left. + S_{l_p}^* S_{l_q} \langle K_p | \Delta V' \, | K_q \rangle_{WS, R_m} \right\} \delta_{\alpha\alpha'} \delta_{\alpha\alpha''} +$$

$$+ S_{l_p}^* \sum_{\beta_1} \left\{ S_{l_q} B_{pq}^{\mathbf{k}*} - \sum_{s_1} B_{ps_1}^{\mathbf{k}*} \langle J_{s_1} | V' \, | X_q \rangle_{WS} \right\} \delta_{\alpha\alpha''} +$$

$$+ S_{l_q} \left\{ S_{l_p}^* B_{qp}^{\mathbf{k}} - \sum_{s_1} B_{qs_1}^{\mathbf{k}} \langle X_p | V' \, | J_{s_1} \rangle_{WS} \delta_{\alpha'\alpha} \right\} +$$

$$+ S_{l_p}^* S_{l_q} \sum_{\beta_1, s_1} \sum_{\beta_2, s_2} B_{ps_1}^{\mathbf{k}*} \langle J_{s_1} | V' \, | J_{s_2} \rangle_{WS} \, B_{qs_2}^{\mathbf{k}} \delta_{\beta_1 \alpha} \delta_{\beta_2 \alpha} . \tag{B196}$$

After having derived the explicit expression of the LCMTO matrix element, we shall comment on a few points relevant to practical calculations.

5.3. COMMENTS ON PRACTICAL CALCULATIONS

The lattice sum in $B_{ps}^{\mathbf{k}}$ is essentially the structure factor D_{LM} of the Green's function method (Section 4) and the layer method (Section 6). Therefore, it can be evaluated by the Ewald method discussed in Section 4.3.

The effect of the core states in Equation (B175) is worth some comments. The potential near the core is spherically symmetric, $\Delta V \to 0$ and ψ_p is orthogonalized to the core states. Therefore, the orthogonalization does not complicate the calculation of $\langle \Phi_p | \Delta V' | X_q \rangle$. It effects the matrix elements $\langle J_p | V'_{MT} | J_q \rangle$, where $V'_{MT}(\mathbf{r}) = V'(\mathbf{r}) - \Delta V(\mathbf{r})$. In these cases, however, the integrals can be computed numerically.

The parameters k_0 and E may in general be treated as variational parameters. For most cases, certain approximations of these quantities may be obtained:

(i) For bound states and for E different from resonance energies all partial waves behave like $\Phi_p = C_{l_p} J_p$ and the corresponding phase shifts are small. From Equation (B178) and with a normalization of $S_{l_p} = -1$ the important terms in Equation (B196) are reduced to a sum of the first term in the large curly bracket and the last term in the equation. Since these two terms are independent of E it is necessary to treat k_0 as a variational parameter to minimize the eigenvalue ε.

(ii) For a fixed k_0, Equation (B196) is roughly a linear function of E except for the case of resonances, where the minimized value of ε should be close to E. Therefore, we may always approximate $E = \varepsilon$ and treat k_0 as parameter which minimizes the energy ε.

An application of the LCMTO-method to the layer compound MoS_2 has been carried out by Kasowski [47]. The results shall be compared with results obtained by other methods in a later chapter.

6. The Layer Transfer-Matrix Method (LTM)

A major difficulty in calculating band structures of transition metal dichalcogenides is caused by their anisotropic crystal structure. The unit cell dimension in these compounds is often several times as large along the $\hat{c}$-axis ($\hat{z}$) as in the plane of the layers. Therefore, a large number of Fourier components in the periodic part of the Bloch states and in the crystal potential will be needed for an expansion in the direction parallel to the $\hat{c}$-axis. This generally results in large size secular equations for obtaining the eigenvalues $E(\mathbf{k})$. To overcome this particular difficulty, Wood and Pendry [48] suggested to apply a 'layer method' to the calculation of band structures of transition metal layer compounds. The method has originally been formulated for the low-energy electron diffraction (LEED) studies on solids.

The essential ingredients of the method involve:

(i) The definition of a basic structural unit consisting of several sheets of different atoms in the $\hat{z}$ direction and extending over a unit cell in the plane of the layers. This structural unit may be smaller along $\hat{z}$ than the crystallographic unit cell.

(ii) A solution of the Schrödinger equation *between* the basic units. The solution can be written in terms of two wavefunctions with equal and opposite crystal momenta in the $\hat{z}$-direction respectively and it can be expanded into a two-dimensional Fourier series in the directions perpendicular to the $\hat{c}$-axis.

(iii) The formulation of the amplitudes of the two wavefunctions in terms of transmission (T) and reflection (R) coefficients characterizing the basic unit.

(iv) The definition of an eigenvalue equation for the $\hat{z}$-component of the crystal momentum.

The potential used in this method is the same as used for APW and Green's function methods. The construction of this potential has been discussed in Section 2.2. The only extra procedure involved in the present approach is to average differently the potential between the layers.

In Section 6.1 we shall formulate the eigenvalue equation. This equation will be expressed in terms of the transmission and reflection coefficients of the basic unit. The expressions of these coefficients will be derived in Section 6.2. Similar to the Green's function method, the coefficients depend on the atomic positions in the basic unit, and can therefore, be expressed in terms of structure factors. However, the formulation given in Section 4 for calculating structure factors for the Green's function method can no longer be used, because in the layer method the $\hat{z}$-direction is treated differently than the other two directions. Expressions defining the structure factors will thus be derived in Section 6.3. For this we shall consider the case of non-overlapping muffin-tin potential only. Finally, in Section 6.4 some mathematical details of the derivation of the structure factors will be presented.

6.1. DERIVATION OF AN EIGENVALUE EQUATION

To derive an eigenvalue equation for the layer method we shall proceed according to the introductory remarks.

(i) Isolating a basic structural unit. In layer compounds, the atoms are stacked in layers along the crystal $\hat{z}$-direction each consisting of several atomic sheets. Depending on stacking sequences the crystallographic unit cell in the $\hat{z}$-direction consists of p layers. For the example of 2H–MoS_2, $(p = 2)$ the corresponding layers are schematically shown in Figure 6. The dashed lines divide the repeating units. Between units the potential is assumed to constant, V_0.

(ii) Obtaining the wavefunction between the repeating units. In the region of constant potential, the solution of the Schrödinger equation can be expressed in terms of two free-electron wavefunctions with opposite momentum.

$$\psi_j(\mathbf{r}) = U_j^{(+)} + U_j^{(-)}. \tag{B197}$$

In the directions parallel to the layer, the two dimensional periodic structure allows to expand the wavefunctions, U_j, in Fourier series. Therefore, for a particular energy E, and for a given set of components of the crystal momentum parallel to the layer k_x and k_y, the wavefunction between the $(j \doteq p)$th and the jth basic units can be written as

$$\psi_j(\mathbf{r}) = \sum_{\mathbf{q}} [U_{j\mathbf{q}}^{(+)} e^{i\mathbf{K}_{\mathbf{q}}^{(+)} \cdot \mathbf{r}} + U_{j\mathbf{q}}^{(-)} e^{-i\mathbf{K}_{\mathbf{q}}^{(-)} \cdot \mathbf{r}}], \tag{B198}$$

where $\mathbf{q}$ is a two-dimensional reciprocal lattice vector. The momenta $\mathbf{K}_{\mathbf{q}}^{(\pm)}$ are given by

$$\mathbf{K}_{\mathbf{q}}^{(\pm)} = \left(k_x + q_x, k_y + q_y, \pm \left[\frac{2m}{\hbar^2} (E - V_0) - \right. \right.$$
$$\left. \left. - (k_x + q_x)^2 - (k_y + q_y)^2 \right]^{1/2} \right),$$

for

$$\frac{2m}{\hbar^2}(E - V_0) > (k_x + q_x)^2 + (k_y - q_y)^2;$$

$$= \left(k_x + q_x, k_y + q_y, \pm i \left[(k_x + q_x)^2 + (k_y + q_y)^2 - \right. \right.$$

$$\left. \left. - \frac{2m}{\hbar^2}(E - V_0) \right]^{1/2} \right)$$

for

$$\frac{2m}{\hbar^2}(E - V_0) < (k_x + q_x)^2 + (k_y + q_y)^2. \tag{B199}$$

Since the unit cell repeats every p layers, the wavefunction $\psi_{j+p}(\mathbf{r})$, between the jth and the $(j + p)$th layers relates to $\psi_j(\mathbf{r})$ by the Bloch condition. Let $\mathbf{d}$ be a lattice vector, then

$$\psi_{j+p}(\mathbf{r}) = e^{ik_z d_z + i\mathbf{k}_t \cdot \mathbf{d}_t} \psi_j(\mathbf{r}), \tag{B200}$$

where d_z and $\mathbf{d}_t$ are the components of $\mathbf{d}$ in the $\hat{z}$-direction and parallel to the plane of the layer respectively. The crystal momentum $\mathbf{k}$ may be written as

$$\mathbf{k} = \hat{z}k_z + \hat{x}k_x + \hat{y}k_y = \hat{z}k_z + \mathbf{k}_t. \tag{B201}$$

(iii) Expressing $U_j^{(\pm)}$ and $U_{j+p}^{(\pm)}$ in terms of the transmission and the reflection coefficients of the basic units.

To facilitate this discussion, we shall use matrix notations. The coefficients of $U_{jq}^{(\pm)}$ appearing in Equations (B197) and (B198) can be written as

$$U_j^{(\pm)} = \begin{pmatrix} U_{jq_1}^{(\pm)} \\ U_{jq_2}^{(\pm)} \\ \vdots \\ U_{jq_n}^{(\pm)} \end{pmatrix}. \tag{B202}$$

With this notation, it is possible to rewrite Equation (B200) as

$$\psi_{j+p}(\mathbf{r}) = U_{j+p}^{(+)} + U_{j+p}^{(-)}$$

and

$$U_{j+p}^{(+)} = e^{i\mathbf{k} \cdot \mathbf{d}} U_j^{(+)}, \qquad U_{j+p}^{(-)} = e^{i\mathbf{k} \cdot \mathbf{d}} U_j^{(-)}. \tag{B203}$$

Now, $U_{j+p}^{(\pm)}$ and $U_j^{(\pm)}$ are interpreted as the incident and the reflected waves of the $(j + p)$th and the jth units respectively. The basic unit between the two constant potential regions can be replaced by a 'black box' characterized by a transfer matrix, Q. The matrix elements of Q are the transmission and the reflection coefficients of the unit. The geometry is shown in Figure 7. Thus a relation between $U_{j+p}^{(\pm)}$ and $U_j^{(\pm)}$ can be obtained using the matrix Q:

$$U_{j+p}^{(+)} = Q^{\mathrm{I}} U_j^{(+)} + Q^{\mathrm{II}} U_{j+p}^{(-)}, \tag{B204}$$

$$U_j^{(-)} = Q^{\mathrm{III}} U_j^{(+)} + Q^{\mathrm{IV}} U_{j+p}^{(-)}, \tag{B205}$$

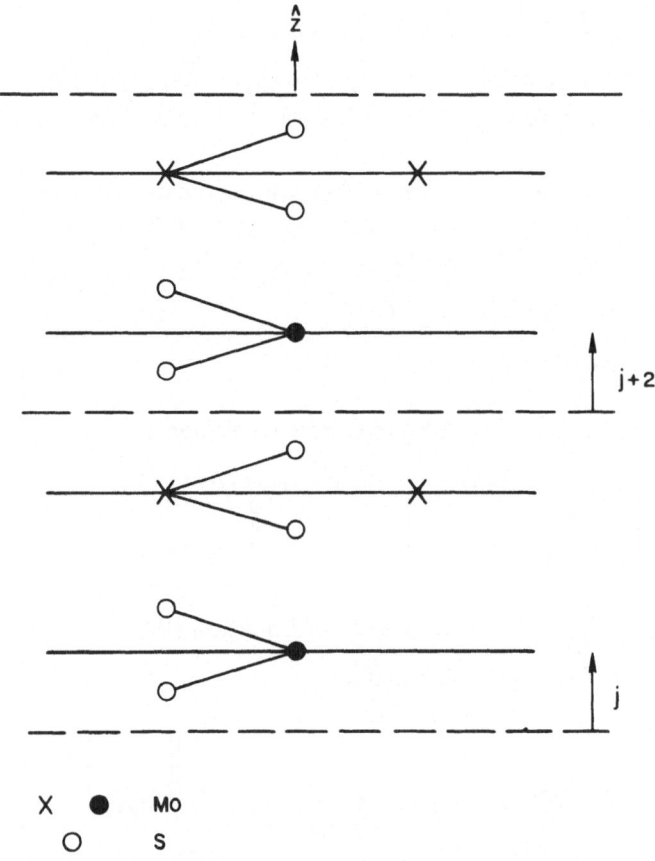

Fig. 6. Example for defining layers (MoS$_2$) in the layer transfer-matrix method.

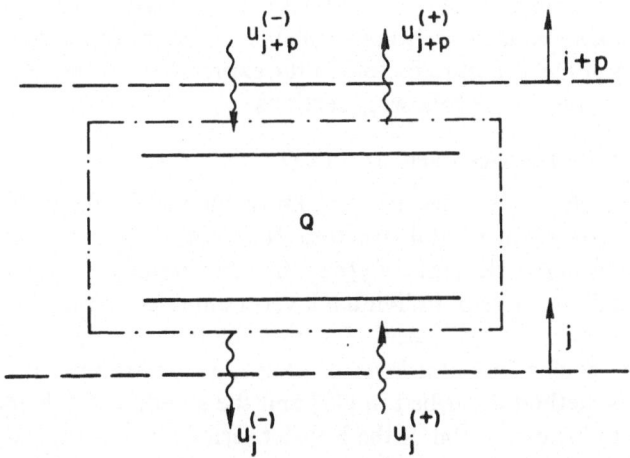

Fig. 7. Schematic representation of incident and reflected waves of a structural unit as used in the layer transfer-matrix method.

where $Q = \begin{pmatrix} Q^{\mathrm{I}} & Q^{\mathrm{II}} \\ Q^{\mathrm{III}} & Q^{\mathrm{IV}} \end{pmatrix}$, Q^{I} and Q^{IV} are the transmission coefficients whereas Q^{II} and Q^{III} are the reflection coefficients. The calculation of these coefficients will be discussed in Section 6.2. From Equations (B203), (B204) and (B205), one can derive an eigenvalue equation.

(iv) Deriving an eigenvalue equation. Substituting the first Bloch condition of Equation (B203) into Equation (B204) and inserting Equation (B204) into the second Bloch condition in Equation (B203), we obtain

$$e^{i\mathbf{k}\cdot\mathbf{d}}\, U_j^{(+)} = Q^{\mathrm{I}} U_j^{(+)} + Q^{\mathrm{II}} U_{j+p}^{(-)},$$

$$e^{i\mathbf{k}\cdot\mathbf{d}}(Q^{\mathrm{III}} U_j^{(+)} + Q^{\mathrm{IV}} U_{j+p}^{(-)}) = U_{j+p}^{(-)}. \tag{B206}$$

The matrix form of Equation (B206) may now be written as

$$e^{i\mathbf{k}\cdot\mathbf{d}} \begin{pmatrix} 1 & 0 \\ Q^{\mathrm{III}} & Q^{\mathrm{IV}} \end{pmatrix} \begin{pmatrix} U_j^{(+)} \\ U_{j+p}^{(-)} \end{pmatrix} = \begin{pmatrix} Q^{\mathrm{I}} & Q^{\mathrm{II}} \\ 0 & 1 \end{pmatrix} \begin{pmatrix} U_j^{(+)} \\ U_{j+p}^{(-)} \end{pmatrix}. \tag{B207}$$

Multiplying $\begin{pmatrix} 1 & 0 \\ Q^{\mathrm{III}} & Q^{\mathrm{IV}} \end{pmatrix}^{-1}$ on both sides of Equation (B207), we obtain an equation of k_z in terms of $e^{i\mathbf{k}\cdot\mathbf{d}}$ for particular E and $\mathbf{k}_t$,

$$\begin{pmatrix} Q^{\mathrm{I}} & Q^{\mathrm{II}} \\ -(Q^{\mathrm{IV}})^{-1} Q^{\mathrm{III}} Q^{\mathrm{I}} & (Q^{\mathrm{IV}})^{-1}(1 - Q^{\mathrm{III}} \; Q^{\mathrm{II}}) \end{pmatrix} \begin{pmatrix} U_j^{(+)} \\ U_{j+p}^{(-)} \end{pmatrix}$$

$$= e^{i\mathbf{k}\cdot\mathbf{d}} \begin{pmatrix} U_j^{(+)} \\ U_{j+p}^{(-)} \end{pmatrix}. \tag{B208}$$

The important feature of Equation (B208) is that the matrix size depends only on the number of *two-dimensional* Fourier components. The complications of the method lie in the calculation of the structure factors in the expressions of the Q's. This will be the topic of the discussions in the following sections.

6.2. CALCULATION OF THE TRANSFER MATRIX Q

To solve Equation (B208) it is necessary to know the matrix elements of Q. The complexity of Q depends on the crystal structure. It is possible to first define Q for just *one* layer. The complete matrix elements of Q can then be obtained from a matrix multiplication of the T- and R-matrices of individual layers contained in the unit cell.

Kambe [49] has applied the Green's function method to calculate the transmission and the reflection coefficients of a layer of atoms. There are several important differences between his method described in [49] and the standard KKR method described in Section 4 due to the expansion of the Fourier series which is carried out only in the plane of the layers. We shall follow Kambe's approach to derive the expressions of T and R of a layer for a single incident plane wave with energy E and momentum $\mathbf{k}_0$. The approach involves: the definition of a wavefunction at the muffin-tin spheres of each atom, the identification of transmitted and reflected waves and the calculation of transmission and reflection coefficients.

Let us start with the usual integral equation of the wavefunction

$$\psi(\mathbf{r}) = \psi^{(0)}(\mathbf{r}) + \int_v G(\mathbf{r}, \mathbf{r}')V(\mathbf{r}')\psi(\mathbf{r}') \, d^3r', \tag{B209}$$

where $\psi^{(0)}(\mathbf{r})$ is an incident wave for scattering states and zero for bound states and where the second term is the scattered wave as given in Equation (B133). v is the volume of the basic unit. The Green's function $G(\mathbf{r}, \mathbf{r}')$ satisfies Equation (B130) with the following two-dimensional boundary conditions:

$$G(\mathbf{r}, \mathbf{r}' + \mathbf{R}_{lt}) = e^{-i\mathbf{k}_t \cdot \mathbf{R}_{lt}} G(\mathbf{r}, \mathbf{r}'), \tag{B210}$$

where $\mathbf{R}_{lt}$ is a lattice vector in the plane of the layer. For convenience, we shall assume that $V(\mathbf{r})$ is composed of non-overlapping muffin-tin potentials centered at the various atoms,

$$V(\mathbf{r}) = \sum_\alpha V_\alpha(\mathbf{r} - \boldsymbol{\tau}_\alpha), \tag{B211}$$

where V_α has the form of Equation (B40) with $V_0 = 0$. $\boldsymbol{\tau}_\alpha$ is the position vector of the αth atom in the layer. To define the wavefunction at the layer boundaries, its values at the surfaces of each particular muffin-tin sphere have to be known.

Let $\mathbf{r} = \mathbf{r}_\alpha + \boldsymbol{\tau}_\alpha$ and $\mathbf{r}' = \mathbf{r}_\beta + \boldsymbol{\tau}_\beta$ where $\mathbf{r}_\alpha$ is measured from the αth atom and is just outside the muffin-tin sphere. $\mathbf{r}_\beta$ is centered at βth atom with its magnitude less than $R_{m\beta}$, the radius of muffin-tin sphere, Figure 8 shows the relations among the vectors. By

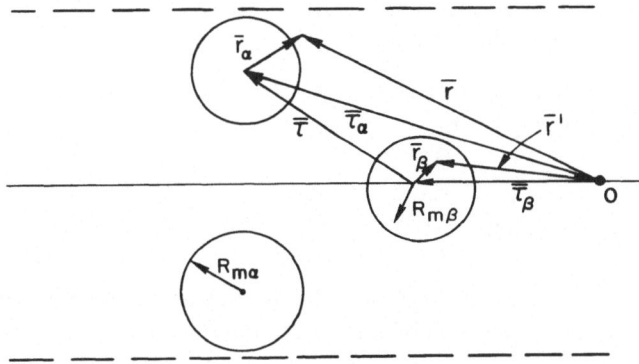

Fig. 8.   Geometry of the position vectors in the layer transfer-matrix method.

substituting Equation (B211) into Equation (B209) and by using the same method for deriving Equation (B143), the wavefunction near the surface of the αth sphere is given by

$$\psi(\mathbf{r}_\alpha + \boldsymbol{\tau}_\alpha) = \psi^{(0)}(\mathbf{r}_\alpha + \boldsymbol{\tau}_\alpha) + \sum_\beta \int dS_\beta [G_{\alpha\beta}(\mathbf{r} - \mathbf{r}') \frac{\partial}{\partial r_\beta} \psi(\mathbf{r}_\beta + \boldsymbol{\tau}_\beta) -$$

$$- \psi(\mathbf{r}_\beta + \boldsymbol{\tau}_\beta) \frac{\partial}{\partial r_\beta} G_{\alpha\beta}(\mathbf{r} - \mathbf{r}')]_{r_\beta = R_{m\beta}}, \tag{B212}$$

where S_β is the surface of the muffin-tin sphere at the βth site, and where

$$G_{\alpha\beta}(\mathbf{r} - \mathbf{r}') = G(\mathbf{r}_\alpha + \boldsymbol{\tau}_\alpha - \mathbf{r}_\beta - \boldsymbol{\tau}_\beta). \tag{B213}$$

Since $\psi^{(0)}(\mathbf{r}_\alpha + \boldsymbol{\tau}_\alpha)$ is assumed to be a single plane wave, it may be expanded in terms of spherical harmonics as:

$$\psi^{(0)}(\mathbf{r}_\alpha + \boldsymbol{\tau}_\alpha) = e^{i\mathbf{k_0}\cdot\boldsymbol{\tau}_\alpha} \sum_{l,m} 4\pi i^l j_l(k_0 r_\alpha) Y^*_{lm}(\theta_{\mathbf{k_0}}, \varphi_{\mathbf{k_0}}) Y_{lm}(\theta_\alpha, \varphi_\alpha), \tag{B214}$$

where $k_0^2 = E$. $\theta_{\mathbf{k_0}}$, $\varphi_{\mathbf{k_0}}$ and θ_α, φ_α are the polar angles of $\mathbf{k_0}$ and $\mathbf{r}_\alpha$ with respect to a coordinate system with $\hat{z} \parallel \hat{c}$ at the αth atom.

In the limit $r_\alpha = R_{m\alpha}$, the integration in Equation (B212) can be carried out by using Equation (B149) with $r > r'$ for the Green's function and letting

$$\psi(\mathbf{r}_\beta + \boldsymbol{\tau}_\beta) = \sum_{l,m} C^\beta_{lm} u_l(r_\beta) Y_{lm}(\theta_\beta, \varphi_\beta). \tag{B215}$$

By comparing the coefficients of $Y_{lm}(\theta_\alpha, \varphi_\alpha)$ on both sides of Equation (B212) after the integration, we obtain,

$$
\begin{aligned}
C^\alpha_{lm} u_l(R_{m\alpha}) = {}& e^{i\mathbf{k_0}\cdot\boldsymbol{\tau}_\alpha} 4\pi i^l j_l(k_0 R_{m\alpha}) Y^*_{lm}(\theta_{\mathbf{k_0}}, \varphi_{\mathbf{k_0}}) - \\
& - \sum_\beta \sum_{l'm'} [A^{\alpha,\beta}_{lm,l'm'} j_l(k_0 R_{m\alpha}) j_{l'}(k_0 r_\beta) r_\beta^2 + k_0 R_{m\alpha}^2 \times \\
& \times \delta_{\alpha\beta} \delta_{ll'} \delta_{mm'} j_l(k_0 r_\beta) n_l(k_0 R_{m\alpha})]_{r_\beta = R_{m\beta}} \times \\
& \times \left[\frac{j'_{l'}(k_0 r_\beta)}{j_{l'}(k_0 r_\beta)} - L_{l'} \right]_{r_\beta = R_{m\beta}} C^\beta_{l'm'} u_{l'}(R_{m\beta}),
\end{aligned}
\tag{B216}
$$

where L_l, j'_l were defined in Section 2. The logarithmic derivatives $L_l(R_{m\beta})$ can be written in terms of phase shifts, η^β_l at the βth site:

$$L_l(R_{m\beta}) = \frac{j'_l(k_0 R_{m\beta}) - \tan\eta^\beta_l n'_l(k_0 R_{m\beta})}{j_l(k_0 R_{m\beta}) - \tan\eta^\beta_l n_l(k_0 R_{m\beta})}, \tag{B217}$$

where

$$n'_l(k_0 r) = \frac{\partial}{\partial r} n_l(k_0 r).$$

Substituting Equation (B217) into Equation (B216) and using the Wronskian of the spherical Bessel functions, Equation (B216) can be simplified as

$$
\begin{aligned}
C^\alpha_{lm} u_l(R_{m\alpha}) = {}& e^{i\mathbf{k_0}\cdot\boldsymbol{\tau}_\alpha} 4\pi i^l j_l(k_0 R_{m\alpha}) Y^*_{lm}(\theta_{\mathbf{k_0}} \varphi_{\mathbf{k_0}}) - \\
& - \sum_\beta \sum_{l'm'} \left[\frac{1}{k_0} A^{\alpha,\beta}_{lml'm'} j_l(k_0 R_{m\alpha}) + \delta_{\alpha\beta} \delta_{ll'} \delta_{mm'} n_l(k_0 R_{m\alpha}) \right] \times \\
& \times \frac{\tan\eta^\beta_{l'}}{j_{l'}(k_0 R_{m\beta}) - \tan\eta^\beta_{l'} n_{l'}(k_0 R_{m\beta})} C^\beta_{l'm'} u_{l'}(R_{m\beta})
\end{aligned}
\tag{B218}
$$

For convenience, we define

$$X^\alpha_{lm} = C^\alpha_{lm} u_l(R_{m\alpha}) \tag{B219}$$

and

$$B_{lm}^{\alpha} = e^{i\mathbf{k}_0 \cdot \boldsymbol{\tau}_{\alpha}} \, 4\pi i^l j_l(k_0 R_{m\alpha}) Y_{lm}^*(\theta_{\mathbf{k}_0}, \varphi_{\mathbf{k}_0}) \tag{B220}$$

then the compact form of Equation (B218) is

$$B_{lm}^{\alpha} = X_{lm}^{\alpha} + \sum_{\beta l'm'} D_{lm,l'm'}^{\alpha\beta} X_{l'm'}^{\beta}$$

where

$$D_{lm,l'm'} = \left[\frac{1}{k_0} A_{lml'm'}^{\alpha,\beta} j_l(k_0 R_{m\alpha}) + \delta_{\alpha\beta}\delta_{ll'}\delta_{mm'} n_l(k_0 R_{m\alpha}) \right] \times$$

$$\times \frac{\tan \eta_{l'}^{\beta}}{j_{l'}(k_0 R_{m\beta}) - \tan \eta_{l'}^{\beta} n_{l'}(k_0 R_{m\alpha})} \tag{B221}$$

Equation (B221) is a set of linear inhomogeneous equations for X_{lm}^{α}. Therefore, if we know $A_{lm,l'm'}^{\alpha\beta}$, Equation (B221) can be solved for the wavefunctions at the muffin-tin spheres. Furthermore, the phase shifts decrease as l increases, so that a finite set of X_{lm}^{α} will be sufficient for a band structure calculation. Sometimes, the last factor in $D_{lm,l'm'}^{\alpha\beta}$ can be very large for certain values of $k_0 R_{m\beta}$ and $\eta_{l'}^{\beta}$, Kambe [49] suggested that this difficulty can be surmounted by renormalizing X_{lm}^{α}. To do this, we redefine

$$X_{lm}^{\alpha} = \frac{X_{lm}^{\alpha}}{\sqrt{4\pi} \, i^l [j_l(k_0 R_{m\alpha}) \cos \hat{\eta}_l^{\alpha} - n_l(k_0 R_{m\alpha}) \sin \hat{\eta}_l^{\alpha}]}, \tag{B222}$$

where $\hat{\eta}_l^{\alpha}$ is the principal value of η_l^{α}, so that its range is from $-\pi$ to π. Equation (B221) can then be rewritten as

$$e^{i\mathbf{k}_0 \cdot \boldsymbol{\tau}_{\alpha}}\sqrt{4\pi} \, Y_{lm}^*(\theta_{\mathbf{k}_0}, \varphi_{\mathbf{k}_0}) = X_{lm}^{\alpha} \cos \hat{\eta}_l^{\alpha} + \sum_{\beta l'm'} \frac{A_{lm,l'm'}^{\alpha,\beta}}{k_0 i^{l-l'}} \sin \hat{\eta}_{l'}^{\beta} X_{l'm'}^{\beta}. \tag{B223}$$

Knowing the wavefunctions at the muffin-tin spheres, we are now in a position to derive transmission and reflection coefficients.

In order to find expressions of the transmitted and the reflected waves of a layer from Equation (B209) Kambe [50] constructed a one-dimensional Green's function which is periodic in the directions perpendicular to the $\hat{z}$-axis by following the general treatment of the one dimensional case [51],

$$G(\mathbf{r}, \mathbf{r}') = \frac{1}{S} \sum_{\mathbf{g}} \frac{1}{2i\Gamma_{\mathbf{g}}} e^{i\Gamma_{\mathbf{g}}|z-z'| + i(\mathbf{k}_t + \mathbf{g})\cdot(\boldsymbol{\rho} - \boldsymbol{\rho}')} \quad \text{for} \quad z \neq z', \tag{B224}$$

where S is the area of the two dimensional unit cell and where

$$\mathbf{r} = \hat{z}z + \boldsymbol{\rho}$$

and

$$\Gamma_{\mathbf{g}} = \sqrt{k_0^2 - |\mathbf{k}_t + \mathbf{g}|^2} \quad \text{for} \quad k_0^2 > |\mathbf{k}_t + \mathbf{g}|^2,$$

$$= i\sqrt{|\mathbf{k}_t + \mathbf{g}|^2 - k_0^2} \quad \text{for} \quad k_0^2 < |\mathbf{k}_t + \mathbf{g}|^2. \tag{B225}$$

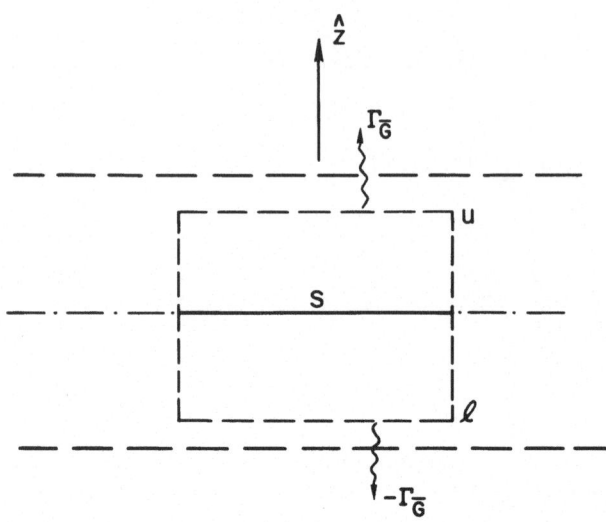

Fig. 9.   Transmitted and reflected waves of a single layer with unit cell area S as defined in the layer transfer-matrix method.

Substituting Equation (B224) into Equation (B209) and setting z equal to z_u at the upper plane u as shown in Figure 9, we obtain

$$\psi(\mathbf{\rho}, z_u) = \psi^{(0)}(\mathbf{\rho}, z_u) + \frac{1}{S} \sum_{\mathbf{g}} \frac{1}{2i\Gamma_{\mathbf{g}}} e^{i\Gamma_{\mathbf{g}} z_u + i(\mathbf{k}_t + \mathbf{g}) \cdot \mathbf{\rho}} \times$$

$$\times \sum_{\alpha} e^{-i\mathbf{k}^{(+)} \cdot \mathbf{\tau}_\alpha} \int \left[e^{-i\Gamma_{\mathbf{g}} z_\alpha - i(\mathbf{k}_t + \mathbf{g}) \cdot \mathbf{\rho}_\alpha} \frac{\partial}{\partial r_\alpha} \psi(\mathbf{r}_\alpha) - \right.$$

$$\left. - \psi(\mathbf{r}_\alpha) \frac{\partial}{\partial r_\alpha} e^{-i\Gamma_{\mathbf{g}} z_\alpha - i(\mathbf{k}_t + \mathbf{g}) \cdot \mathbf{\rho}_\alpha} \right]_{r_\alpha = R_{m\alpha}} dS_\alpha$$

$$= \psi^{(0)}(\mathbf{\rho}, z_u) + \frac{1}{S} \sum_{\mathbf{g}} \frac{1}{2i\Gamma_{\mathbf{g}}} e^{i\Gamma_{\mathbf{g}} z_u + i(\mathbf{k}_t + \mathbf{g}) \cdot \mathbf{\rho}} \sum_{\alpha} \psi_s(R_{m\alpha}, -\Gamma_{\mathbf{g}}),$$

$$(B226)$$

where $\psi_s(R_{m\alpha}, -\Gamma_{\mathbf{g}})$ is the amplitude of the scattered wave from the αth atom, and $\mathbf{k}^{(+)} = \hat{z}\Gamma_{\mathbf{g}} + (\mathbf{k}_t + \mathbf{g})$.

With the assumption that a plane wave with amplitude $U_{j\mathbf{g}}^{(+)}$ is incident on the lower boundary (denoted by l in Figure 9) of the layer and that $\Gamma_{\mathbf{g}}$ is real, the amplitude of the gth transmitted wave at the upper boundary (u in Figure 9) is given by

$$U_{j+1,\mathbf{g}}^{(+)} = U_{j,\mathbf{g}}^{(+)} + \frac{1}{2iS\Gamma_{\mathbf{g}}} \sum_{\alpha} \psi_s(R_{m\alpha}, -\Gamma_{\mathbf{g}}). \tag{B227}$$

By expanding the exponentials of ψ_s in terms of spherical harmonics and using the solution of ψ given by Equation (B221), it is shown that

$$U_{j+1,\mathbf{g}}^{(+)} = U_{j,\mathbf{g}}^{(+)} + \frac{i2\pi}{S\Gamma_\mathbf{g}k_0} \sum_\alpha \sum_{l,m} e^{-i\mathbf{k}^{(+)}\cdot\mathbf{\tau}_\alpha} \sqrt{4\pi} \sin \hat{\eta}_l^\alpha \times$$

$$\times Y_{lm}(\theta_{\mathbf{k}^{(+)}}, \varphi_{\mathbf{k}^{(+)}}) X_{lm}^\alpha(U_{j\mathbf{g}}^{(+)}), \tag{B228}$$

where $X_{lm}^\alpha(U_{j\mathbf{g}}^{(+)})$ denotes the wavefunction at the αth muffin-tin sphere with the amplitude of the incident wave as $U_{j\mathbf{g}}^{(+)}$.

Similarly, the wavefunction at the lower boundary l can be written as

$$\psi(\mathbf{\rho}, z_l) = \psi^{(0)}(\mathbf{\rho}, z_l) + \frac{1}{S} \sum_\mathbf{g} \frac{1}{2i\Gamma_\mathbf{g}} e^{-i\Gamma_\mathbf{g}z_l + i(\mathbf{k}_t + \mathbf{g})\cdot\mathbf{\rho}} \times$$

$$\times \sum_\alpha e^{-i\mathbf{k}^{(-)}\cdot\mathbf{\tau}_\alpha} \int \left[e^{i\Gamma_\mathbf{g}z_\alpha - i(\mathbf{k}_t + \mathbf{g})\cdot\mathbf{\rho}_\alpha} \frac{\partial}{\partial r_\alpha} \psi(\mathbf{r}_\alpha) - \right.$$

$$\left. - \psi(\mathbf{r}_\alpha) \frac{\partial}{\partial r_\alpha} e^{i\Gamma_\mathbf{g}z_\alpha - i(\mathbf{k}_t + \mathbf{g})\cdot\mathbf{\rho}_\alpha} \right] dS_\alpha,$$

$$= \psi^{(0)}(\mathbf{\rho}, z_l) + \frac{1}{S} \sum_\mathbf{g} \frac{1}{2i\Gamma_\mathbf{g}} e^{-i\Gamma_\mathbf{g}z_l + i(\mathbf{k}_t + \mathbf{g})\cdot\mathbf{\rho}} \sum_\alpha \psi_s(R_{m\alpha}, \Gamma_\mathbf{g}),$$

where

$$\mathbf{k}^{(-)} = -\hat{z}\Gamma_\mathbf{g} + (\mathbf{k}_t + \mathbf{g}). \tag{B229}$$

The amplitude of the gth reflected wave is given by

$$U_{j\mathbf{g}}^{(-)} = \frac{2\pi i}{Sk_0\Gamma_\mathbf{g}} \sum_\alpha \sum_{lm} e^{-i\mathbf{k}^{(-)}\cdot\mathbf{\tau}_\alpha} \sqrt{4\pi} \sin \hat{\eta}_l^\alpha \times$$

$$\times Y_{lm}(\theta_{\mathbf{k}^{(-)}}, \varphi_{\mathbf{k}^{(-)}}) X_{lm}^\alpha(U_{j\mathbf{g}}^{(+)}). \tag{B230}$$

From Equations (B228) and (B230), expressions of Q^{I} and Q^{III} for the layer can be obtained. By using the same procedures discussed above and assuming a plane wave with amplitude $U_{j+1,\mathbf{g}}^{(-)}$ incident onto the layer from the upper boundary u, Q^{II} and Q^{IV} of the layer can be calculated. If the basic unit contains p layers, the resulting Q can be expressed in terms of matrix multiplications of the Q's for individual layers.

If $\Gamma_\mathbf{g}$ is imaginary, the spherical harmonics in the expansion of the plane wave should be written as [49]

$$Y_{lm}(\theta_{\mathbf{k}^{(\pm)}}, \varphi_{\mathbf{k}^{(\pm)}}) = \sqrt{\frac{2l+1}{4\pi} \frac{(l-|m|)!}{(l+|m|)!}} \times$$

$$\times (\mp i)^{|m|} P_l^{|m|}(\pm i |\Gamma_\mathbf{g}|k_0) e^{im\varphi_{\mathbf{k}^{(\pm)}}}, \tag{B231}$$

where $P_l^{|m|}$ is the associated Legendre polynomial. This change should also be considered in calculating X_{lm}^α for the structure factors.

6.3. FORMULATION OF THE STRUCTURE FACTORS

The derivation of the expressions of the structure factors, $A_{lm,l'm'}^{\alpha,\beta}$, can largely be simplified, if none-overlapping muffin-tin potentials are used, as it has been done in [48]. In this case, the complication of treating the singular behavior of the Green's function appears only in the diagonal structure factor, $A^{\alpha,\,\alpha}_{lm,l'm'}$. Therefore, we shall divide the following discussions into two parts:

(i) derivation of the off-diagonal structure factor, $A^{\alpha,\,\beta}_{lm,l'm'}$, $\alpha \neq \beta$; and

(ii) derivation of the diagonal structure factor, $A^{\alpha,\,\alpha}_{lm,l'm'}$.

(i) In Figure 10, we have shown the configuration of the non-overlapping muffin-tin

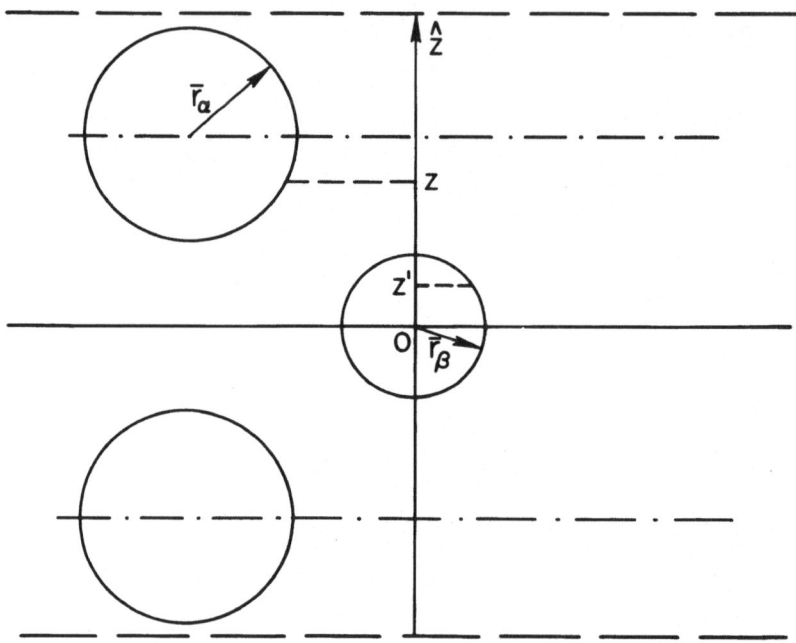

Fig. 10.   Geometrical configuration of non-overlapping muffin-tin potentials as defined in the layer transfer-matrix method.

potentials in a layer of e.g. MoS_2. In this configuration z is never equal to z' and the Green's function is regular for all $\mathbf{r}_\alpha$ and $\mathbf{r}_\beta$. Equation (B224) can then be used,

$$G(\mathbf{r}, \mathbf{r}') = G(\mathbf{r}_\alpha, \mathbf{r}_\beta) = \frac{1}{2iS} \sum_{\mathbf{g}} \frac{1}{\Gamma_{\mathbf{g}}} e^{i\mathbf{k}(\pm)\cdot(\mathbf{r}_\alpha - \mathbf{r}_\beta + \tau_{\alpha\beta})}, \tag{B232}$$

where $\tau_{\alpha\beta} = \tau_\alpha - \tau_\beta$. For $z > z'$, Equation (B149) without the second term in the curly bracket can be used for $G(\mathbf{r}_\alpha, \mathbf{r}_\beta)$, so

$$A^{\alpha,\,\beta}_{lm,l'm'} = \frac{1}{j_l(k_0 r_\alpha)j_{l'}(k_0 r_\beta)} \sum_{\mathbf{g}} \frac{e^{i\mathbf{k}(+)\cdot\tau_{\alpha\beta}}}{2iS\Gamma_{\mathbf{g}}} \int d\Omega_\alpha\, e^{i\mathbf{k}(+)\cdot\mathbf{r}_\alpha} \times$$
$$\times\, Y^*_{lm}(\theta_\alpha, \varphi_\alpha) \int d\Omega_\beta\, e^{-i\mathbf{k}(+)\cdot\mathbf{r}_\beta}\, Y_{l'm'}(\theta_\beta, \varphi_\beta), \tag{B233}$$

where Ω_α, Ω_β are the solid angles subtended at τ_α and τ_β. The integrals can simply be evaluated by expanding the plane waves in terms of spherical harmonics. Equation (B233) is then reduced to

$$
A^{\alpha,\beta}_{lm,l'm'} = \frac{8\pi^2}{S} i^{l-l'-1} \sum_{\mathbf{g}} \frac{1}{\Gamma_{\mathbf{g}}} e^{i\mathbf{k}(+)\cdot\tau_{\alpha\beta}} \times
$$

$$
\times Y^*_{lm}(\theta_{\mathbf{k}(+)}, \varphi_{\mathbf{k}(+)}) Y_{l'm'}(\theta_{\mathbf{k}(+)}, \varphi_{\mathbf{k}(+)}). \tag{B234}
$$

This expression is valid only when $\Gamma_{\mathbf{g}}$ is real. For imaginary $\Gamma_{\mathbf{g}}$, Equation (B231) should be used, then Y^*_{lm} is replaced by Y_{l-m}. Note that the last two terms are not the same because of the complex arguments in the associated Legendre polynomial in Y_{l-m}.

For $z < z'$ and real $\Gamma_{\mathbf{g}}$, the corresponding structure factor is given by

$$
A^{\alpha,\beta}_{lm,l'm'} = \frac{8\pi^2}{S} i^{l-l'-1} \sum_{\mathbf{g}} \frac{1}{\Gamma_{\mathbf{g}}} e^{i\mathbf{k}(-)\cdot\tau_{\alpha\beta}} Y^*_{lm}(\theta_{\mathbf{k}(-)}, \varphi_{\mathbf{k}(-)}) \times
$$

$$
\times Y_{l'm'}(\theta_{\mathbf{k}(-)}, \varphi_{\mathbf{k}(-)}). \tag{B235}
$$

Equations (B234) and (B235) define the off-diagonal structure factors to be used in Equation (B223) depending on whether $z > z'$ or $z < z'$.

To derive an expression for the diagonal structure factor $A^{\alpha,\alpha}_{lm,l'm'}$, we use Equation (B168). The D_{LM}'s satisfy the following equation (from Equation (B163))

$$
G(R) = -\frac{1}{4\pi} \frac{\cos k_0 R}{R} + \sum_{L,M} D_{LM} j_L(k_0 R) \times
$$

$$
\times Y_{LM}(\Theta, \Phi), \qquad R \leq 2R_{m\alpha}. \tag{B236}
$$

The Green's function given in Equation (B224) is no longer a convenient form because it is valid for $z \neq z'$ only. Therefore, the task is to derive an alternative form of the Green's function so that its singular behavior can properly be taken into account. By examining the asymptotic behavior of the Hankel function H, it may be shown that Equation (B224) can be rewritten as follows:

$$
G(R) = G^{\alpha\alpha}(\mathbf{r}, \mathbf{r}') = \frac{1}{2iS} \sqrt{\frac{\pi}{2}} \sum_{\mathbf{g}} \left(\frac{|z - z'|}{\Gamma_{\mathbf{g}}} \right)^{1/2} \times
$$

$$
\times H^{(1)}_{-1/2}(\Gamma_{\mathbf{g}} |z - z'|) e^{i(\mathbf{k}_t + \mathbf{g})\cdot(\boldsymbol{\rho} - \boldsymbol{\rho}')}, \tag{B237}
$$

where $G^{\alpha\alpha}(\mathbf{r}, \mathbf{r}') = G(\mathbf{r}_\alpha + \tau_\alpha, \mathbf{r}'_\alpha + \tau_\alpha)$ and $H^{(1)}_{-1/2}$ is the Hankel function of the first kind. This equation in turn, suggests a possible integral representation of $G(R)$. It will be shown in Section 6.4 that a useful form of the Green's function is given by

$$
G^{\alpha\alpha}(\mathbf{r}, \mathbf{r}') = -\frac{1}{2\pi S} \sqrt{\frac{\pi}{2}} \sum_{\mathbf{g}} e^{i(\mathbf{k}_t + \mathbf{g})\cdot(\boldsymbol{\rho} - \boldsymbol{\rho}')} \int_{\omega}^{\infty e^{i\varphi_{\mathbf{g}}}} \times
$$

$$
\times \zeta^{-1/2} \exp\left[\tfrac{1}{2}\left(\Gamma_{\mathbf{g}}^2 \zeta - \frac{|z - z'|^2}{\zeta} \right) \right] d\zeta - \frac{1}{4\pi^2} \sqrt{\frac{\pi}{2}} \times
$$

$$\times \sum_{n}^{|\mathbf{R}_{nt}| \neq 0} e^{-i\mathbf{k}_t \cdot \mathbf{R}_{nt}} \int_{1/\omega}^{\infty} \zeta^{-1/2} \times$$

$$\times \exp\left[-\tfrac{1}{2}\left(|\mathbf{r} - \mathbf{r}' + \mathbf{R}_{nt}|^2\, \zeta - \frac{k_0^2}{\zeta} \right) \right] d\zeta - \frac{1}{4\pi^2} \times$$

$$\times \sqrt{\frac{\pi}{2}} \int_{1/\omega}^{\infty} \zeta^{-1/2} \exp\left[-\tfrac{1}{2}\left(|\mathbf{r} - \mathbf{r}'|^2\, \zeta - \frac{k_0^2}{\zeta} \right) \right] d\zeta, \qquad \text{(B238)}$$

where ω is a real number with $\infty > \omega > 0$. The phase angle φ_g determining the contour of integration in the complex ζ plane equals π for $\Gamma_g^2 > 0$ and 0 for $\Gamma_g^2 < 0$. $\mathbf{R}_{nt}$ is a lattice vector in the plane of the layer.

From Equation (B236), we obtain

$$D_{LM} = D_{LM}^{(1)} + D_{LM}^{(2)} + D_{00} \underset{R \to 0}{=} \frac{1}{j_L(k_0 R)} \int \left[G(\mathbf{R}) + \frac{\cos(k_0 R)}{4\pi R} \right] Y_{LM}^*(\Theta, \Phi)\, d\Omega,$$
$$\text{(B239)}$$

where

$$D_{LM}^{(1)} \underset{R \to 0}{=} -\frac{1}{j_L(k_0 R)} \frac{1}{2\pi S} \sqrt{\frac{\pi}{2}} \sum_{\mathbf{g}} \int d\Omega\, Y_{LM}^*(\Theta, \Phi)\, e^{i(\mathbf{k}_t + \mathbf{g}) \cdot (\boldsymbol{\rho} - \boldsymbol{\rho}')} \times$$

$$\times \int_{\omega}^{\infty e^{i\varphi_{\mathbf{g}}}} \zeta^{-1/2} \exp\left[\tfrac{1}{2}\left(\Gamma_g^2 \zeta - \frac{|z - z'|^2}{\zeta} \right) \right] d\zeta,$$

$$D_{LM}^{(2)} \underset{R \to 0}{=} \frac{-1}{j_L(k_0 R)} \frac{1}{4\pi^2} \sqrt{\frac{\pi}{2}} \sum_{n}^{|R_{nt}| \neq 0} e^{i\mathbf{k}_t \cdot \mathbf{R}_{nt}} \int d\Omega\, Y_{LM}^*(\Theta, \Phi) \times$$

$$\times \int_{1/\omega}^{\infty} \zeta^{-1/2} \exp\left[-\tfrac{1}{2}\left(|\mathbf{R} + \mathbf{R}_{nt}|^2\, \zeta - \frac{k_0^2}{\zeta} \right) \right] d\zeta,$$

$$D_{00} \underset{R \to 0}{=} \frac{\sqrt{4\pi}}{j_0(k_0 R)} \left[\cos\frac{(k_0 R)}{4\pi R} - \frac{1}{4\pi^2} \sqrt{\frac{\pi}{2}} \int_{1/\omega}^{\infty} \zeta^{-1/2} \exp\left[-\tfrac{1}{2}\left(R^2 \zeta - \frac{k_0^2}{\zeta} \right) \right] d\zeta \right].$$

In the last expression, the integration over the solid angle has been carried out. Following the discussions that will be given in the next section we obtain

$$D_{LM}^{(1)} = -\frac{1}{Sk_0} \frac{(-1)^L i^{|M|+1}}{2^L} \left[(2L+1)(L+|M|)!(L-|M|)! \right]^{1/2} \times$$

$$\times \sum_{\mathbf{g}} e^{-iM\Phi_{\mathbf{k}_t + \mathbf{g}}} \sum_{m=0}^{(L-|M|)/2} \times$$

$$\times \frac{\left(\dfrac{1}{k_0} |\mathbf{k}_t + \mathbf{g}| \right)^{L-2m} \left(\dfrac{1}{k_0} \Gamma_{\mathbf{g}} \right)^{2m-1}}{m! \left[\tfrac{1}{2}(L - |M| - 2m) \right]! \left[\tfrac{1}{2}(L + |M| - 2m) \right]!} \times$$

$$\times \; \Gamma(\tfrac{1}{2} - m, \exp{(-i\pi\omega/2\Gamma_{g}^{2})}) \quad \text{for} \quad L - |M| \text{ even},$$

$$= 0 \quad \text{for} \quad L - |M| \text{ odd}, \tag{B240}$$

where $\Phi_{\mathbf{k}_t + \mathbf{g}}$ is the polar angle of $\mathbf{k}_t + \mathbf{g}$, and $\Gamma(\tfrac{1}{2} - m, \exp{(-i\pi\omega/2\Gamma_{g}^{2})})$ is an incomplete gamma function.

$$D_{LM}^{(2)} = -\frac{k_0}{4\pi} \frac{(-1)^L (-1)^{(L-|M|)/2}}{2^{2L}\left(\dfrac{L-|M|}{2}\right)!\left(\dfrac{L+|M|}{2}\right)!}\; [(2L+1) \times$$

$$\times \; (L-|M|)!(L+|M|)!]^{1/2} \sum_{n}^{|\mathbf{R}_{nt}| \neq 0} e^{-i\mathbf{k}_t \cdot \mathbf{R}_{nt} - iM\Phi_{\mathbf{R}_{nt}}} \times$$

$$\times \; (k_0 R_{nt})^L \int\limits_{0}^{k_0^2(\omega/2)} u^{-3/2 - L}\, e^{u - k_0^2(R_{nt}^2/4u)}\, \mathrm{d}u \quad \text{for} \quad L - |M| \text{ even},$$

$$= 0 \quad \text{for} \quad L - |M| \text{ odd}, \tag{B241}$$

where $\Phi_{\mathbf{R}_{nt}}$ is the polar angle of $\mathbf{R}_{nt}$. The integral does not have a closed form, it may be carried out by numerical integration.

$$D_{00} = -\frac{k_0}{2\pi}\left[2 \int\limits_{0}^{\sqrt{k_0^2\omega/2}} e^{t^2}\, \mathrm{d}t - \frac{e^{k_0^2\omega/2}}{\sqrt{k_0^2\omega/2}}\right]. \tag{B242}$$

Substituting Equations (B240), (B241) and (B242) into Equation (B239), D_{LM} can be evaluated. Finally, $A_{lm,l'm'}^{\alpha,\alpha}$ can be obtained from Equation (B168). In practice, ω is chosen to be $S/2\pi$.

TABLE 1

Structure factors D_{LM} for $L = 0, 1, 2$ used in the layer transfer-matrix (LTM) method (after [50])

L, M	$D_{LM}^{(1)}$
0, 0	$-\dfrac{i}{S}\sum_{g}\dfrac{1}{\Gamma_{g}}\,\Gamma(\tfrac{1}{2}, \tfrac{1}{2}e^{-i\pi\omega\Gamma_{g}^{2}})$
1, 0	0
1, ± 1	$-\dfrac{1}{Sk_0}\sum_{g} e^{\mp i\Phi_{\mathbf{k}_t+\mathbf{g}}}\dfrac{\sqrt{6}}{-k_0}\dfrac{|\mathbf{k}_t + \mathbf{g}|}{\Gamma_{g}}\,\Gamma(\tfrac{1}{2}, \tfrac{1}{2}e^{-i\pi\omega\Gamma_{g}^{2}})$
2, 0	$-\dfrac{1}{Sk_0}\sum_{g} i\dfrac{\sqrt{5}}{2}\left\{\dfrac{|\mathbf{k}_t + \mathbf{g}|^2}{k_0\Gamma_{g}}\,\Gamma(\tfrac{1}{2}, \tfrac{1}{2}e^{-i\pi\Gamma_{g}^{2}}) + \dfrac{\Gamma_{g}}{k_0}\,\Gamma(-\tfrac{1}{2}, \tfrac{1}{2}e^{-i\pi\omega\Gamma_{g}^{2}})\right)$
2, ± 1	0
2, ± 2	$-\dfrac{1}{Sk_0}\sum_{g} e^{\mp i2\Phi_{\mathbf{k}_t+\mathbf{g}}}\left(-i\dfrac{\sqrt{15}}{2}\right)\dfrac{|\mathbf{k}_t + \mathbf{g}|^2}{2k_0\Gamma_{g}}\,\Gamma(\tfrac{1}{2}, \tfrac{1}{2}e^{-i\pi\omega\Gamma_{g}^{2}})$

6.4. MATHEMATICAL DETAILS

In this section, we shall derive the integral form of the Green's function given in Equation (B238).

Equation (B237) leads us to consider the following series:

$$
G_s(\mathbf{r} - \mathbf{r}') = \frac{1}{2iS} \sqrt{\frac{\pi}{2}} \sum_{\mathbf{g}} \left(\frac{|z - z'|}{\Gamma_{\mathbf{g}}} \right)^{s/2} H^{(1)}_{-s/2} (\Gamma_{\mathbf{g}} |z - z'|) \times
$$
$$
\times \, e^{i(\mathbf{k}_t + \mathbf{g}) \cdot (\mathbf{\rho} - \mathbf{\rho}')}, \tag{B243}
$$

where s is a complex number, and $G_s(\mathbf{r} - \mathbf{r}')$ is an analytic function of s for all values of $|\mathbf{r} - \mathbf{r}'|$ if the real part of s, Re(s), is greater than 2. We shall show that by using an integral representation of $H^{(1)}_{-s/2}$ and applying Riemann's method [52] (for the Riemann integral of the zeta-function) it is possible to analytically continue s to the value 1.

Let's start with an integral representation of the Hankel function given by Watson [53]

$$
\left(\frac{|z - z'|}{\Gamma_{\mathbf{g}}} \right)^{s/2} H^{(1)}_{-s/2}(\Gamma_{\mathbf{g}} |z - z'|) =
$$
$$
= \frac{1}{i\pi} \int_0^{\infty e^{i\varphi_{\mathbf{g}}}} \zeta^{(s/2)-1} \exp\left[\frac{1}{2}\left(\Gamma_{\mathbf{g}}^2 \zeta - \frac{|z - z'|^2}{\zeta} \right) \right] \mathrm{d}\zeta, \tag{B244}
$$

where $\Gamma_{\mathbf{g}}$ is treated as a complex number defined in Equation (B225), and $\varphi_{\mathbf{g}} = \pi - 2\,\mathrm{Arg}\,\Gamma_{\mathbf{g}}$ (Arg means the argument of a complex number). Substituting Equation (B244) into Equation (B243), G_s is rewritten as

$$
G_s(\mathbf{r} - \mathbf{r}') = -\frac{1}{2\pi S} \sqrt{\frac{\pi}{2}} \sum_{\mathbf{g}}^{\Gamma_{\mathbf{g}}^2 > 0} e^{i(\mathbf{k}_t + \mathbf{g}) \cdot (\mathbf{\rho} - \mathbf{\rho}')} \int_{0^+}^{-\infty} \zeta^{(s/2)-1} \times
$$
$$
\times \exp\left[\frac{1}{2}\left(\Gamma_{\mathbf{g}}^2 \zeta - \frac{|z - z'|^2}{\zeta} \right) \right] \mathrm{d}\zeta - \frac{1}{2\pi S} \sqrt{\frac{\pi}{2}} \sum_{\mathbf{g}}^{\Gamma_{\mathbf{g}}^2 < 0} \times
$$
$$
\times \, e^{i(\mathbf{k}_t + \mathbf{g}) \cdot (\mathbf{\rho} - \mathbf{\rho}')} \int_{0^+}^{\infty} \zeta^{(s/2)-1} \times \exp\left[\frac{1}{2}\left(\Gamma_{\mathbf{g}}^2 \zeta - \frac{|z - z'|^2}{\zeta} \right) \right] \mathrm{d}\zeta. \tag{B245}
$$

The first term on the right hand side corresponds to $\mathrm{Arg}\,\Gamma_{\mathbf{g}} = 0$, $\varphi_{\mathbf{g}} = \pi$, while the second term arises from $\mathrm{Arg}\,\Gamma_{\mathbf{g}} = \pi/2$, $\varphi_{\mathbf{g}} = 0$. From Equation (B225) the summation with $\Gamma_{\mathbf{g}}^2 > 0$ can involve only finite terms. Therefore, the summation and the integration can be interchanged. By using Equation (B225) and by subdividing the integral into 0^+ to ω and into ω to $-\infty$, the first term (now defined as $G_s^I(\mathbf{r} - \mathbf{r}')$) is

$$
G_s^I(\mathbf{r} - \mathbf{r}') = -\frac{1}{2\pi S} \sqrt{\frac{\pi}{2}} \left\{ \int_{0^+}^{\omega} \zeta^{(s/2)-1} \exp\left(\frac{\zeta}{2} k_0^2 - \frac{1}{2\zeta} |z - z'|^2 \right) \sum_{\mathbf{g}}^{\Gamma_{\mathbf{g}}^2 > 0} \times \right.
$$
$$
\times \exp\left(\frac{\zeta}{2}(\mathbf{k}_t + \mathbf{g})^2 + i(\mathbf{k}_t + \mathbf{g}) \cdot (\mathbf{\rho} - \mathbf{\rho}') \right) \mathrm{d}\zeta +
$$

$$+ \int\limits_{\omega}^{-\infty} \zeta^{(s/2)-1} \exp\left(\frac{-|z-z'|^2}{2\zeta}\right) \times$$

$$\times \sum_{\mathbf{g}}^{\Gamma_{\mathbf{g}}^2>0} e^{1/2\Gamma_{\mathbf{g}}^2\zeta + i(\mathbf{k}_t+\mathbf{g})\cdot(\mathbf{\rho}-\mathbf{\rho}')}\, d\zeta \Bigg\}, \tag{B246}$$

where ω is a complex number with $\mathrm{Re}(\omega) > 0$, $|\omega| < \infty$.

The second term in Equation (B245) involves the summation of an infinite number of terms. However, by following Rieman's method, an interchange of summation and integration is possible if $\mathrm{Re}(s) > 4$ and if $\mathrm{Re}(\zeta)$ is always greater than 0 on the contour. Similar to Equation (B246), the integration can be broken into two ranges: 0^+ to ω and ω to ∞. Let's define this second term as $G_s^{\mathrm{II}}(\mathbf{r}-\mathbf{r}')$. Then,

$$G_s^{\mathrm{II}}(\mathbf{r}-\mathbf{r}') = -\frac{1}{2\pi S}\sqrt{\frac{\pi}{2}}\Bigg\{\int\limits_{0^+}^{\omega} \zeta^{(s/2)-1} \sum_{\mathbf{g}}^{\Gamma_{\mathbf{g}}^2<0} \times$$

$$\times \exp\left[\tfrac{1}{2}\left(\Gamma_{\mathbf{g}}^2\zeta - \frac{|z-z'|^2}{\zeta}\right) + i(\mathbf{k}_t+\mathbf{g})\cdot(\mathbf{\rho}-\mathbf{\rho}')\right]d\zeta +$$

$$+ \int\limits_{\omega}^{\infty} \zeta^{(s/2)-1} \sum_{\mathbf{g}}^{\Gamma_{\mathbf{g}}^2<0} \times$$

$$\times \exp\left[\tfrac{1}{2}\left(\Gamma_{\mathbf{g}}^2\zeta - \frac{|z-z'|^2}{\zeta}\right) + i(\mathbf{k}_t+\mathbf{g})\cdot(\mathbf{\rho}-\mathbf{\rho}')\right]d\zeta. \tag{B247}$$

The first part in the above equation can be combined with the first term in Equation (B246), so that the sum is now valid for all $\mathbf{g}$. This sum may be converted to a sum over the lattice vectors by the two dimensional theta transformation [54]. Furthermore, by changing $\zeta \to 1/\zeta$ in the $\int_0^\omega$ integration, the final form of $G_s(\mathbf{r}-\mathbf{r}')$ is

$$G_s(\mathbf{r}-\mathbf{r}') = -\frac{1}{2\pi S}\sqrt{\frac{\pi}{2}}\Bigg\{\int\limits_{\omega}^{-\infty} \zeta^{(s/2)-1} \sum_{\mathbf{g}}^{\Gamma_{\mathbf{g}}^2>0} \times$$

$$\times \exp\left[\tfrac{1}{2}\left(\Gamma_{\mathbf{g}}^2\zeta - \frac{|z-z'|^2}{\zeta}\right) + i(\mathbf{k}_t+\mathbf{g})\cdot(\mathbf{\rho}-\mathbf{\rho}')\right]d\zeta +$$

$$+ \int\limits_{\substack{\omega \\ \mathrm{Re}(\zeta)>0}}^{\infty} \zeta^{(s/2)-1} \sum_{\mathbf{g}}^{\Gamma_{\mathbf{g}}^2<0} \times$$

$$\times \exp\left[\tfrac{1}{2}\left(\Gamma_{\mathbf{g}}^2\zeta - \frac{|z-z'|^2}{\zeta}\right) + i(\mathbf{k}_t+\mathbf{g})\cdot(\mathbf{\rho}-\mathbf{\rho})\right]d\zeta\Bigg\} -$$

$$- \frac{1}{4\pi^2}\sqrt{\frac{\pi}{2}} \int\limits_{\substack{1/\omega \\ \mathrm{Re}(\zeta)>0}}^{\infty} \zeta^{-s/2} \times$$

$$\times \sum_{n} \exp\left(-\frac{\zeta}{2}\,|\mathbf{r}-\mathbf{r}'+\mathbf{R}_{nt}|^2 + \frac{k_0^2}{2\zeta} - i\mathbf{k}_t\cdot\mathbf{R}_{nt}\right)\mathrm{d}\zeta, \qquad (B248)$$

where $\mathbf{R}_{nt}$ is a two-dimensional lattice vector which lies in the layer. Using again the Riemann method, Equation (B248) is analytic for all value of s if $|\mathbf{r}-\mathbf{r}'| \neq 0$. Therefore, it is the analytic continuation of Equation (B243) for $\mathrm{Re}(s) \leq 2$. Since $\mathrm{Re}(\omega) > 0$, we may remove the restriction of $\mathrm{Re}(\zeta) > 0$. Furthermore, the upper limits of the first two integrals can be written compactly in terms of the phase angle $\varphi_\mathbf{g}$. So, the useful form of $G(\mathbf{r}-\mathbf{r}')$ as $s \to 1$ is

$$G(\mathbf{r}-\mathbf{r}') = -\frac{1}{2\pi S}\sqrt{\frac{\pi}{2}}\sum_{\mathbf{g}} e^{i(\mathbf{k}_t+\mathbf{g})\cdot(\boldsymbol{\rho}-\boldsymbol{\rho}')} \int_{\omega}^{\infty e^{i\varphi_\mathbf{g}}} \zeta^{-1/2} \times$$

$$\times \exp\left[\tfrac{1}{2}\left(\Gamma_\mathbf{g}^2\zeta - \frac{|z-z'|^2}{\zeta}\right)\right]\mathrm{d}\zeta - \frac{1}{4\pi^2}\sqrt{\frac{\pi}{2}}\sum_{n}^{|\mathbf{R}_{nt}|\neq 0} \times$$

$$\times e^{-i\mathbf{k}_t\cdot\mathbf{R}_{nt}} \int_{1/\omega}^{\infty} \zeta^{-1/2} \exp\left[\frac{-1}{2}\left(|\mathbf{r}-\mathbf{r}'+\mathbf{R}_{nt}|^2\,\zeta - \frac{k_0^2}{\zeta}\right)\right]\mathrm{d}\zeta -$$

$$-\frac{1}{4\pi^2}\sqrt{\frac{\pi}{2}}\int_{1/\omega}^{\infty} \zeta^{-1/2} \exp\left[\frac{-1}{2}\left(|\mathbf{r}-\mathbf{r}'|^2\,\zeta - \frac{k_0^2}{\zeta}\right)\right]\mathrm{d}\zeta. \qquad (B249)$$

To show how Equations (B240), (B241) and (B242) were derived, we shall consider the three terms in Equation (B239) separately.

(i) The calculation of $D_{LM}^{(1)}$. Since $|z-z'|$ is a function of Θ, it is convenient to expand $e^{(-1/2\zeta)|z-z'|^2}$ in a power series. Thus

$$\int_{\omega}^{\infty e^{i\varphi_\mathbf{g}}} \zeta^{-1/2}\, e^{(-1/2\zeta)|z-z'|^2}\, e^{1/2\Gamma_\mathbf{g}^2\zeta}\, \mathrm{d}\zeta$$

$$= \sum_{m} \frac{1}{m!}\left[-\tfrac{1}{2}\,|z-z'|^2\right]^m \int_{\omega}^{\infty e^{i\varphi_\mathbf{g}}} \zeta^{-1/2-m}\, e^{(1/2)\Gamma_\mathbf{g}^2\zeta}\, \mathrm{d}\zeta. \qquad (B250)$$

The last integral can be carried out [55],

$$\int_{\omega}^{\infty e^{i\varphi_\mathbf{g}}} \zeta^{-1/2-m}\, e^{(1/2)\Gamma_\mathbf{g}^2\zeta}\, \mathrm{d}\zeta = \left(e^{i\pi}\frac{\Gamma_\mathbf{g}^2}{2}\right)^{m-1/2}\Gamma\left(\tfrac{1}{2}-m,\, e^{-i\pi(\omega/2)\Gamma_\mathbf{g}^2}\right), \qquad (B251)$$

where

$$\Gamma\left(\tfrac{1}{2}-m,\, e^{-i\pi(\omega/2)\Gamma_\mathbf{g}^2}\right)$$

is the incomplete gamma function.

To carry out the angular integration, we start with the Φ-integration first. The Φ-dependent terms in the integrand are Y^*_{LM} and the exponential factor $e^{i(\mathbf{k}_t + \mathbf{g}) \cdot (\boldsymbol{\rho} - \boldsymbol{\rho}')}$. It is known [56] that

$$\int_0^{2\pi} d\Phi \exp[-iM\Phi + i \, |\mathbf{k}_t + \mathbf{g}| \, R \sin \Theta \cos (\Phi_{\mathbf{k}_t + \mathbf{g}} - \Phi)]$$

$$= 2\pi i \, |M| \exp(-iM\Phi_{\mathbf{k}_t + \mathbf{g}}) \, J_{|M|}(|\mathbf{k}_t + \mathbf{g}| \, R \sin \Theta), \tag{B252}$$

where $\Phi_{\mathbf{k}_t + \mathbf{g}}$ is the polar angle of $\mathbf{k}_t + \mathbf{g}$. $J_{|M|}$ is the Bessel function of order $|M|$.

Next, the Θ integration will be carried out. Since $|z - z'| = \cos \Theta$, a typical term of $|z - z'|^{2m}$ in (B250) gives

$$\int_{-1}^{1} d\mu \, P_L^{|M|}(\mu) \mu^{2m} J_{|M|}(|\mathbf{k}_t - \mathbf{g}| \, R\sqrt{1 - \mu^2}), \tag{B253}$$

where $\mu = \cos \Theta$. $P_L^{|M|}$ is the associated Legendre polynomial. This polynomial can be expressed either in the form $\mu^{(L-|M|)}f(\mu^2)$ or $\mu^{(L-|M|)}f(1/\mu^2)$. Therefore, the integral is zero if $L - |M|$ is an odd integer. For even $L - |M|$, the Bessel function in Equation (B253) is first expanded in a power series [56]

$$J_{|M|}(|\mathbf{k}_t + \mathbf{g}| \, R\sqrt{1 - \mu^2}) = \sum_{m=0} \frac{(-1)^m[\frac{1}{2} \, |\mathbf{k}_t + \mathbf{g}| \, R\sqrt{1 - \mu^2}]^{2m}}{m!(|M| + m)!} \times$$

$$\times [\tfrac{1}{2}|\mathbf{k}_t + \mathbf{g}| \, R\sqrt{1 - \mu^2}]^{|M|} \tag{B254}$$

then the integration of Equation (B253) can be carried out term by term. We anticipate to take the limit $R \to 0$, such that $j_L(k_0 R)$ in the denominator approaches R^L. The lowest power of R in Equation (B254) for non-vanishing of the integral is $2m + |M| = L$. Hence, it is necessary to sum m from 0 to $(L - |M|)/2$. By collecting all the numerical factors, we obtain,

$$D_{LM}^{(1)} = -\frac{1}{Sk_0} \frac{(-1)^L i^{|M|+1}}{2^L} [(2L + 1)(L + |M|)!(L - |M|)!]^{1/2} \times$$

$$\times \sum_{\mathbf{g}} \exp(-iM\Phi_{\mathbf{k}_t + \mathbf{g}}) \times \sum_{m=0}^{1/2(L-|M|)} \times$$

$$\times \frac{\left(\frac{1}{k_0} \, |\mathbf{k}_t + \mathbf{g}|\right)^{L-2m} \left(\frac{1}{k_0} \Gamma_{\mathbf{g}}\right)^{2m-1}}{m![\frac{1}{2}(L - |M| - 2m)]![\frac{1}{2}(L + |M| - 2m)]!} \times$$

$$\times \Gamma\left(\tfrac{1}{2} - m, \, e^{-i\pi(\omega/2)\Gamma_{\mathbf{g}}^2}\right), \quad \text{for} \quad L - |M| \cdot \text{even},$$

$$= 0 \quad \text{for} \quad L - |M| \text{ odd} \tag{B240}$$

Examples of $D_{LM}^{(1)}$ for $L = 0, 1, 2$ are listed in Table 1.

(ii) The calculation of $D^{(2)}_{LM}$. Similar to the procedures given in (i), we expand the exponential $\exp(-\frac{1}{2}|\mathbf{R} + \mathbf{R}_{nt}|^2\,\zeta)$ to find the explicit function of Θ and Φ,

$$\exp\left(-\frac{\zeta}{2}|\mathbf{R} + \mathbf{R}_{nt}|^2\right) = \exp\left[-\frac{\zeta}{2}(R^2 + R^2_{nt})\right]\exp[i(i\zeta\mathbf{R}\cdot\mathbf{R}_{nt})]$$

$$= \exp\left[-\frac{\zeta}{2}(R^2 + R^2_{nt})\right] \times$$

$$\times 4\pi \sum_{L,M} i^L j_L(i\zeta R R_{nt}) Y^*_{LM}(\Theta_{\mathbf{R}_{nt}}, \Phi_{\mathbf{R}_{nt}}) Y_{LM}(\Theta, \Phi),$$

$$(B255)$$

where $\Theta_{\mathbf{R}_{nt}}, \Phi_{\mathbf{R}_{nt}}$ are the polar angles of $\mathbf{R}_{nt}$. The angular integrations in $D^{(2)}_{LM}$ of Equation (B239) can now be carried out. Only one term in Equation (B255) will survive because of the orthonormality condition of the Y_{LM}'s. Since $\Theta_{\mathbf{R}_{nt}}$ is 90 degree with respect to a fixed coordinate system, $Y^*_{LM}(\Theta_{\mathbf{R}_{nt}}, \Phi_{\mathbf{R}_{nt}})$ vanishes when $L - |M|$ is an odd integer, so as $D^{(2)}_{LM}$. As $R \to 0$, we expand both $j_L(i\zeta R_{nt})$ in Equation (B255) and $j_L(k_0 R)$ in Equation (B239) in power series, and take the leading terms. By setting $u = k^2_0/2\zeta$ in the ζ-integration and expressing Y^*_{LM} in terms of $e^{-iM\Phi_{R_{nt}}}$ and $P_L^{|M|}$, the final form of $D^{(2)}_{LM}$ is

$$D^{(2)}_{LM} = -\frac{k_0}{4\pi} \frac{(-1)^L(-1)^{1/2(L-|M|)}}{2^{2L}\frac{(L-|M|)}{2}!\frac{(L+|M|)}{2}!} \times$$

$$\times [(2L+1)(L-|M|)!(L+|M|)!]^{1/2} \times$$

$$\times \sum_n^{|\mathbf{R}_{nt}|\neq 0} \exp(-i\mathbf{k}_t\cdot\mathbf{R}_{nt} - iM\Phi_{\mathbf{R}_{nt}})(k_0 R_{nt})^L \int_0^{(k^2_0\omega)/2} \times$$

$$\times u^{-3/2-L}\exp\left(u - \frac{k^2_0 R^2_{nt}}{4u}\right)\mathrm{d}u, \quad \text{for} \quad L - |M| \text{ even,}$$

$$= 0 \quad \text{for} \quad L - |M| \text{ odd.} \tag{B244}$$

(iii) The calculation of D_{00}. The integration of ζ can be performed after the factor $e^{k^2_0/2\zeta}$ in Equation (B239) is expanded in a power series. Setting $\eta = \omega\zeta$, then the integral of D_{00} is

$$\int_{1/\omega}^\infty \zeta^{-1/2}\exp\left[\frac{-1}{2}\left(R^2\zeta - \frac{k^2_0}{\zeta}\right)\right]\mathrm{d}\zeta =$$

$$= \sum_m \frac{1}{m!}\left(\frac{k^2_0}{2}\right)^m \omega^{m-1/2}\int_1^\infty \eta^{-(1/2+m)}\exp\left(-\frac{R}{2\omega}\eta\right)\mathrm{d}\eta. \tag{B256}$$

The last integral can be expressed in terms of the incompleted gamma function [57]

$$\int_{1}^{\infty} \eta^{-(1/2+m)} \exp\left(-\frac{R^2}{2\omega}\eta\right) d\eta = \left(\frac{R^2}{2\omega}\right)^{m-\frac{1}{2}} \Gamma\left(\frac{1}{2}-m, \frac{R^2}{2\omega}\right). \tag{B257}$$

From [54], p. 135, the incomplete gamma function can be written as

$$\Gamma\left(\frac{1}{2}-m, \frac{R^2}{2\omega}\right) = \Gamma(\frac{1}{2}-m) - \sum_{n=0}^{\infty} \frac{(-1)^n}{\frac{1}{2}-m+n} \left(\frac{R^2}{2\omega}\right)^{1/2-m+n} \frac{1}{n!}. \tag{B258}$$

Substituting Equations (B256), (B257) and (B258) into Equation (B239) we get,

$$D_{00} = \frac{\sqrt{4\pi}}{j_0(k_0 R)} \left\{ \frac{\cos k_0 R}{4\pi R} - \frac{1}{4\pi^2} \sqrt{\frac{\pi}{2}} \left[\sum_m \frac{1}{m!} \left(\frac{k_0^2}{2}\right)^m \left(\frac{R}{\sqrt{2}}\right)^{2m-1} \Gamma(\frac{1}{2}-m) - \right. \right.$$
$$\left. \left. - \sum_m \frac{1}{m!} \left(\frac{k_0^2\omega}{2}\right)^m \sqrt{\frac{1}{\omega}} \times \sum_{n=0}^{\infty} \frac{(-1)^n}{n!} \frac{1}{\frac{1}{2}-m+n} \left(\frac{R^2}{2\omega}\right)^n \right] \right\}. \tag{B259}$$

As $R \to 0$, $j_0(k_0 R) \to 1$, and the first term in the square bracket approaches $1/4\pi R$ which cancels $[(\cos k_0 R)/4\pi R]$ $R \to 0$. Therefore,

$$D_{00} = -\frac{k_0}{4\pi} \sum_{m=0}^{\infty} \frac{(k_0^2\omega/2)^{m-1/2}}{m!(m-\frac{1}{2})}. \tag{B260}$$

Using relations on p. 134, and 135 and 147 of [57], the final expression of D_{00} is

$$D_{00} = -\frac{k_0}{2\pi} \left[2 \int_{0}^{\sqrt{k_0^2\omega/2}} e^{t^2} dt - \frac{e^{k_0^2\omega/2}}{\sqrt{k_0^2\omega/2}} \right]. \tag{B242}$$

An application of the layer-method to the layer-compound MoS_2 has been carried out by Wood and Pendry [48]. The results shall be compared with results obtained by other methods in a later chapter.

7. The Orthogonalized-Plane-Wave Method (OPW)

The Orthogonalized Plane-Wave-method (OPW), though one of the 'established' band-structure calculation methods [58], has only recently been applied to layer-type crystals [59, 60]. The reason for this late and limited use is certainly connected to its inherent convergence problems, which may nowadays be overcome by using modern high-speed computers. In fact, the OPW method, if carried out self-consistently as in [60, 61] represents probably one of the most accurate approaches to a bandstructure. Some basic ideas of the OPW method have already been discussed in Section 3 as being the fundaments of the pseudopotential method. Nevertheless we shall present here a complete discussion of the method in spite of repeating some of the facts. Section 7.1 contains the general formulation of the OPW method and the derivation of its matrix elements. In Section 7.2 we shall focus our attention on some special problems connected to the self-consistent calculation of the crystal potential. Finally in Section 7.3 a combined OPW tight binding scheme, which has recently been proposed for layer type crystals shall be reviewed.

7.1. GENERAL FORMULATION OF THE OPW METHOD

The mathematical description of a valence wavefunction in a periodic crystal may naturally be in terms of a Fourier series. A simple planewave series, however, would only very poorly converge because of the rapid oscillations in the wavefunction in the vicinity of the atomic cores. To improve convergence, Herring [27] proposed to ortho-gonalize the plane waves to all tightly bound core wavefunctions. The core wavefunction contributions now present in an OPW should simulate the strongly varying part of the valence wavefunction in the core regions, whereas outside the cores the overall behavior of the Bloch function is well described by plane waves.

In the original OPW formalism, the electronic states of the crystal are thus divided into core and valence states which are treated rather differently. The tightly-bound core states are assumed not to overlap for atoms centered at different sites and to see a spherically symmetric potential. They may thus be written in the atomic form:

$$a_{nlm}(\mathbf{r}) = R_{nl}(r) Y_{lm}(\theta, \varphi) \tag{B261}$$

with nlm indicating the usual atomic quantum numbers. To obtain translational invariance, a Bloch state of the form

$$\psi_{nlm}(\mathbf{k}, \mathbf{r}) = \frac{1}{\sqrt{N}} \sum_{n=1}^{N} e^{i\mathbf{k}\cdot\mathbf{R}_n} a_{nlm}(\mathbf{r} - \mathbf{R}_n) \tag{B262}$$

may be constructed. It is assumed that N one-atomic unit cells, each of volume Ω form the crystal.

The valence state wavefunction is expanded in a modified Fourier series of plane waves, each of which has the form

$$\varphi(\mathbf{k} + \mathbf{G}, \mathbf{r}) = \frac{1}{\sqrt{N\Omega}} e^{i(\mathbf{k}+\mathbf{G})\cdot\mathbf{r}}. \tag{B263}$$

These plane waves have now to be orthogonalized to the core states according to

$$\text{OPW}(\mathbf{k} + \mathbf{G}, \mathbf{r}) = \varphi(\mathbf{k} + \mathbf{G}, \mathbf{r}) - \sum_{t} \langle \psi_t(\mathbf{k}, \mathbf{r}) \mid \varphi(\mathbf{k} + \mathbf{G}, \mathbf{r}) \rangle \psi_t(\mathbf{k}, \mathbf{r}), \tag{B264}$$

where t stands for the triplet of quantum numbers n, l, m.

If, as assumed the core states do not show any overlap between neighboring atoms, the orthogonality matrix element has the simple form:

$$O_t(\mathbf{k} + \mathbf{G}) = \langle \psi_t(\mathbf{k}, \mathbf{r}) \mid \varphi(\mathbf{k} + \mathbf{G}, \mathbf{r}) \rangle =$$

$$= i^l \frac{4\pi}{\sqrt{\Omega}} Y_{lm}^*(\theta_{\mathbf{k}+\mathbf{G}}, \varphi_{\mathbf{k}+\mathbf{G}}) \int_0^{\infty} j_l(|\mathbf{k} + \mathbf{G}|\cdot r) R_{nl}(r) r^2 \, dr, \tag{B265}$$

where $j_l(x)$ is the spherical Bessel function of order l. In the expression above it is assumed that the atom is centered at the origin, for an arbitrary atomic position $\boldsymbol{\tau}$ a structure factor $e^{-i\mathbf{G}\cdot\boldsymbol{\tau}}$ has to be inserted.

The total valence wavefunction $\Phi(\mathbf{k}, \mathbf{r})$ which is expanded in terms of OPW's has then the form:

$$\Phi(\mathbf{k}, \mathbf{r}) = \sum_{\mathbf{G}} A_{\mathbf{G}}(\mathbf{k}) \, \text{OPW}(\mathbf{k} + \mathbf{G}, \mathbf{r}). \tag{B266}$$

The expansion coefficients $A_{\mathbf{G}}(\mathbf{k})$ are calculated by minimizing the Hamiltonian expectation value. This leads to a secular equation

$$\sum_{\mathbf{G}'} [H_{\mathbf{G}\mathbf{G}'}(\mathbf{k}) - S_{\mathbf{G}\mathbf{G}'}(\mathbf{k})E(\mathbf{k})]A_{\mathbf{G}'}(\mathbf{k}) = 0 \tag{B267}$$

with Hamiltonian matrix elements

$$H_{\mathbf{G}\mathbf{G}'}(\mathbf{k}) = \langle \, \text{OPW}(\mathbf{k} + \mathbf{G}, \mathbf{r})| \, H \, |\text{OPW}(\mathbf{k} + \mathbf{G}', \mathbf{r})\rangle \tag{B268}$$

and overlap matrix elements arising from the non-orthogonal basis

$$S_{\mathbf{G}\mathbf{G}'}(\mathbf{k}) = \langle \, \text{OPW}(\mathbf{k} + \mathbf{G}, \mathbf{r}) \, | \, \text{OPW}(\mathbf{k} + \mathbf{G}', \mathbf{r})\rangle. \tag{B269}$$

The general eigenvalue equation of Equation (B267) may be solved by the Choleski decomposition method as discussed in Section 1 in connection with the tight-binding method.

As next step we evaluate the matrix elements of Equations (B268) and (B269).

For a Hamiltonian of the form

$$H = -\nabla^2 + V(\mathbf{r}) \tag{B270}$$

one obtains the following matrix elements

$$H_{\mathbf{G}\mathbf{G}'}(\mathbf{k}) = (\mathbf{k} + \mathbf{G})^2 \delta_{\mathbf{G},\mathbf{G}'} + V(\mathbf{G}' - \mathbf{G})$$
$$- \sum_t O_t^*(\mathbf{k} + \mathbf{G})O_t(\mathbf{k} + \mathbf{G}')E_t, \tag{B271}$$

where $V(\mathbf{G}' - \mathbf{G})$ is the Fourier transform of the crystal potential, E_t is a core state energy and where the orthogonality matrix elements O_t are defined in Equation (B265). For the overlap matrix elements one finds accordingly

$$S_{\mathbf{G}\mathbf{G}'}(\mathbf{k}) = \delta_{\mathbf{G}\mathbf{G}'} - \sum_t O_t^*(\mathbf{k} + \mathbf{G})O_t(\mathbf{k} + \mathbf{G}'). \tag{B272}$$

The potential Fourier transforms appearing in Equation (B271) are defined by

$$V(\mathbf{r}) = \sum_{\mathbf{G}} V(\mathbf{G}) \, e^{i\mathbf{G}\cdot\mathbf{r}}$$

with

$$V(\mathbf{G}) = \frac{1}{\Omega} \int V(\mathbf{r}) \, e^{-i\mathbf{G}\cdot\mathbf{r}} \, d\mathbf{r}. \tag{B273}$$

The various terms appearing in the crystal potential may be studied in the model of isolated atoms suggested in [58]. The model uses isolated-atom core states and potentials. The corresponding atomic charges are then placed into the crystal lattice and the resulting potential is calculated. Results based on that procedure, however, have to be improved by empirical adjustments to agree with experimental data. On the other hand the model serves as a good starting point for self-consistent calculations. In particular, the model crystal potential is obtained from solving Poisson's equations of the form

$$\nabla^2 V_n(\mathbf{r}) = 8\pi \rho_n(\mathbf{r}) = 8\pi Z \sum_{\mathbf{R}_n} \delta(\mathbf{r} - \mathbf{R}_n) \tag{B274}$$

for the nuclear charges and of the form

$$\nabla^2 V_c(\mathbf{r}) = -8\pi \rho_{el}(\mathbf{r}) = -8\pi \sum_{\mathbf{R}_n} \rho_{tot}^{at}(\mathbf{r} - \mathbf{R}_n) \tag{B275}$$

for the atomic electronic charges. The exchange potential may be approximated by the usual local $\rho^{1/3}$ term

$$V_X(\mathbf{r}) = -6 \left[\frac{3}{8\pi} \right]^{1/3} \sum_{\mathbf{R}_n} [\rho_{tot}^{at}(\mathbf{r} - \mathbf{R}_n)]^{1/3}, \tag{B276}$$

where some ambiguity remains for the order of executing the sum over $\mathbf{R}_n$ and taking the cube root. The Fourier transform of the total potential is then given by

$$V(\mathbf{G}) = \frac{8\pi}{\Omega} \frac{1}{|\mathbf{G}|^2} \left(-Z + 4\pi \int_0^\infty \rho_{tot}^{at}(r) j_0(|\mathbf{G}|r) r^2 \, dr \right) -$$

$$- \frac{6}{\Omega} (24\pi^2)^{1/3} \int_0^\infty [\rho_{tot}^{at}(r)]^{1/3} j_0(|\mathbf{G}|r) r^2 \, dr. \tag{B277}$$

The $|\mathbf{G}| = 0$ term may be found by expanding the Bessel function for small arguments:

$$V(0) = \frac{8\pi}{\Omega} \left[-\frac{1}{|\mathbf{G}|^2} \rho_{ion} - \frac{2\pi}{3} \int_0^\infty \rho_{tot}^{at}(r) r^4 \, dr \right] -$$

$$- \frac{6}{\Omega} (24\pi^2)^{1/3} \int_0^\infty [\rho_{tot}^{at}(r)]^{1/3} r^2 \, dr. \tag{B278}$$

In Equation (B278) the first term

$$- \frac{8\pi}{\Omega |\mathbf{G}|^2} \rho_{ion} = - \frac{8\pi}{\Omega |\mathbf{G}|^2} \left[-Z + 4\pi \int_0^\infty \rho_{tot}^{at}(r) r^2 \, dr \right] \tag{B279}$$

represents the Madelung potential. In contrast to the pseudopotential method, where the $|\mathbf{G}| = 0$ term merely represents a shift of the energy scale, in the OPW method this term is of importance, whenever self-consistent potentials have to be constructed. We shall return to a discussion of $V(|\mathbf{G}| = 0)$ in the next section.

We shall briefly discuss some of the convergence problems of the OPW method. The convergence of the plane wave expansion in Equation (B266) depends mainly on three factors [61]:

(i) The presence or absence of core wavefunctions in the OPW. If core states of a certain type are absent in an OPW, the series expansion of the wavefunction becomes a pure plane wave series and consequently converges very slowly. As an example, this is cited in [61]: Valence s-states, for which core s-states exist converge fully with some 700 OPW's, whereas valence p-states, for which no core states exist would only strictly converge after several thousands of plane waves. Similar problems have, of course, been encountered in the pseudopotential method, where non-local model potentials had to be inserted.

(ii) The relative core sizes of atoms in the unit cell. To exactly represent the effect of core states, orthogonality integrals O_t as given in Equation (B265) have to be evaluated. The integrals O_t represent the radial Fourier transforms of core states. Since the scale of the lattice $\mathbf{G}$-vectors is determined by the unit cell dimensions, the relative size of core state to unit cell dictates the number of $\mathbf{G}$-vectors to be included.

(iii) The amount of localization of the plane wave part in the valence band wavefunction. This convergence problem usually is the least severe of the mentioned three problems. It appears identically in the pseudopotential formalism. For highly localized wavefunctions appearing in strongly covalent or in highly ionic compounds the problem is more severe than for e.g. metals with spread out valence – and conduction wavefunctions.

In general, the necessary number of OPW's to be included can be inferred from comparing localization lengths of e.g. core states or bonds with the characteristic wavelength $\lambda = 2\pi/|\mathbf{G}|$ of an OPW. Often symmetry reductions using symmetrized OPW's (for their construction see Chapter A) are necessary to circumvent the above mentioned convergence problems.

7.2. SELF-CONSISTENT OPW CALCULATIONS

Self-consistent charge rearrangements may lead to important changes in the electronic structure of crystals. Changes in gap energies up to several eV are not unusual and in particular for ionic compounds correct charge transfer may only be determined self-consistently. It is, however, clear that any self-consistent calculation represents an enormous computational effort which explains its relatively rare use for complicated bandstructure calculations.

In a self-consistent OPW calculation [60, 61] a characteristic cycle as shown in Figure 11 is iteratively performed until relative convergence of eigen energies (or potentials) is reached. The cycle is started with a model calculation as discussed in the previous section. Valence wavefunctions are then calculated at selected (high symmetry) $\mathbf{k}$-points and the change densities are averaged over the Brillouin zone. The valence change density is then spherically symmetrized about each atomic site to evaluate new core wavefunctions for this new valence charge environment. Then the total new crystal potential (not spherically symmetric) is computed from the valence charge density and the new core charge density. New core-plane wave orthogonality coefficients are evaluated and the valence band structure is re-calculated at the selected $\mathbf{k}$-points. The convergence test of this band structure determines termination or repetition of the cycle. We now briefly describe the formulation of the charge densities used in the calculations. The total charge density may be written as the sum of the terms

$$\rho(\mathbf{r}) = \rho_n(\mathbf{r}) + \rho_c(\mathbf{r}) + \rho_v(\mathbf{r}) \tag{B280}$$

comprising the nuclear, core and valence electron charges respectively.

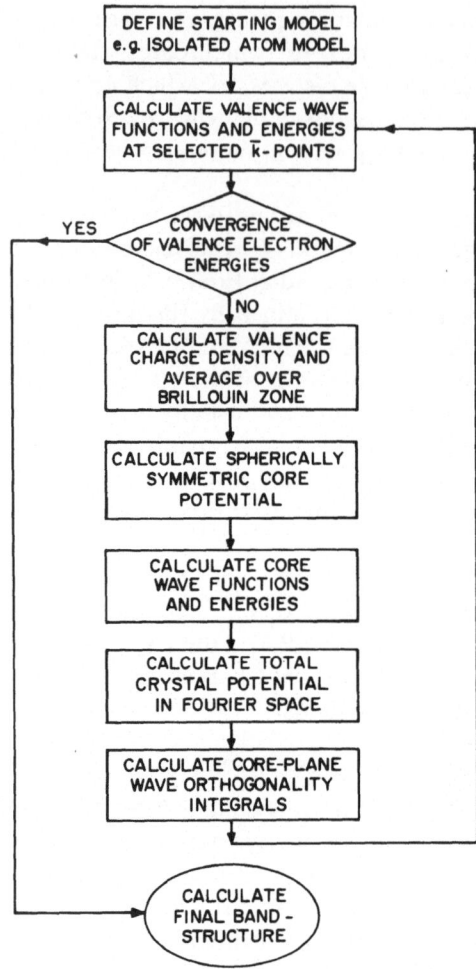

Fig. 11. Block-diagram for calculating a band structure using the self-consistent OPW method.

(1) The valence charge density

$$\rho_v(\mathbf{r}) = \sum_{\text{states}} \sum_{\mathbf{k}} |\Phi(\mathbf{k}, \mathbf{r})|^2 \tag{B281}$$

contains according to Equations (B264) and (B266) combinations of plane wave and core terms. It may be written as

$$\rho_v(\mathbf{r}) = \rho_{pw}(\mathbf{r}) + \rho_{vc}(\mathbf{r}), \tag{B282}$$

where

$$\rho_{pw}(\mathbf{r}) = \sum_{\mathbf{k},\text{bands}} \sum_{\mathbf{G},\mathbf{G}'} \frac{1}{\Omega} A_{\mathbf{G}}^*(\mathbf{k}) A_{\mathbf{G}'}(\mathbf{k}) \, e^{i(\mathbf{G}-\mathbf{G}')\cdot\mathbf{r}}$$

$$= \sum_{\mathbf{G}} \rho_{pw}(\mathbf{G}) \, e^{i\mathbf{G}\cdot\mathbf{r}} \tag{B283}$$

is that part of the valence charge density involving plane wave parts only and where

$$\rho_{vc}(\mathbf{r}) = \sum_{\mathbf{R}_n} \rho_{vc}^{at}(\mathbf{r} - \mathbf{R}_n) \tag{B284}$$

is the part of the valence charge density involving products of core wavefunctions with each other and with plane waves. Equation (B284) is conveniently decomposed into a superposition of spherically symmetric charges ρ_{vc}^{at} centered at each atomic site.

(ii) The core charge density

$$\rho_c(\mathbf{r}) = \sum_{\mathbf{R}_n} \rho_c^{at}(\mathbf{r} - \mathbf{R}_n) \tag{B285}$$

may easily be evaluated in a spherically symmetric form from the atomic core wavefunctions (Equation (B261)).

(iii) The nuclear charge density is given by:

$$\rho_n(\mathbf{r}) = \sum_{\mathbf{R}_n} Z\delta(\mathbf{r} - \mathbf{R}_n). \tag{B286}$$

While ρ_c, ρ_n and the second part ρ_{vc} of ρ_v are already in spherically symmetric forms, $\rho_{pw}(\mathbf{r})$ as given by Equation (B283) has to be spherically averaged for the purpose of calculating new core states. (For the calculation of valence states it is not spherically averaged.) Thus for the core state calculation one sets:

$$\rho_{pw}^{at}(\mathbf{r} - \mathbf{R}_n) = \sum_{\mathbf{G}} \rho_{pw}(\mathbf{G})\, j_0(|\mathbf{G}|\ |\mathbf{r} - \mathbf{R}_n|). \tag{B287}$$

The total (spherically symmetric) potential seen by the core electrons then has the form:

$$V^{at}(\mathbf{r}) = -\frac{2Z}{r} + \frac{2}{r}\int_0^r [\rho_c^{at}(r') + \rho_{vc}^{at}(r')]4\pi r'^2 \mathrm{d}r' +$$

$$+ 2\int_r^\infty [\rho_c^{at}(r') + \rho_{vc}^{at}(r')]4\pi r'\, \mathrm{d}r' + V_{pw}^{at}(r) -$$

$$- 6\left(\frac{3}{8\pi}\right)^{1/3} [\rho_c^{at}(r) + \rho_{vc}^{at}(r) + \rho_{pw}^{at}(r)]^{1/3} + V_{\mathrm{Mad}}. \tag{B288}$$

In Equation (B288) the first three terms define the electrostatic potentials arising from the nuclear charge, the core charge and the core-plane wave contribution of the valence charge. These terms are followed by

$$V_{pw}^{at}(r) = 8\pi \sum_{\mathbf{G}} \frac{\rho_{pw}^{at}(r)}{|\mathbf{G}|^2} j_0(|\mathbf{G}|r) \tag{B289}$$

representing the potential due to the plane wave part of the valence electrons and by the complete exchange potential. The last term in (B288) contains the Madelung corrections which are non-zero only for ionic compounds. For an atom of type x in an xy ionic compound this potential becomes

$$V_{\text{Mad}}^{x} = \sum_{\substack{y \text{ atoms} \\ \text{in unit cell}}} M_{xy}\left[Z_y - 4\pi \int_0^\infty [\rho_c^y(r) + \rho_{vc}^y(r)] \, r^2 \, dr \right] -$$

$$- \frac{4\pi}{3} \rho_{pw}(|\mathbf{G}| = 0)r^2, \tag{B290}$$

where the first term is a constant Madelung-type energy originating from the only partially compensated nuclear potentials and where the second term originates from the remaining, spread out electronic charge, warranting charge neutrality in the unit cell. The Madelung constants, M_{xy}, defined for unit charge may be evaluated using an Ewald-type lattice summation [61] to be

$$M_{xy} = \frac{8\pi\eta N_{\text{cell}}}{\Omega} - \sum_y \frac{2}{|\mathbf{R}_x - \mathbf{R}_y|} \, \text{erf}\left(\frac{|\mathbf{R}_x - \mathbf{R}_y|}{2\sqrt{\eta}} \right) -$$

$$- \frac{8\pi}{\Omega} \sum_{\mathbf{G} \neq 0} \frac{e^{-|\mathbf{G}|^2 \cdot \eta}}{|\mathbf{G}|^2} \sum_y e^{-i\mathbf{G} \cdot (\mathbf{R}_x - \mathbf{R}_y)}. \tag{B291}$$

Above, erf denotes the complementary error function and N_{cell} the number of atoms per unit cell. The summation of atoms extends over the whole crystal and is controlled by the convergence parameter η.

In the total potential (Equation (B288)) new core states are calculated to derive new orthogonality integrals. It remains to construct the Fourier transform of the total potential as it acts upon the valence states. One finds:

$$V(\mathbf{G}) = \frac{8\pi}{\Omega|\mathbf{G}|^2}\left[-Z + 4\pi \int_0^\infty (\rho_c(r) + \rho_{vc}(r)) \, j_0(|\mathbf{G}|r) \, r^2 \, dr \right] -$$

$$- \frac{6(24\pi^2)^{1/3}}{\Omega} \int_0^\infty (\rho_c(r) + \rho_{vc}(r))^{1/3} \, j_0(|\mathbf{G}|r) \, r^2 \, dr +$$

$$+ \frac{8\pi}{\Omega|\mathbf{G}|^2} \rho_{pw}(\mathbf{G}) - 6\left(\frac{3}{8\pi} \right)^{1/3} \frac{1}{\Omega} [\rho_{\text{tot}}^{1/3} - (\rho_c + \rho_{vc})^{1/3}](\mathbf{G}). \tag{B292}$$

The first term in Equation (B292) is the Fourier transform of the spherical Coulomb potential, followed by the Fourier transform of the spherical part of the exchange potential, the plane wave part of the valence electron Coulomb potential and the exchange potential due to the non-spherical part of the charge density. The Fourier transform of this latter part has to be evaluated numerically by integrating over a dense mesh of points in the unit cell. The singular terms, appearing for $\mathbf{G} = 0$ may be treated as before the Madelung-type corrections.

With the total valence electron potential given in Equation (B292) new valence states are calculated completing one self-consistency cycle. To reach fast convergence several smoothing or relaxing procedures may be used in calculating the new input potential for cycle n. Usually, the input potential for cycle n is taken to be a linear combination of input and output potential of cycle $n - 1$, etc.

A version of the self-consistent OPW method described above has been applied by Krusius *et al.* [60] to the transition metal layer compound TiS_2. The results of this calculation shall be presented in a later chapter and be compared to results obtained by other methods.

7.3. COMBINED OPW-TIGHT BINDING CALCULATIONS

An inherent difficulty in calculating the electronic energy band structure of layer compounds is the structural anisotropy. The weak interaction between layers is a result of the bonding and the closed shell configuration of the valence electrons within the layers. This situation led Bromley *et al.* [3] to treat the layer compounds as a two-dimensional crystals and led Wood *et al.* [48] to formulate the layer transfer-matrix method. In this section, we discuss another layer-adapted method suggested by Tsukada *et al.* [59]. The essential ideas of the method are described as follows:

Since the wavefunction of the valence electrons are less localized in the plane of the layers ($\hat{x}$, $\hat{y}$ or $\hat{a}$, $\hat{b}$ directions) than in the direction perpendicular to the layers ($\hat{z}$ or $\hat{c}$), it is reasonable to approximate the wavefunction in a mixed representation. A plane wave expansion in the $\hat{x}$, $\hat{y}$ directions may in turn be treated as localized tight-binding wavefunction along the $\hat{z}$-axis. Tsukada *et al.* modified the plane wave approximation along $\hat{x}$, $\hat{y}$ by using orthogonalized plane waves (OPW) as basis functions to improve convergence. Therefore, this approach may be called combined orthogonalized plane wave and tight binding method.

The general concept of the OPW method is discussed in Section 7.1 of this section. To elaborate the details of the combined orthogonalized plane wave and tight binding method, we divide the following discussions into: (i) the application of the OPW expansion for a two-dimensional monolayer; (ii) the formulation of a tight-binding method using the solutions of (i) to calculate the three dimensional bandstructure of a layer compound; and (iii) the formulation of the crystal potential.

(i) The solution of a monolayer problem:

We first examine the expansion of the wavefunction into plane waves with expansion coefficients as functions of z. These coefficients will turn out to satisfy a set of linearly coupled differential equations.

Consider the one-electron Hamiltonian of the monolayer.

$$H_m(\boldsymbol{\rho}, z) = \frac{p_{\parallel}^2}{2m} + \frac{p_z^2}{2m} + \sum_l \sum_{\alpha=1}^{\alpha_m} V(\boldsymbol{\rho} - \boldsymbol{\rho}_l - \boldsymbol{\tau}_\alpha, z), \tag{B293}$$

where $p_{\parallel}$ and p_z are the electronic momenta parallel and perpendicular to the plane of the layers, $\boldsymbol{\rho}$ is the two-dimensional position vector of the electron with respect to chosen origin, $\boldsymbol{\rho}_l$ is the lattice vector in the layer and $\boldsymbol{\tau}_\alpha$ is the position vector of the αth atom in the unit cell. z is the coordinate normal to the layer. To find a proper trial wavefunction for solving Equation (293) we expand the wavefunction into a set of plane waves.

$$\Phi_{n\mathbf{k}}(\mathbf{r}) = \frac{1}{\sqrt{S}} \sum_j e^{i(\mathbf{k} + \mathbf{G}_j) \cdot \boldsymbol{\rho}} F_{n\mathbf{k}, j}(z), \tag{B294}$$

where n is the band index, $\mathbf{k}$ is the two-dimensional crystal momentum of the electron, $\mathbf{G}_j$ is the two-dimensional reciprocal lattice vector and s is the area of the monolayer. $F_{nk,j}(z)$ are the expansion coefficients. By substituting Equation (B294) into (B293) and by scalar multiplying the plane wave $(1/\sqrt{s})\, e^{i(\mathbf{k}+\mathbf{G}_j)\cdot\boldsymbol{\rho}}$, we obtain

$$\left\{ -\frac{\hbar^2}{2m}\frac{\partial^2}{\partial z^2} + \frac{\hbar^2}{2m}(\mathbf{k}+\mathbf{G}_j)^2\, F_{nk,j}(z) + \sum_{j\neq j'} U(\mathbf{G}_{j'}-\mathbf{G}_j, z)\, F_{nk,j'}(z) \right\}$$

$$= E_{nk} F_{nk,j}(z) \tag{B295}$$

and

$$U(\mathbf{G}_{j'}-\mathbf{G}_j) = \frac{1}{S}\sum_{\alpha} e^{i(\mathbf{G}_{j'}-\mathbf{G}_j)\cdot\tau_\alpha} \int d\boldsymbol{\rho}\, e^{i(\mathbf{G}_{j'}-\mathbf{G}_j)\cdot\boldsymbol{\rho}}\, V(\boldsymbol{\rho}, z), \tag{B296}$$

where S is the area of the two-dimensional unit cell. The sum over α in Equation (B296) is just the two-dimensional structure factor, and the integral gives the two-dimensional Fourier transform of the potential. Equation (B295) represents a set of linear coupled equations for $F_{nk,j}(z)$.

For non-transition metal layer compounds, the potential characterizing the s and p valence states, $V(\boldsymbol{\rho})$, is expected to vary rather smoothly as a function of $\boldsymbol{\rho}$. Therefore, one may use the following first-order approximation to decouple the set of Equation (B295):

$$|U(0, z)| \geqslant |U(\mathbf{G}_{j'}-\mathbf{G}_j, z)| \quad \text{for} \quad j' \neq j. \tag{B297}$$

Furthermore, the solution $F_{nk}(z)$ thus obtained, depends on $\mathbf{k}$ only through the constant term $\hbar^2 k^2/2m$. It is thus possible to simply express $F_{nk}(z)$ in terms of the eigenfunctions of a one-dimensional Schrödinger equation:

$$-\frac{\hbar^2}{2m}\frac{\partial^2}{\partial z^2} f_\lambda(z) + U(0, z) f_\lambda(z) = E_\lambda f_\lambda(z), \tag{B298}$$

where λ labels the eigenvalues in increasing order. With the plane waves $(\mathbf{k}+\mathbf{G}_j)$ and f_λ we are able to solve variationally the eigenvalue problem of Equation (B298) by using the following trial wavefunction.

$$\Psi_{nk}^p(\mathbf{r}) = \frac{1}{\sqrt{s}}\sum_j \sum_\lambda C_{nk,j\lambda}\, e^{i(\mathbf{k}+\mathbf{G}_j)\cdot\boldsymbol{\rho}}\, f_\lambda(z), \tag{B299}$$

where the superscript p means that the expansion at the right hand side is in terms of a plane wave basis, $C_{nk,j\lambda}$ are the expansion coefficients. The corresponding secular equation can be easily derived:

$$\det |\langle j'\lambda'|\, H\, |j\lambda\rangle - E_{nk}\delta_{j'j}\delta_{\lambda'\lambda}| = 0 \tag{B300}$$

and

$$\langle j'\lambda'|\, H\, |j\lambda\rangle = \left\{ \frac{\hbar^2}{2m}(\mathbf{k}+\mathbf{G}_{j'})^2 + \varepsilon_{\lambda'} \right\}\delta_{j'j}\delta_{\lambda'\lambda} +$$

$$+ (1-\delta_{j'j})U(\mathbf{G}_{j'}-\mathbf{G}_j, \lambda'\lambda), \tag{B301}$$

where

$$U(\mathbf{G}_{j'} - \mathbf{G}_j, \lambda'\lambda) = \int dz \, f_{\lambda'}^*(z) \, U(0, z) f_\lambda(z).$$

We shall see how Equations (B300) and (B301) are modified if the basis functions are OPW's rather than simple plane waves. The OPW corresponding to $\mathbf{k} + \mathbf{G}_j$ can be written as

$$\varphi_{\mathbf{k}+\mathbf{G}_j,\lambda}(\mathbf{r}) = \frac{1}{\sqrt{S}} e^{i(\mathbf{k}+\mathbf{G}_j)\cdot\boldsymbol{\rho}} f_\lambda(z) - \sum_c O_\lambda(\mathbf{k} + \mathbf{G}_j) \, a_{c\mathbf{k}}(\mathbf{r}), \qquad (B302)$$

where $a_{c\mathbf{k}}$ is the Bloch form of the core wavefunctions. The sum runs over all core states. $O_\lambda(\mathbf{k} + \mathbf{G}_j)$ is the orthogonal integral:

$$O_\lambda(\mathbf{k} + \mathbf{G}_j) = \frac{1}{\sqrt{S}} \int d^3 r \, a_{c\mathbf{k}}^*(r) \, e^{i(\mathbf{k}+\mathbf{G}_j)\cdot\boldsymbol{\rho}} f_\lambda(z). \qquad (B303)$$

In terms of OPW's the trial wavefunction in Equation (B299) is modified to

$$\Psi_{n\mathbf{k}}^0(\mathbf{r}) = \sum_{j,\lambda} C_{n\mathbf{k},\lambda} \varphi_{\mathbf{k}+\mathbf{G}_j,\lambda}(\mathbf{r}). \qquad (B304)$$

The superscript 0 stands for the expansion in terms of OPW's. The secular equation in Equation (B300) has the same form, whereas Equation (B301) is also modified due to the presence of the second term in Equation (B302):

$$\langle j'\lambda' | H | j\lambda \rangle^0 = \langle j'\lambda' | H | j\lambda \rangle + \sum_c (E_{n\mathbf{k}} - E_c) \, O_{\lambda'}^*(\mathbf{k} + \mathbf{G}_{j'}) \, O_\lambda(\mathbf{k} + \mathbf{G}_j), \qquad (B305)$$

where the first term on the right is just Equation (B301). E_c is the energy of the core states.

The eigenvalues and eigenfunctions of the monolayer can be easily obtained by diagonalizing Equation (B300) with matrix elements specified in Equation (B305), and with the potential discussed in the previous section. These solutions can then be used to calculate the band structure in three dimensions as follows.

(ii) Let us stack N units of different monolayers with the period C. The one-electron Hamiltonian can then be written as

$$H = \frac{p_\parallel^2}{2m} + \frac{p_z^2}{2m} + \sum_{r=0}^{N} \sum_{s=1}^{s_r} \sum_l \sum_\alpha V(\boldsymbol{\rho} - \boldsymbol{\rho}_l - \boldsymbol{\tau}_{\alpha s}, z - \tau_s - rC), \qquad (B306)$$

where $(\boldsymbol{\tau}_{\alpha s}, \tau_s)$ is the coordinate of the αth atom in the sth layer measured with respect to a lattice vector $(\boldsymbol{\rho}_l, rC)$. To calculate the eigenvalues of Equation (B306) we linearly combine the wave functions given in Equation (B304).

$$\Psi_{n\mathbf{k},s}(\mathbf{r}) = \frac{1}{\sqrt{N}} \sum_{r=0}^{N} r^{ik_z \, rC} \Psi_{n\mathbf{k}}^0(\boldsymbol{\rho}, z - \tau_s - rC) \quad \text{for} \quad s = 1, \cdots, s_r, \qquad (B307)$$

where k_z is the component of the crystal momentum in the $\hat{c}$-direction. The resulting secular equation has the following form:

$$\det |H_{ss'} - E_{n\mathbf{k}}| = 0. \qquad (B308)$$

The matrix element, $H_{ss'}$, consists of diagonal and multi-center terms:

$$H_{ss'} = E_s \delta_{ss'} + H_{ss'}^{(2)} + H_{ss'}^{(3)} \tag{B309}$$

and

$$H_m (\boldsymbol{\rho}, z - \tau_s) \Psi_{nk}^0 (\boldsymbol{\rho}, z - \tau_s) = E_s \Psi_{nk}^0 (\boldsymbol{\rho}, z - \tau_s), \tag{B310}$$

where H_m is the Hamiltonian given in Equation (B293). In practice, the three-center integrals $H_{ss'}^{(3)}$ are neglected and the two-center contributions are given by:

$$H_{ss'}^{(2)} = \frac{1}{N} \sum_{r,r'} \sum_{j,j'} \sum_{\lambda,\lambda'} \sum_\alpha e^{-ik_z C(r'-r)} C_{nkj'\lambda'}^* C_{nkj\lambda}\, e^{i(\mathbf{G}_j - \mathbf{G}_{j'})\cdot\boldsymbol{\tau}_{\alpha s}} \times$$

$$\times \{ \int\int f_{\lambda'}^*(z - \tau_{s'} - r'C)\, U(\mathbf{G}_j - \mathbf{G}_{j'},$$

$$z - \tau_s - rC)\, f_\lambda(z - \tau_s - rC)\, \mathrm{d}z +$$

$$+ \int f_{\lambda'}^*(z - \tau_s - rC)\, U(\mathbf{G}_j - \mathbf{G}_{j'},$$

$$z - \tau_{s'} - r'C)\, f_\lambda(z - \tau_s - rC)\, \mathrm{d}z \}, \tag{B311}$$

where

$$U(\mathbf{G}_j - \mathbf{G}_{j'}, z - \tau_s - rC) = \int \mathrm{d}\boldsymbol{\rho}\, e^{i(\mathbf{G}_j - \mathbf{G}_{j'})\cdot\boldsymbol{\rho}}\, V(\boldsymbol{\rho}, z - \tau_s - rC).$$

Compared to the multi-center integrals discussed in Section 1, the calculations of $H_{ss'}^{(2)}$ are simpler because these integrals are all one-dimensional. The core states do not contribute since they are highly localized. Finally, the Cholesky decomposition method may be used to diagonalize Equation (B308).

(iii) So far we discussed the procedures for solving the one-electron eigen equation for a layer compound by the combined OPW and tight-binding method, the problem of constructing a rigorous OPW potential has been treated in the previous section. The use of the trial wavefunction, $f_\lambda(z)$, in the present method introduces further complications. In principle, the solutions of Equations (B298) have to include the continuum case as well, which increases the complexity of the problem. One way to circumvent this difficulty is to use the bound solutions of Equation (B298) only and to introduce additional empirical terms in the potential. Tsukada et al. [59] suggested to use an empirical model potential from the outset. For example, to calculate the π band structure of graphite, a potential of the form

$$V(\boldsymbol{\rho}, z) = \frac{Ae^2}{r}\, e^{-r/R} \tag{B312}$$

is proposed where $r = \sqrt{\rho^2 + z^2}$ and where A and R are treated as disposable parameters. The results of this graphite calculation will be discussed in a later section.

8. Conclusions

In this chapter we have discussed seven different fundamental methods which have been applied to calculate electronic band structures of layer compounds. A natural question may be at this point: is there any simple criterion that may be used for selecting one of these methods for a particular problem? Since the answer is difficult to be stated in a few words, we shall rather present the following summary about versatilities, involved computational efforts, precisions and special limiting features of the discussed methods. This way, we hope to provide some information to the reader who may be new in the field and is in the process of choosing a suitable method of calculation for his particular problem.

(i) *The Tight-binding method.*

The method is based on the picture that electronic states in a solid do not significantly deviate from a superposition of free atomic states. This assumption qualifies the method to be used mainly for molecules, semiconductors and insulators. Moreover, the difficulties met in calculating multi-center integrals from first-principles have generally restricted its use among solid state physicists, though the method is popular among molecular chemists. Several *ab-initio* bandstructure calculations, however, have been done for layered structures and results for e.g. graphite or boron nitride are available. More *ab-initio* type calculations are expected using improved integration schemes.

The semi-empirical or empirical approach to the tight-binding matrix elements, on the other hand, has been rather successful. This approach has recently been extended to treat the electronic properties of amorphous semiconductors and semiconductors surfaces, i.e. to more complicated systems. The success of the empirical approach is certainly based on its simple mathematical form and its capability to yield semi-quantitative results. The agreement between theory and experiment may in favorable cases be of the order of 0.1 to 0.5 eV, in particular if only valence bands are considered. The matrix elements in these cases are determined either by reference to molecular and atomic spectroscopic data or by directly using experimental results of the solid under consideration. Conduction states are generally poorly described. Qualitative conduction band structures may be obtained by implementing excited atomic states into the basic set.

As for all empirical approaches one has to choose judiciously among the available experimental information. A good prescription is to systematically analyze the experimental results not only for the crystal under study but also for crystals having similar electronic configurations or crystal-structure, in order to detect trends in the electronic system. Once the tight-binding matrix elements are parametrized, energy eigenvalues can easily be calculated. The size of the Hamiltonian matrices ranges typically between four and ten times the number of atoms per unit cell. These features make the empirical tight-binding method also an attractive tool for valence band studies and for k-space interpolations.

(ii) *The APW, KKR, LCMTO and LTM methods.*

Though systematic differences in the approaches exist for these methods, their results, advantages and shortcomings are rather similar. This of course is due to the muffin-tin approximation used as their basic potential form. The advantages of this potential form are simplicity and computational economy, its shortcomings are connected to the rather insufficient description of valence electron potentials in covalent, open (or layer-type) structures. Various improvement schemes to account for non-muffin-tin corrections have been suggested and used. The LCMTO method, in fact, has specifically been devised to treat non-muffin-tin corrections. Concerning computational effort, the four methods are rather equivalent (smaller matrices in one or the other case may be obtained at the expense of more complicated matrix elements) and standard for first-principles calculations (i.e. well above empirical calculations). Self-consistent treatments are possible in principle, though of limited usefulness because of the underlying muffin-tin potential approximation. No such treatment has to our knowledge yet been reported for a layer crystal. With respect to the characteristic structural anisotropy of layer structures the LTM method seems to show certain formal advantages, which, however, may be over-compensated by its more complicated mathematics. More prototype calculations are certainly necessary to establish judgement. Advantage of the APW and KKR methods is its wide, common usage for various kinds of bandstructure problems. Experiences gained in these calculations are certainly very useful in studying layer materials. The four methods have been applied to study transition metal layer compounds yielding rather similar results. The general overall agreement with experimental data is better than 1 eV, which is very remarkable for first-principles calculations based on ad hoc-type crystal potentials.

(iii) *The OPW and EP methods.*

Fundamentally, both the OPW and the EP methods are derived from the same physical picture. Their characteristics, however, are strongly different. While the OPW method is a strict first-principles method (without the limiting muffin-tin potential approximation), the EP method is basically an empirical method, though self-consistent modifications have recently been developed and applied to surface calculations.

Among the first-principles methods, the OPW scheme represents the most rigorous approach. The price for this quality, in particular, if carried out self-consistently has to be paid in terms of computational effort. Thus, mostly because of convergence problems (matrix sizes of several hundreds are often necessary) the OPW method has not been widely used yet for layer crystal bandstructure calculations. The existing, self-consistent calculations for TiS_2, are probably not unaffected by these problems.

Its derivative, the EPM does not suffer from these extreme convergence problems. This advantage, of course, goes on the expense of loosing the first-principles character and becoming an empirical method. Though empirical, the EPM yields highly accurate bandstructures, due to the sound underlying physical concept. With simple local pseudopotentials the method's application range from perfect crystals-metals, semi-conductors and insulators with non-transition elements to surfaces and defects. Using non-local pseudopotentials the method has successfully been extended to transition metals and their compounds, though these cases seem to uncover some limits of applicability. An important advantage which the EPM has over all other methods is connected to the simplicity of its wavefunctions. Based on the use of simple plane waves, matrix elements, charge densities etc. are very easily calculated. Thus, spectral quantities as dielectric functions and its derivatives can directly be evaluated and compared to experiment. The EPM certainly represents the most widely applied bandstructure method. Its usefulness in studying new materials is known. Difficulties characteristic of the method have to do with its modest convergence behavior which on the other hand is well compensated by the simplicity of the matrix-elements, and with its empirical character. This latter difficulty is essentially absent in the self-consistent version of the method, which probably shall soon be used to calculate electronic structures of layer crystals.

This short summary has certainly not given a direct answer to the question of an appropriate method for a particular problem; it may, however, have shed some extra light on the pro's and contra's connected with the various methods.

References

[1] F. Bloch: *Z. Physik* **52** (1929), 555.
[2] For example, D. E. Ellis and G. S. Painter: in *Computational Methods in Band Theory*, ed. by P. M. Marcus, J. F. Janak, and A. R. Williams, Plenum Press, New York 1971, p. 271.
[3] R. A. Bromley and R. E. Murray: *J. Phys. C: Solid State Phys.* **5** (1972), 738.
[4] J. C. Slater and G. F. Koster: *Phys. Rev.* **94** (1954), 1498.
[5] L. F. Mattheiss: *Phys. Rev. Letters* **30** (1973), 784.
[6] F. Herman and S. Skillman: *Atomic Structure Calculations*, Prentice Hall, New Jersey 1963.
[7] J. Robinson, F. Bassani, G. Knox and J. R. Schrieffer: *Phys. Rev. Letters* **9** (1962), 215; and J. Pendry: *J. Phys. C: Solid State Phys.* **4** (1971), 427.
[8] E. Clementi: *IBM J. Res. Developm.* **9** (1965), 2 and suppl.; and R. E. Watson and A. J. Freeman: *Phys. Rev.* **124** (1961), 1117.
[9] R. F. Steward: *J. Chem. Phys.* **52** (1967), 431.
[10] F. Bassani and G. P. Parravicini: *Nuovo Cimento* **50B** (1967), 95.
[11] J. C. Phillips: in *Solid State Phys.*, ed. by H. Ehrenreich and F. Seitz, D. Turnbull, Academic Press, New York 1966, Vol. 18, p. 55.
[12] J. H. Wilkinson: *The Algebraic Eigenvalue Problem*, Clarendon Press, London 1965, p. 229.
[13] P. O. Löwdin: *J. Chem. Phys.* **18** (1950), 365.
[14] J. C. Slater: *Phys. Rev.* **51** (1937), 846.
[15] L. F. Mattheiss: *Phys. Rev.* **B8** (1973), 3719.
[16] P. O. Löwdin: *Adv. Phys.* **5** (1956), 1.
[17] L. F. Mattheiss: *Phys. Rev.* **181** (1969), 987.
[18] L. F. Mattheiss, J. H. Wood and A. C. Switendick: in *Methods in Computational Physics*, ed. by B. Alder, S. Fernbach, and M. Rotenberg, Academic Press, New York 1968, Vol. 8, p. 98.
[19] P. D. DeCicco: *Phys. Rev.* **153** (1967), 931.
[20] H. Schlosser and P. M. Marcus: *Phys. Rev.* **131** (1963), 2529.
[21] R. S. Leigh: *Proc. Phys. Soc. London* **A69** (1956), 388.
[22] M. L. Cohen and V. Heine: in *Solid State Physics*, ed. by H. Ehrenreich, F. Seitz and D. Turnbull, Academic Press, New York 1970, Vol. 24, p. 37.
[23] C. Y. Fong and M. L. Cohen: *Phys. Rev. Letters* **24** (1970), 306.
[24] J. A. Appelbaum and D. R. Hamann: *Phys. Rev,* **B8** (1973), 1777.
[25] M. Schlüter, J. R. Chelikowsky, S. G. Louie, and M. L. Cohen: *Phys. Rev.* **B12** (1975), 4200.
[26] J. C. Phillips and L. Kleinman: *Phys. Rev.* **116** (1959), 287; L. Kleinman and J. C. Phillips: *Phys. Rev.* **116** (1959), 880; ibid **117** (1960), 460; ibid **118** (1960), 1153; ibid **125** (1962), 819.
[27] C. Herring: *Phys. Rev.* **57** (1940), 1169.
[28] M. H. Cohen and V. Heine: *Phys. Rev.* **122** (1961), 1821.
[29] P. O. Löwdin: *J. Chem. Phys.* **19** (1951), 1396.
[30] D. Brust: *Phys. Rev.* **134** (1964), A1337.
[31] C. Y. Fong and M. L. Cohen: *Phys. Rev.* **B5** (1972), 3095.
[32] M. Schlüter: *Nuovo Cimento* **13B** (1973), 313.
[33] G. Weisz: *Phys. Rev.* **149** (1966), 504.
[34] B. R. Watts: *J. Phys. C: Solid State Phys.* **6** (1973), 3605.
[35] M. J. G. Lee and L. M. Falicov: *Proc. Roy. Soc.* **A304** (1968), 319.
[36] C. Y. Fong and M. L. Cohen: *Phys. Rev. Letters* **32** (1974), 720.
[37] A. O. E. Animalu and V. Heine: *Phil. Mag.* **12** (1965), 1249.
[38] J. Korringa: *Physica* **13** (1947), 392.
[39] W. Kohn and N. Rostoker: *Phys. Rev.* **94** (1954), 1111; also B. Segall and F. S. Ham: in *Methods in Computational Physics*, ed. by B. Alder, S. Fernbach and M. Rotenberg, Academic Press, New York 1968, Vol. 8, p. 251.

[40] For example, B. Segall: *Phys. Rev.* **125** (1962), 109.

[41] H. W. Myron and A. J. Freeman: *Phys. Letters* **44A** (1973), 167.

[42] B. Segall: *Phys. Rev.* **105** (1957), 108.

[43] P. Ewald: *Ann. Physik* **64** (1921), 253.

[44] O. K. Andersen: in *Computational Methods in Band Theory*, ed. by P. M. Marcus, J. F. Janak, and A. R. Williams, Plenum Press, New York 1971, p. 178.

[45] O. K. Andersen and R. V. Kasowski: *Phys. Rev.* **B4** (1971), 1064.

[46] R. V. Kasowski and O. K. Andersen: *Solid State Comm.* **11** (1972), 799.

[47] R. V. Kasowski: *Phys. Rev. Letters* **30** (1973), 1175.

[48] K. Wood and J. B. Pendry: *Phys. Rev. Letters* **31** (1973), 1400.

[49] K. Kambe: in *Computational Methods in Band Theory*, ed. by P. M. Marcus, J. F. Janak, and A. R. Williams, Plenum Press, New York 1971, p. 409; *Z. Naturforsch.* **23a** (1968), 1280.

[50] K. Kambe: *Z. Naturforsch.* **22a** (1967), 322.

[51] P. M. Morse and H. Feshbach: *Method of Theoretical Physics*, McGraw-Hill, New York 1953, Vol. 1, p. 891.

[52] E. T. Whittaker and G. N. Watson: *A Course of Modern Analysis*, Cambridge Univ. Press, 4th ed., Cambridge 1958, p. 273.

[53] G. N. Watson: *The Theory of Bessel Functions*, Cambridge Univ. Press, 2nd ed., Cambridge 1944, p. 178.

[54] P. Epstein: *Math. Ann.* **56** (1903), 615.

[55] K. Kambe: *Z. Naturforsch.* **22a** (1967), 422.

[56] P. M. Morse and H. Feshbach: *Method of Theoretical Physics*, McGraw-Hill, New York 1953, Vol. 2, pp. 1321 and 1371.

[57] A. Erdelyi: *Higher Transcendental Functions*, McGraw-Hill, New York 1953, Vol. 2, p. 134.

[58] F. Herman, R. L. Kortum, C. D. Kuglin, J. P. Van Dyke and S. Skillman: in *Methods in Computational Physics*, ed. by B. Alder, S. Fernbach, and M. Rotenberg, Academic Press, New York 1968, Vol. 8, p. 193.

[59] M. Tsukada, K. Nakao, Y. Uemura, and S. Nagai: *J. Phys. Soc. Japan* **32** (1972), 54.

[60] P. Krusius, J. von Boehm, and H. Isomäki: *J. Phys. C: Solid State Phys.* **8** (1975), 3788.

[61] R. N. Euwema, D. J. Stuckel, and T. C. Collins: in *Computational Methods in Band Theory*, ed. by P. M. Marcus, J. F. Janak, and A. R. Williams, Plenum Press, New York 1971, p. 82.

ELECTRONIC STRUCTURE OF SOME LAYER COMPOUNDS

C. Y. FONG

Dept. of Physics, University of California, Davis, Calif. 95616, U.S.A.

M. SCHLÜTER

Bell Laboratories, Murray Hill, N.J. 07974, U.S.A.

1. Transition Metal Layer Compounds

The transition metal dichalcogenides MX_2, including group IVB to group VIB transition metal ions, crystallize in layered structures. Many physical properties of these layer compounds are dominated by the two-dimensional structural characteristics. More or less strong anisotropies are reported for mechanical, electrical and optical properties [1]. Moreover, intercalation compounds of the form I_xMX_2, have been prepared where I is an atom or molecule which occupies the interstitial sites between adjacent sandwiches. The semiconducting group IVB and VIB compounds are found to exhibit metallic conductivity when intercalated with alkali atoms. The metallic group VB compounds intercalate with organic molecules which may increase the inter-layer distance by as much as 50 Å. Most of the metallic compounds are superconductors with low to medium T_c with the superconducting property persisting even in the intercalated phases [2]. According to the discussions in Chapter A.3 and similar to non-transition metal compounds the materials can be grouped into *two* major classes depending on the crystal field symmetry around the metallic ions. The two types of structures, the octahedral and the trigonal prismatic structures, shall be discussed separately in the following subsections.

1.1. COMPOUNDS WITH OCTAHEDRAL ARRANGEMENT OF THE CATIONS

The standard octahedral arrangement (combined with the most simple stacking sequence), also commonly found in non-transition metal layer compounds, is the CdI_2 type structure (also referred to as C6-type). Most of the group IVB transition metal compounds and also some cases of the group VB dichalcogenides at high temperature (1000 K) belong to this class. In their fundamental review article Wilson and Yoffe [1] presented extensive tabulations about this class. Some of the crystals which have been studied theoretically, their lattice constants and the corresponding references to theoretical investigations are listed in Table 1. A common approximation made in most of the calculations for this class of crystals is to assume the following atomic arrangement within the unit cell.

$$M: (0, 0, 0,)$$

$$X: \pm(\tfrac{1}{3}, \tfrac{2}{3}, z) \quad z = \tfrac{1}{4}. \tag{C1}$$

This would correspond to ideal octahedral symmetry for an ideal c/a ratio of 1.63. Ideal octahedral symmetry of the cation site would still be possible for $c/a \neq 1.63$ if

TABLE 1

List of transition metal compounds with octahedral coordination of the cation. The various theoretical approaches are indicated.

Crystals D_{3d}^3	Lattice constants a(Å)	c/a	Schematic approach	ETB	APW	KKR	OPW	EPM
TiS₂	3.405	1.67	Wilson-Yoffe [1] White-Lucovsky [12]	Murray-Yoffe [16]		Myron-Freeman [14]	Krusius et al. [13]	
TiSe₂	3.535	1.698	WY, WL	MY				
TiTe₂	3.766	1.724	WY, WL	MY				
ZrS₂	3.662	1.590	WY, WL	Murray, Bromley and Yoffe [20]				
ZrSe₂	3.770	1.628	WY, WL	MBY				
ZrTe₂	3.950	1.678	WL					
HfS₂	3.635	1.606	WY, WL	MBY	Mattheiss [22]			Fong et al. [21]
HfSe₂	3.748	1.643	WY, WL	MBY				
1T–TaS₂	3.365	1.751	WL		Mattheiss [22]			

$z = 1/\sqrt{6}\, c/a$. Crystal structure refinements have been done only in few cases so that precise values for the positional parameter z are generally unavailable. The coordinates given in Equation (C1) are taken along the hexagonal basis vectors (a_1, a_2, a_z). The symmorphic space group associated with the structure is D_{3d}^3 and has been discussed in full detail in Chapter A,1.3.2. A perspective view of the structure and its unit cell is presented in Figure 1. This crystal structure generally represents the high temperature phase which is in particular the case for 1T–TaS₂. In a recent review article Wilson *et al.* [3] summarize numerous low-temperature phase transitions of this compound. Large variations of crystal lattice parameters are reported. These variations are probably due to non-stoichiometric effects and the coexistence of different phases.

As seen from Table 1, most of the theoretical investigations in this class of crystals with octahedrally coordinated cations are limited to the group IVB transition metal dichalcogenides (except 1T–TaS₂).

Several semi-empirical band structure and ligand field models have been proposed to describe the electronic states in these MX_2 layer compounds. These schemes heavily depend on the input of experimental results. We shall therefore first summarize some important properties of these crystals deduced from various experimental data. A discussion of the features of the schematic band structure and ligand field approaches will follow. We then discuss in detail the results of the various explicit band structure calculations. Since density of states curves which may be compared to photoemission data or other optical response functions derived from these calculations are in general not available, comparison between theory and experiment will be of more qualitative

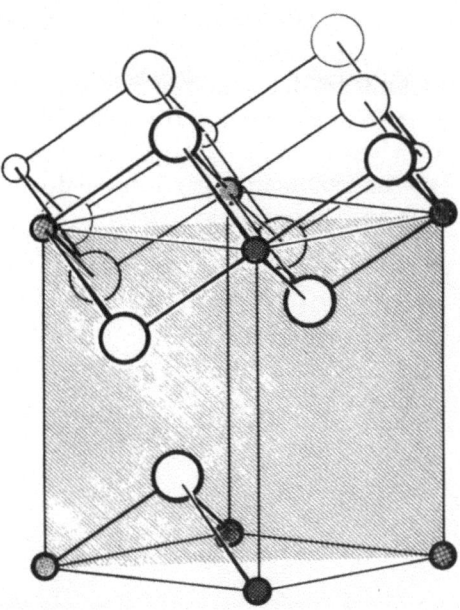

Fig. 1. Perspective view of the CdI$_2$-type structure. The unit cell contains one formula unit and has been chosen to extend over two half layers (from [60]).

nature. Finally, the group VB compound 1T–TaS$_2$ shall be considered in some detail.

A comprehensive review of the experimental aspects of transition metal layer compounds up to 1969 has been presented in [1]. In other volumes of this series and in [3] more recent experiments are reviewed. We shall at this point summarize some important properties deduced from experiments which are relevant to the empirical theoretical approaches.

The measured properties of the Ti dichalcogenides are quite different from those obtained for other group IVB compounds. Electrical measurements suggest that TiS$_2$ is either a small gap semiconductor ($E_g \leq 0.1$ eV) or a metal [4]. The 'metallic' behavior has also been found in X-ray [5] and optical studies [6, 7]. Similar free carrier behavior in the low energy region was also reported for TiTe$_2$.

Measurements of electrical [8] and optical [6, 7] properties on Zr and Hf compounds indicate the existence of a semiconducting gap. The fundamental energy gaps are indirect and range from 1.13 eV to 1.96 eV. Exceptions may again be the heavier anion compounds ZrTe$_2$ and HfTe$_2$. On the basis of this information several empirical, schematic approaches to the band structures of this class of crystals have been proposed.

A lowest order model to the electronic structure of the MX$_2$ transition metal compounds is obtained by analyzing the ligand-field splitting of the cation d-levels in the octahedral environment and of the anion s, p-levels in the trigonal environment [9, 10].

The neutral atomic configurations of the chalcogenides S, Se and Te are $(3s)^2(3p)^4$, $(4s)^2(4p)^4$ and $(5s)^2(5p)^4$ respectively. The group IVB transition metals have $(nd)^2((n+1)s)^2$ configurations in their neutral state where n ranges from 3 for Ti to 5 for Hf, while the group VB transition metal Ta, has the configuration $(5d)^3(6s)^2$. In ideal octahedral environment the metal d-states split into a doubly degenerate level e_g, associated with the d_{xz} and d_{yz} orbitals and into a three-fold degenerate level t_{2g} associated with the d_{z^2}, $d_{x^2-y^2}$ and d_{xy} orbitals.

Further symmetry reductions are due to deviations from the ideal octahedral environment which includes e.g. effects originating from the finite thickness of the layers. These reductions lead to the space group D_{3d}^3. For completeness and for later more quantitative arguments we shall enumerate here the irreducible representations of D_{3d}^3 corresponding to the cation and anion valence states.

$$
\text{cation} \quad d \quad
\begin{cases}
\Gamma_3^+(d_{xz}, d_{yz}) \\
\Gamma_3^+(d_{x^2-y^2}, d_{xy}) \\
\Gamma_1^+(d_{z^2})
\end{cases}
$$

$$
\text{anion } s \quad \Gamma_1^+, \Gamma_2^-(s) \tag{C2}
$$

$$
\text{anion } p \quad
\begin{cases}
\Gamma_1^+, \Gamma_2^-(p_z) \\
\Gamma_3^+, \Gamma_3^-(p_x, p_y).
\end{cases}
$$

The $\pm$ representations refer to the inversion parity behavior of the *two* anions per layer unit cell. The complete hybridization possibilities of orbitals may be deduced from Equation (C2). These expressions are valid for the D_{3d} symmetry of $\mathbf{k} = 0$ states in the crystal. For general $\mathbf{k}$-points no symmetry restrictions on hybridization are imposed.

On the basis of the simple octahedral field induced splitting of the d-levels, it was suggested in [1] that the geometrical configuration of the crystal is most suitable for the two e_g states to hybridize with the p-states of the chalcogenides to form together with the cation s-levels the covalent contribution to the bonds, whereas the t_{2g} states should form the basis of non-bonding bands. The anion p and the cation d hybrids thus constitute the bonding and antibonding orbitals (σ, σ^*) separated by an energy gap of several eV. Due to differences in ionicity the cation d-character should be rather found in the antibonding σ^*-states, whereas the anion p-character should be more strongly present in the bonding σ-states. Complete separation of course means dehybridization and ionic bonding. The cation t_{2g} non-bonding orbitals are located somewhere in the bonding-antibonding gap. In this covalent picture a fundamental gap would appear for the group IVB compounds between the occupied bonding p-d hybrids and the empty non-bonding cation d-orbitals of predominant t_{2g} character. The resulting schematic band structures are shown in Figure 2. All energies are measured with respect to the top of the valence bands. The suggested *indirect* energy gaps are indicated by arrows. The indirect nature of the gaps is deduced from experiments and may also be deduced from the behavior of p- and d-states in an octahedral crystal field. Based on experimental information the gaps were estimated to range from 0.25 eV for $TiTe_2$ to 1.95 eV for HfS_2. The width of the uppermost valence bands was proposed to be around 1 eV (which is presumably too small) whereas the lower non-bonding conduction bands should have a width of about 2 eV.

It is emphasized that the Ti compounds are thus considered as semiconductors instead of metals as indicated by experimental data. Non-stoichiometric effects are suggested to be responsible for the observed metallic behavior [1, 11].

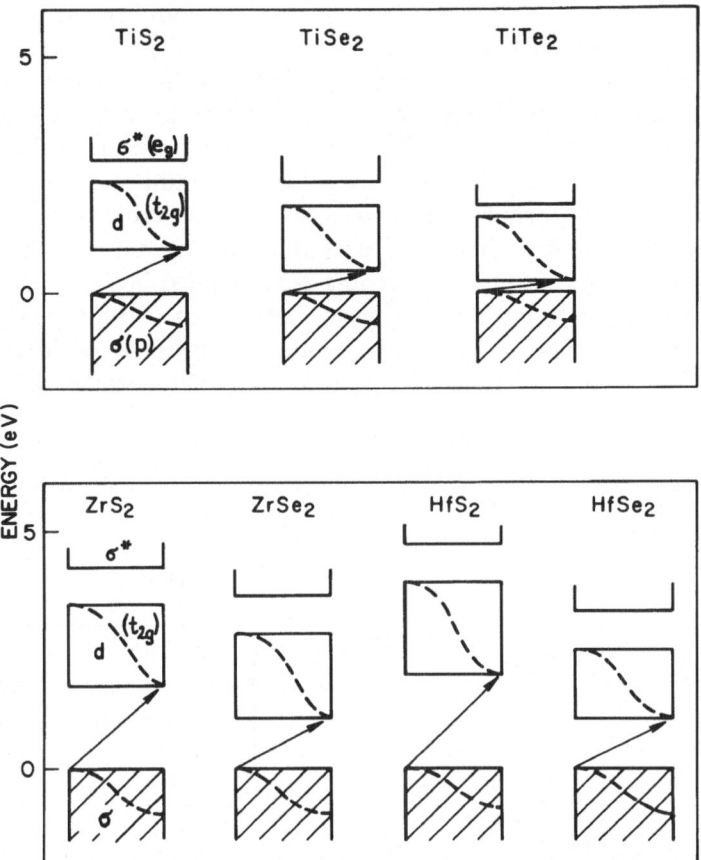

Fig. 2.   Schematic band structure of 7 group IVB transition metal dichalcogenides as proposed in [1].

This covalent picture is contrasted by an ionic picture which is based on the standard model used in crystal chemistry to first fill the anion valence shells and then consider the remaining cation electrons to determine the electrical and optical properties. A filling of the anion valence shells in the group IVB chalcogenides empties all cation s and d-levels. In this case all five d-levels should appear as empty non-bonding bands above the filled anion p-bands. In contrast to the semiconducting situation obtained by complete charge transfer, incomplete transfer would result in metallic behavior. The ionic picture has been chosen in another schematic approach which is based on stability arguments of the crystal structure [12]. Assuming ionic crystals with varying degree of ionicity the essential contributions for determining the crystal structure stability are the Madelung energy and the repulsive overlap energy. In this picture the total potential energy per unit cell can be written as

$$V(r) = -\frac{\alpha(Ze)^2}{r} + \frac{C\lambda}{r^n},\tag{C3}$$

where α is the Madelung constant, Z the valency of the ion and r the separation between the nearest neighbor ions. C denotes the coordination number and λ is an empirical parameter. In a crude model the repulsive part of the potential may be accounted for by treating the ions as hard spheres. Then $V(r)$ depends only on the Madelung term. Structural characteristics and ionic radii thus determine the relative stability through α and r. By examining thirty-seven different crystals, it was found [12] that *two* critical values of the ratio (r_+/r_-) of positive and negative ionic radii exist. These critical values of (r_+/r_-), 0.65 and 0.33 form border lines between three stable types of structures for compounds of the form MX_2. The structures are characterized by eight, six, and four-fold coordination of the cations. The results are compiled in Figure 3. All transition metal compounds fall into the middle region, favoring the CdI_2-type structure. Their trend to increased covalency, however, is indicated by their positions close to the $r_+/r_- = 0.33$ line which separates them from four-fold, tetrahedrally coordinated

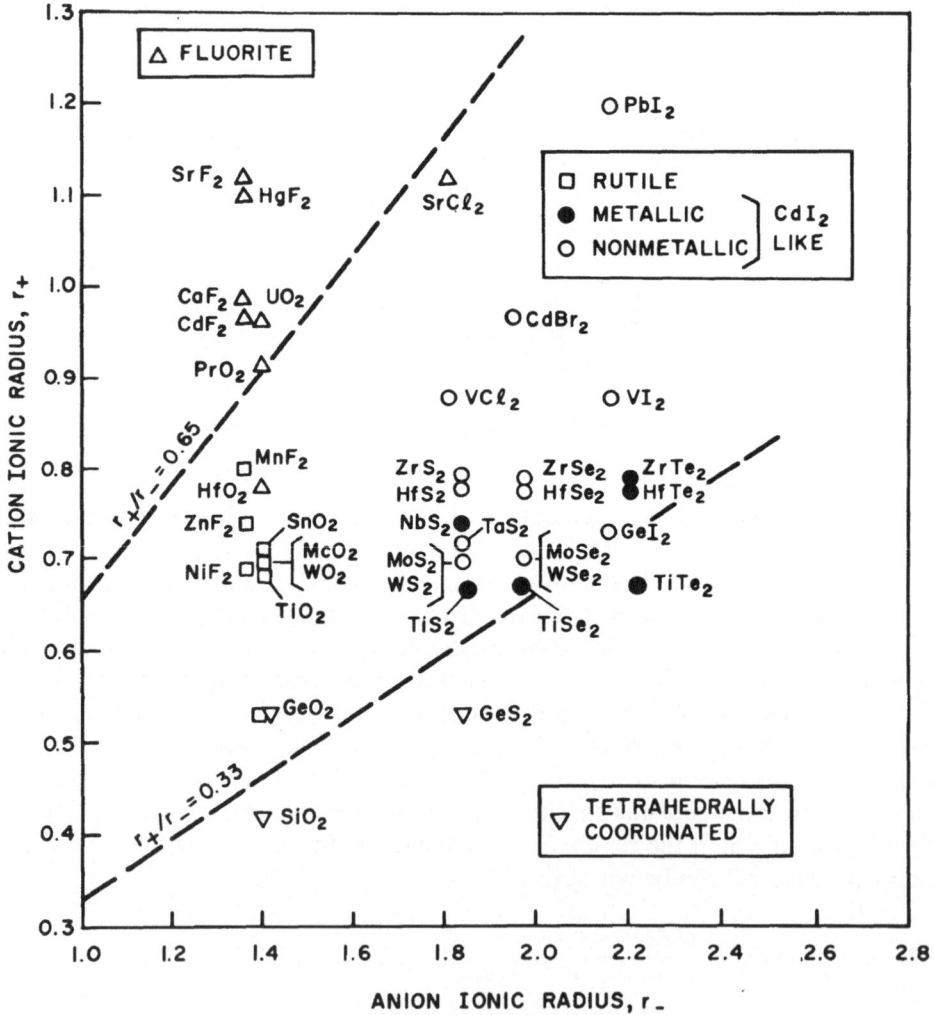

Fig. 3.   Classification of MX_2 compounds based on ionic radii and crystal structures according to [12].

compounds. Taking into account this varying degree of ionicity which is simply described by the r_+/r_- ratio, schematic band structures were devised as shown in Figure 4. It is emphasized that the Ti-compounds and the heavier Zr and Hf chalcogenides are proposed to be *metallic*, in contrast to the suggestion of [1].

More realistic band structure calculations have been performed since. In the following we shall discuss the various compounds separately proceeding from the Ti to the Hf compounds.

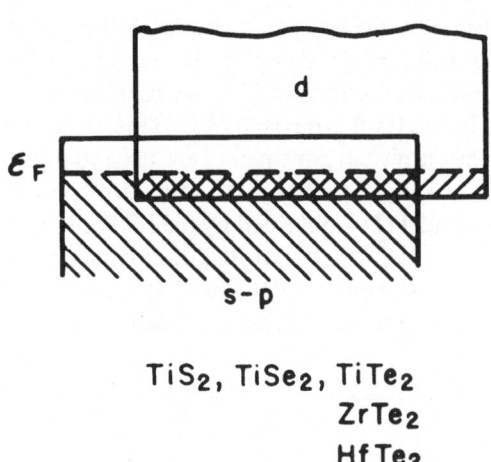

$$\text{TiS}_2, \text{TiSe}_2, \text{TiTe}_2$$
$$\text{ZrTe}_2$$
$$\text{HfTe}_2$$

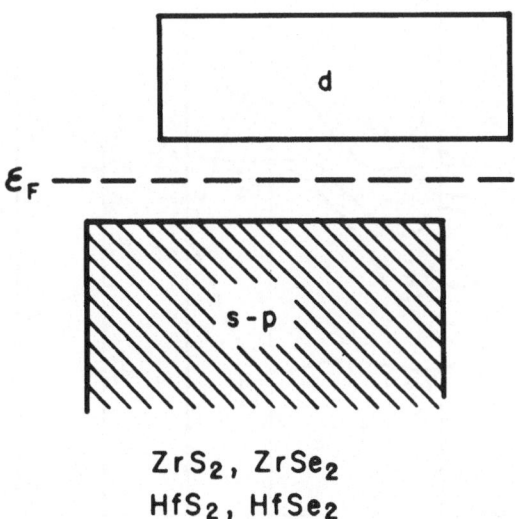

$$\text{ZrS}_2, \text{ZrSe}_2$$
$$\text{HfS}_2, \text{HfSe}_2$$

Fig. 4.   Schematic band structure of 9 group IV transition metal dichalcogenides as proposed in [12].

1.1.1. *The Group IVB Transition Metal Chalcogenides*

The band structure of TiS_2 has been studied by the orthogonalized plane wave (OPW) method [13], the Green's function (KKR) method [14], by first-principles tight binding (TB) methods [14, 15] and by an empirical tight binding (ETB) method [16]. The ETB method has also been applied to the cases of $TiSe_2$ and $TiTe_2$. In the TB calculations of [15] results for various molecules TiS, TiS_2 and TiS_6 as well as band structure results of TiS_2 for $\mathbf{k} = 0$ are reported. All methods start with or use the 'neutral atom' configurations of $(4s)^2(3d)^2$ for Ti and $(3s)^2(3p)^4$ for S. Two different values of α for the local $\alpha\rho^{1/3}$ exchange interaction have been used. In the TB, KKR [14] and OPW [13] calculations both cases of $\alpha = 1$ and $\alpha = \frac{2}{3}$ are explored whereas the TB calculation of [15] is based on $\alpha = 1$. The OPW calculations of [13] are carried out until selfconsistency of the potential is reached. In the ETB approach of [16] the Ti-compounds are assumed to be semiconducting and the empirical parameters are used to fit a few energy gaps to the structures in optical spectra [6]. The band structures of TiS_2 obtained by the KKR method [14] and of $TiSe_2$ and $TiTe_2$ obtained by the ETB method [16] are plotted along some high symmetry lines in the hexagonal Brillouin zone and displayed in Figures 5, 6, and 7 respectively. The TiS_2 results of [13] and [16] are qualitatively similar to the results of [14] and are not presented explicitly. In all band structures the lowest two anion

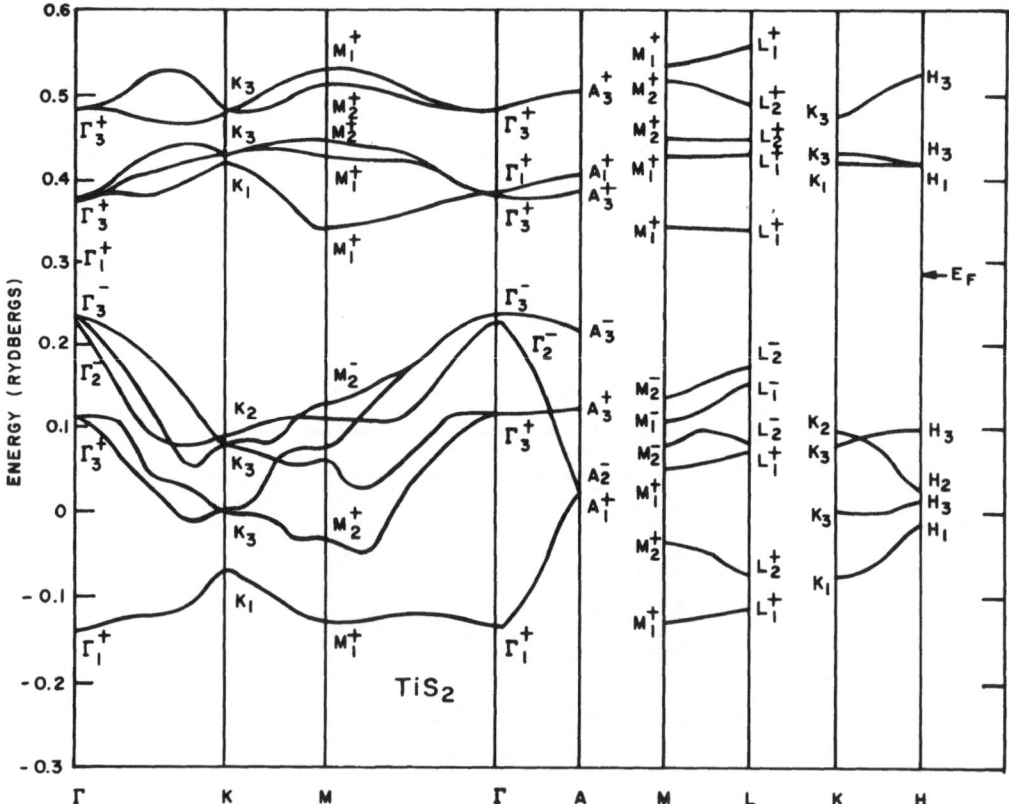

Fig. 5.   Band structure of TiS_2 as obtained from KKR calculations of [14].

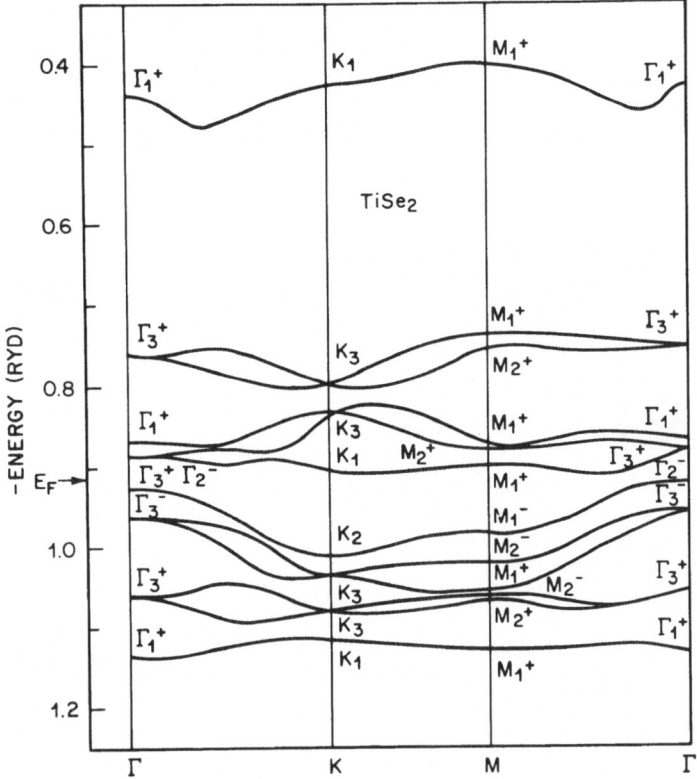

Fig. 6. Band structure of TiSe$_2$ as obtained from ETB calculations of [16].

s-bands are not shown. Except for TiTe$_2$ a semiconducting gap is found between the six valence bands (primarily chalcogenide p-states) and the five split off conduction bands (primarily non-bonding Ti d-states). It is emphasized that the first-principles calculations [13, 14, 15] suggest semiconducting properties for TiS$_2$. Recent self-consistent TB calculations [14] based on numerical basis sets seem to yield a very small or even vanishing gap for TiS$_2$. Some preliminary results are indicated in Table 2.

The top of the valence bands predicted by the KKR, OPW and the ETB methods has Γ_3^- or Γ_2^- symmetry respectively. The small splitting between Γ_3^- and Γ_2^- shows the influence of small deviations from the ideal octahedral environment on p-like orbitals. (p_x, p_y transform like Γ_3^- whereas p_z transforms like Γ_2^-). The Γ_3^+ and Γ_1^+ states which correspond to symmetric combinations of the anion p-orbital centered at the two atoms and which also allow for cation d-admixture are found at somewhat lower energies. Five states with Γ_3^+ and Γ_1^+ symmetry are found above the Fermi level and clearly constitute the e_g and t_{2g} cation d-levels. For the KKR results these bands are closely grouped together and are followed by a large gap to higher energies. The lower t_{2g} d-levels may be interpreted as non-bonding [1, 9]. The situation seems to favor the ionic picture over the covalent picture. In the OPW results of [13] no clear grouping of the conduction bands is seen. The smallest difference in energy between the valence and conduction bands gives the fundamental indirect gap. The calculated values of these gaps, the widths of the lowest group of non-bonding d-bands and of the occupied

TABLE 2

Comparison between experimental and theoretical results for TiS$_2$

Authors energy (eV) features	TiS$_2$ Theory				Experiment				
	Wilson Yoffe [1]	Krusius et al. $\alpha=\frac{2}{3}$ [13]	Murray Yoffe [16]	Myron Freeman $\alpha=\frac{2}{3}$ [14]	Greenaway Nitsche [6]	Beal Knight Liang [7]	Lee Said Davis Lim [17]	Thompson Pisharody Koehler [4]	Fischer [5]
Indirect gap	0.93	1.5	0.83	1.45 (0.32)[a] (-0.05)[b]	0.83		0.7	metallic	metallic
		$(\Gamma_3^- \to L_1^+)$ 3.38	$(\Gamma_2^- \to L_1^+)$ 1.139	$(\Gamma_3^- \to M_1^+)$ 1.97					
Minimum direct gap		$L_2^- \to L_1^+$	$(\Gamma_2^- \to \Gamma_1^+)$ (forbidden for $\mathbf{E}\perp\hat{c}$)	$(\Gamma_3^- \to \Gamma_3^+)$					
Width of the lower nonbonding d bands	1.43	~6 overlap	0.85	1.44					
Width of the topmost p bands	0.68	~2.5	1.27 $(K_3 \to \Gamma_2^-)$	1.69 $(K_2 \to \Gamma_3^-)$					
$L_2^- \to L_1^+$		3.38	1.22	2.18	1.16	1.04			1.2
$\Gamma_2^- \to \Gamma_3^+$		3.75	1.22	2.06					
$\Gamma_3^- \to L_1^+$		3.69	1.73	1.958		1.64			
$A_3^- \to A_3^+$		~4.20	2.04	2.205	1.95(E_1)	2.01			2.0
$M_1^- \to M_1^+$		~3.70	2.07	3.09					
$M_1^- \to M_1^+$		~7.0	2.55	4.2	2.2(E_2)	2.29			2.2
$\Gamma_3^- \to \Gamma_3^+$		-	3.26	3.29	3.35(E_3)	3.37			3.4

[a] Recent results obtained in a SCF calculation using a numeric basis set and $\alpha = \frac{2}{3}$. Order of Γ_3^- and Γ_2^- valence bands interchanges.

[b] As [a] but with $\alpha = 1$.

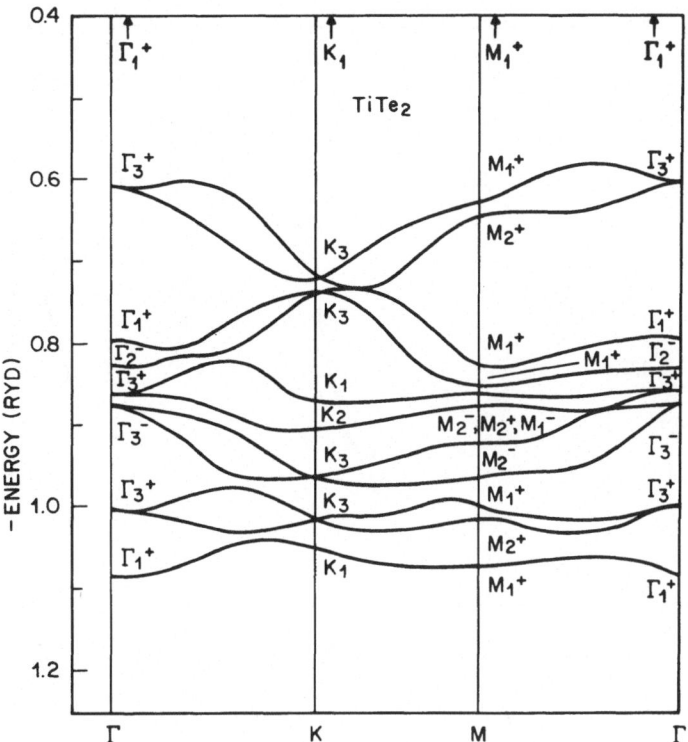

Fig. 7.   Band structure of $TiTe_2$ as obtained from ETB calculations of [16].

p-bands are presented in Table 2. The dispersion of the d-conduction bands increases from TiS_2 to $TiTe_2$ whereas the width of the p-valence bands decreases. In general the d-bands show less dispersion than the p-band which indicates weak hybridization effects and the more localized nature of the d-states. In fact, the d-levels of the transition metals show strong localization which may only approximately be described by band-pictures.

To facilitate comparisons between the theoretical band calculations and optical spectra, we reproduce in Figures 8 and 9 experimental absorption spectra of TiS_2 and $TiSe_2$ obtained in [7]. The spectra consist of two main absorption structures between 0 and 3.5 eV. In TiS_2, the low energy structure is centered around 2.3 eV with a width of 1.5 eV, while the corresponding structure in $TiSe_2$ of comparable width is shifted towards lower energies and centered at about 1.6 eV. A similar shift of the high energy structure is observed. The band structures of TiS_2 and $TiSe_2$ presented before, qualitatively account for these features in the spectra. The low energy structure can be identified as due to transitions from the upper p-bands into the lower triplet of non-bonding d-bands followed by a window due to the separation between the lower non-bonding and the upper non-bonding (or antibonding) d-bands. The higher energy absorption structure can consequently be attributed to transitions between the bonding p-bands and the upper d-bands. The shifting in energy of the measured spectra of TiS_2 and $TiSe_2$ may qualitatively be explained by relative shifts of the d-bands in the band structures. In [7] an attempt was made to correlate fine structures in the spectra with

156 CHAPTER C

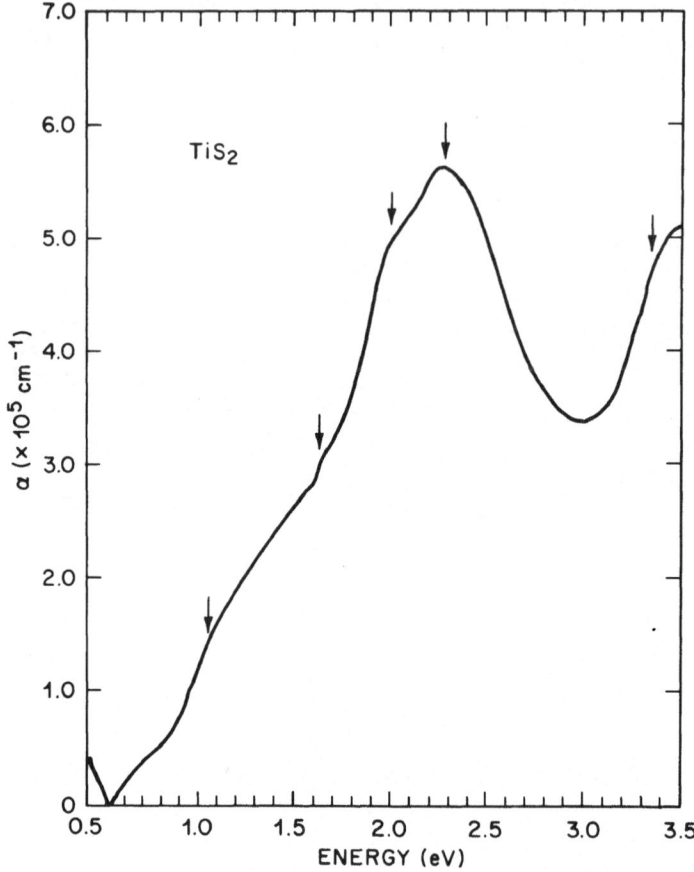

Fig. 8.   Optical absorption spectrum of TiS₂ as measured in [7].

critical points derived from the ETB band structures. These identifications with a few minor modifications are given in Tables 2 and 3.

Although most of the present theoretical results seem to suggest that stoichiometric compounds of TiS_2 and $TiSe_2$ are *semiconductors*, some doubts remain as to the precision of these calculations. More selfconsistent first-principles calculations in the spirit of the OPW calculations of [13] will be useful to understand the contradictions between the theories and the experimental data. The results of [13] compared to non-selfconsistent calculations show that the relative positions of the *p* and *d* bands can differ as much as 3 eV. The need for selfconsistent treatment seems also to be indicated by the observation that all reported non-selfconsistent band structures (this will also be valid for other transition metal layer compounds to be discussed) indicate more or less ionic results, whereas they were based on neutral atom configurations. Moreover, the localization of the 3*d*-levels produces strong correlation effects which are not accounted for in standard band schemes. Modified band structure schemes as proposed for the transition metal oxides [18] may therefore be of need to clarify the situation.

Further uncertainties result from the neglect of muffin-tin corrections in the KKR calculations. No muffin-tin approximation has to be used in the OPW calculations.

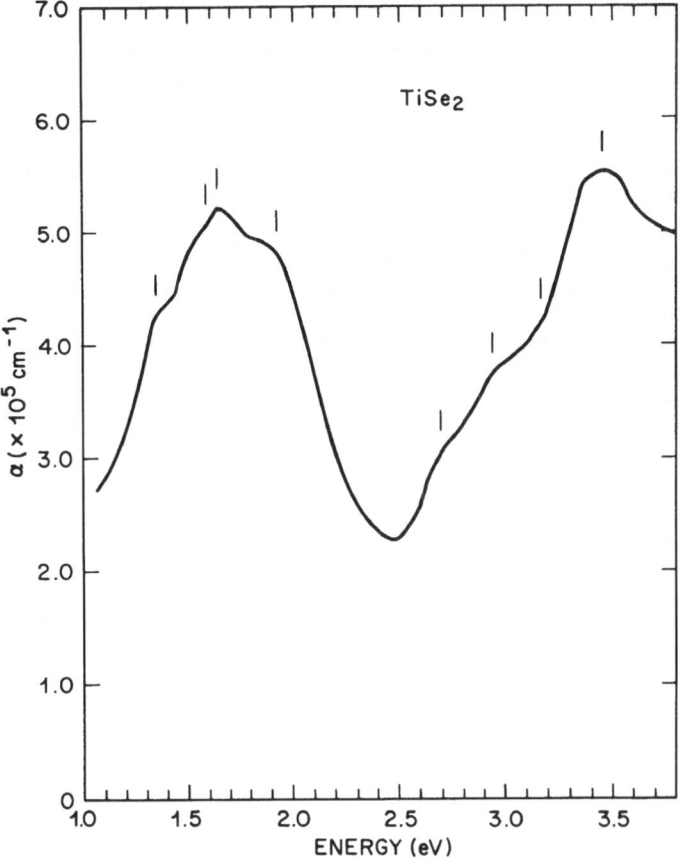

Fig. 9.   Optical absorption spectrum of TiSe$_2$ as measured in [7].

The empirical tight binding calculations on the other hand suffer from the neglect of interlayer coupling. As already pointed out earlier, interlayer coupling, though only weakly affecting the ensemble of valence electrons, can have quite drastic effects on certain particular electronic levels. Thus large dispersions of the $\Gamma_2^- - A_2^-$ and $\Gamma_1^+ - A_1^+$ bands along Δ in Figure 4 can occur compared to the relatively flat behavior of other bands. Obviously, the above mentioned bands correspond to chalcognide p_z-type orbitals which point into the space between layers.

The selfconsistent OPW results of [13] provide real space valence charge density distributions. The total valence charge of TiS$_2$ is shown in Figure 10 in a (110) plane containing sulphur and titanium atoms of two adjacent layers and in Figure 11 in a (001) plane containing three titanium atoms of one layer. The Ti–S bond is found to have sizeable covalent character. The ionic contribution of the selfconsistent OPW valence charges can be accounted for by equivalent effective charges of 1.75 electrons and -0.88 electrons at the Ti and S sites respectively [13]. These quantities are illustrated for TiS$_2$ in a so called charge transfer function in Figure 12. The charge transfer function shows the difference between the selfconsistent OPW charge density and isolated atom valence charge densities along the Ti–S bond.

TABLE 3

Comparison between experimental and theoretical results for $TiSe_2$

| | TiSe$_2$ | | | |
| | Theory | | Experiment | |
Authors energy (eV) features	Wilson Yoffe [1]	Murray Yoffe [16]	Greenaway Nitsche [6]	Beal Knight Liang [7]
Indirect gap	0.496	0.17 ($\Gamma_2^- \to \Sigma_1$)		
Minimum direct gap		0.544 ($\Gamma_2^- \to \Gamma_3^+$)		
Width of the lower nonbonding d-bands	1.36	1.26		
Width of the topmost p-bands	0.62	1.05 ($K_2 \to \Gamma_2^-$)		
$M_1^- \to M_1^+$		1.19	1.55 (E_1)	
$\Gamma_3^- \to \Gamma_1^+$ with s.o.		1.35 1.67		1.58
$M_1^- \to M_1^+$		1.56	1.85 (E_2)	1.63
$\Gamma_2^- \to \Gamma_3^+$		2.3		
$\Gamma_3^- \to \Gamma_3^+$		2.82	2.85 (E_3)	2.94

In the case of $TiTe_2$, a significant overlapping between the top of the valence bands and bottom parts of the non-bonding conduction bands is obtained by the ETB method. Therefore, metallic behavior is suggested for this crystal. A comparison of the overall widths of the p valence bands in Figures 6 and 7, suggests that the larger sizes of the Te ions cause more overlap between the p-orbitals and thus broaden the corresponding bands. A similar broadening is observed for the 'non-bonding' d-levels which indicates a decrease in their non-bonding character. The present ETB results seem to be consistent with the banding scheme of [12].

The reflectivity of $TiTe_2$ has been measured in [6] and is shown in Figure 13. Corresponding to the low energy absorption structures in TiS_2 and $TiSe_2$ (Figures 8 and 9), the E_2 peak is shifted down to 1.31 eV. The window is observed in the sulphide and selenide cases at 3.0 eV and 2.5 eV and is smeared out in $TiTe_2$. By reference to Tables 2 and 3 this result indicates that the $M_1^- \to M_1^+$ transition has about 1.3 eV in $TiTe_2$. No energy gap separates the various groups of d-states. The relatively low reflectivity in the range $3.0\,\text{eV} \leq E < 7.0\,\text{eV}$ suggests a smaller joint-density of states between the valence and conduction states. All these features can be qualitatively accounted for by the band structure shown in Figure 7. In Table 4 we summarize these results together with

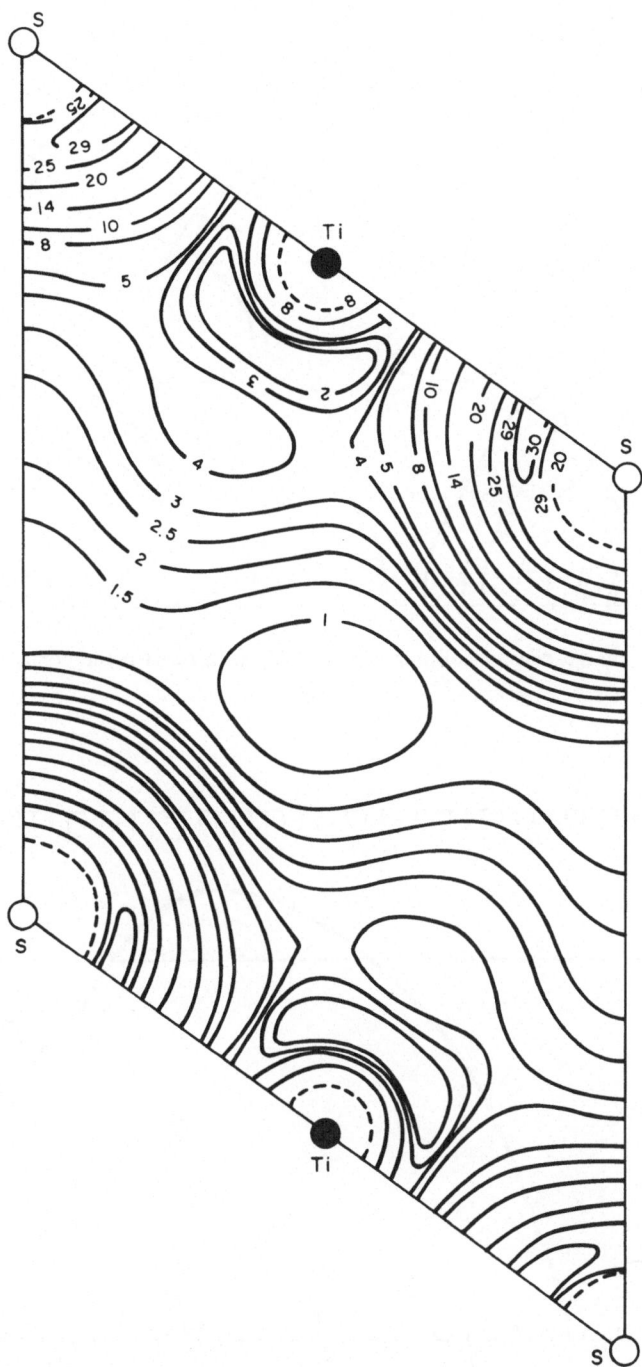

Fig. 10.   Valence charge density contour map of TiS$_2$ as obtained from the self-consistent OPW calculations of [13] shown in a (110) plane extending over two layers. The contours are given in units of 10^{-3} electrons per a_0^3.

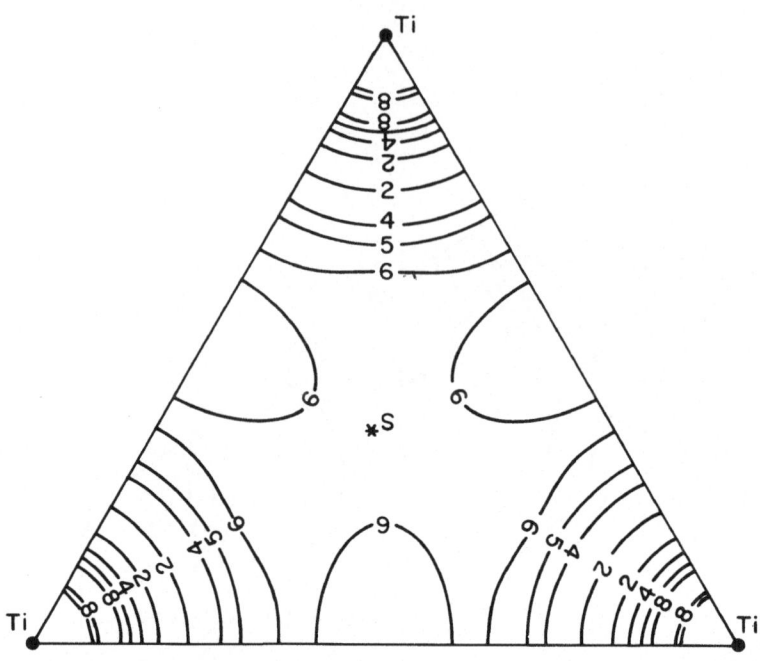

Fig. 11.   Valence charge density contour map of TiS_2 in a (001) plane containing the Ti atoms (from [13]).

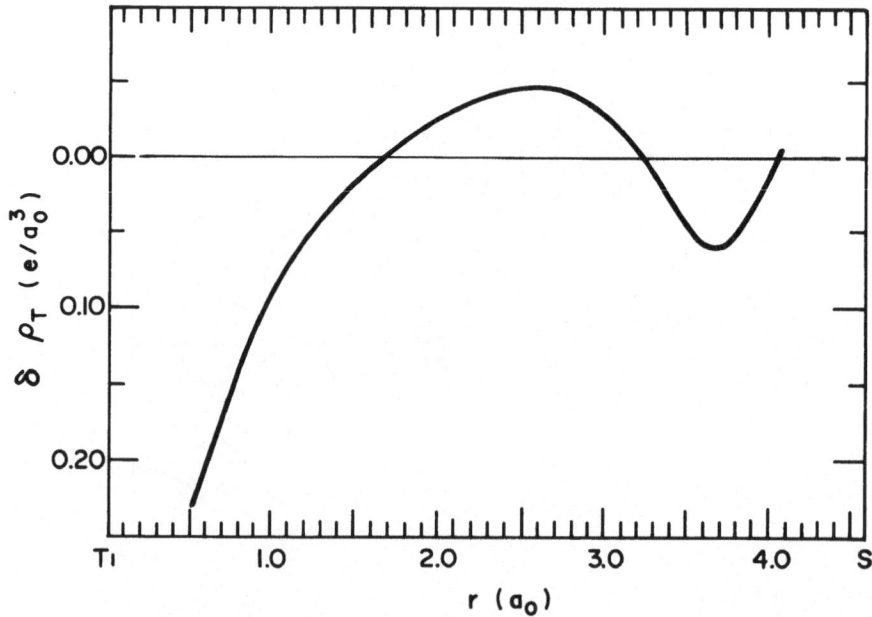

Fig. 12.   Charge density transfer function $\delta\rho_T$ for TiS_2 along the Ti–S bond ([13]). The definition of $\delta\rho_T$ is given in the text; contributions from the core states are neglected.

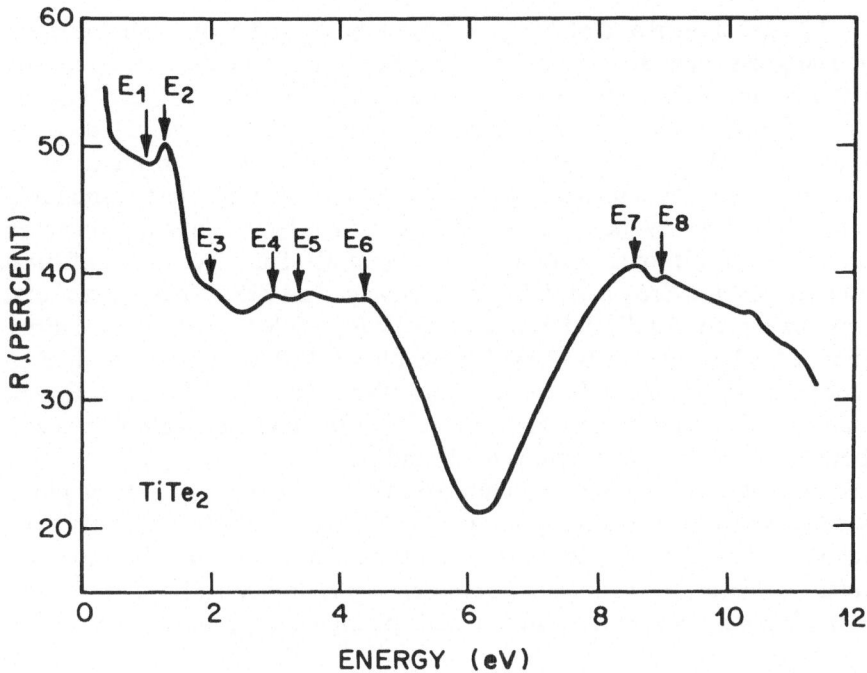

Fig. 13. Reflectivity spectrum of TiTe₂ as measured in [6].

TABLE 4

Comparison between experimental and theoretical results for TiTe₂

	TiTe₂		
	Theory		Experiment
Authors energy (eV) features	Wilson Yoffe [1]	Murray Yoffe [16]	Greenaway Nitsche [6]
Indirect gap	0.25	metallic	
Width of the lower nonbonding d-bands	1.36	1.99	
Width of the uppermost p-bands	0.62	overlap between d-p bands	
$M_1^- \rightarrow M_1^+$		1.0	$1.00(E_1)$
$\Gamma_3^- \rightarrow \Gamma_1^+$		1.17	
$M_1^- \rightarrow M_1^+$		1.33	$1.30(E_2)$
$\Gamma_3^- \rightarrow \Gamma_3^+$		3.66	$2.00(E_3)$

identifications of critical point transitions with the fine structures in the reflectivity. The Fermi energy has tentatively been placed between the Γ_3^- and the second Γ_3^+ bands.

The only existing band structure calculations on ZrS₂ and ZrSe₂ were carried out by the ETB method [20]. Results for the band structures are shown in Figures 14 and 15. The parameters used in the calculation were determined by fitting the energies at

$L(L_1^- \to L_{1,3}^+$ not shown) and at $\Gamma(\Gamma_3^- \to \Gamma_1^+)$ to the E_1 and E_2 peaks in the reflectivities of [6]. The gross qualitative features of the two band structures are quite similar to those of TiS_2 and $TiSe_2$. The lowest six bands are the bonding p-like bands separated from the next higher non-bonding d-bands by nearly 2 eV. The second (pair) of d-bands is considerably split off and approaches the high s-type band. It might thus also be viewed as antibonding band, in the spirit of the discussion in [1]. The non-bonding bands would thus be formed by the t_{2g} triplet of d-bands. The band configuration nevertheless indicates considerable ionic character in the bonding. The exact values of the gap are 1.84 and 1.22 eV for ZrS_2 and $ZrSe_2$ respectively. The ZrS_2-value agrees reasonably well with the experimentally determined 1.68 eV [6], while the $ZrSe_2$-value differs from the estimated value of [1] by 0.17 eV. The widths of the lower non-bonding d-bands as well as of the uppermost valence bands are listed in Tables 5 and 6. Except some quantitative differences in energies, the results of the ETB calculations are consistent with both schematic band structures of [1] and [12].

To compare to optical data, in Figure 16 the reflectivity [6] of ZrS_2 is plotted as a typical spectrum for the two compounds. There is an average 0.5 eV shift of the peaks towards lower energies in $ZrSe_2$. More recent absorption spectra [7] show more fine structures. However, their range in photon energy is limited to 1.5 to 4.3 eV; thus the important features such as the window and the high energy absorption peak are not

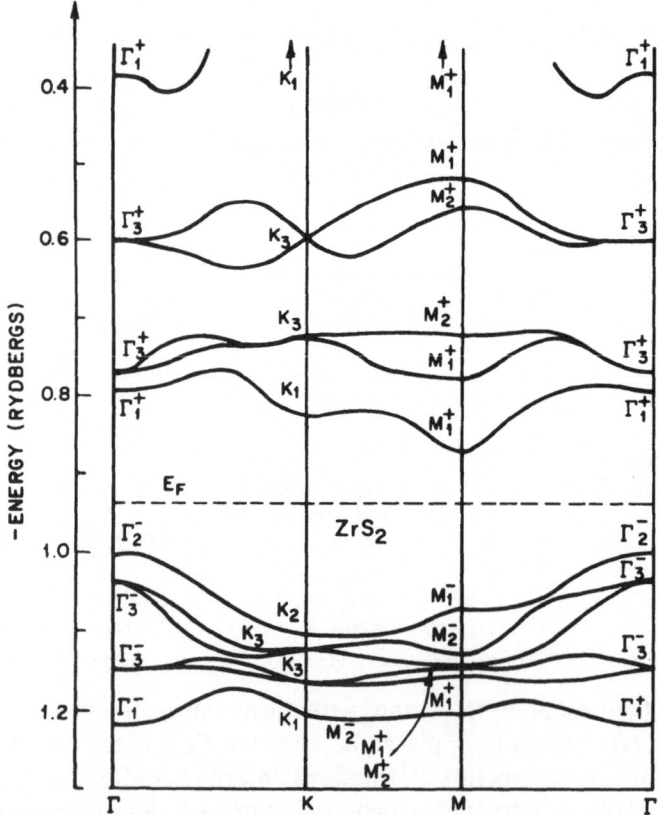

Fig. 14. Band structure of ZrS_2 as obtained from ETB calculations of [20].

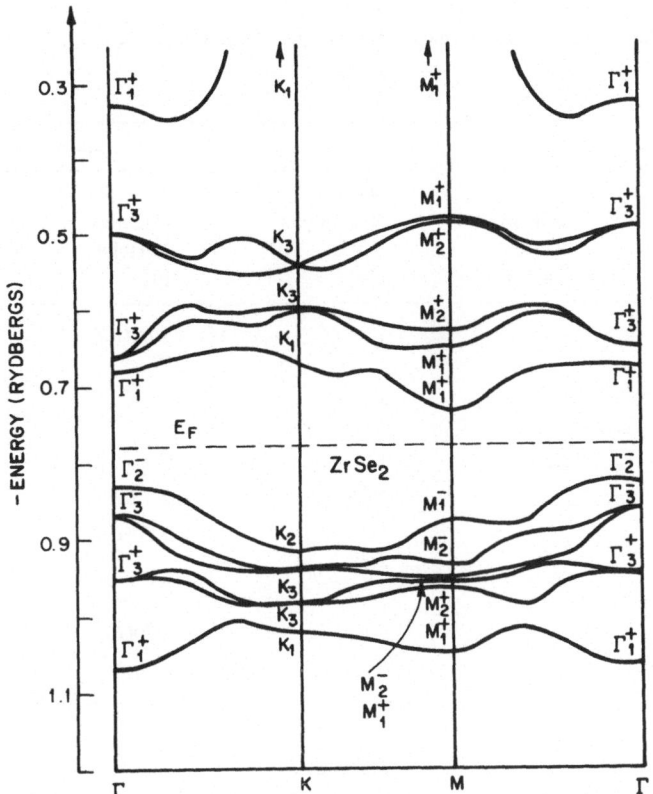

Fig. 15. Band structure of ZrSe₂ as obtained from ETB calculations of [20].

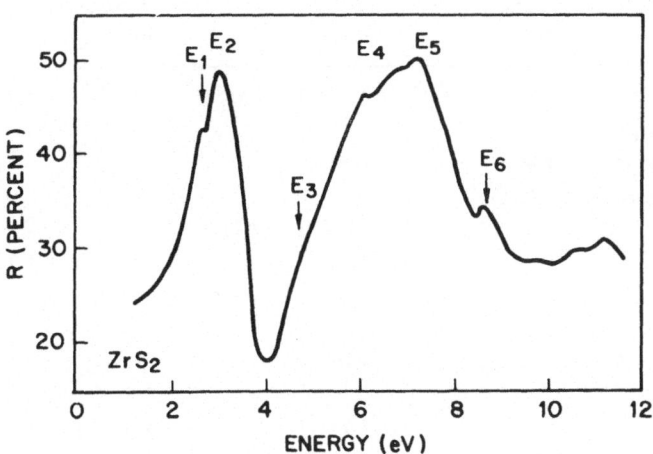

Fig. 16. Reflectivity spectrum of ZrS₂ as measured in [6].

TABLE 5

Comparison between experimental and theoretical results for ZrS_2

| | ZrS$_2$ | | | |
| | Theory | | Experiment | |
Authors energy (eV) features	Wilson Yoffe [1]	Murray Bromley Yoffe [20]	Greenaway Nitsche [6]	Beal Knights Liang [7]
Indirect gap	1.74	1.84 $(\Gamma_2^- \rightarrow M_1^+)$	1.68(E_0)	
Width of the lower nonbonding d-bands	1.74	2.25		
Width of the uppermost p-bands	0.99	1.36 $(K_2 \rightarrow \Gamma_2^-)$		
$M_1^- \rightarrow M_1^+$		2.72	2.75(E_1)	2.73
$\Gamma_2^- \rightarrow \Gamma_3^+$		3.3	3.00(E_2)	2.99
$\Gamma_3^- \rightarrow \Gamma_1^+$		3.35		
$M_1^- \rightarrow M_1^+$		4.08		3.93
$\Gamma_2^- \rightarrow \Gamma_3^+$		5.44	4.70(E_3)	
$\Gamma_3^- \rightarrow \Gamma_3^+$		6.01	6.10(E_4)	

TABLE 6

Comparison between experimental and theoretical results for $ZrSe_2$

| | ZrSe$_2$ | | | |
| | Theory | | Experiment | |
Authors energy (eV) features	Wilson Yoffe [1]	Murray Bromley Yoffe [20]	Greenaway Nitsche [6]	Beal Knights Liang [7]
Indirect gap	1.05	1.22 $(\Gamma_2^- \rightarrow M_1^+)$		
Width of the lower nonbonding d-bands	1.80	1.90		
Width of the uppermost p-bands	0.93	1.16 $(K_2 \rightarrow \Gamma_2^-)$		
$M_1^- \rightarrow M_1^+$		1.96	2.00(E_1)	1.97
$\Gamma_2^- \rightarrow \Gamma_3^+$		2.3	2.60(E_2)	2.36
$\Gamma_3^- \rightarrow \Gamma_1^+$		2.58		
$M_1^- \rightarrow M_1^+$		3.10		
$\Gamma_2^- \rightarrow \Gamma_3^+$		4.49	4.30(E_3)	
$\Gamma_3^- \rightarrow \Gamma_3^+$		5.03	5.50(E_4)	

included. Assignment of the structure in the observed spectra and a comparison with the schematic banding models are given in Tables 5 and 6. As compared to the Ti dichalcogenides, the respective contributions to the E_2 peak originate from different transitions. However, both structures involve transition between the bonding p-like and non-bonding d-like bands. Thus the consistency in the assignments of structures observed for Ti and Zr compounds is maintained.

The series of ETB calculations on group IVB transition metal dichalcogenides [19, 20] also includes results for HfS_2 and $HfSe_2$. In addition the EPM method [21] and the APW method [22] are applied to HfS_2. The APW calculations of [22] involve ad-hoc crystal potentials that are derived from atomic Hartree-Fock-Slater charge densities [23]. Neutral atomic configurations are assumed and the Slater $\alpha = \frac{2}{3}$ exchange parameter is used. Muffin-tin corrections outside the spheres are taken into account by perturbation.

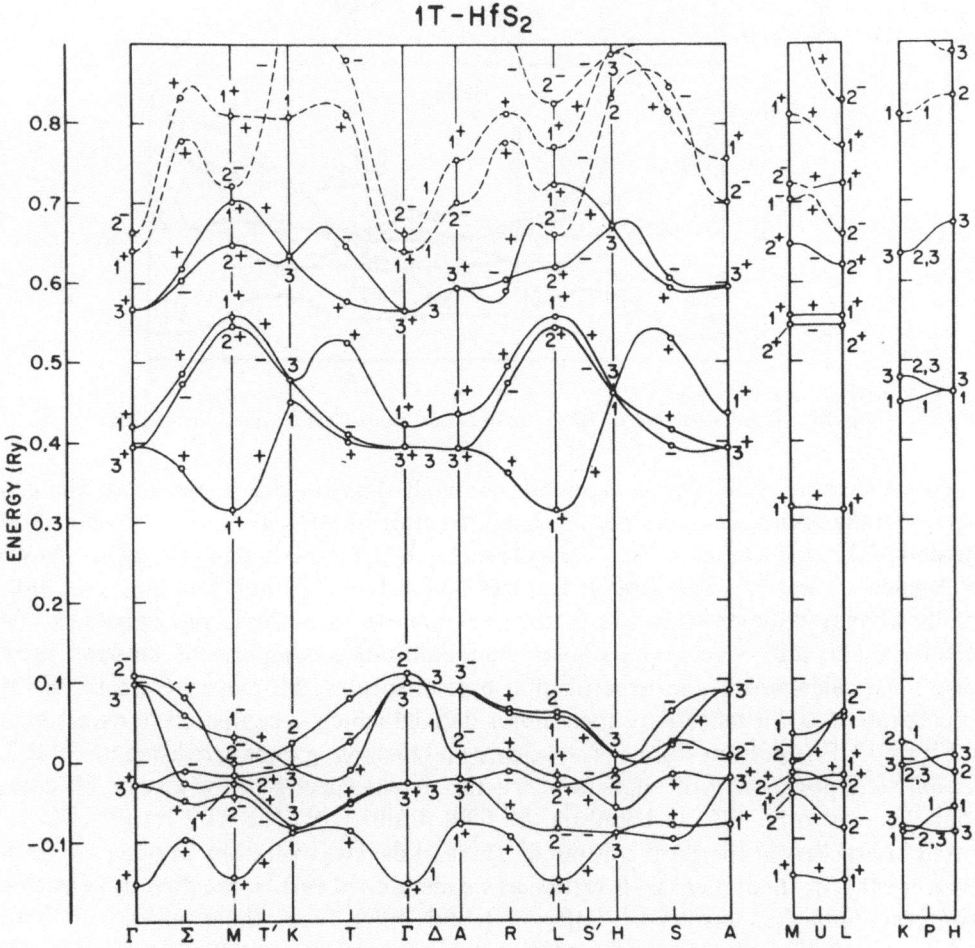

Fig. 17.   Band structure of HfS_2 as obtained from the APW calculations of [22].

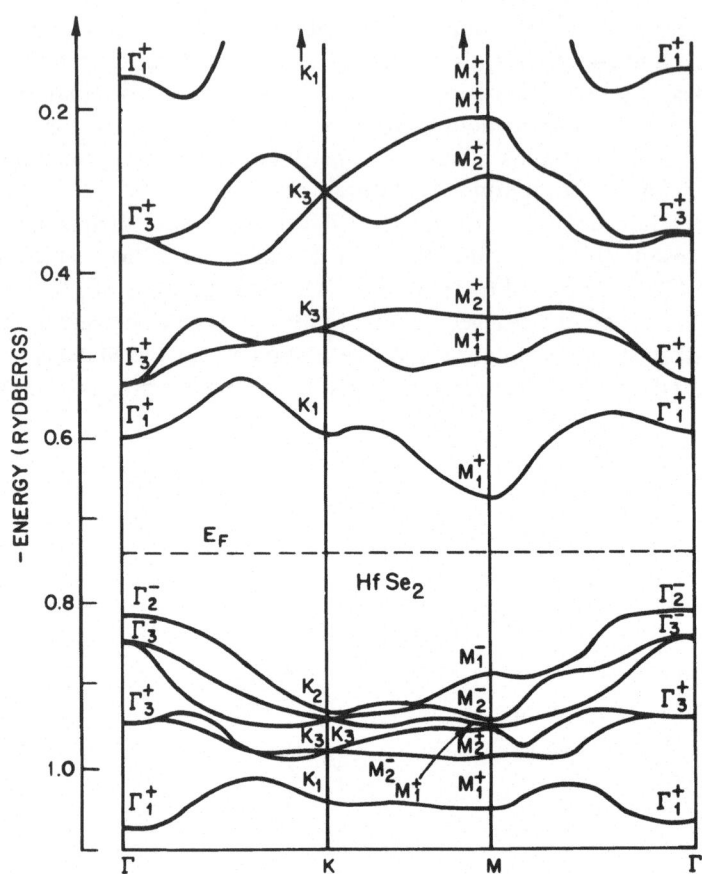

Fig. 18.   Band structure of HfSe$_2$ as obtained from ETB calculations of [20].

No calculation on HfTe$_2$ is reported; we shall thus restrict ourselves to the discussion of the sulfide and selenide. The band structure of HfS$_2$ as obtained by the APW method of [22] and that of HfSe$_2$ as calculated by the ETB method of [19, 20] are shown in Figures 17 and 18. The general features of the bonding, non-bonding, and antibonding bands in these two band structures resemble those of the Ti and Zr sulfides and selenides. This allows to make similar semiquantitative comparisons between these theoretical calculations and experimental optical spectra. Moreover, from the EPM calculation of [21] a reflectivity spectrum is derived which is compared to experiment in Figure 19. Energetic positions of structures and their assignment are listed in Table 7. A consistent assignment of important structure in the spectra of the Zr and Hf compounds seems to be difficult based on the ETB results [20]. The ETB results of HfS$_2$ (listed in brackets in the third column of Table 7) deviate overall by about 1 eV from experiment. The theoretical ETB values, as we mentioned earlier, are directly extracted from the published results of HfS$_2$ [20] and differ from those given in [7] by 0.7 eV. If the d-bands are rigidly shifted down by 1.0 eV, energy gaps as given in Table 7 (without brackets) are obtained. Except for the $M^- \rightarrow M_1^+$ transition the agreement between the ETB results and experiment is now greatly improved. A similar

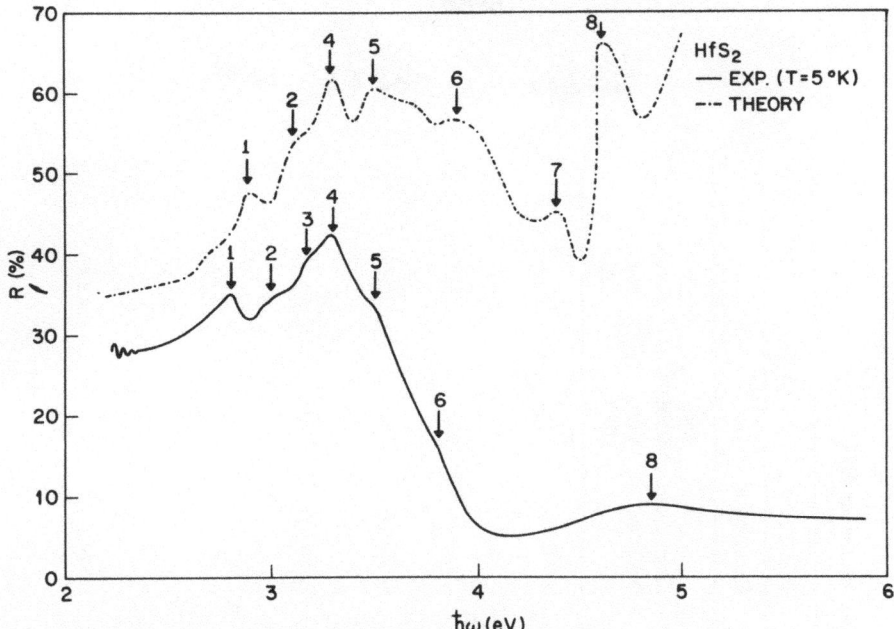

Fig. 19. Experimental and theoretical (EPM) reflectivity of HfS₂ as given in [21].

shift of about 0.7 eV is expected in $HfSe_2$ (Table 8). The reflectivities [6] of the two Hf compounds show some features (relative width of the E_1, E_2 structures) not in agreement with transmission spectra [7] which may question any attempt of detailed assignment.

If, however, the tentative identifications of spectral structures to the results of [20] are correct, the APW calculation seems to give a separation between the p and the non-bonding d bands 1 eV wider than experimentally obtained. As also seen in the other 3-dimensional calculations (e.g. for TiS_2) interlayer interactions have little effect on the non-bonding d-states in HfS_2. The p_z bands ($\Gamma_1^+ - A_1^+$; $\Gamma_2^- - A_2^-$), however, are largely influenced by this interaction. Furthermore, the APW results for HfS_2 show considerable hybridization between the metal s, p states and the second group of metal d-states. Hybridization effects of this kind are found in the OPW results but not in the KKR results for TiS_2. Such differences may originate from the different atomic binding energies of the s and d states for the elements Ti and Hf. From atomic calculations [23] we find the difference in $s - d$ binding energy for Ti to be 2.31 eV which is about three times more than that for Hf.

From the discussions given above, one realizes that comparison between theory and experiment for the group IVB transition metal dichalcogenides is generally still in a rather qualitative stage. Since calculations of reflectivities including the **k**-dependent matrix elements are available only in the case of HfS_2 (EPM) [21], only limited quantitative conclusions can be drawn at the time from the data presented in Tables 2 to 8. More theoretical work is certainly needed to get a better understanding of physical properties of these compounds.

TABLE 7

Comparison between experimental and theoretical results for HfS$_2$

HfS$_2$

Authors energy (eV) features	Theory				Experiment		
	Wilson Yoffe [1]	Fong et al. [21]	Murray Bromley Yoffe [20]	Mattheiss [22]	Greenaway Nitsche [6]	Beal Knights Liang [7]	Fong et al. [21]
Indirect gap	1.98	1.80 ($\Gamma_2^- \to M_1^+$)	2.08 (2.70)	2.77 ($\Gamma_2^- \to L_1^+$)	1.96(E_0)		
Width of the lower nonbonding d-bands	1.98	4.00 ($L_1 \to L_1$)	3.14 ($M_1^+ \to M_2^+$)	3.31 ($M_1^+ \to M_1^+$)			
Width of the uppermost p-bands	0.87	3.1 ($H_1 \to \Gamma_2^-$)	1.56 ($M_2^- \to \Gamma_2^-$)	1.54 ($M_2^- \to \Gamma_2^-$)			
$L_1^- \to L_1^+$		2.90 (volume $8 \to 9$ near Δ)	2.81 (3.54)	3.55	2.90(E_1)	2.88	2.82
$\Gamma_3^- \to \Gamma_1^+$		3.30 (volume $8 \to 9$ near R)	3.08 (4.08)	4.38	3.35(E_2)	3.20	3.30
$M_1^- \to M_1^+$		3.50	4.85 (5.85)	7.12			3.50
		3.90 (volume $7,8 \to 9$)					3.80
$\Gamma_2^- \to \Gamma_3^+$		4.60 (volume)	5.80 (6.80)	6.22	5.00(E_3)		4.90
$\Gamma_3^- \to \Gamma_3^+$			6.21 (7.21)	6.38			

TABLE 8

Comparison between experimental and theoretical results for $HfSe_2$

| | $HfSe_2$ | | | |
| | Theory | | Experiment | |
Authors energy (eV) features	Wilson Yoffe [1]	Murray Bromley Yoffe [20]	Greenaway Nitsche [6]	Beal Knights Liang [7]
Indirect gap	1.05	1.90	$1.13(E_0)$	
Width of the lower nonbonding d-bands	1.43	3.10		
Width of the uppermost p-bands	0.93	1.60		
$M_1^- \to M_1^+$		2.96	$2.20(E_1)$	2.28
$\Gamma_3^- \to \Gamma_1^+$		3.40	$2.80(E_2)$	2.76
$M_1^- \to M_1^+$		5.20		
$\Gamma_2^- \to \Gamma_3^+$		6.25	$4.40(E_3)$	
$\Gamma_3^- \to \Gamma_3^+$		6.73		

In the following we shall present a few simple concluding remarks about relationships deduced from the theoretical and experimental results obtained for the group IVB transition metal dichalcogenides.

It is of interest to study the effect of atomic size on the physical properties of this group of crystals. It is argued that the metallic behavior of $TiTe_2$ may be caused by the larger ionic size of Te as compared to the other two chalcogenides. In Figure 20(a) the indirect fundamental gap of the group IVB transition metal sulfides and selenides is plotted as a function of the anion radii [24]. The energies of the Ti and Zr series are taken from the ETB calculations of [20] whereas experimental values [6] are used for the Hf-series. As a general trend, the fundamental gap decreases as the size of the anion increases. The effect is about equally strong for all group IVB compounds and mainly results from a broadening of the anion p-bands. The width of the non-bonding d-bands may in a similar way be correlated to the cation size.

Since the core of Ti does not have any d-states the $3d$ electrons experience a stronger attractive potential in the vicinity of the nucleus than do the $4d$ and $5d$ electrons Zr and Hf respectively. A plot of the width of the lower non-bonding d-bands versus the cationic radii (core sizes) provides information about the strength of the potential acting on the d-electrons. In Figure 20(b) the lower d-band widths, obtained by the ETB method, are plotted against the cationic radii of the transition metal ions as given in [24]. Obviously,

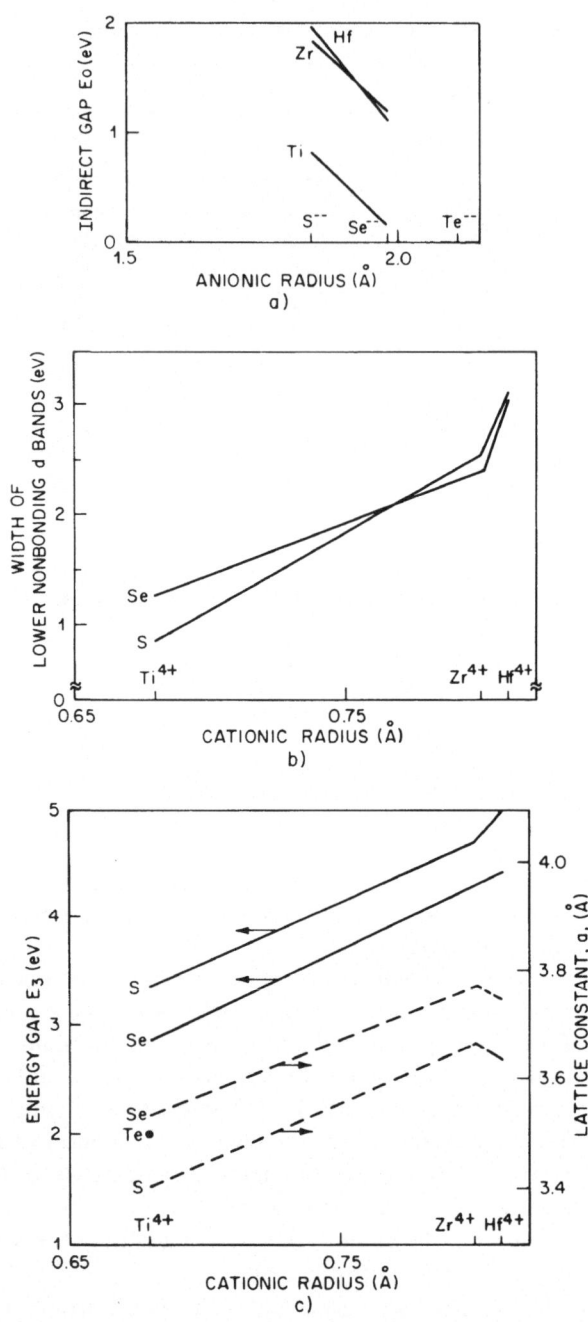

Fig. 20.   (a) Plot of calculated indirect energy gaps of the group IVB transition metal compounds vs ionic radii of the cholcogenides. (b) Plot of calculated widths of the lower non-bonding d-bands vs cation radii for group IVB transition metal sulfides and selenides. (c) Plot of the experimental energies of the E_3 structure vs cation radii for the group IVB transition metal sulfides and selenides.

the strength of the potential acting on the d-states follows the order: Ti > Zr > Hf. However, the d-band width of the Ti-compounds seems to be too large. The plot of Figure 20b also seems to indicate the limit of validity of standard band structure calculations for compounds containing $3d$ transition metals because of the strong correlation effects associated with localization [18].

In all these compounds, one of the mechanisms responsible for the bonding is the overlap-covalency effect between the cation d-states and the anion p-states. To obtain a measure of the relative amount of ionicity involved in the bonding, an averaged bonding-antibonding gap combined with the bond length may be considered. Comparing the theoretical results with the experimental data, the structure E_3 (Table 3) was identified as resulting from the transition $\Gamma_3^- \to \Gamma_3^+$ representing the minimum energy gap between the bonding and antibonding bands. In Figure 20c we plot for the sulfides and selenides the values of E_3 [6] vs the cation radii (solid lines). The dotted lines show the corresponding lattice constants, as a function of the ionic radii. The plots seem to suggest that the overlap-covalency effect contributes equally to the bonding in the Ti-compounds and the Zr and Hf-compounds. No difference in ionicity between the Ti and the Zr, Hf dichalcogenides would therefore imply a constant energy of the E_3 structure. However, from the 1.5 eV shift in the position of the E_3 structure one may thus conclude that the Hf and Zr compounds are more ionic than the Ti-dichalcogenides, which is consistent with the results of [12] shown in Figure 2, and with a list of electro-negativity differences for various structures [25] (see Table 9).

It is further of interest to compare the electronic structure of the semiconducting transition metal dichalcogenides with octahedral coordination to that of *non-*transition metal compounds with the same crystal structure, such as SnS_2 and $SnSe_2$. The most striking difference, of course, appears in the conduction bands. In Sn dichalcogenides the character of the conduction bands is $s - p$ like, whereas in the transition metal compounds, the d-states of the cation constitute the lowest conduction bands. Theoretical valence electron charge distributions (which shall be presented later in this section) of the non-transition metal semiconductors SnS_2 and $SnSe_2$ show that the bonding states at the chalcogenides form to a large extent sp^3 type orbitals, with three lobes pointing to the neighboring metallic ions and one lobe forming a lone pair in the van der Waals region. The OPW charge densities for TiS_2, in contrast, do not show this pronounced behavior, though significant covalent contributions to the bonding are reported in [13]. As we shall see later, charge distributions around the anions in trigonal prismatic $NbSe_2$ also have somewhat different configurations. It would therefore be of interest, to investigate the effect of the presence of (empty) transition metal d-states on the bonding properties of crystals with trigonal prismatic or octahedral coordination. Ultimately, the results could serve as a check on the various bonding models proposed in [1, 9–12].

1.1.2. *The Group VB Transition Metal Dichalcogenide* TaS_2

TaS_2 is found in *both* types of coordination and numerous polytypes of stacking sequences. The trigonal prismatic 2H–TaS_2 will be discussed in the next section. The octahedral 1T–TaS_2 is stable for $T \geq 1050$ K but also may be found at lower temperature by quenching. Measurements of resistivity and magnetic susceptibility on 1T–TaS_2 have revealed two phase transitions at $T = 200$ K and 315 K [26]. The metallic behavior

TABLE 9

List of electronegativity differences (ΔX), fractional ionic bond character (f_i), structure (s), lattice parameters (a, c) and bond lengths (d) for various layer compounds after [25]

	ΔX	f_i	S	a (Å)	c (Å)	$d(MX)$[a] (Å)
1. HfS$_2$	1.2	0.30	O	3.635	5.837	2.56
2. ZrS$_2$	1.1	0.26	O	3.662	5.813	2.56
3. HfSe$_2$	1.1	0.26	O	3.748	6.159	2.66
4. TiS$_2$	1.0	0.22	O	3.405	5.690	2.42
5. ZrSe$_2$[b]	1.0	0.22	O	3.770	6.138	2.66
6. 2H–NbS$_2$	0.9	0.19	T	3.31	2 × 5.945	2.47
7. 2H–TaS$_2$	0.9	0.19	T	3.315	2 × 6.05	2.48
1T–TaS$_2$	0.9	0.19	O	3.36	5.90	2.44
8. TiSe$_2$	0.9	0.19	O	3.535	6.004	2.54
9. VS$_2$[b]	0.8	0.15	O	3.29	5.66	2.37
10. 2H–WS$_2$	0.8	0.15	T	3.154	2 × 6.181	2.41
11. 2H–NbSe$_2$	0.8	0.15	T	3.442	2 × 6.27	2.595
12. 2H–TaSe$_2$	0.8	0.15	T	3.437	2 × 6.362	2.59
1T–TaSe$_2$	0.8	0.15	O	3.477	6.272	2.55
13. HfTe$_2$[b]	0.8	0.15	O	3.949	6.651	2.82
14. 2H–MoS$_2$	0.7	0.12	T	3.160	2 × 6.147	2.42
15. VSe$_2$[b]	0.7	0.12	O	3.352	6.104	2.47
16. 2H–WSe$_2$	0.7	0.12	T	3.286	2 × 6.488	2.51
17. ZrTe$_2$[b]	0.7	0.12	O	3.950	6.630	2.82
18. 2H–MoSe$_2$	0.6	0.09	T	3.288	2 × 6.460	2.49
19. TiTe$_2$	0.6	0.09	O	3.766	6.491	2.717
20. NbTe$_2$	0.5	0.06	'O'			
21. TaTe$_2$	0.5	0.06	'O'			
22. VTe$_2$	0.4	0.04	O	3.6	6.45	2.67
23. WTe$_2$	0.4	0.04	'O'			
24. 2H–MoTe$_2$	0.3	0.02	T	3.517	2 × 6.983	2.73

[a] In only a few cases are accurate bond distances available from refined crystal structures. The z parameter obtained for the chalgogens (z chalcogen–z metal) in the trigonal prismatic compounds varied in these cases from 0.130 to 0.132. Where refinements were not available we used 0.131. For octahedral coordination, the z value, $\frac{1}{4}$, was used.
[b] Metal rich.

of the crystal has been observed for temperatures above 315 K. Recent experimental evidence [2] seems to indicate that the phase transition at 315 K can be attributed to a distortion of the periodic structure and a formation of superlattices induced by charge density waves. These charge density waves in turn seem to be closely related to the topology of the Fermi surface. Room temperature transmission spectra [1] and photo-emission data [27] suggest that 1T–TaS$_2$ is metallic or a small gap semiconductor. The occupied d-band width obtained by photoemission is about 1.5 eV and the sulphur $3p$ valence band width is about 6–7 eV. There may be a slight overlap between the bonding p-bands and the d-bands. Shown in Figure 21 are the results of a band structure calculation based on the APW method [22] referring to the high temperature phase of 1T–TaS$_2$. Compared with the group IVB transition metal compounds, qualitative features of the

1T–TaS$_2$

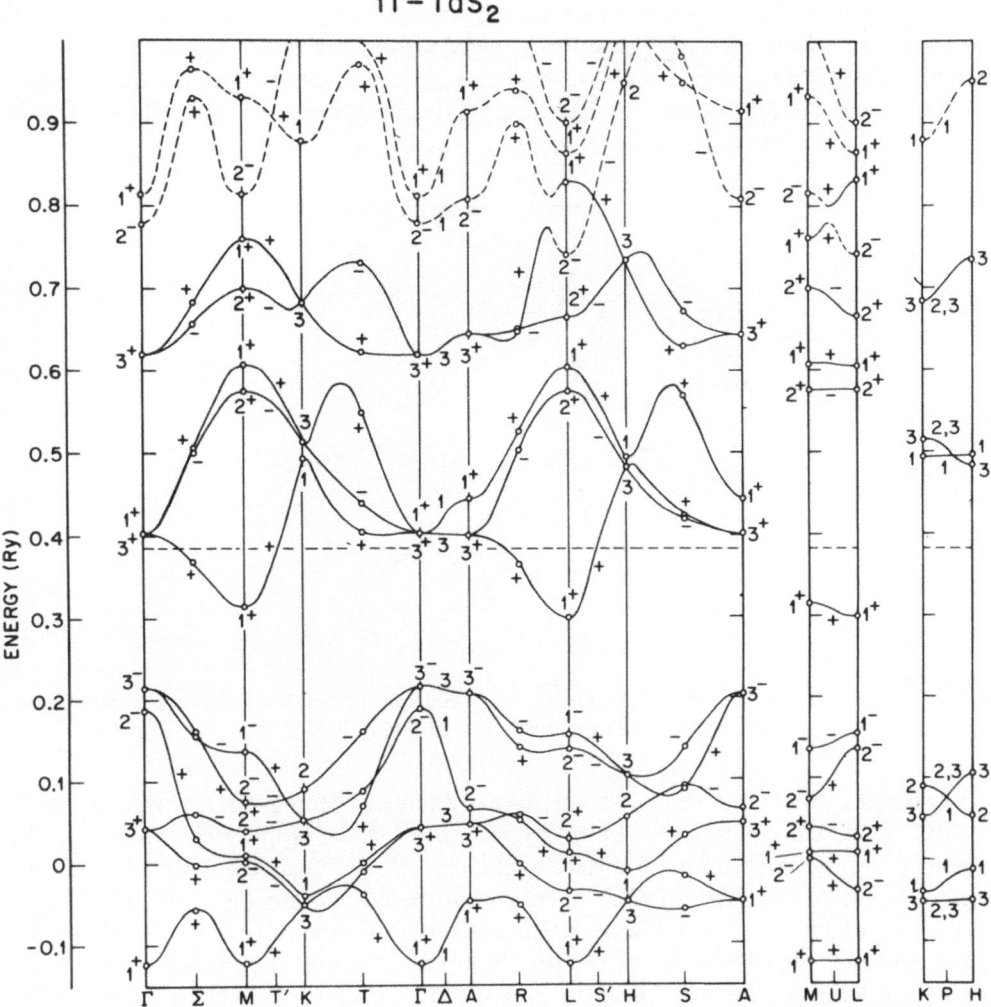

Fig. 21.   Band structure of 1T–TaS$_2$ as obtained from APW calculations of [22].

band structures are very similar. Due to an extra electron in 1T–TaS$_2$, the non-bonding
d-bands are partially occupied. The Fermi energy, E_f is indicated by the dashed line in
Figure 21. Metallic behavior is predicted from the band structure. Very similar results
are obtained in KKR calculations for TaS$_2$ and TaSe$_2$ [28]. The width of the occupied
d-states in the APW calculation is 1.16 eV (1.19 eV for KKR) which differs from the
experimental value by about 0.44 eV. However, a gap of 1.17 eV (APW) and 2.5 eV
(KKR) between the p and d states below E_f is found in the calculations which does not
agree with experiment [27]. The calculated p-band width is 4.57 eV (APW) which is
about 1.5 eV narrower than the photoemission results. The Fermi surface determined
from the APW [22] band structure plotted in the ΓKM plane in the hexagonal Brillouin
zone is shown in Figure 22. Since the energies along ML, below E_f are almost dis-
persionless the Fermi surface sections around M extend nearly cylindrically along

$\pm k_z$. The cylinders are spanned by a vector $\mathbf{q}_0$ indicated in Figure 22. It is argued that the wave vector dependent susceptibility $x(\mathbf{q})$ may show divergent behavior for $\mathbf{q} = \mathbf{q}_0$ which would drive a charge density wave with $\mathbf{q}_0$ period. More detailed qualitative studies are presented in [28]. Recent observations of these phenomena and qualitative theoretical explanations are compiled in a review article in [3].

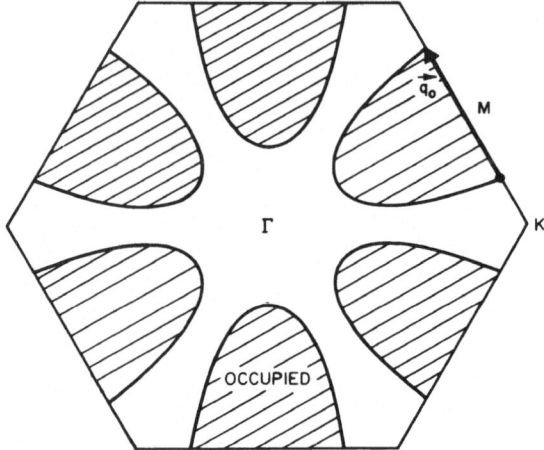

Fig. 22.   Schematic plot of Fermi surface of $1T–TaS_2$ intersecting the ΓKM plane as obtained from the APW bandstructure of [22] shown in Figure 21.

A detailed analysis in terms of atomic-like orbitals has been given for the d-bands in $1T–TaS_2$ [22]. The splitting of the $5d$ bands into non-overlapping sub-bands can be explained on the basis of crystal field arguments [1] given at the beginning of this section. In the calculations of [22] the lower sub-band has thus been identified as t_{2g} with d_{z^2}, d_{xy} and $d_{x^2-y^2}$ type orbitals. Because of the close proximity of the metal $6s - 6p$ bands, stronger hybridization occurs for the upper two-fold degenerate e_g sub-band originating from d_{xz} and d_{yz} orbitals. A Slater-Koster type LCAO fit to the APW d-bands of $1T–TaS_2$ as given in [22] is shown in Figure 23. In Figure 23a the ligand-field levels are related to the LCAO density of states curves for the corresponding sub-bands, neglecting interband hybridization and interlayer interaction. In Figure 23b the effect of interband hybridization is included. Finally, in Figure 23c the results for three-dimensional $1T–TaS_2$ are given. One readily observes the minor role played by the interlayer interaction in the case of the TaS_2 d-bands.

Direct experimental evidence on the band structure of $1T–TaS_2$ and $1T–TaSe_2$ is available from angular resolved photoemission studies [29, 30, 31]. In Figure 24 photoelectron spectra are shown for $1T–TaS_2$ taken at various polar angles θ of emission. The azimuthal angle φ was set to select only those photoelectrons propagating in a plane spanned by the ΓMLA points in the hexagonal Brillouin zone. The observed energy shifts of structure as a function of θ can directly be related to the band structure from which the photoelectrons originate. In a simple model which neglects final state effects and assumes $\mathbf{k}$-conserving optical transitions as well as $k_{\parallel}$ ($\parallel$ to surface of emission) conservation of the photoelectrons, $E(k_{\parallel})$ spectra may be plotted [29, 30].

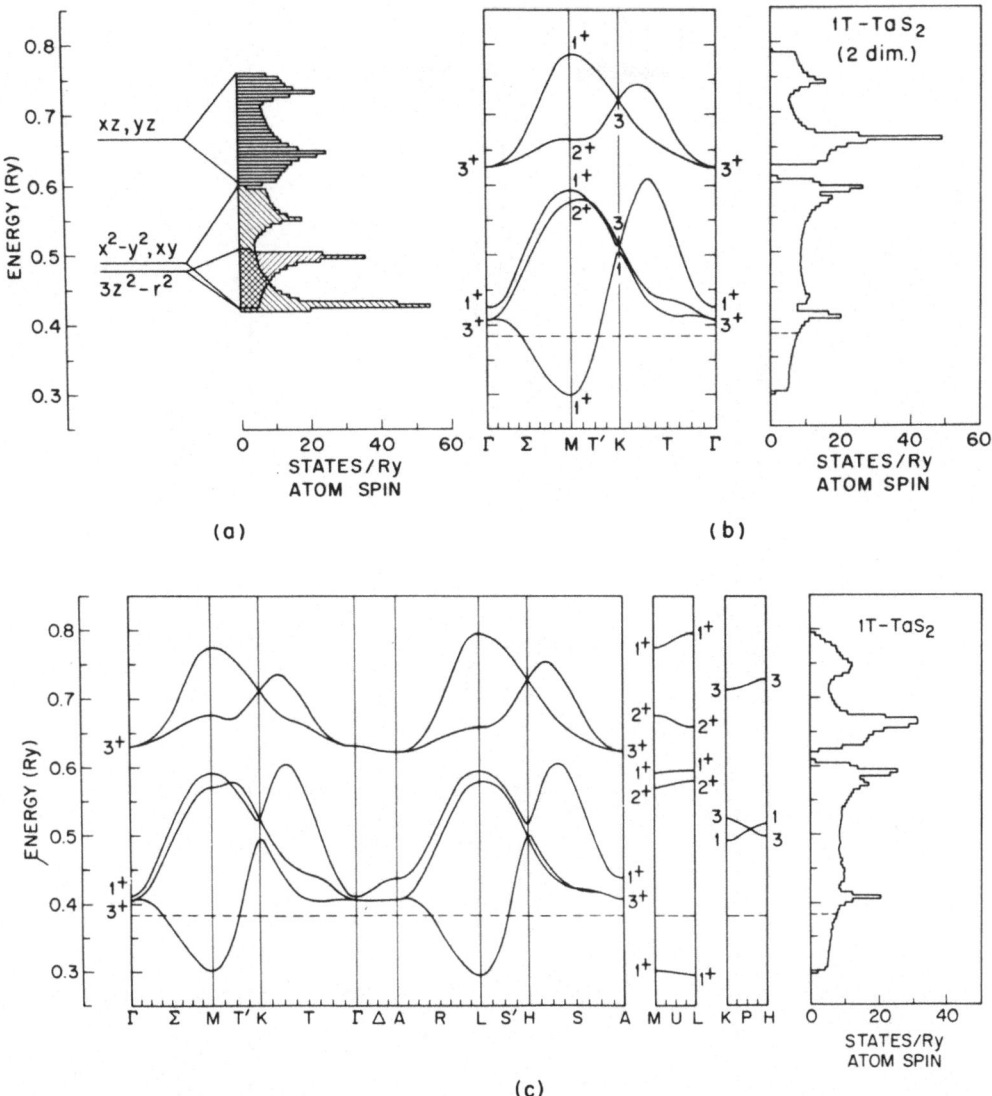

Fig. 23. (a) Ligand-field levels and LCAO density of states histograms for unhybridized d-bands in 1T–TaS$_2$. LCAO results for 1T–TaS$_2$ with hybridization first neglecting (b) and then including (c) layer interactions (from [22]).

The incident light is of sufficiently low energy for electrons to remain within the first Brillouin zone and to avoid umklapp processes. Assuming further $k_\perp$-independence of E (negligible interlayer interaction) an experimental band structure as shown in Figure 25 (dots and circles) may be derived. In Figure 25c the theoretical APW-results for 1T–TaS$_2$ of [22] are displayed which show reasonable agreement with the experimental data if the $p - d$ gap is reduced by about 0.8 eV.

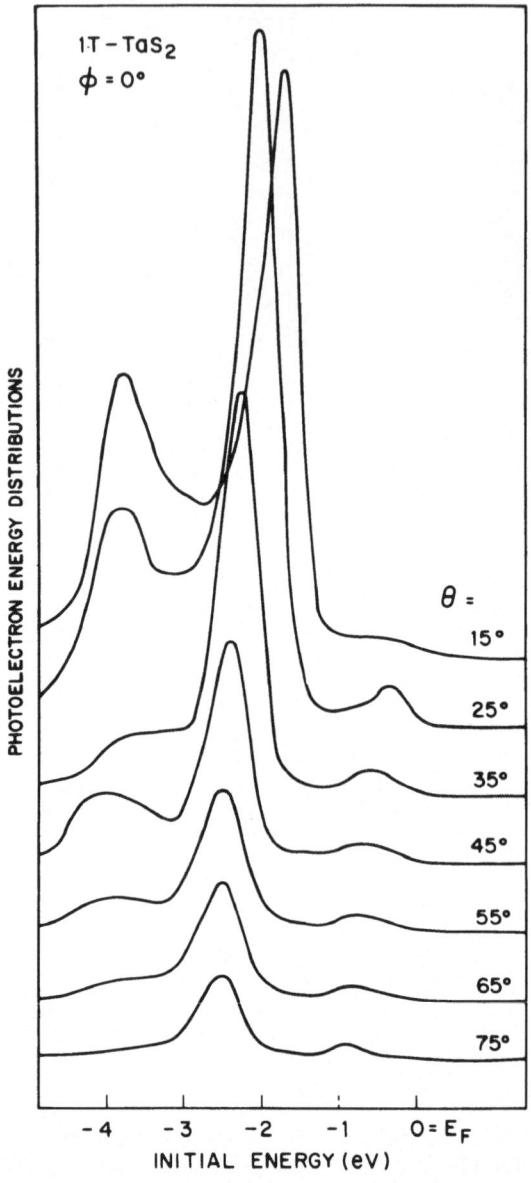

Fig. 24.   UV-photoelectron spectra on 1T–TaS$_2$ at various polar angles of emission (after [29]).

Photoemission studies for fixed polar angle θ and variable azimuthal angle φ may yield information about orbital symmetries [30]. Experimental results for $\theta = 55°$ and variable φ are given in Figure 26. The outer radial plot (open circles) corresponds to photoemission intensities as a function of φ for photoelectrons with binding energies between 0 and −4 eV (i.e. mainly the anion p-electrons). The three-fold rotational symmetry of the crystal structure is clearly demonstrated. The inner curve (full circles) represents the intensity of photoelectrons with binding energies between 0 and

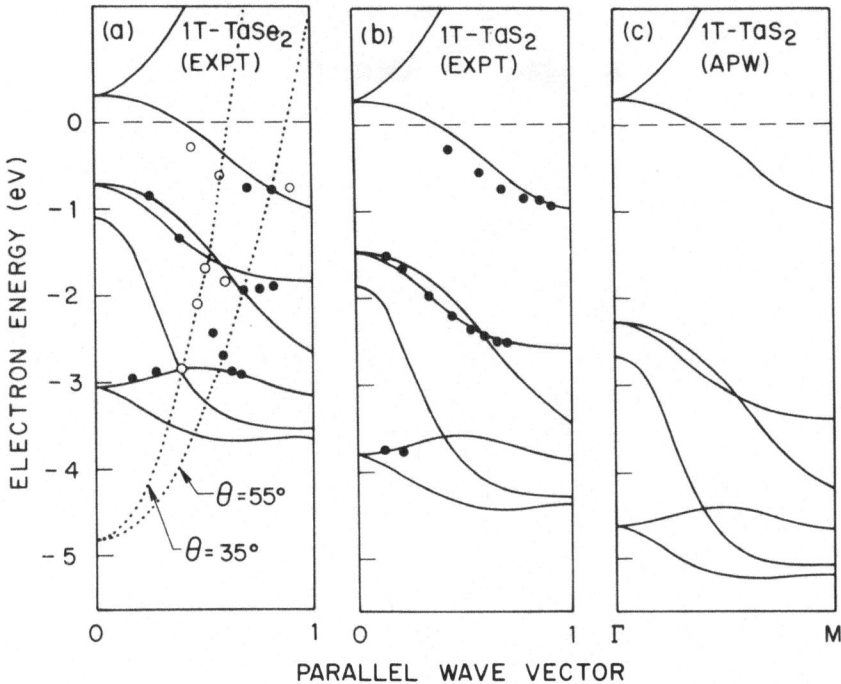

Fig. 25. Electron energy versus parallel wave vector: (a) experimental data for 1T–TaSe₂ (full and open circles) after [29]; (b) experimental data for 1T–TaS₂ (full circles) after [29]; (c) APW results of 1T–TaS₂ after [22].

-1 eV (i.e. the cation d-electrons). Referring these results to the orientation of the crystal one concludes that the atomic (non-bonding) Ta d-orbitals must pile up their *momentum* densities in directions between the nearest neighbor anions rather than along anion-cation bonds [30].

These examples show, how angular-resolved photoemission experiments can be used to obtain detailed information on the momentum distribution of valence electrons. It is, however, not clear yet how much of the observed anisotropic pattern results from the properties of the final states in the optical transition and how much is due to initial state orbital effects. It is desirable to find an energy range in which it is reasonable to approximate the final states by plane waves. In that case the photoemission intensity is proportional to the momentum density of the initial state wave function. Related information is available from Compton scattering and positron annihilation studies, which also measure some section of the electronic momentum density. Layer compounds may be suitable objects for such studies because of the quasi two-dimensional properties of their electronic states.

1.2. COMPOUNDS WITH TRIGONAL PRISMATIC ARRANGEMENT OF THE CATIONS

Layer structures with trigonal prismatic cation-coordination are found for group VB and group VIB transition metal dichalcogenides. The transition to the octahedrally coordinated structure occurs for Ta compounds which exist in both structural types. Unlike the octahedrally coordinated structures, trigonal prismatic structures contain

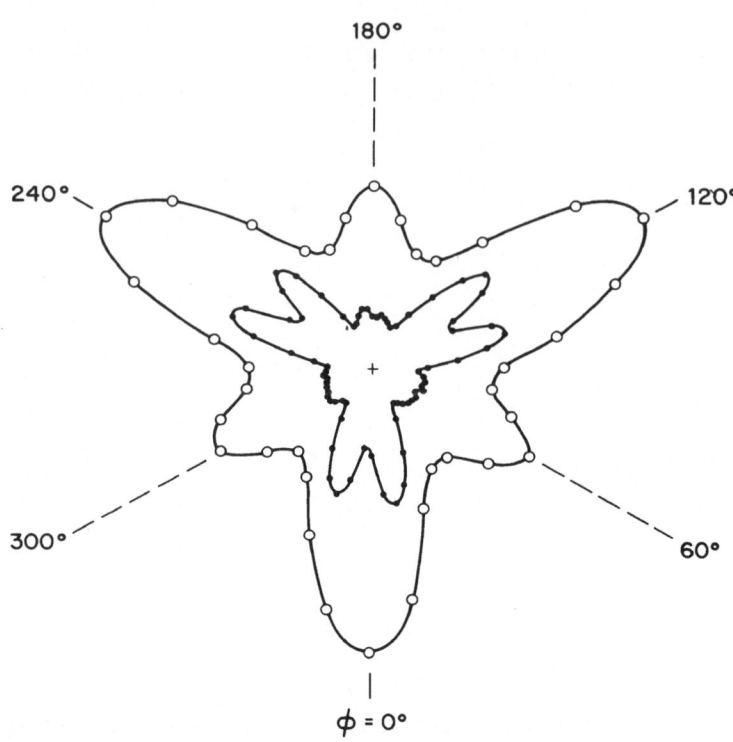

Fig. 26. Radial plots of the azimuthal dependence at $\theta = 55°$ of the total UV photoemission intensity of 1T–TaSe$_2$ (open circles) and on a 5 times expanded scale the Ta–d emission intensity (full circles) (after [30]).

at least two layers per hexagonal unit cell which complicates the discussion of the electronic structure. For some polytypes (D_{3d}^5) a three-layer hexagonal unit cell may be replaced by a one-layer rhombohedral unit cell (see Chapter A). The structural complications for trigonal prismatic layers, of course, originate from the close packing of anions of neighboring layers.

The most common polytype, the two-layer (2H) hexagonal structure, belongs to the non-symmorphic space group D_{6h}^4, which has extensively been discussed in Chapter A. Due to the trigonal prismatic symmetry, the D_{6h}^4 structures may further be subdivided into two distinct polytypes corresponding to different stacking sequences. The two structure types (labelled C7 and C27 in Structurberichte) are schematically shown in Figure 27. While the C27 type is met in the group VB compounds, the C7 structure is mostly found for group VIB compounds.

The atomic positions for the C27 structure are specified by

$$M: \pm(0, 0, \tfrac{1}{4})$$
$$X: \pm(\tfrac{1}{3}, \tfrac{2}{3}, \tfrac{1}{8}); \; \pm(\tfrac{1}{3}, \tfrac{2}{3}, \tfrac{3}{8}) \tag{C3}$$

whereas for the C7 structure they are given by

$$M: \pm(\tfrac{1}{3}, \tfrac{2}{3}, \tfrac{1}{4})$$
$$X: \pm(\tfrac{1}{3}, \tfrac{2}{3}, -\tfrac{3}{8}); \; \pm(\tfrac{1}{3}, \tfrac{2}{3}, -\tfrac{1}{8}) \tag{C4}$$

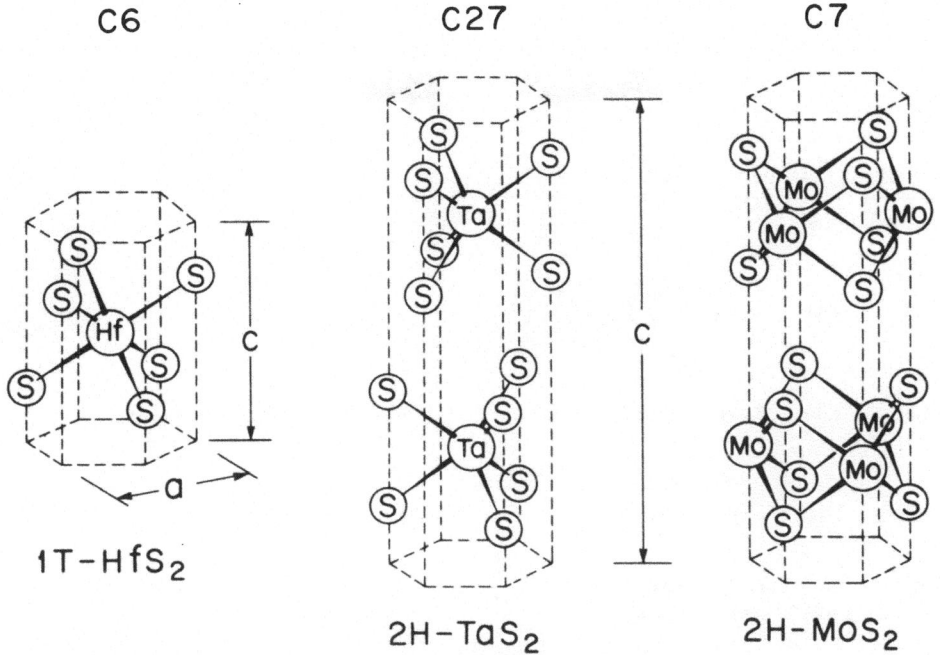

C6   C27   C7

1T-HfS$_2$

2H-TaS$_2$ 2H-MoS$_2$

Fig. 27. Symmetric unit cells for the C27 (2H–TaS$_2$) and C7 (2H–MoS$_2$) crystal structures. Both structures have the space group D_{6h}^4. The C6 (1T–HfS$_2$) structure with D_{3d}^3 symmetry is shown for reference (after [22]).

A list of crystals, their lattice parameters and references to theoretical investigations are given in Table 10 for the (C27) group VB compounds and in Table 11 for the (C7) group VIB compounds.

We shall first summarize important experimental results which are relevant for band structure calculations. Electrical measurements [26, 32] over a wide temperature range show that the group VB compounds are good conductors and undergo superconducting transitions for $T < 10$ K in contrast to their 1T-phases which undergo metal-insulator transitions. If organic molecules, such as pyridine, are intercalated in the 2H structure, the system remains metallic and T_c usually rises. Hall effect measurements for the group VB compounds suggest that both electrons and holes are present at room temperature [32, 33]. From magnetothermal oscillations the form of the Fermi surface in NbSe$_2$ has been deduced to be pancake shaped [34]. UV photoemission measurements have been carried out for NbSe$_2$ [35, 36] and MoTe$_2$ [35], for TaSe$_2$ [31] and for MoS$_2$ [36, 37, 38]. The density of states deduced from these photoemission measurements shall be compared to the theoretical results in the following individual discussions. The optical properties of NbSe$_2$ have been studied in detail in [39] and [40] whereas for the Mo compounds recent transmission and absorption spectra up to 4 eV are presented in [41]. Earlier experimental data up to 1969 are compiled in the review article in [1].

Ligand-field theories [9, 10] predict in the trigonal field a splitting of the five-fold degenerate d-orbitals similar to the octahedral field into e_g and t_{2g} levels with an *additional* splitting of the t_{2g} level into singly degenerate d_{z^2} and doubly degenerate ($d_{x^2 - y^2}$,

TABLE 10

List of group VB transition metal compounds with trigonal prismatic coordination of the cation. The various theoretical approaches are indicated.

Crystals D_{6h}^4(C27)	Lattice constants a(Å)	c/a	Theoretical approaches Schematic approach	APW	EPM	LCMTO
NbS$_2$	3.31	3.592	Wilson, Yoffe [1] Huisman et al. [10] White, Lucovsky [12]			Kasowski [42]
NbSe$_2$	3.44	3.628	WY HJHJ WL	Mattheiss [22]	Fong-Cohen [43]	
2H–TaS$_2$	3.316	3.64	WY HJHJ WL	Mattheiss [22]		
2H–TaSe$_2$	3.436	3.694	WY HJHJ WL			

TABLE 11

List of group VIB transition metal compounds with trigonal prismatic coordination of the cation. The various theoretical approaches are indicated.

Crystals D_{6h}^4(C7)	Lattice constants a(Å)	c/a	Theoretical approaches Schematic approach	APW	ETB	LCMTO	LTM
MoS$_2$	3.16	3.89	Wilson, Yoffe [1] Huisman et al [10] White, Lucovsky [12]	Mattheiss [22]	Bromley et al. [51] Edmondson [52]	Kasowski [42]	Wood, Pendry [54]
MoSe$_2$	3.288	3.924	WY HJHJ WL		BMY		
WS$_2$	3.154	3.92	WY HJHJ WL		BMY		
WSe$_2$	3.286	3.952	WY HJHJ WL		BMY		
MoTe$_2$	3.517	3.968	WY HJHJ WL		BMY		

d_{xy}) levels. The trigonal field thus reduces the spherical symmetry further than the octahedral field. The complete symmetry classification of cation and anion levels in the D_{6h}^4 group is given in Equation (C5):

$$\Gamma_6^+, \Gamma_5^-(d_{xz}, d_{yz})$$

cation d $\quad \Gamma_5^+, \Gamma_6^-(d_{x^2-y^2}, d_{xy})$

$$\Gamma_1^+, \Gamma_4^-(d_{z^2})$$

anion s $\quad \Gamma_1^+, \Gamma_2^-, \Gamma_3^+, \Gamma_4^-(s)$

anion p $\quad \Gamma_1^+, \Gamma_2^-, \Gamma_3^+, \Gamma_4^-(p_z)$

$$\Gamma_5^+, \Gamma_6^-, \Gamma_5^-, \Gamma_6^+(p_x, p_y). \tag{C5}$$

Since the unit cell contains two MX_2 molecules, the orbitals transform according to several (two for M and four for X) representations. The $\pm$ signs refer to spatial inversion.

Similar arguments as presented in Section 1 for octahedrally coordinated compounds are used to establish a simple bonding picture for the trigonal prismatic compounds [1]. The doubly degenerate states d_{xz}, d_{yz} are supposed to hybridize with the p-states of the chalcogenides to form the covalent bond. Due to the differences in electronegativity these states should predominantly be found in the antibonding conduction bands. The $d_{x^2-y^2}$, d_{xy} nonbonding d-states are expected to be at higher energy than d_{z^2} orbitals. The difference between the group VB and group VIB compounds can be seen to originate from the different degree of (rigid) band filling. Thus the group VB compounds are expected to be metallic, whereas the group VIB compounds should exhibit semiconducting properties, provided the lowest non-bonding d_{z^2}-band is flat and separated from the next higher non-bonding d-bands. This band ordering agrees well with that proposed in [10] but is opposite to that of [9].

Alternative band schemes based on information obtained from spectroscopic infrared studies [12] have been suggested for the group VIB dichalcogenides. In these schemes the group VIB compounds are supposed to be strongly covalent materials in which the metal d-states hybridize with its s and p states to form d^4sp orbitals. These orbitals combine with the anion p orbitals in trigonal prismatic arrangements to form bonding and antibonding bands. Non-bonding dp^2 hybrids form the first conduction band in this picture. The group VB Ta compounds are assumed to follow the ionic picture developed for the octahedrally coordinated compounds. In Figure 28 the schematic band models of [1, 10, 12] are shown in (a), (b), and (c) respectively. The most striking difference in the resulting energy band schemes is the appearance of a gap in the model of [1] (Figure 28a) between the lowest (half occupied or filled) d_{z^2}-states and the lower valence bands.

Our next object is to discuss the results of more realistic band structure calculations for the group VB and VIB transition metal dichalcogenides and to compare these results to existing experimental data.

1.2.1. *The Group VB Transition Metal Dichalcogenides*

Theoretical investigations of the electronic properties of this class of crystals have been advanced by several recent band structure calculations. Thus, the method of linear

182

CHAPTER C

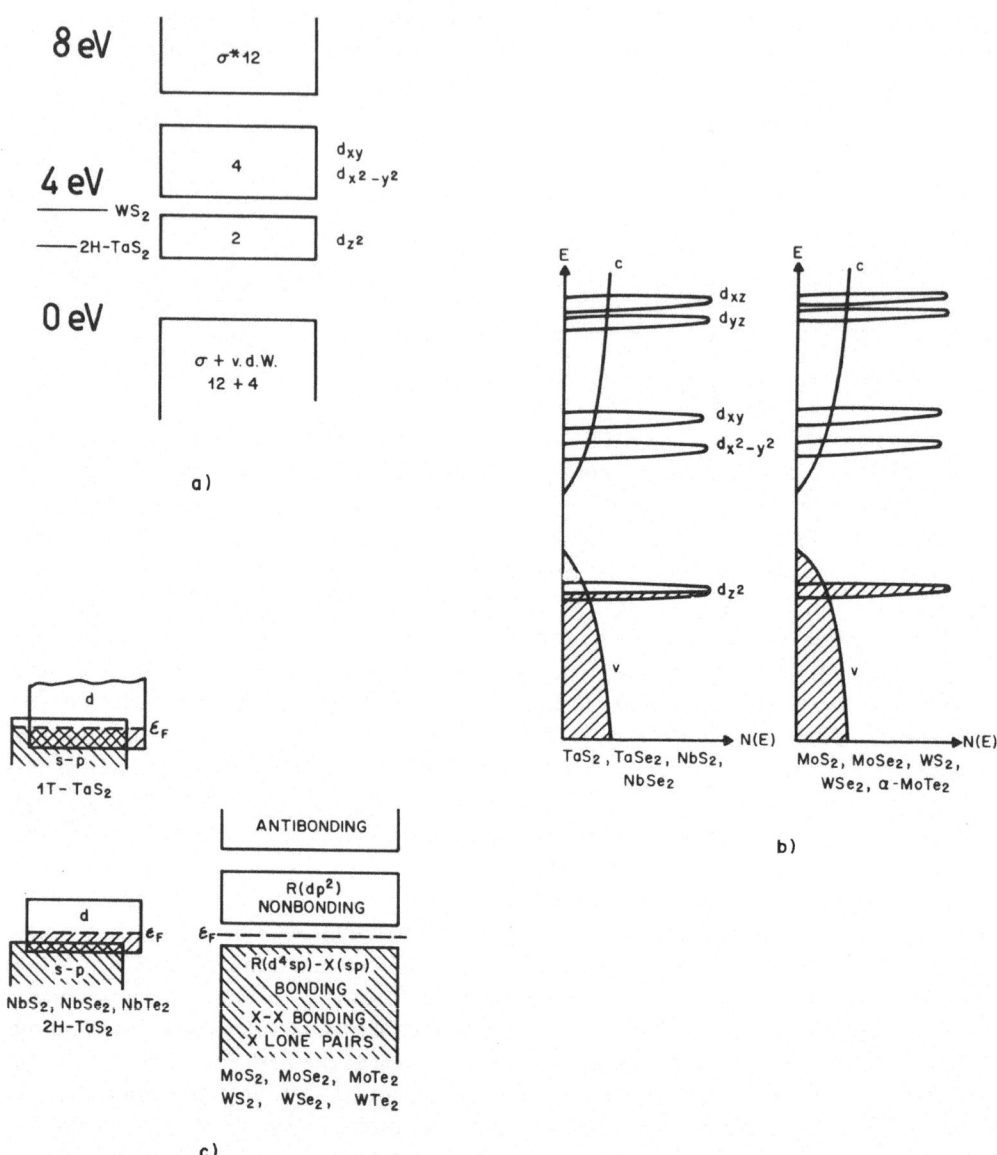

Fig. 28. Band schemes for compounds with trigonal prismatic coordination proposed in [1] (a), (note the new energy scale, enlarged by a factor of two with respect to [1], in [10] (b) and in [12] (c).

combination of muffin-tin orbitals (LCMTO) has been applied to NbS_2 [42], the augmented plane wave method (APW) has been applied to $NbSe_2$ and $2H–TaS_2$ [22] and the non-local empirical pseudopotential method (EPM) has been applied to $NbSe_2$ [43]. Both, the LCMTO and the APW calculations were based on neutral atomic configurations and Slater's $\alpha = 1$ value for the exchange potential. Non-muffin-tin corrections are naturally included in the LCMTO method and have been treated by perturbation in the APW method. The EPM used photoemission [28, 35] and optical [39, 40] data to adjust the potential form factors.

The LCMTO band structure of NbS_2 is represented in Figure 29 showing the upper part of the bonding anion p-like bands and the metal d-bands only. The APW results for $2H–TaS_2$ and $NbSe_2$ resemble closely each other; thus only the $2H–TaS_2$ band structure is presented in Figure 30. The EPM band structure of $NbSe_2$ is presented in Figure 31. In all cases the low lying anion s-states are not shown.

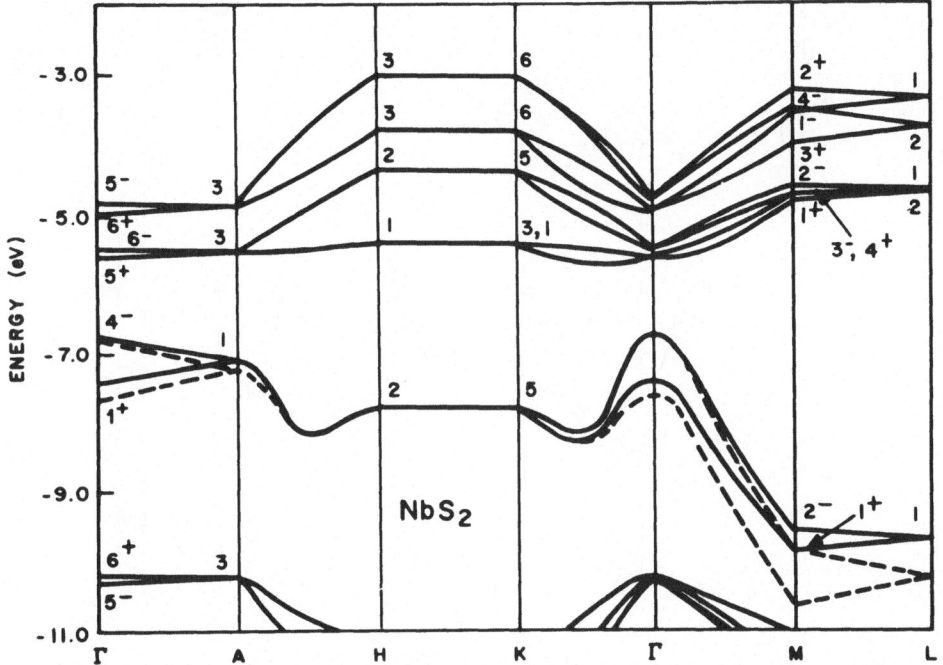

Fig. 29.   Band structure of NbS_2 as obtained from the LCMTO calculations of [42].

Although the methods used for these calculations are rather different, qualitatively the results resemble strongly each other. The filled valence bands primarily derive from the anion s (not shown) and p orbitals. The average separation of these bands is about 12–15 eV. Above the anion $s − p$ valence bands two split-off bands are found. According to the simple crystal-field models of [1] and [10] these states should correspond to the two d_{z^2} orbitals of the two layers per unit cell. Detailed symmetry analysis [43] and LCAO type fits to the band structure [22], however, show that these flat bands are a result of strong hybridization between the d_{z^2} and $d_{xy}, d_{x^2-y^2}$ sub-bands. Thus the states at Γ are of d_{z^2} character, whereas at K they are of $d_{xy}, d_{x^2-y^2}$ character. LCAO results [22] for MoS_2 shall be shown later demonstrating the various steps of hybridization and banding which lead to the flat d-band and a $d − d$ gap.

The two flat bands (often referred to as d_{z^2} bands) are occupied by two electrons and thus intersected by the Fermi energy. The theoretical results thus show metallic behavior

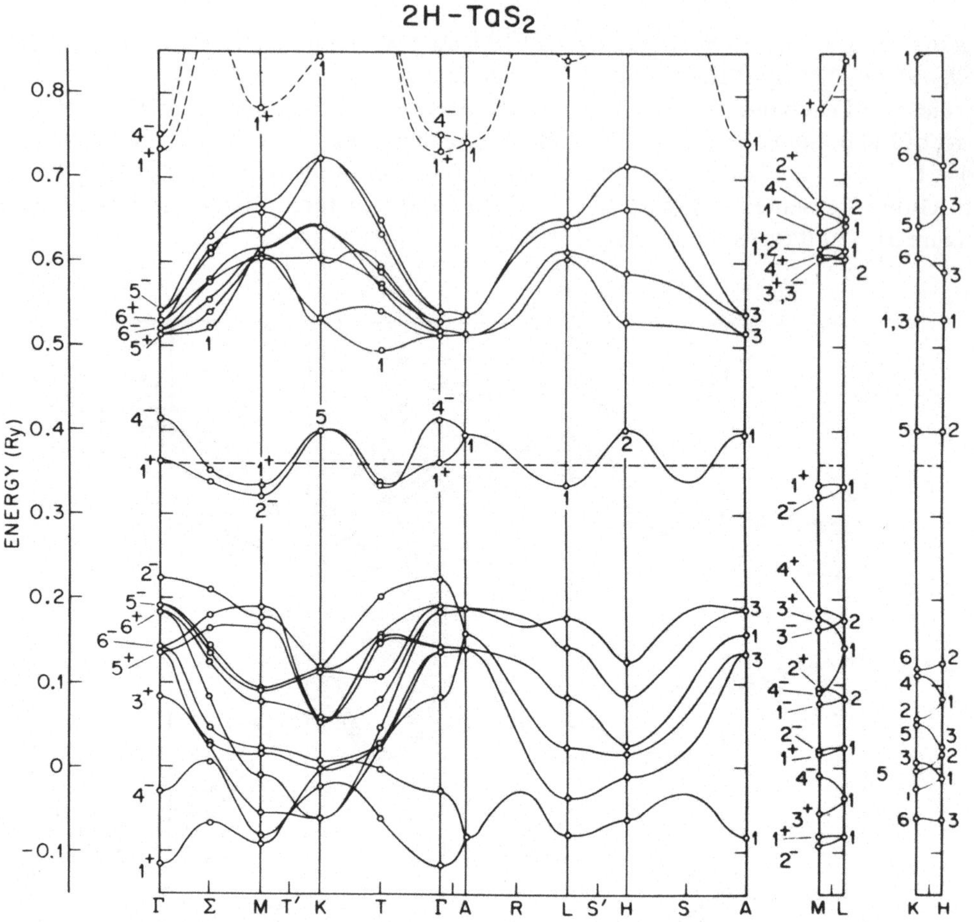

2H–TaS$_2$

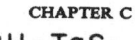

Fig. 30.   Band structure of SH–TaS$_2$ as obtained from the APW calculations of [22].

consistent with experiment. As seen from Figures 29, 30 and 31 the carriers are composed of electrons and holes.

There are several quantitative differences between the results of the three calculations applied to TaS$_2$, NbS$_2$ and NbSe$_2$. Some of them are:

(i) In the LCMTO results for NbS$_2$ shown in Figure 29, the top of the p-like bonding bands is of Γ_6^+ symmetry and derived from p_x, p_y states of the chalcogenides, whereas that obtained by both the APW method and the EPM is of Γ_2^- symmetry and like for other layer-compounds is associated with the p_z states of the anions. The difference in energy between the two states is due to the anisotropic crystal field.

(ii) There is about 0.36 eV overlap between the top of the p-states (Γ_6^+) and the bottom of the d states (M_1^+) in the LCMTO calculation of NbS$_2$ whereas in the APW calculations a gap of 1.28 eV for 2H–TaS$_2$ and of 0.64 eV for NbSe$_2$ was obtained. The corresponding gap in the EPM calculation of NbSe$_2$ is 0.3 eV. No drastic change is expected if the S-atoms are replaced by the Se atoms (except an eventual increase in

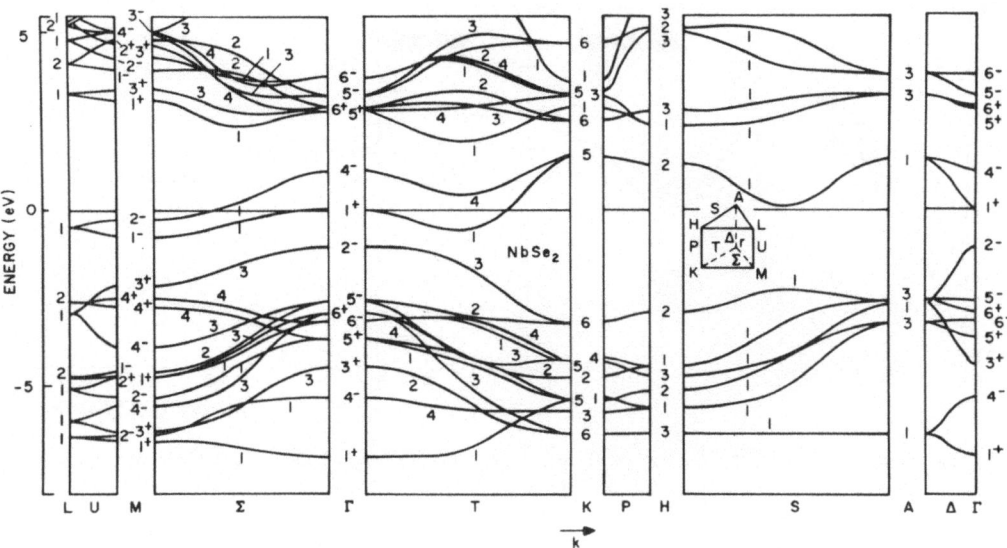

Fig. 31.   Band structure of $NbSe_2$ as obtained from the EPM calculations of [43].

overlap). Thus, among the three calculations, the LCMTO results shown in Figure 29 for NbS_2 give the best agreement with the density of states deduced from photoemission experiments [36]. In [42] the $p - d$ overlap was tentatively attributed to non-muffin-tin contributions in the potential. This is in contrast to the results of [22] where similar corrections have been considered and found lead to a lowering of the energies of the bonding p-states with respect to the non-bonding d-states resulting in an increase of the $p - d$ gap.

(iii) The calculated widths of the partially filled d-bands are 2.2 eV in NbS_2 [42], 0.51 eV [22] and 0.9 eV [43] in $NbSe_2$, and 0.85 eV in $2H–TaS_2$ [22]. Compared with the measured width of about 1 eV for $NbSe_2$ the results obtained by the EPM give the best agreement.

(iv) The APW results differ from the other two calculations in the d-band region along the T direction, where T_1 and T_4 cross. This happens in both $2H–TaS_2$ and $NbSe_2$.

(v) APW and EPM results show the top of the d_{z^2}-bands to be K_5 whereas the LCMTO result shows it to be Γ_4^-. Differences in the topology of the Fermi surface are expected.

The tight-binding method has been used in [22] as an interpolation scheme for the APW results to calculate a density of states for the conduction bands of TaS_2. These results (shown in Figure 32) provide some insight and allow semi-quantitative comparison between theory and experiment. In Figure 32, the maximum of the density-of-states for TaS_2 obtained by the APW-tight-binding interpolation scheme coincides with the Fermi level. A similar result was obtained for $NbSe_2$. The reason for the coincidence is due to the fact that the Fermi energy just passes through the flat region of the interpolated band structure (Figure 32) along T. This suggests that the maximum in

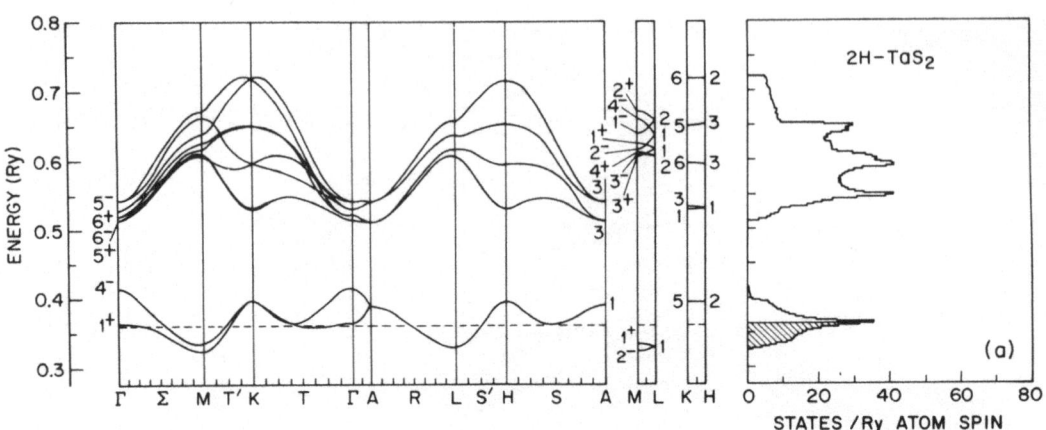

Fig. 32.   Conduction band structure and density of states of 2H–TaS$_2$ calculated from a LCAO interpolation scheme (after [22]).

the conduction band density-of-states of the non-interpolated direct APW calculation on 2H–TaS$_2$ (Figure 30) may occur at -0.32 eV with respect to E_F. The corresponding structure in NbSe$_2$ may be at -0.32 eV and two structures in the bonding states are estimated to be at -1.59 eV and -4.3 eV. The Fermi energy in the EPM result of [43] was estimated to be 0.2–0.3 eV above Γ_1^+ (Figure 31). Sections of the partially filled T_1 band and other bands with energies around -2.4 eV and -5 eV may give maxima in the density of states at -0.6 eV, -2.7 eV and -5.2 eV below E_F. Both sets of theoretical energies agree reasonably with photoemission data [35, 36]. The values of the density of states at E_F were calculated in [22] for NbSe$_2$ and 2H–TaS$_2$ based on the LCAO interpolation results. The values are 3.0 states/eV-spin and 2.3 states/eV-spin for the Nb and Ta atoms respectively. The Nb value agrees with the 2.8 states/eV-spin deduced from specific heat measurement [44] whereas low temperature (<10 K) heat capacity data [45] give a value of 1.8 states/eV-spin for TaS$_2$. Structural phase transition, however, have influenced this value.

Due to the crossing of the T_1 and T_4 of the d_{z^2} bands in NbSe$_2$ in [22], interband transitions are expected to start near zero photon energy. On the other hand, the EPM result shown in Figure 31 should give a minimum for interband transitions at 0.6 eV near Σ. The band structure of Figure 31 shows sections of parallel bands separated by about 2 eV in the Σ and T region ($\Sigma_3 - \Sigma_1$ and $T_3 - T_4$). These volume contributions may give rise to the measured structure, in the interband optical spectrum in this region of energy. A summary of the details of the various band structures is given in Table 12.

To understand the effect of intercalation from the band structure point of view, an APW band structure [22] has been done for NbSe$_2$ with the interlayer distance increased by about 1.5 Å. The main consequences of the reduced interlayer interaction are:

(i) The bonding anion p-state Γ_2^- drops between the anion p_x, p_y states Γ_5^-, Γ_6^+ and Γ_5^+, Γ_6^- due to the changes in the anisotropy of the crystal field. Furthermore, all energies become nearly doubly degenerate, and the width of the bands along the crystal $\hat{c}$-direction is greatly diminished indicating a decoupling of the two layers per unit cell.

TABLE 12

Comparison between experimental and theoretical results for some group VB transition metal compounds

		Lowest s-states of the Chalcogenides	$p-d$ gap	Width of the partially filled d_{z^2} band	Interband transition onset
NbS$_2$ Theory	Kasowsky [42]	–	−0.36 (overlap)	2.2	< 0.1
NbSe$_2$ Theory	Mattheiss [22]	− 13.81 eV[a]	0.64	0.51	≈ 0
	Fong-Cohen [43]	− 15.00	0.30	0.90	0.60
NbSe$_2$ Experiment	McMenamin-Spicer [36]		− 0.20 (overlap)	1.0	
	Williams-Shepherd [35]			0.70	
	Liang [40]				1.41
2H–TaS$_2$ Theory	Mattheiss [22]	− 14.42	1.28	0.85	0.11

[a] Relative to the d_{z^2} (Γ_1^+) band.

(ii) Similar effects are found for the non-bonding d bands which result in changes in the topology of the Fermi surface in the vicinity of Γ.

(iii) The $p - d$ energy gap increases as a consequence of the decreased interlayer interaction. We like to mention that all these effects may be reversed if pressure is applied to the crystal. It is assumed that hydrostatic pressure primarily affects the interlayer separation.

Another attempt to illustrate the effect of intercalation has been made in the LCMTO scheme for NbS$_2$ [42]. In order to treat the interlayer region accurately, the LCMTO calculation for normal non-intercalated NbS$_2$ has been done by assuming six atomic centers and two additional non-atomic centers in the interlayer region. The empty site potentials result from the overlap of the charge tails from neighboring sites. The results are shown as full lines in Figure 29. The effect of intercalation is simulated by excluding the potential contributions of the two non-atomic interlayer centers. The effect of this modification on the d-bands is shown as broken line in Figure 29. As a result the d-band width increases in contrast to the APW results [22] and the overlap with the p-bands becomes more important. The geometry of the Fermi surface also changes. Both APW and LCMTO results show that the Fermi surface is changed even without the donation of electrons from the intercalated molecules. Therefore, the question, whether the variation of the superconducting transition temperature T_c is due to the change of the geometry of the Fermi surface or to the donation of electrons remains unanswered.

The geometry of the Fermi surface of the group VB transition metal compounds is subject of several theoretical studies. In a simple tight-binding approach [45] for un-hybridized (d_{z^2}) bands the energy may be expressed as a linear function of the intra (D) and interlayer (δ) bonding strength [45]. With a hypothetical value of $\delta/D = 0.75$

proposed in [45] the Fermi surface contains hole surfaces around Γ and K. This model, however, may be questionable because of the assumption of unhybridized (d_{z^2}) bands. As mentioned, APW results show strong hybridization effects between the d_{z^2} and $d_{x^2-y^2}$, d_{xy} states [22]. Also, the value of $\delta/D = 0.75$ may be overestimated [22]. Both APW and EPM results do not show the existence of a hole surface around Γ but allow for it around K. In the realistic three dimensional case, the shape of the Fermi surface is more complicated. In Figures 33(a) and (b), sketched Fermi surfaces are shown as obtained from the seventeenth and eighteenth bands of the EPM band structure of NbSe$_2$. The shaded regions are occupied by electrons. For the seventeenth band, almost cylindrical hole surfaces are found along $K-H$ and spherical hole surfaces appear around A. The APW and LCMTO show similar geometries for this case [46]. For the eighteenth band, the Fermi surface may best be described by cylindrical electron surfaces along $M-L$. The theoretical results do not exhibit a simple pancake shape as suggested on experimental grounds in [34]. This disagreement may be related to the structural phase transition at 40 K.

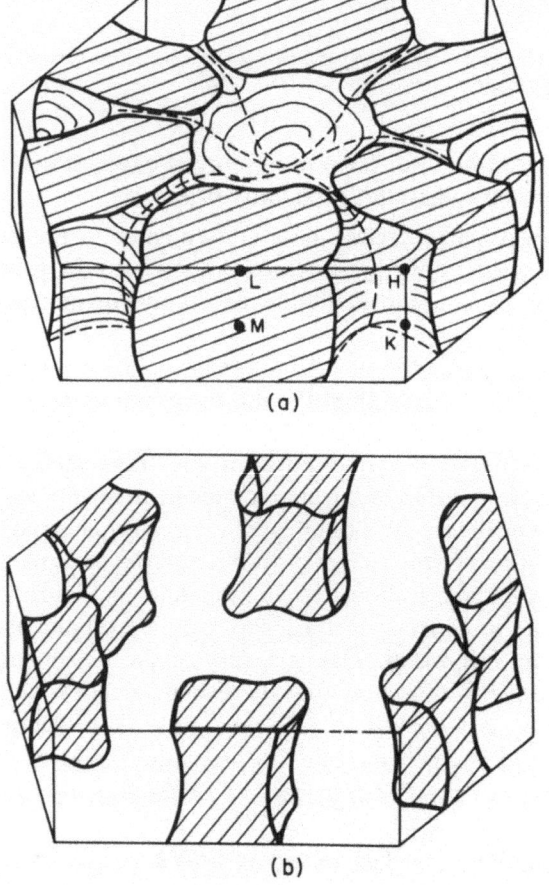

(a)

(b)

Fig. 33.   Sketch of the Fermi surfaces for the 17th (a) and the 18th (b) bands of NbSe$_2$ based on the EPM band structure of [43].

The bonding properties of the group VB transition metal layer compounds have been studied schematically in [1]. It is suggested that each chalcogenide possesses lone pair electrons which extend into the region between the layers. The cations supply four electrons each to fill the bonding states of the anions. Additional electrons of the cations also contribute to non-bonding orbitals. Realistic valence charge distributions for $NbSe_2$ have been calculated from the band structure results of [43]. Charge densities obtained from the pseudo wave functions averaged over six 'special' k-points in the BZ [47] are shown in Figure 34(a) and (b). The method defining 'special' k-points shall be described later in the subsection on non-transition metal layer compounds.

Figure 34(a) shows the total charge distribution of the sixteen filled valence bands in a vertical plane DEFG which contains one Nb atom and two Se atoms within one layer. The actual location of the plane with respect to the unit cell is shown in the lower left corner of the figures. The dotted circles in the figures around the atoms show the core radii in which the charge is expected to be wrong because of its derivation from pseudo wave functions. The closed shell type distribution of the charge around the anions illustrates clearly the characteristics of the filled bands. Most of the charge is concentrated around the chalcogenides with its maxima between the Se and Nb ions. Both these figures are similar to the calculated distributions in SnS_2 and $SnSe_2$ which are presented in the second part of this section. However, the present distribution of the bonding charge seems to have less sp^3 type character at the anions indicated by the only weakly appearing 'lone pair' charge between the layers.

According to the directed valence bond theory [24], the d and s states of the transition metal ion under the trigonal prismatic environment should hybridize with the higher unoccupied p states of the same ion to form six equivalent cylindrical bonds pointing towards the chalcogenides. The mixing of the p-states should be at least 10%. These dsp^4 type orbitals and the p-orbitals of the anions should form the covalent bonding charge of the crystal. To investigate these ideas a projection operator of the $5p$-states centered at the Nb ions has been applied to the valence wavefunction to find the extent of the $5p$-hybridization in the sixteen bands at Γ, M, K. The results indicate no significant contribution from the $5p$ states. Although the calculated charge distribution shown in Figure 34(a) does not exactly agree with either the simple bonding picture of [1] nor with the directed bonding theory of [24], the shapes of the contours seem to be consistent with the molecular orbital pictures of [10]. The results seem to show that simple chemical bonding pictures may easily be overinterpreted. Figure 34(b) shows the charge density of the partially filled conduction bands in the plane OABC which contains the Nb ions at the corners. The chalcogenides are directly above and below the center of the triangular contour with magnitude equal to 2. Unlike the uniform charge distribution in simple metals, such as Al [48] and even the ones in noble metals [49] where the partially filled conduction bands have strong $d - sp$ hybridization, most of the charge density shown in Figure 34(b) is concentrated near the Nb ions. This fact results from the pure non-bonding d-characters of these bands. The shapes of the contours indicate that the charge results from mixtures of d_{xy} and $d_{x^2-y^2}$ states of the metallic ion. As we mentioned earlier, the presence of the d_{xy}, $d_{x^2-y^2}$ states in the lower conduction band is a result of strong $d - d$ hybridization [22]. Pure d_{z^2} states should not lead to any charge in the OABC plane. The maximum value of the contours for these partially filled conduction bands is about 14% of the maximum value for the filled valence bands.

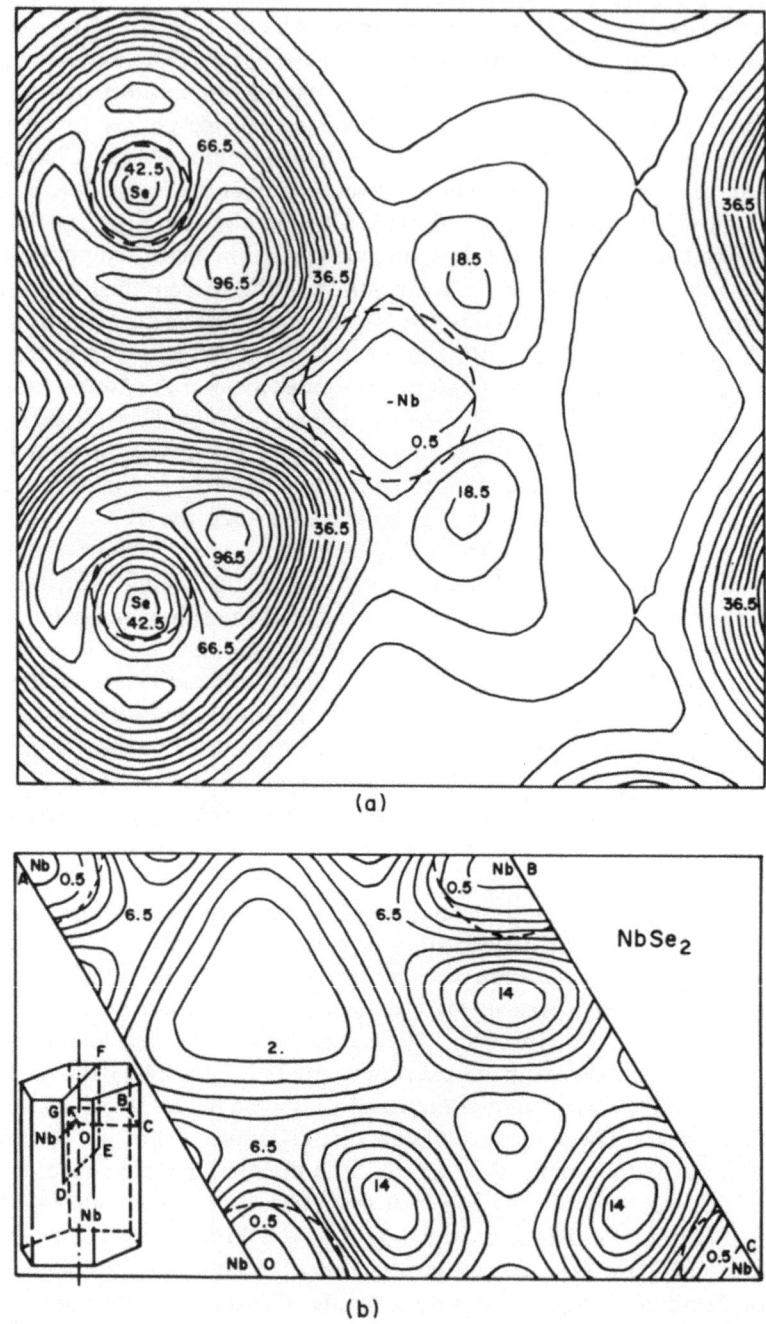

(a)

(b)

Fig. 34.   EPM charge distribution of the filled sixteen valence bands of $NbSe_2$ in the plane DEFG in the hexagonal unit cell (a) and charge distribution of the partially filled d_{z^2} bands in the plane OABC (b). The planes are indicated in the insert. The contours are given in units of 2 electrons per unit cell per band (after [43]).

The absolute charge values should be considered as an approximation since the distribution was calculated from only six 'special' k-points inside BZ which represents a rough sampling of the Fermi surface. However, the characteristic features of the charge distributions in a d-band metal are expected to be well accounted for.

Finally we shall give some comments on the differences between 1T–TaS$_2$ and 2H–TaS$_2$. As mentioned, TaS$_2$ is the only compound which exists with both octahedral and trigonal prismatic coordination of the cations. The most simple polytypes are the 1T and the 2H forms respectively.

In comparing the gross features of the LCAO band structures [22] of 1T–TaS$_2$ (Figure 23) and 2H–TaS$_2$ (Figure 32) the most striking difference appears in the hybridization of the non-bonding d-bands. In the 1T phase the d_{z^2} and $d_{x^2-y^2}$, d_{xy} derived bands overlap strongly without forming a '$d - d$ hybridization gap' (see Figure 23). In contrast to that the trigonal prismatic crystal field of the 2H phase allows an extreme $d - d$ hybridization at certain k-points which results in a split-off pair of bands. (There are two bands because of the two layers per unit cell (see Figure 32).) This split-off pair of bands has sometimes (inexactly) been attributed to pure d_{z^2} states [45]. The strong hybridization as well as the fact that two layers form the unit cell lead to drastic differences in the Fermi-surfaces. Whereas the 1T phase exhibits a simple cylindrical single sheet electron surface along $M - L$, the 2H phase shows a complicated three sheets electron and hole surfaces similar to the results of 2H–NbSe$_2$ shown in Figures 33(a) and (b). From the LCAO results and also from inspection of Figures 21 and 30 one finds a higher density of states at the Fermi level for the 2H phase than for the 1T phase.

Recent angular dependent photoemission results [31] showed a marked difference between the 1T and the 2H phase, based on the following structural differences. In the 1T phase the unit cell in the $\hat{c}$-direction extends over only *one* sandwich of three-fold rotational symmetry. The trigonal (1T) symmetry is hence preserved for the complete crystal. In the 2H phase each sandwich again has three-fold rotational symmetry. However, the stacking is such that the orientations of successive sandwiches alternate between two positions turned by 180°. Two sandwiches are required to define the unit cell and the crystal has hexagonal symmetry. (Note that some of the hexagonal symmetry operations interchange the two layers.) The photoemission results obtained on the 1T phase are shown as radial plots in Figure 35 (right hand side). The outer curve (open circles) represents the variation with ϕ of the total emission (i.e. all valence bands between 0 and about -5 eV) at $\theta = 55°$. The inner curve (full circles) represents the φ dependence of the Ta d-emission (i.e. valence bands between 0 and about -1 eV). The three-fold rotational symmetry is clearly recognizable for this 1T phase.

Figure 35 (left hand side) shows the corresponding results for the 2H phase. As expected for the 2H polytype, the symmetry of the emission is close to six-fold. However, in the Ta d-emission some residual three-fold symmetry can be observed. This lower symmetry may be attributed to the finite attenuation length of the photoelectrons. The three-fold symmetry of the outermost sandwich overpowers the overall six-fold crystal symmetry. From the quantitative deviation of the emission pattern from six-fold symmetry an approximate attenuation length of $l = 13$ Å for photoelectrons with $E = 9.4$ eV above the Fermi level has been derived [31].

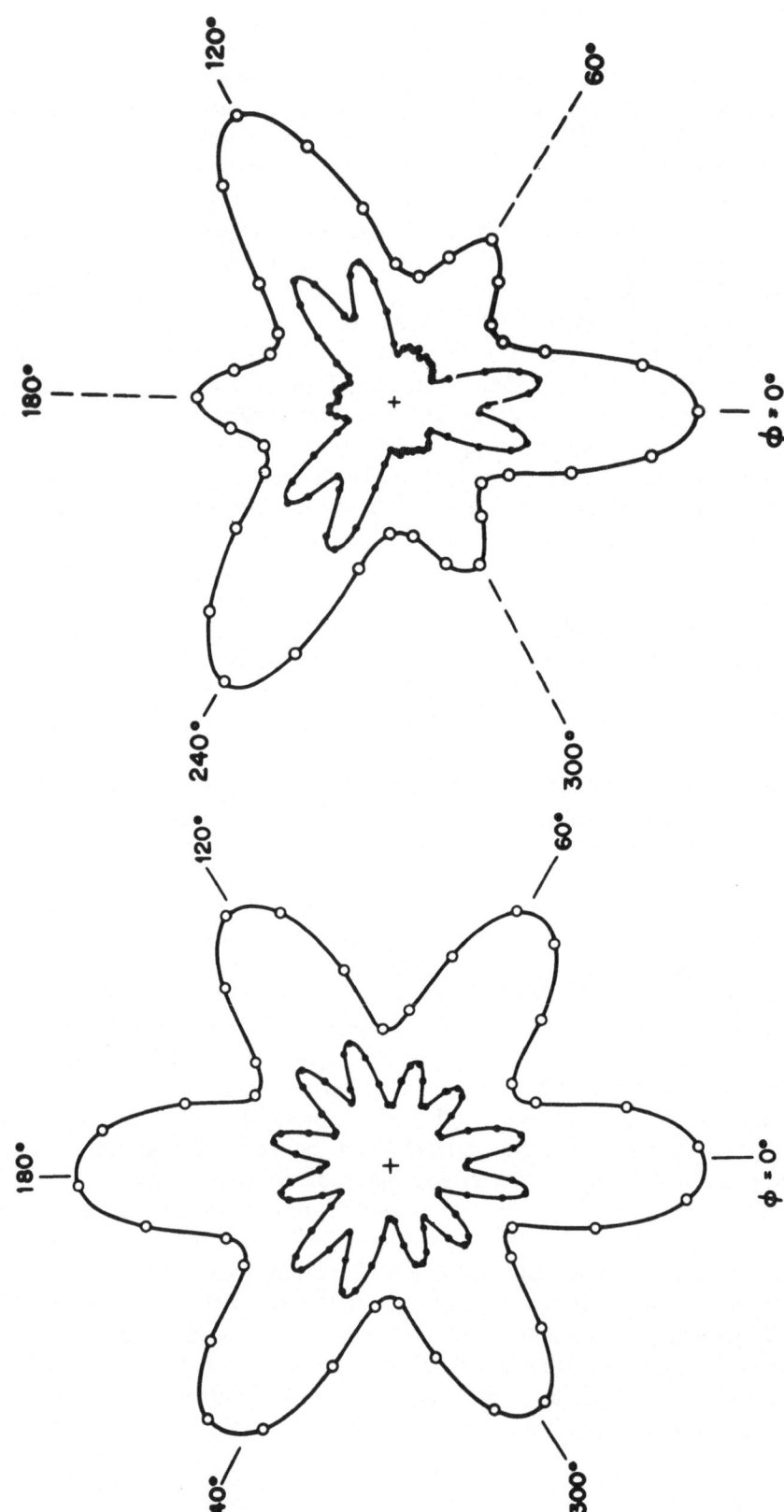

Fig. 35. Radial plots of the azimuthal dependence at $\theta = 55°$ of the total UV photoemission intensity of 1T–TaSe$_2$ (open circles) and on a 5 times expanded scale the Ta–d emission intensity (full circles) (right hand plot) and Radial plots of the azimuthal dependence at $\theta = 55°$ of the total UV photoemission intensity of 2H–TaSe$_2$ (open circles) and on a 10 times expanded scale the Ta–d emission intensity (full circles) (left hand plot) (after [31]).

1.2.2. *The Group VIB Transition Metal Dichalcogenides*

Molybdenum sulphide is one of the most intensively investigated layer compounds. It exists in nature as the mineral molybdenite and can therefore be easily obtained. MoS_2 and several other group VIB transition metal layer compounds are included in the fundamental review article of [1]. The optical properties of these compounds have been studied very thoroughly. This is in contrast to their electrical properties which are still not well understood. It was concluded in [1] that the group VIB dichalcogenides are diamagnetic small gap semiconductors. Recent transmission and absorption spectra [41] taken on a large series of group VIB dichalcogenides confirm the general semiconducting

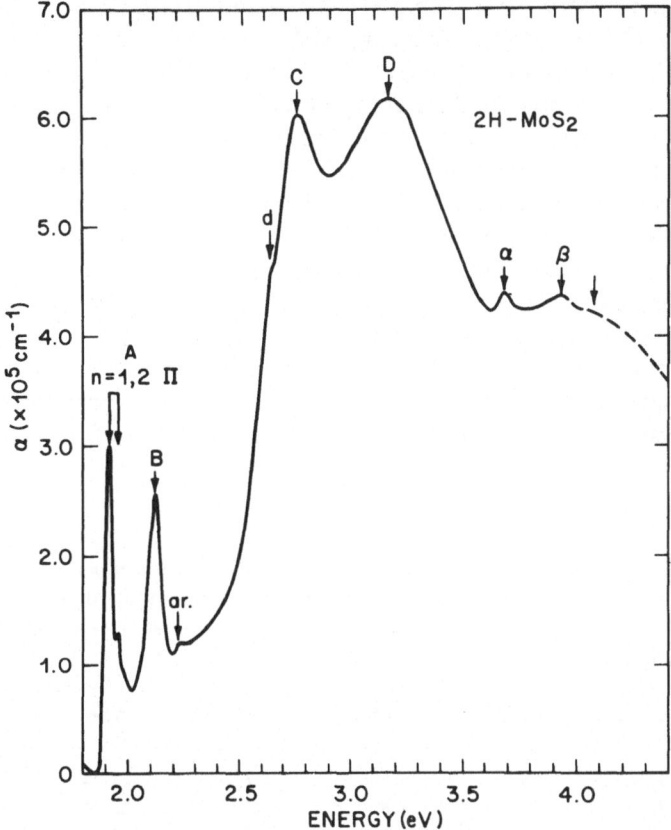

Fig. 36.   Optical absorption spectrum of 2H–MoS₂ as measured in [41].

behavior. A representative spectrum of MoS_2 is displayed in Figure 36. The general shape of the various spectra may be summarized as follows, starting at the low energy side.

(i) Two very sharp absorption peaks, labelled *A* and *B* in Figure 36 are identified as excitonic transitions. Two subsidiary peaks (*A'* and *B'*) may be seen in the heavier (Se and Te) chalcogenides. The exciton binding energies of the *A* structure are estimated to be 50 meV, 67 meV and 50 meV in MoS_2, $MoSe_2$ and WSe_2 respectively.

(ii) These excitonic structures are followed by a broad strong absorption band exhibiting fine structures, labelled as *d*, *C* and *D*.

(iii) Two weaker structures, labelled α and β in Figure 36, are present in MoS_2.

The positions of these structures for various compounds are listed in Table 13.

Photoemission measurements [36, 37] allow to establish a band picture of the valence bands. The uppermost d-band is about 2 eV wide with its maximum density of states about 1.5 eV below the top of the valence bands. The maximum of the p-like valence bands is located at 3 eV below the top. Finally, the nature of the wave function at the top of the valence bands has been suggested to have d_{z^2} character, based on electron paramagnetic resonance measurements on As-doped MoS_2 [50].

Most of the theoretical studies on the group VIB transition metal dichalcogenides have been done for MoS_2 (see Table 11). Exceptions are ETB calculations [51] on a series of MoS_2, $MoSe_2$, $MoTe_2$, WS_2 and WSe_2. As representative band structure of this series we show in Figure 37 the ETB results for MoS_2. The calculations of [51] are done similarly to those presented for group IVB transition metal layer compounds [20]. The single layer approximation is used and the empirical tight-binding parameters are fitted to reproduce optical transition energies as listed in Table 13. Finally, interlayer interaction is considered in a perturbation scheme.

The general features of the ETB band structures are determined by the $\Gamma_3^- - \Gamma_3^-$ gap between bonding p-like orbitals and non-bonding or antibonding d-orbitals. The absorption complex (A, B, A', B') is ascribed to this $p - d$ gap. The perturbations of spin-orbit and interlayer interactions are proposed to yield the fine structure of the absorption complex. Thus, both valence and conduction bands are supposed to be split by spin-orbit interaction which should give rise to allowed transitions resulting in the two main structures A and B. Additional inter-layer splitting of the valence bands but

TABLE 13

Important optical structure (reported in [41]), in the experimental spectra of some group VIB transition metal compounds

Crystals energy (eV) features	MoS_2	$MoSe_2$	α-$MoTe_2$	WSe_2
A	1.910	1.598	1.120	1.694
B	2.112	1.864	1.431	2.174
A'		2.122	1.790	2.260
B'		2.320	2.046	2.625
d	2.630	$\begin{cases} 2.550 \\ 2.660 \end{cases}$	$\begin{cases} 2.280 \\ 2.494 \end{cases}$	$\begin{cases} 2.935 \\ 3.095 \end{cases}$
C	2.760	$\begin{cases} 2.836 \\ 3.001 \end{cases}$	$\begin{cases} 2.600 \\ 2.820 \end{cases}$	$\begin{cases} 3.200 \\ 3.470 \end{cases}$
D	3.175	$\begin{cases} 3.001 \\ 3.210 \end{cases}$	$\begin{cases} 2.670 \\ 2.935 \end{cases}$	$\begin{cases} 3.590 \\ 3.755 \end{cases}$
α	3.685			
β	3.930			
E		3.96	$\begin{cases} 3.640 \\ 3.810 \end{cases}$	$\begin{cases} 3.930 \\ 4.190 \end{cases}$
F				4.380

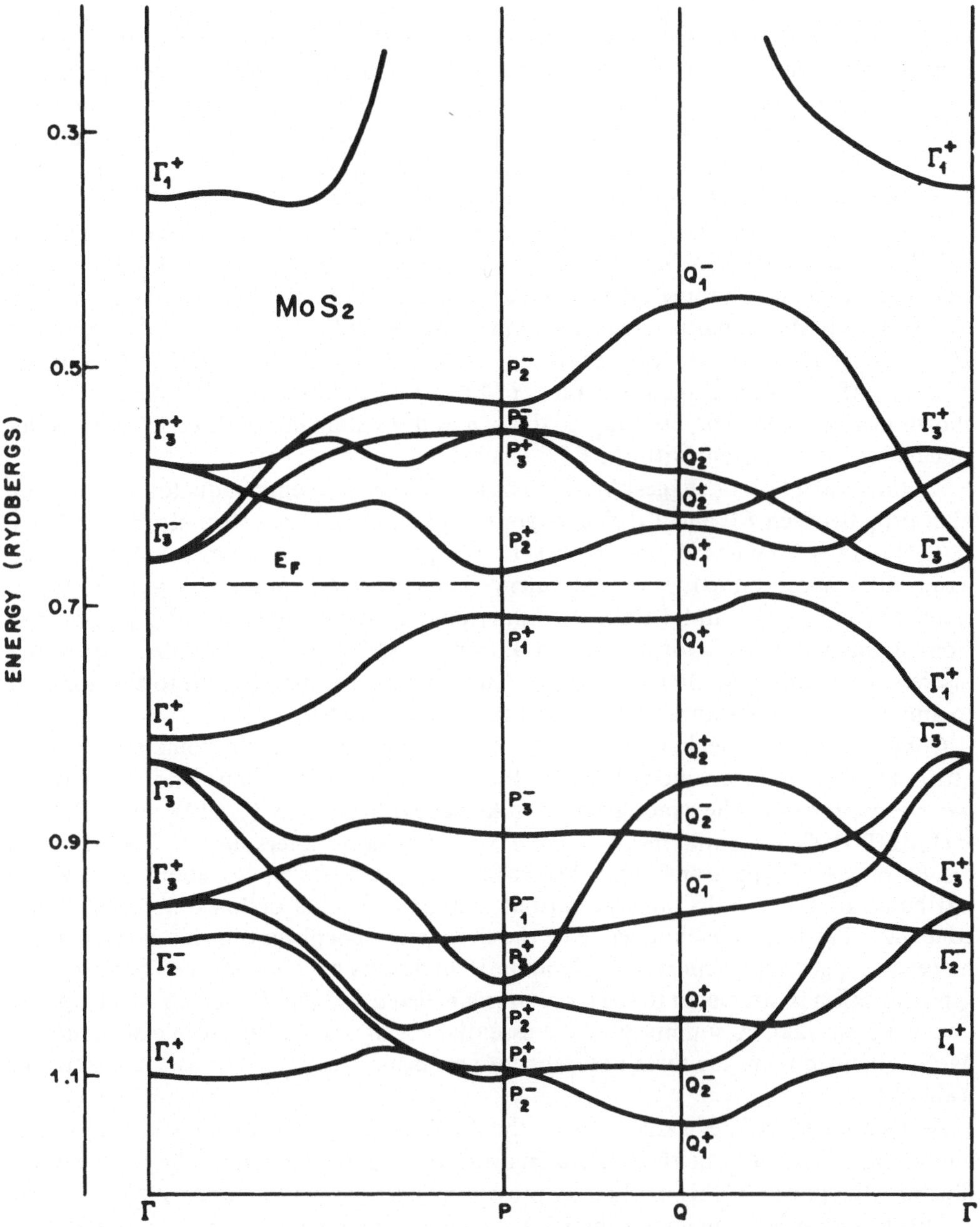

Fig. 37.   Band structure of a single layer MoS_2 as calculated by the ETB method in [51].

not of the conduction bands should result in the repeated structures A' and B'. Transitions at the zone boundary should contribute to the structures C, D. It is emphasized that in this ETB picture of [51, 52, and 53] the uppermost occupied d-band (Γ_1^+) does not play any role in the assignment of the main optical structures. This band is separated from the p-like valence bands by several tenths of an eV at Γ and 1 eV to 2 eV at the zone boundary. The conduction bands are of pure cationic-character. Some

$d - d$ hybridization between these conduction bands and the filled valence d-bands occurs but it is generally weak resulting in small indirect gaps. (MoTe$_2$ is even predicted to be semimetallic.) Generally, this tight-binding picture confirms the schematic band pictures of [1]. It may be criticized, since it uses weak perturbations such as spin-orbit or interlayer interactions to render symmetry forbidden transitions ($\Gamma_3^- - \Gamma_3^-$) allowed to account for the strong optical structures (A, B, A', B').

In addition to these ETB calculations, several independent three-dimensional calculations are reported for MoS$_2$. The APW [22], LCMTO [42] and LTM [54] methods have been applied. All three calculations are based on the muffin-tin potential concept. Corrections to this potential are included in the APW and LCMTO but not in the LTM calculations. The essential features of the resulting band structures are quite comparable. Therefore only the band structure of MoS$_2$ obtained in the LTM scheme of [54] is shown in Figure 38. The ordering of the s, p and d-bands in the three first-principles calculations is consistent with the ETB results.

All three calculations suggest MoS$_2$ to be an indirect gap semiconductor with the gap appearing between Γ_4^- and T_2. The magnitude of this gap varies strongly from 0.25 eV (LTM) to over 1 eV (APW and LCMTO). The position of the topmost d-like valence band with respect to the lower p-like valence bands varies from a 0.8 eV gap (APW) to small overlap (LTM and LCMTO). Experimental absorption [41] and photoemission [36] data seem to favor the LCTMO over the APW and LTM results. Part of the discrepancies among the first-principles calculations are presumably due to the different treatments of the non-muffin-tin part of the crystal potential.

Concerning the optical spectra, in all three studies the A, B excitonic structure is attributed to $d - d$ transitions between the topmost d-valence-band and the bottom d-conduction band. The exact location in k-space differs from Γ (APW and LTM) to A (LMCTO). The essential result of $d - d$ type transitions differs sharply from the ETB results of $p - d$ type transitions. Moreover, all structures up to about 3.5 eV are attributed to $d - d$ transitions at various symmetry points or lines in the Brillouin zone [42]. The large absolute value of the absorption coefficient in this energy range, however, suggests the importance of volume joint density of states effects. The first $p - d$ type transition is attributed to the structures labelled α and β in Figure 36 which appear above 3.5 eV [22]. A summary of the essential results on the electronic structure of MoS$_2$ obtained from the three independent first-principles calculations is presented in Table 14.

As mentioned before, LCAO fits to the APW results are presented in [22] which allow a detailed study of the various hybridization and interaction effects. We shall illustrate the power of this method by presenting detailed results for MoS$_2$.

The important parameters of the LCAO method are given as $\varepsilon_{\alpha\beta}(\tau_m)$ which describe the interaction between orbital α at the origin and orbital β at a distance τ_m. The parameters are determined by fitting the LCAO band structure to the APW results at some high symmetry k-points. Physically, the parameters $\varepsilon_{\alpha\beta}(\tau_m)$ are chosen to characterize the intra- and interlayer interactions between various d-orbitals as well as the overlap-covalency interactions between these d-orbitals and the s- and p-orbitals of neighboring anions. The centers of gravity of the various crystal-field split d-sub-bands are characterized by the quantities $\varepsilon_{\alpha\alpha}(0)$. Results for various stages of approximation are shown in Figures 39(a) to (c). The results are based on the discrete levels of the $4d$ states of an

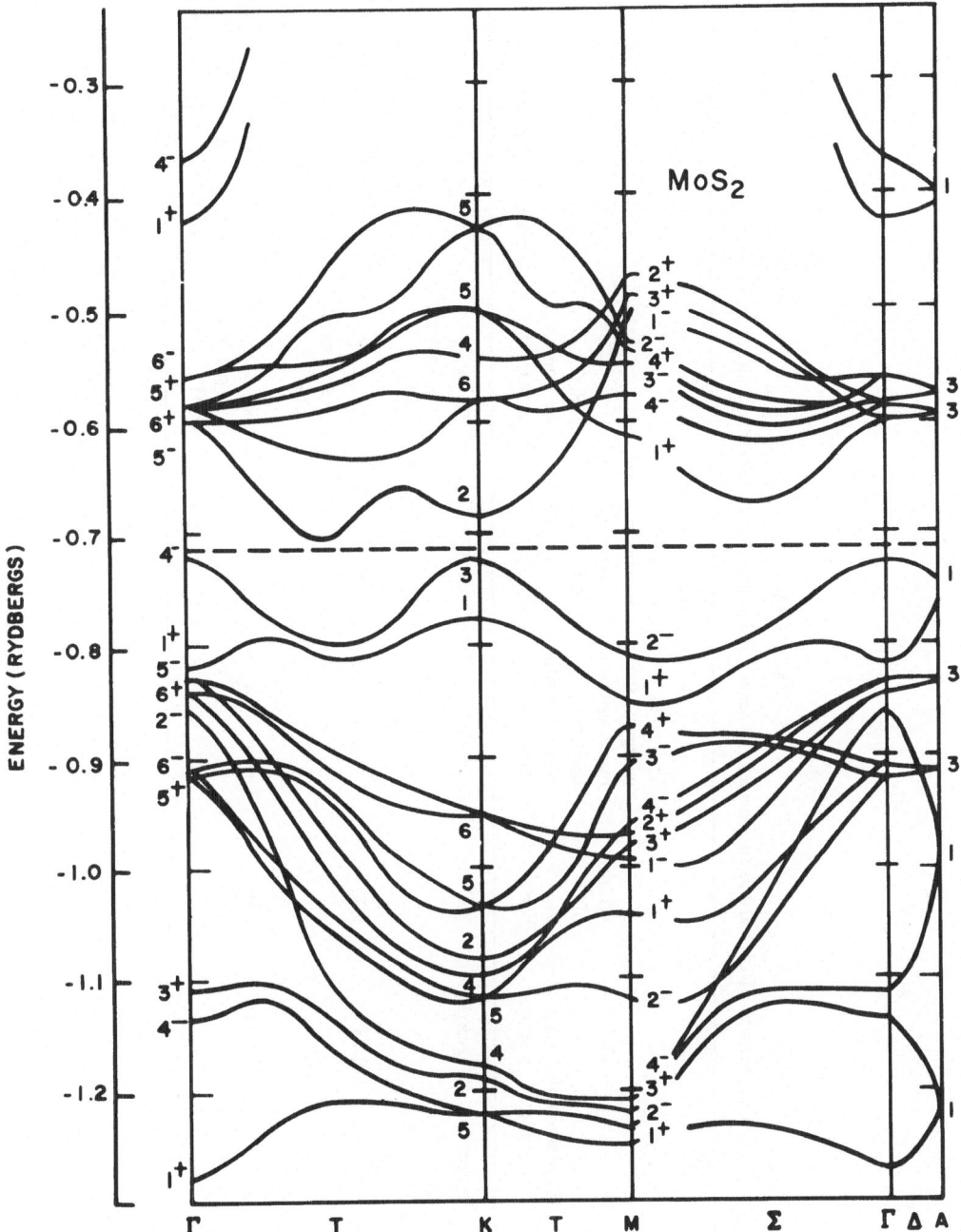

Fig. 38.   Band structure of 2H–MoS₂ as calculated by the LTM method in [54].

TABLE 14

Comparison of the various first-principles theoretical results for the electronic structure of MoS_2

Features energy (eV) authors	Fundamental energy gap	$E(d_{z^2})$ $-E$ (bonding p-states)	Width of d_{z^2} bands	Estimated energy of the maximum of the d_{z^2} density-of-states (with respect to Γ_4^-)	Estimated energy of the maximum of the p-like bonding density-of-states (with respect to Γ_4^-)	AB excitonic structure	C-peak	D-peak	$\alpha'\beta'$ structure
Wood Pendry [54]	Indirect $\Gamma_4^- \rightarrow T_1$ 0.25	Overlap	1.84	−1.20	−3.50	$\Gamma_4^- \rightarrow \Gamma_5^+$ + spin orbit 1.87			
Mattheiss [22]	Indirect $\Gamma_4^- \rightarrow T_1$ 1.1	0.75	1.00	−1.00	−3.90	$\Gamma_4^- \rightarrow \Gamma_5^+$ + spin orbit 1.78			
Kasowski [42]	Indirect $\Gamma_4^- \rightarrow T_1$ 1.23	Overlap	2.40	−1.07	–	$A_1 \rightarrow A_3$ + spin orbit 2.03	along Δ near A $(d \rightarrow d)$	along P near H $(d \rightarrow d)$	along Δ near A $(p \rightarrow d)$

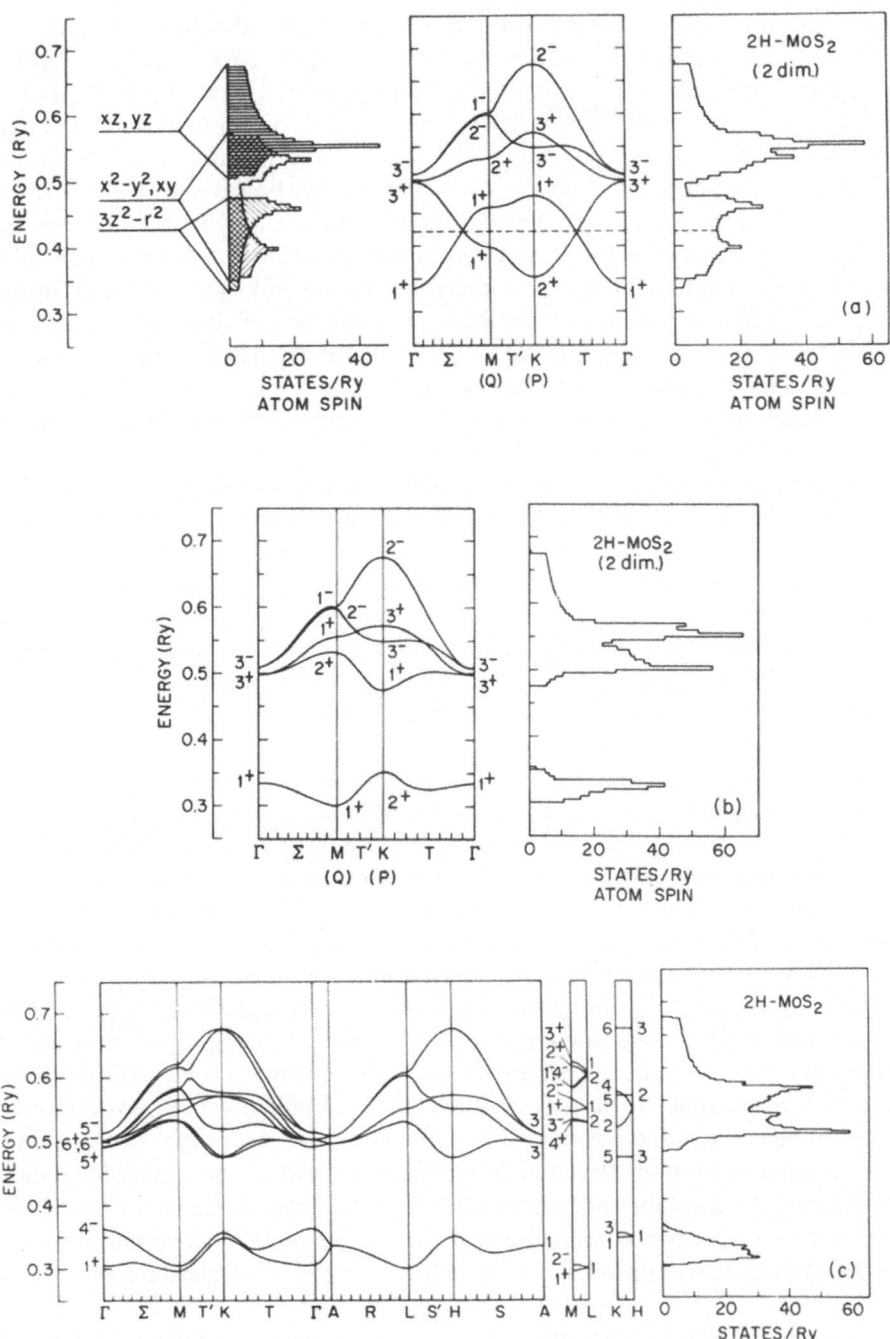

Fig. 39.   (a) Ligand-field levels of MoS$_2$, LCAO bandstructure without hybridization and interlayer interaction, and the corresponding density-of-states. (b) LCAO band structure of 2H–MoS$_2$ showing the additional effect of interlayer interaction (after [22]).

isolated Mo atom split in the presence of a trigonal crystal field. Solid state effects broaden these levels into bands which are shown in Figure 39(a), interband hybridization effects are neglected in this two-dimensional picture. The strong overlap between d_{z^2} and $d_{x^2-y^2}$, d_{xy} sub-bands should be emphasized. The inclusion of interband hybridization removes the overlap and creates a large $d - d$ hybridization gap (Figure 39(b)). This feature is strongly present in the 2H structure, not however in the 1T structure. The lower d-sub-band may falsely be interpreted as purely d_{z^2}-like because of its small width. Finally, the inclusion of interlayer coupling results in a relatively small splitting of the two-dimensional band structure at certain regions in **k**-space. (Figure 39(c)). No splitting occurs at the top face of the Brillouin zone due to time reversal symmetry. (See the discussion of the space group D_{6h}^4 in Chapter A). These detailed LCAO-type calculations illustrate very clearly the role of $d - d$ hybridization in determining the electronic properties of transition metal layer compounds with trigonal prismatic coordination of the cations.

In concluding this section we shall give a general comparison between the qualitative empirical band schemes and the results of the recent more realistic calculations. Especially the arrangement and order of the various d- sub-band deserves comments.

The principal elements of the three ligand-field type approaches [1, 9, 10] are a valence band complex, consisting mainly of anion s- and p-orbitals, a high-lying conduction band complex, consisting mainly of cation s- and p-orbitals and a complex of cation d-orbitals between them. The differences between the three approaches appear in the position, relative order and eventual overlap of these metal d-orbitals with respect to the conduction hands.

In the work of [1], the d_{z^2} and $d_{x^2-y^2}$, d_{xy} orbitals contribute to anion-cation bonding and merge with the aforementioned conduction band complex.

In the work of [9], the order (with increasing energy) of the d-orbitals which are supposed to fall into the valence conduction complex gap is as follows: $d_{x^2-y^2}$, d_{xy}, then d_{z^2} and then d_{xz}, d_{yz}. With respect to [1] the lower two d- sub-bands are reversed.

In the work of [10] the order of the d-orbitals is proposed as in [1]. However, it is suggested that the d_{z^2} sub-band overlaps with the anion p-type valence bands and that both $d_{x^2-y^2}$, d_{xy} and d_{xz}, d_{yz} sub-bands overlap with the cation s and p-type conduction bands. The reason for the d_{z^2}-valence band overlap, proposed in [10] was to account for the anomalous change with temperature of the Hall constant of both $NbSe_2$ and $TaSe_2$ from n type to p type that had been observed in [32, 33]. Later, however, it was shown [56] that the anomalous conductivity was associated with a structural phase transition and that the d_{z^2}-valence band overlap is no longer necessary. In all three cases the bands are rigidly filled by the available electrons which results in metallic behavior for the group VB compounds and in semiconducting behavior for the group VIB compounds.

In a fourth schematic approach, based on infrared spectroscopic data and on crystal stability arguments [12] a distinction between the metallic group VB and the semiconduction group VIB compounds is made. A strong ionic configuration in which metal s- and d-electrons are transfered to fill the anion valence shells is proposed for the group VB metals. In contrast a strongly covalent configuration with a bonding antibonding combination of d^4sp cation hybrids and sp anion hybrids is proposed for the semiconducting group VIB compounds. Between these bonding-antibonding bands, a non-bonding band is placed formed by cation dp^2 hybrids.

The main result of all quantitative band structure calculations is that strong $d - d$ hybridization effects do not allow to define d-sub-bands in the spirit of the empirical ligand-field methods. In particular d_{z^2} and $d_{x^2-y^2}$, d_{xy} orbitals hybridize strongly throughout the Brillouin zone. The order of the centers of gravity of the original (unhybridized) sub-bands coincide fairly well with the band scheme proposed in [1]. Even though the ordering of the d-states given in [10] seems to be correct, no strong overlap with valence and conduction bands has been found. The calculations also indicate that all group VB and group VIB compounds may be viewed in a partially ionic picture as proposed in [12] for the group VB compounds. On the other hand the strongly covalent scheme proposed in [12] for the group VIB compounds does not seem to agree well with the calculations. One may extrapolate the results of the charge density calculations of $NbSe_2$ [43] to MoS_2. It suggests that the strong cation p-admixture in the occupied bands proposed in the covalent scheme of [12] is *not* indicated by these calculations.

The question of relative stability of the octahedral versus the trigonal prismatic arrangement of transition metal layer compounds has recently been studied on the basis of electronegativity differences, lattice constants and atomic radii [25]. In Table 9 a complete list has been presented correlating electronegativity differences (ΔX), fractional ionic bond character (f_i), type of coordination (s), lattice parameters (a, c) and anion-cation bond lengths (d). It can be seen from this table that the simple argument, stating that covalency favors trigonal prismatic coordination whereas electrostatics favors octahedral coordination cannot be used to divide the two structures. A rather striking, if not complete, separation however, has been found in [25] by plotting the compounds on an effective ionic radius ratio r^+/r^- versus ionicity (see Figure 40). An effective separation of trigonal prismatic from octahedral coordination can be made by drawing a gently curving line. It is emphasized in [25] that the choice of particular ionic radii is not critical to the result.

It is suggested in [25] that among the IVB-, VB-, and VIB-layered dichalcogenides, where geometry permits, the compounds adopt the trigonal prismatic coordination. Those with electronic configurations d^n where n is 3 or greater, all adopt the octahedral structure. If these are placed on the radius ratio vs ionicity plot (Figure 40), they fall in both regions. This is not precluded by geometry. It is suggested that compounds falling in the upper, trigonal-prismatic, region (i.e., compounds that are not too ionic and have a large enough radius ratio, r^+/r^-) have the option to adopt the trigonal prismatic structure, whereas those in the lower region do not. Compounds falling in either region can adopt the electrostatically favored octahedral structure. One must look beyond geometry to determine why those compounds (d^n, $n < 3$) adopting the trigonal prismatic structure do so despite the electrostatic considerations favoring octahedral coordination. Additional information can be gained from band calculation e.g. [22]. In these calculations a hybridization gap is found opening up in the d-manifold of trigonal prismatic compounds that substantially favors this structure in MoS_2. As this gap splits off the lower two bands from the rest, configurations d^n ($n > 2$) would probably not be so stabilized in accord with the absence of the trigonal structure in these compounds. Another consequence of this argument is that d^0 compounds would not have this driving force toward the trigonal prismatic structure, in accord with the octahedral structures adopted by all group IV non-transition metal and group IVB transition metal compounds.

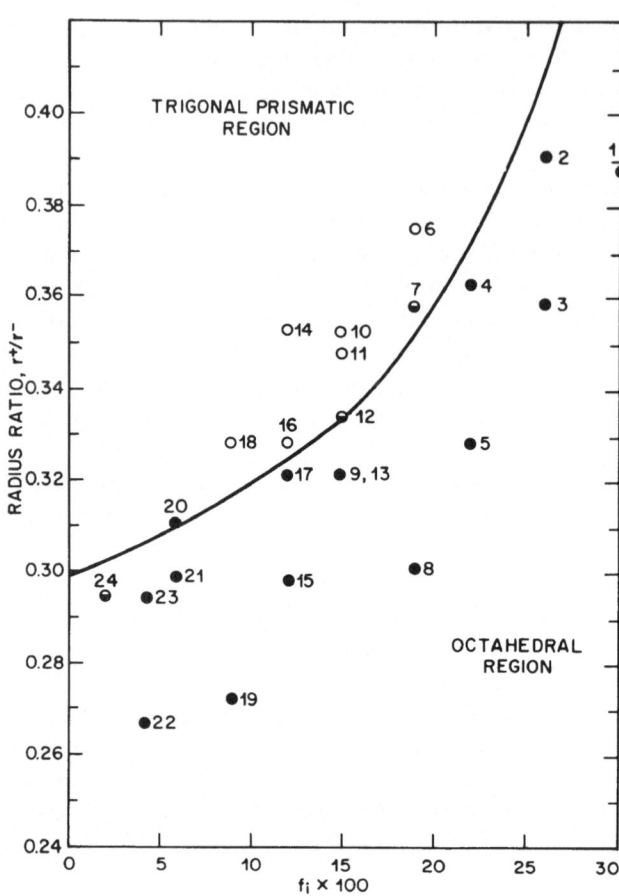

Fig. 40.   Ionic radius ratio r^+/r^- versus the fractional ionic character f_i of the metal-chalcogen bond. Open circles represent compounds in which the cation coordination is trigonal prismatic; solid circles, octahedral. The half solid circles (7, 12, 24) represent compounds in which both configurations can be obtained at room temperature. The compounds can be identified by reference to Table 9 (after [25]).

We may note that various different explanations of the relative stability of octahedral versus trigonal prismatic structures have been proposed. (See [18] and references in [25].) In particular, in [18] a completely different picture is presented involving electro-static crystal field stabilization versus electron spin-correlation energies. In fact, correlation energies are of substantial strength (order of eV's) for the lighter 3-d transition metals.

2. Non-Transition Metal Layer Compounds

In continuation of the first part of this chapter we shall discuss the electronic prop-erties of several layer materials involving *non*-transition metal ions. The electronic properties of transition metal dichalcogenides are strongly characterized by partially filled d-bands in the vicinity of the Fermi-level. While these d-bands are of non-bonding character, the chemical bonding is achieved predominantly by s- and p-electrons. The absence of any metal d-states among the valence states should thus not change the bonding properties considerably. In fact, a large number of layer compounds involving non-transition metal ions is found to exist in similar crystal structures. The materials to be discussed here crystallize in layered structures in which the atoms are tightly bound in two-dimensionally extended sandwiches being only several atomic layers thick. The sandwiches are stacked on top of each other and held together by relatively weak interactions. Thus, a large number of polytypes corresponding to different stacking sequences exists among these crystals. In the following discussion of the electronic properties only the most simple stacking sequences shall be considered.

It is mainly the bonding and the atomic arrangement within the layers which determine the valence band structure of these materials. Interlayer interactions may globally be considered as perturbations on the isolated sandwich system. This is in particular true for properties (like e.g. mechanical properties) which depend on the electronic screening of the entire ensemble of valence electrons. It may be a poor approximation for the detailed structure of individual electron states, which may considerably be influenced by interlayer interactions.

The structures to be discussed are stabilized by strong covalent $s - p$ bonds within the layers resulting in relatively large gap semiconductors. This semiconducting behavior persists even in the presence of weak resonance-like bonding between the layers. It may however reduce the gap considerably as in the case of Bi_2Te_3. The non-transition metal layer compounds thus represent a class of 'exotic' semiconductors whose galvanometric and optical properties have been the subject of intensive studies over the last several years. Detailed reviews of experimental results are presented in other volumes of this series. Here our attention shall be focused on a small but relatively representative number of layer compounds which have been investigated theoretically in the past. Due to major differences in structure and electronic configuration a close comparison as for the various groups of transition metal layer compounds is not possible. However, similarly to the transition metal layer compounds the materials to be discussed here may be grouped into classes characterized by either octahedral or trigonal prismatic coordination of the cations within the layers. An additional group, the (planar) hexagonal network materials, i.e. graphite and boron nitride shall also be included.

Experimental results shall be discussed along with the theoretical results as they are relevant to the calculations.

2.1. COMPOUNDS WITH OCTAHEDRAL ARRANGEMENT OF THE CATIONS

Similar to the 1T-type group IVB transition metal compounds a number of materials involving octahedrally coordinated non-transition metal ions crystallize in the simple CdI_2 type structure. In this structure the unit cell contains one layer consisting of three atomic sheets in the sequence of anion-metal-anion.

A perspective view of the structure and the unit cell has been given in Figure 1. The cations in the middle of each sandwich are octahedrally surrounded by six anions which form top and bottom sheets of the sandwiches. Each anion is thus bound to three cations within the same sandwich in addition to a weak bonding to the adjacent sandwich. The theoretical results of three materials, SnS_2, $SnSe_2$ and PbI_2 crystallizing in this structure shall be presented. Two different structures which derive from the

204   CHAPTER C

CdI$_2$-type are found for Bi$_2$Te$_3$ and for BiI$_3$ respectively. These more complex structures, in which the cations, however, remain octahedrally coordinated shall be described
later. In Table 15 we list several materials with octahedrally coordinated cations for
which theoretical studies have been undertaken.

The calculations of the CdI$_2$-type materials are based on a structure with positional
parameter $z = \frac{1}{4}$. The atomic positions within the unit cell are given by Equation (C1).
The symmorphic space group associated with the structure is D_{3d}^3 and has been discussed in full detail in Chapter A, Section 1.

TABLE 15

Table of non-transition metal layer-compounds showing octahedral coordination of the cation, for
which theoretical studies are undertaken

Crystals	Lattice constants		Theoretical approaches		
	a(Å)	c/a	EPM	ETB	APW
SnS$_2$ D_{3d}^3	3.64	1.61	Au-Yang, Cohen [58] Fong, Cohen [59] Aymerich et al. [61] Schlüter, Schlüter [60]	Murray, Williams [62]	
SnSe$_2$ D_{3d}^3	3.81	1.61	Au-Yang, Cohen [58] Fong, Cohen [59] Aymerich et al. [61] Schlüter, Schlüter [60]	Murray, Williams [62]	
PbI$_2$ D_{3d}^3	4.56	1.53	Schlüter, Schlüter [94]	Doni et al. [92] [93]	
Bi$_2$Te$_3$ D_{3d}^5	4.39	6.95[a]	Katsuki [77] Borghese, Donato [76] Togei, Miller [79]		Lee, Pincherle [75]
BiI$_3$ D_{3d}^1	7.50[b]	0.92	Schlüter, Cohen, Kohn, Fong [103]		

[a] The hexagonal unit cell of Bi$_2$Te$_3$ contains three layers. A rhombohedral cell with one layer may be
chosen.
[b] BiI$_3$ has a $\sqrt{3}$ times larger a-dimension than the usual CdI$_2$ type structures.

2.1.1. The Tin-Dichalcogenides SnS$_2$ and SnSe$_2$

The semiconducting properties of the tin chalcogenides have been predicted in [57]
assuming tin to form resonating sp^3d^2 bonds. The chemical bonding shall be discussed
later in this spirit on the basis of quantitative calculations. Several band structure calculations have been performed for SnS$_2$ and SnSe$_2$. The first calculations, based on the
empirical pseudopotential method (EPM), were reported in [58]. The group-theoretical
analysis of this work contained several errors. Later a corrected version of these
calculations has been published [59]. An independent EPM calculation has been carried

out at the same time [60] giving essentially identical results. While these calculations were based on empirically determined potential form factors, other pseudopotential calculations of SnS_2 and the SnS_xSe_{2-x} mixed crystal series [61] have been reported using screened atomic model potentials. Recently, the empirical tight-binding method (ETB) has been applied to calculate band structures and photoemission electron distributions of SnS_2 and $SnSe_2$ [62]. This represents an unusual amount of theoretical work done on one type of material. The overall agreement among the results of the various theoretical approaches thus reassures the general quality of the calculations. In Figure 41 we present EPM band structures for SnS_2 (upper plot) and $SnSe_2$ (lower

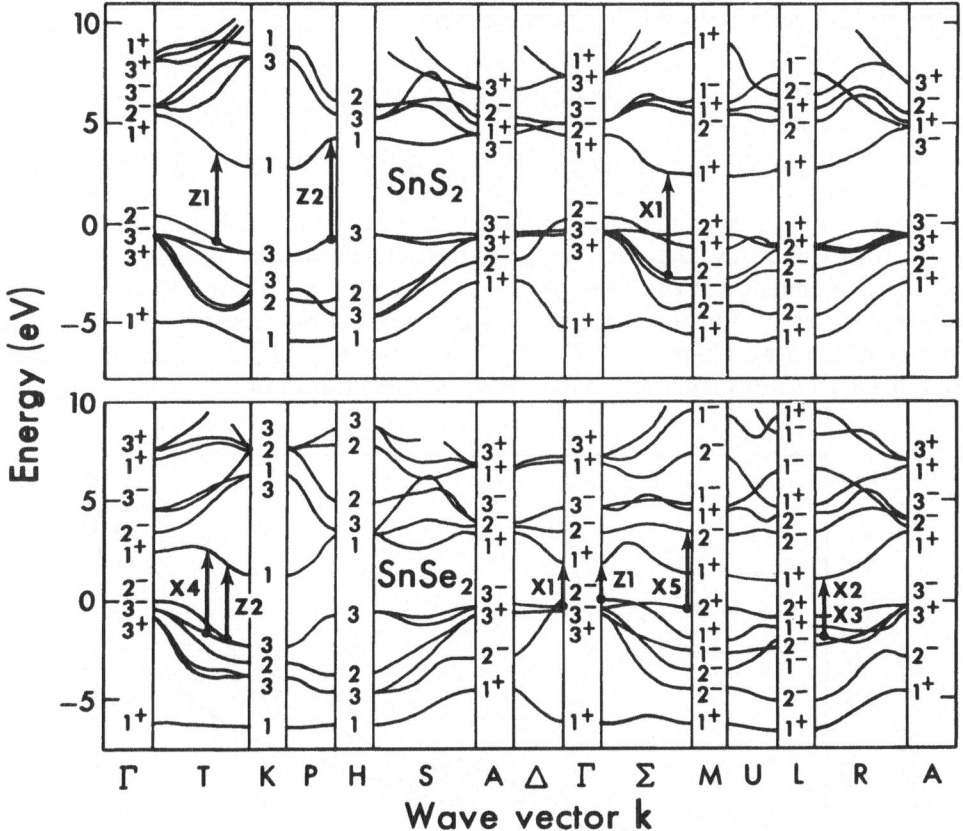

Fig. 41.   Pseudopotential band structure of SnS_2 and $SnSe_2$ along some high symmetryl ines in the hexagonal Brillouin zone. The locations of some main optical transitions are indicated and refer to Tables 16, 17 (after [63]).

plot). The results for SnS_2 are taken from [59] whereas those for $SnSe_2$ are obtained from [60]. Though the band structures of [59] and [60] show reasonable overall agreement and also quantitative agreement in the immediate vicinity of the fundamental gap, the derived spectra can differ somewhat for higher energy optical transitions. We thus chose the two band structures proposed in [63] to most correctly represent optical reflectivity spectra.

Both materials exhibit strong indirect gaps around 1 eV (SnSe$_2$) and 2 eV (SnS$_2$). The exact values have been fitted to experimental data [64] and are listed in [59] and [60]. The gaps appear between the states Γ_2^- and M_1^+. Group theoretical arguments show that this indirect transition is forbidden for polarization $E\perp c$ which is in accord with experimental findings. Less pronounced indirect gaps have been proposed [61, 62] on the basis of model pseudopotentials. In [61] a virtual crystal potential has been devised to describe the mixed crystal series SnS$_x$Se$_{2-x}$. In this version of the pseudopotential method the one electron potential is obtained by renormalizing the dielectric screening of model potentials or potentials used in other compounds by means of a Penn-type semiconductor dielectric function [65]. For the solid solution SnS$_x$Se$_{2-x}$ the potential Fourier transform used in [61] has thus the form:

$$V(q)^x = \frac{1}{\varepsilon_x(q)\Omega_x}\{S_{Sn}(q)\Omega_{GT}\varepsilon_{GT}(q)V_{GT}(q) +$$

$$+ S_{S,Se}(q)[x\Omega_{ZnS}\varepsilon_{ZnS}(q)V_S(q) + (2-x)\Omega_{ZnSe}\varepsilon_{ZnSe}(q)V_{Se}(q)]\}, \quad (C6)$$

where q depends on the composition via the lattice constant which is assumed to vary linearly with x. Ω_x, Ω_{GT}, Ω_{ZnS}, and Ω_{ZnSe} are the unit cell volumes of the mixed crystal, grey tin, zinc sulphide and zinc selenide respectively. ε_a, S_a and V_a are the corresponding Penn-type dielectric functions, structure and atomic form factors. Results for the variation of the lowest gaps $\Gamma_2^- - \Gamma_1^+$, $\Gamma_2^- - M_1^+$ and $\Gamma_2^- - L_1^+$ as a function of composition are given in Figure 42. The energies of these transitions change non-linearly

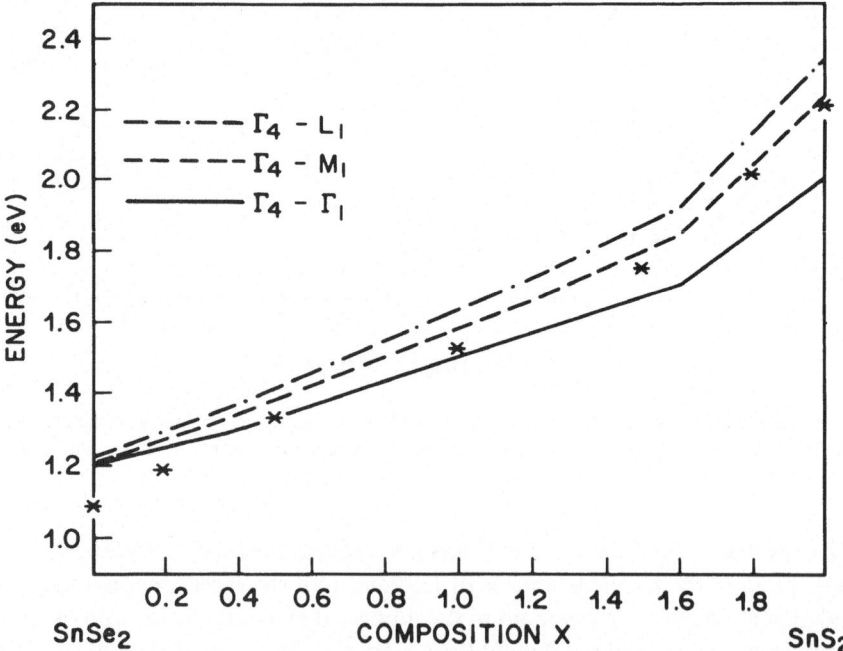

Fig. 42. Variation of the lowest energy gaps in the mixed crystal series SnS$_x$Se$_{2-x}$ with composition as calculated in [61]. Experimental data of [61] are also indicated (asterisks).

with x as a consequence of non-linear variation of the potential. Experimental transmission data of the lowest gap [61] show very similar non-linear behavior as indicated in the figure.

The ETB calculations of [62], though giving similar overall results differ from the band structures shown in Figure 41 in several ways:

(i) The top of the valence bands is found at M and the lowest conduction band at Γ, thus creating a 'reversed' indirect gap.

(ii) A relatively wide (1–3 eV) gap is formed between the first and second conduction bands which should lead to a window in the reflectivity of these materials around 4 eV. This, however, is in contrast to recent reflectivity data [63].

(iii) The low-lying chalcogenide s-states are found at 1–2 eV lower energies than reported in the EPM calculations. This fact seems to be in better agreement with X-ray photoemission (XPS) data [66].

In general, the ETB results seem qualitatively to be correct, larger quantitative uncertainties, however, are expected to occur among the conduction bands.

We shall now discuss the optical properties of SnS_2 and $SnSe_2$ for energies ranging between 3 eV and 6 eV. The theoretical analysis is based on the computation of the imaginary part of the frequency-dependent dielectric function $\varepsilon_2(\omega)$. The calculation of $\varepsilon_2(\omega)$ requires the knowledge of the band structure throughout the Brillouin zone since it may be written as

$$\varepsilon_2(\omega) = \frac{e^2\hbar}{\pi m^2}\frac{1}{\omega^2}\sum_{v,c}\int d\mathbf{k}\ \delta(E_c(\mathbf{k}) - E_v(\mathbf{k}) - \hbar\omega)\ \times$$

$$\times\ |\langle U_{v\mathbf{k}}|\ \nabla\cdot\hat{e}\ |U_{c\mathbf{k}}\rangle|^2, \tag{C7}$$

where the U's are the periodic parts of the valence and conduction band wavefunctions and $E_v(\mathbf{k})$ and $E_c(\mathbf{k})$ are the energies of these states. $\hat{e}$ is the polarization vector of the incident light. The integration over $\mathbf{k}$-space in Equation (C7) may be written as

$$\int d\mathbf{k} = \int \frac{d\mathbf{s}}{|\nabla_\mathbf{k}\omega(\mathbf{k})|}, \tag{C8}$$

where s is a surface of constant interband energy $\hbar\omega(\mathbf{k}) = E_c - E_v$. Besides matrix element effects the structure in $\varepsilon_2(\omega)$ originates from van Hove singularities [67], at critical points where $\nabla_\mathbf{k}\omega(\mathbf{k}) = 0$. These critical points can be classified according to symmetry as minimum M_0, saddle points M_1 and M_2, and maximum M_3.

The interband energies $E_c - E_v$, the dipole matrix elements $\langle U_{v\mathbf{k}}|\ \nabla\cdot\hat{e}\ |U_{c\mathbf{k}}\rangle$ and the energy gradients $\nabla_\mathbf{k}\omega(\mathbf{k})$ are obtained from eigenvalues and eigenvectors of the pseudo-Hamiltonian at a number of mesh points in the irreducible part ($\frac{1}{12}$) of the Brillouin zone. The error arising from calculating the dipole matrix elements by using pseudowave-functions rather than crystal wave functions is estimated to be of the order of 10–20%. The $\mathbf{k}$-space integration is performed using the Gilat-Raubenheimer scheme [68]. In order to compare the theoretical results directly to experimental reflectivity measurements, $R(\omega)$ has to be derived from the $\varepsilon_2(\omega)$ spectrum. This is done by first performing a Kramers-Kronig integration over $\varepsilon_2(\omega)$ resulting in the real part $\varepsilon_1(\omega)$ of the frequency dependent dielectric function

$$\varepsilon_1(\omega) = 1 + \frac{2}{\pi} \int\limits_0^\infty \frac{\omega' \varepsilon_2(\omega')}{\omega'^2 - \omega^2} \, d\omega'. \tag{C9}$$

This equation implies that $\varepsilon_2(\omega)$ is known explicitly over the entire frequency range. An analytic tail is chosen to replace the calculated $\varepsilon_2(\omega)$ at higher energies. A tail function

$$F(\omega) = \frac{\beta\omega}{(\omega^2 + \gamma^2)^2} \tag{C10}$$

is usually used with β and γ as empirical parameters to be determined by requiring continuity of Equation (C10) and $\varepsilon_2(\omega)$ at some cut-off frequency ω_c. $\varepsilon_1(\omega)$ together with $\varepsilon_2(\omega)$ allow the calculation of the reflectivity $R(\omega)$ for normal incidence by use of the formulas

$$\varepsilon_1(\omega) = n(\omega)^2 - k(\omega)^2,$$

$$\varepsilon_2(\omega) = 2n(\omega)k(\omega),$$

$$R(\omega) = \frac{[n(\omega) - 1]^2 + k(\omega)^2}{[n(\omega) + 1]^2 + k(\omega)^2}. \tag{C11}$$

The theoretical reflectivity calculated this way may be compared to experimental data. Even though some uncertainty in the absolute values of $R(\omega)$ usually exists resulting from complicated experimental situations and from an insufficient description of the high energy part of $\varepsilon_2(\omega)$ by Equation (C10), the position of structure in the spectra may directly be compared.

In Figures 43 and 44 we present the calculated and experimental reflectivities for

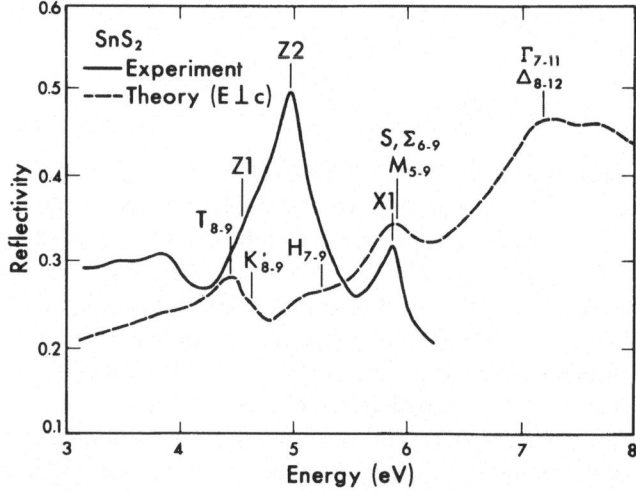

Fig. 43.   Comparison of the experimental (5 K) and calculated reflectivity spectra of SnS$_2$ (see Table 16) (after [63]).

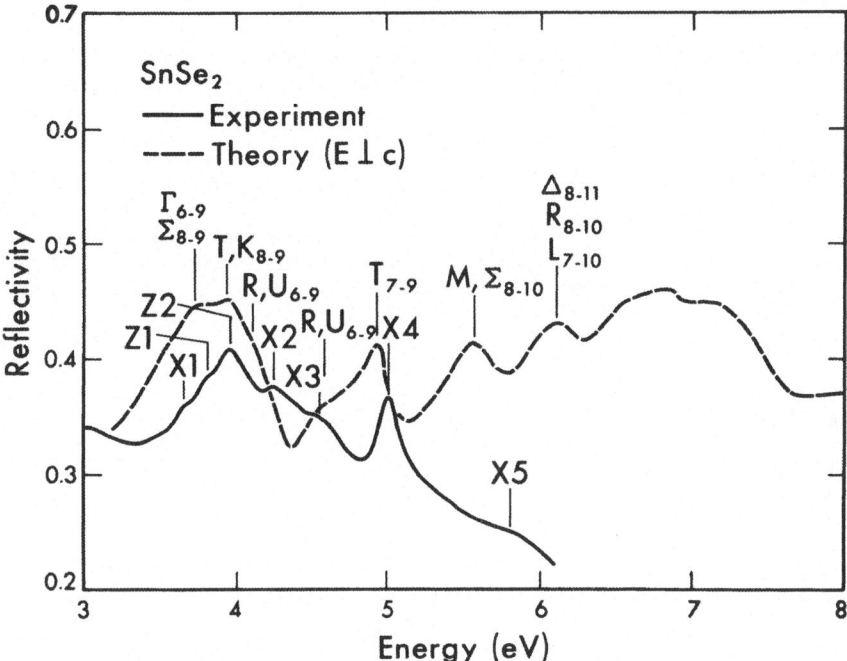

Fig. 44. Comparison of the experimental (5 K) and calculated reflectivity spectra of SnSe₂ (see Table 17) (after [63]).

SnS$_2$ and SnSe$_2$ [63]. The calculations are based on the band structures of Figure 41. In both cases the experimentally determined dielectric response consists of *two* main peaks. In SnSe$_2$ more fine structure as compared to SnS$_2$ appears. The first peak is more spread out coupled to an intensity decrease. The various structures found in the reflectivity spectra of SnS$_2$ and SnSe$_2$ are listed in Tables 16 and 17 respectively.

Structures forming the first main peak in SnS$_2$ around 5 eV correspond to transitions which take place along the lines T and P, including the high symmetry points K and H, between the topmost valence bands (they are degenerate along P) and the lowest conduction band. These transitions are marked as $Z1$ and $Z2$ in the band structure of Figure 41. Inspection of the corresponding wavefunctions shows that in these transition electrons are excited from sulphur p_z-like orbitals into sulphur and tin s- and p-like orbitals. In contrast to these transitions, the second large peak in the reflectivity of SnS$_2$ around 5.8 eV correspond to transitions in which electrons from sulphur p_{xy}-like orbitals are promoted into sulphur and tin s- and p-like orbitals. These transitions are labelled by $X1$ in Figure 41 and take place along the lines Σ and S between lower valence bands and the first conduction band. These findings are summarized in Figure 43 and Table 16. The obvious disagreement in amplitudes between theory and experiment of the 5 eV structure in Figure 43 may be due to the existence of excitonic character in these transitions [63].

Roughly speaking one may attribute the two main peaks to transitions from sulphur p-like orbitals into sulphur and tin s- and p-like orbitals. The splitting of about 0.9 eV between the two peaks is caused by the splitting of the sulphur p-states in the anisotropic

TABLE 16

Listing of energies for the main structures in the experimental reflectivity spectrum of SnS_2. Theoretical assignments are also indicated referring to Figures 41, 43 (after [63]).

					SnS_2				
Experiment: 300 K [63]	2.32	2.82	3.32	3.57	4.37	4.89	5.09	5.74	6.04
							5.41		
Experiment: 5 K [63]	2.5	2.61	3.37	3.61	4.17	4.98	5.22	5.58	
		2.70		3.80	4.53		5.32	5.85	
		2.81		3.92					
		2.90							
		2.95							
Labelling in Figures 41, 43					$Z1$	$Z2$		$X1$	
Theory: Energy		2.70		3.90	4.49	4.65	5.40	5.91	
Assignment [63]		M, L_{8-9}		Σ_{8-9}	T_{8-9}	K_{8-9}	H_{7-9}	Σ, S_{6-9}	
					Sulphur (p_z)			Sulphur (p_{xy})	
					Sulphur (s), tin (s)			Sulphur (s), tin (s)	

TABLE 17

Listing of energies for the main structures in the experimental reflectivity spectrum of $SnSe_2$. Theoretical assignments are also indicated referring to Figures 41, 44 (after [63]).

				$SnSe_2$						
Experiment: 300 K [63]	1.92	2.88	3.52		3.79		4.09	4.27	4.76	5.58
								4.46	5.06	
Experiment: 77 K [63]		2.99	3.56	3.72	3.85		4.14	4.41	4.88	5.23
									5.02	5.66
5 K	1.92	3.05	3.60	3.75	3.88	3.95	4.19	4.45	4.91	5.29
			3.66		3.91	3.99	4.30	4.55	5.06	5.71
						4.05				
Labelling in Figures 41, 44			$X1$	$Z1$	$Z2$		$X2$	$X3$	$X4$	$X5$
Theory: Energy			3.70	3.70		3.95	4.15	4.55	4.90	5.55
Assignment [63]			Γ_{6-9}	Σ_{8-9}		T, K_{8-9}	R, U_{6-9}	R, U_{6-9} Γ_{7-9}	M_{8-10}	

crystal field. Similar splittings which show the p_{xy}-levels at lower energy than the p_z-levels are also found in other layer compounds like GaSe and GaS which will be discussed later in this chapter. It is interesting to note that the two main peaks originate from transitions which take place at locations in k-space which are close to 'special' k-points [47]. These 'special' k-points are entirely determined by the crystal space group symmetry and represent an optimum zero order approximation for evaluating quantities which require k-space integration. We shall briefly describe this very elegant method which is mainly used to calculate valence electron charge densities. Its application to the evaluation of dielectric response functions is suggested by the present and other results but has not quantitatively been investigated yet.

Let us suppose we want to evaluate a quantity $\rho(\mathbf{r}, \mathbf{k})$ which is a periodic function in $\mathbf{k}$-space. Due to this periodicity it may be expanded in a (real-space) Fourier series

$$\rho(\mathbf{r}, \mathbf{k}) = \rho_0(\mathbf{r}) + \sum_{m=1}^{\infty} \rho_m(\mathbf{r}) \, e^{i\mathbf{k}\cdot\mathbf{R}_m}, \tag{C12}$$

where $\mathbf{R}_m$ are lattice vectors in real space. From $\rho(\mathbf{r}, \mathbf{k})$ we construct a function

$$\tilde{\rho}(\mathbf{r}, \mathbf{k}) = \frac{1}{M} \sum_{\{T\}} \rho(\mathbf{r}, T\mathbf{k}) \tag{C13}$$

which has the complete symmetry of the Bravais lattice. $\{T\}$ represents the set of M point group symmetry operations of the Bravais lattice.

We may now express $\tilde{\rho}(\mathbf{r}, \mathbf{k})$ in the form

$$\tilde{\rho}(\mathbf{r}, \mathbf{k}) = \rho_0(\mathbf{r}) + \sum_{m=1}^{\infty} \rho_m(\mathbf{r}) A_m(\mathbf{k}) \tag{C14}$$

with

$$A_m(\mathbf{k}) = \frac{1}{M} \sum_{\{T\}} e^{i\mathbf{k}\cdot T\mathbf{R}_m}. \tag{C15}$$

Equation (C15) associates each $A_m(\mathbf{r})$ with a particular star of real space lattice vectors $\mathbf{R}_m$. If we carry out the $\mathbf{k}$-space integration over the Brillouin zone formally we, of course, obtain

$$\rho(\mathbf{r}) = \rho_0(\mathbf{r}). \tag{C16}$$

Using Equation (C14) this result is equivalent to

$$\rho(\mathbf{r}) = \tilde{\rho}(\mathbf{r}, \mathbf{k}) - \sum_{m=1}^{\infty} \rho_m(\mathbf{r}) A_m(\mathbf{k}). \tag{C17}$$

In other words, the integrated quantity $\rho(\mathbf{r})$ is equal to the (lattice-symmetric) value $\tilde{\rho}(\mathbf{r}, \mathbf{k})$ at one particular $\mathbf{k}$-point plus corrective terms. The object of the method is to make as many of these corrective terms $A_m(\mathbf{k})$ equal to zero as possible by an appropriate choice of the particular point $\mathbf{k}$. The method may be improved by use of more than one $\mathbf{k}$-points. Those N $\mathbf{k}$-points, $\{\mathbf{k}_i\}$, and their weighting factors α_i, are then determined by

$$\sum_{i=1}^{N} \alpha_i A_m(\mathbf{k}_i) = 0, \qquad m = 1, 2, 3 \ldots,$$

$$\sum_{i=1}^{N} \alpha_i = 1, \tag{C18}$$

where m runs over as many stars as possible. Since the expansion coefficients $\rho_m(\mathbf{r})$ drop rapidly in magnitude with increasing m we should have to have good approximation

$$\rho(\mathbf{r}) = \sum_{i=1}^{N} \alpha_i \rho(\mathbf{r}, \mathbf{k}_i). \tag{C19}$$

In general the solutions of Equation (C18) are not unique; several procedures to solve them and to find the 'special' k-points in the irreducible part of the Brillouin zone are given in [47]. For a hexagonal lattice one finds for $N = 1$ (one k-point) either $\mathbf{k} = (\frac{1}{3}, 0, \frac{1}{4})$ or $\mathbf{k}$ (0.1904, 0.1904, $\frac{1}{4}$) which both satisfy Equation (C18) for the first $\{\mathbf{R}_m\}$ star along z and for the first star in the plane $z = 0$. These two points are located in the $T - S$ plane and $\Sigma - R$ plane respectively of the hexagonal Brillouin zone shown in Figure 3 in Chapter A.

The main transitions determining the optical spectrum of SnS_2 are found to originate exactly from these regions. In other words, a reasonable approximation to the dielectric response of SnS_2 between 2 and 6 eV may be obtained by evaluating the band structure at only two 'special' k-points.

The interpretation of the $SnSe_2$ spectra given in [63] leads to conclusions similar to those obtained for SnS_2. To illustrate the results, charge density contour maps of initial and final states involved in typical transitions are presented in [63]. The contour maps are obtained by plotting and quantity $\rho(\mathbf{r}, \mathbf{k}) = |\psi_i(\mathbf{r}, \mathbf{k})|^2$ as a function of position in a particular plane in real space. The plane chosen is a (110) plane (see Figure 1) which starts and ends in the middle of two $SnSe_2$ layers and contains the cation-anion bonds. In addition an optical transition function is defined

$$\mathbf{F}_{if}(\mathbf{r}, \mathbf{k}) = \psi_i(\mathbf{r}, \mathbf{k})\mathbf{p}\psi_f(\mathbf{r}, \mathbf{k}) \tag{C20}$$

between initial and final states where $\mathbf{p}$ represents the momentum operator. If integrated over a real space

$$\int \mathbf{F}_{if}(\mathbf{r}, \mathbf{k}) \, d\mathbf{r} = \mathbf{M}_{if}(\mathbf{k}) \tag{C21}$$

which is the usual transition dipole matrix element describing the strength of the transition. A real space plot of $\mathbf{F}_{if}(\mathbf{r}, \mathbf{k})$ before integration yields additional information regarding the real space localization and polarization of transitions. By examining the transition functions one may discern between atomic-like (intra-ionic), charge transfer-like (inter-ionic) or bond-like (bonding-antibonding) transitions.

To visualize the results for $SnSe_2$ concerning transitions between crystal field split anion p-levels and mixed anion-cation conduction bands, two representative transitions are investigated.

As first transition one may take $Z1$ (see Figures 41, 44) representing an excitation of selenium p_z-like electrons into selenium and tin s- and p-like orbitals (Figure 45). As seen from the transition function (Figure 45(c)) the optical excitation is strongly localized around the anions and exhibits $\mathbf{E}\|c$ polarization dependence. Note the strong, clear selenium p_z-like character of the valence band wavefunction.

As second representative transition one may choose the M_{8-10} transition which occurs at slightly higher energy than the experimental $X4$ peak (see Figure 44). Both the T_{7-9} transition which probably corresponds to the $X4$ peak and the M_{8-10} transition are of comparable character with the latter being more pronounced. In Figure 46 charge density contour maps of initial and final states are displayed together with the corresponding transition function. Again, the excitation is strongly localized around the anions.

Additional, independent experimental information on the valence electron structure is obtained from ultraviolet (UPS) and X-ray (XPS) photoelectron spectroscopy. Recently, the tin dichalcogenides SnS_2 and $SnSe_2$ have been investigated by both

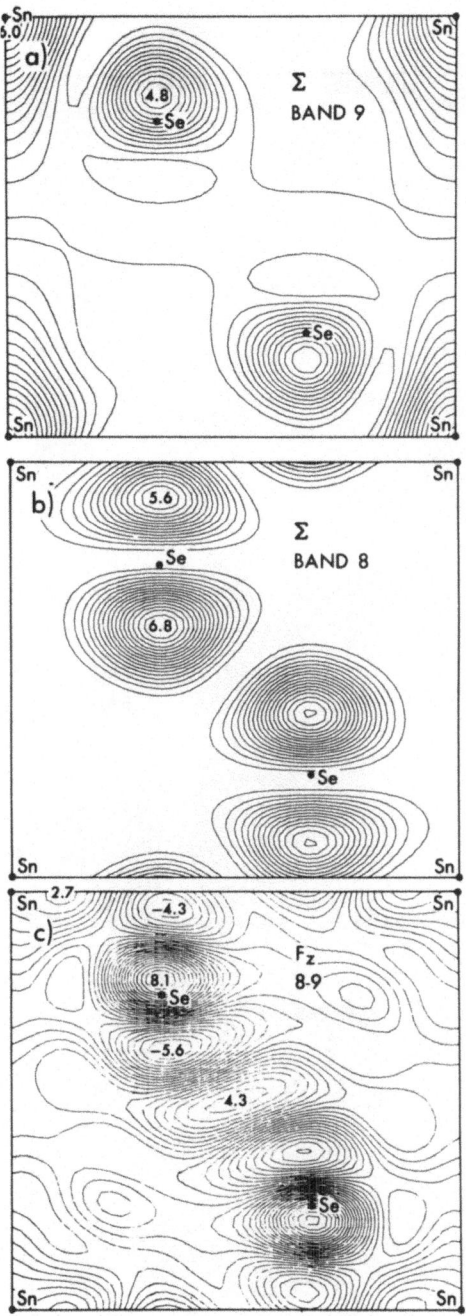

Fig. 45.   Charge density contour plots of conduction band state (a) and valence band state (b) at the line Σ in the Brillouin zone of $SnSe_2$. The transition between these states, which gives rise to the Z1, Z2 structures indicated in Figure 44 and in Table 17, is chosen as representative for anion p_z-like transitions. The transition function F_z (c) which indicates the real space location (around the anions) of the transition is defined in the text. All contour plots are displayed in a (110) plane extending over two half layers of $SnSe_2$. The charge density contours are given in units of electrons per unit cell. The transition function contours are given in atomic units (after [63]).

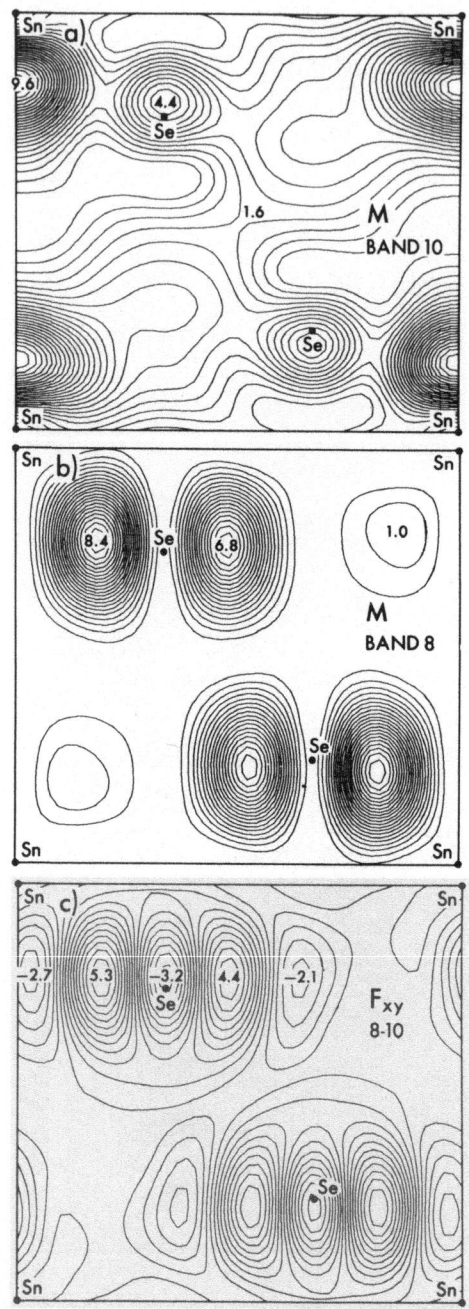

Fig. 46.   Charge density and transition function contour plots for the transition X5 between states in bands 8 and 10 at the point M in the hexagonal Brillouin zone of SnSe$_2$. The transition is chosen representative for anion p_{xy}-like transitions. Units are as in Figure 45 (after [63]).

methods [66, 69]. Whereas XPS data may be used to study the overall density of states of all valence electrons, low energy UPS data (up to 10 eV) yield information about both, valence and conduction band distributions. The UPS data of [66] have been studied theoretically on the basis of the ETB calculation of [62]. To compare the experimental photoelectron energy distribution curves with theory, two approaches have been followed.

(i) The concept of non-direct transitions which simply results in a convolution of valence and conduction band density of states. Thus the number of electrons emitted from initial states of energy E, when light of energy $\hbar\omega$ is used, is given by

$$D^{n-d}(E, \hbar\omega) = N_c(E + \hbar\omega) \, N_v(E), \tag{C22}$$

where the individual density of states is given by

$$N(E) = \frac{1}{(2\pi)^3} \int d\mathbf{k} \, \delta(E - E(\mathbf{k})). \tag{C23}$$

(ii) The direct transition model which results in the energy distribution of the joint density of states

$$D^{\mathrm{dir}}(E, \hbar\omega) = \frac{1}{(2\pi)^3} \sum_{v,c} \int d\mathbf{k} \, \delta(E_c(\mathbf{k}) - E_v(\mathbf{k}) - \hbar\omega) \, \delta(E - E_v(\mathbf{k})). \tag{C24}$$

This second approach is generally believed to give a better approximation to the measured EDC's. In either case, any matrix elements effects, surface effects and effects related to the finite escape length of photoelectrons are neglected.

In Figures 47(a) and (b) we show a comparison between experimental EDC's and theoretical distributions based on the direct transition model and the band structures of [62]. Though the quantitative agreement is only fair, there is qualitative agreement in the progression from one peak, through two, to three peaks as $\hbar\omega$ increases. The limited quality of the tight-binding conduction bands may be the reason for the limited quantitative correspondence.

Finally the XPS data of [66] may be compared to complete density of states calculations of the valence bands. This is done in Figure 48 in which experimental XPS results for $SnSe_2$ [66] are compared with EPM density of states results [70], based on the calculations of [60]. The theoretical density of states curve is computed from Equation (C23) using the Gilat-Raubenheimer k-space integration scheme [68] as discussed before. Good agreement is found between the theoretical and experimental curves except for the position of the low-lying anion s-states around -12 eV. Though the EPM value is probably somewhat too high in energy, the experimental position of the Se $4s$ level is not clearly determined by the XPS data due to the simultaneous excitation of Sn $4d$ electrons by the Al $K_{\alpha_3\alpha_4}$ X-ray satellite, which overlaps in energy with the Se $4s$ electrons excited by the main Al $K_{\alpha_1\alpha_2}$ X-ray line. A comparison of the experimental values with theoretical ETB and EPM values is listed in Table 18.

We shall use the valence electron density of states of Figure 48 in connection with charge density contour maps derived for various groups of bands to discuss the nature

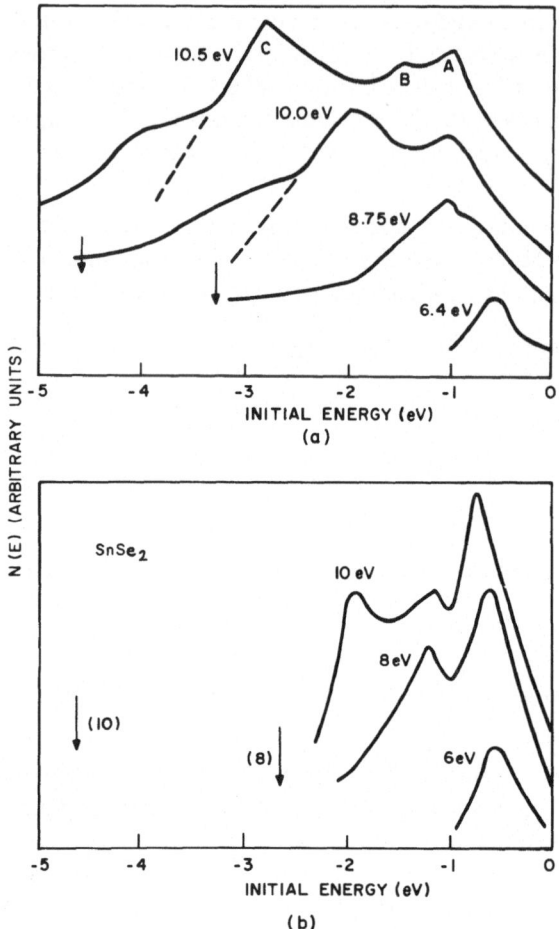

Fig. 47. (a) Photoelectron energy distribution curves (EDC) for SnSe$_2$, cleaved in ultra high vacuum and (b) Calculated (ETB) energy distributions of the joint density of states (after [69]).

of chemical bonding in SnSe$_2$ (and SnS$_2$) [70]. For this purpose we define a band charge density

$$\rho_n(\mathbf{r}) = e \sum_{\mathbf{k}} |\psi_n(\mathbf{r}, \mathbf{k})|^2 = e \sum_{\mathbf{k}} \rho_n(\mathbf{r}, \mathbf{k}) \tag{C25}$$

with $\mathbf{k}$ running over all $\mathbf{k}$-states in the Brillouin zone. The total charge density is then given by summing over all valence bands

$$\rho(\mathbf{r}) = \sum_n \rho_n(\mathbf{r}) = e \sum_{\mathbf{k}} \rho(\mathbf{r}, \mathbf{k}). \tag{C26}$$

Since usually the wavefunctions of $\psi_n(\mathbf{r}, \mathbf{k})$ are evaluated for $\mathbf{k}$-points within the irreducible part of the Brillouin zone only, $\rho(\mathbf{r}, \mathbf{k})$ has to be symmetrized according to the identical representation of the full crystal space group. This symmetrization procedure is completely analogous to the one necessary to obtain symmetrized basis functions of

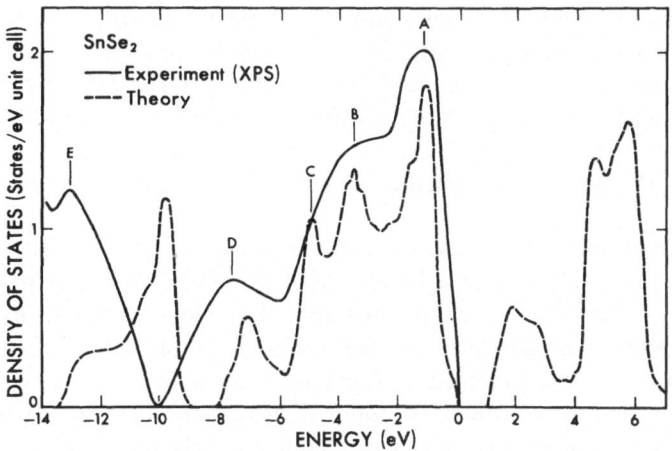

Fig. 48. Calculated density of states for SnSe$_2$, based on the EPM calculations of [63] compared to experimental photoemission data of [66].

TABLE 18

Listing of energies for the main peaks in the experimental photoemission spectrum of SnSe$_2$ (after [66]) compared to theoretical density of states calculations (see Figure 48).

	SnSe$_2$			
Structure	Experiment [66]	EPM [59]	ETB [62]	Predominant nature of states
A	− 1.0	− 1.0	− 1.0	
B	− 3.6	− 3.6	− 3.0	Se(p)
C	–	− 5.0	− 4.0	
D	− 7.7	− 7.1	− 7.0	Sn(s)
E	− 13.0[a]	$\begin{cases} -10.0 \\ -12.6 \end{cases}$	− 13.0	Se(s)

[a] The Sn $4d$ states excited by the Al $K_{a_3 a_4}$ satellite overlap in this region.

the Hamiltonian which has been described in length in Chapter A. It remains to carry out the **k**-summation in Equation (C26) over the irreducible part of the Brillouin zone. This summation was replaced in [70] by a summation over three 'special' **k**-points, which were discussed in connection with the reflectivity spectra. The coordinates of the three 'special' **k**-points with respect to the hexagonal coordinate system are

$$\mathbf{k}_1 = \left(\tfrac{1}{3}, \tfrac{1}{9}, \tfrac{1}{4}\right)$$

$$\mathbf{k}_2 = \left(\tfrac{2}{9}, \tfrac{2}{9}, \tfrac{1}{4}\right)$$

$$\mathbf{k}_3 = \left(\tfrac{4}{9}, \tfrac{1}{9}, \tfrac{1}{4}\right) \tag{C27}$$

218   CHAPTER C

with a weighting factor of $\frac{1}{3}$ for each point. The three points satisfy Equations (C18) for the first eight stars in $SnSe_2$ except for the two starts with no $x - y$ dependence and even z-coordinates $\{R_4\} = \{\begin{smallmatrix}0&0&2\\0&0&2\end{smallmatrix}\}$ and $\{R_{14}\} = \{\begin{smallmatrix}0&0&4\\0&0&4\end{smallmatrix}\}$. The choice of these k-points for $SnSe_2$ which has D_{3d}^3 symmetry has the further advantage that all three points lie on symmetry planes. The points therefore need to be considered in $\frac{1}{24}$ of the Brillouin zone only in spite of the fact that the irreducible part extends of $\frac{1}{12}$ of the Brillouin zone for D_{3d}^3.

The band structures in Figure 41 show that the sixteen electrons per SnS_2 or $SnSe_2$ unit cell fill the lowest eight valence bands which are separated by a semiconducting gap from the empty conduction bands. According to the general principle of crystal chemistry, namely the filling of the anion subshells with available cation electrons, the eight filled bands in $SnSe_2$ are expected to originate from anion s- and p-orbitals. This is essentially confirmed by the calculated total (EPM) charge distribution, which is shown in Figure 49. It can be seen from this figure that most of the charge is located around the

Fig. 49.   Total valence charge density of $SnSe_2$. The units and the location of the displayed plane are chosen as in Figure 45 (after [70]).

anion sites. A comparison of the experimental anion-cation bond length which is 2.57 Å for $SnSe_2$ and 2.67 for SnS_2 with the sum of covalent (2.43 Å and 2.57 Å) or ionic (2.55 Å and 2.69 Å) radii also favors the ionic picture. The total charge, however, is not exactly centered around the atomic positions thus indicating an incompletely filled anion subshell. In fact, total filling of the anion valence shells would ionize the tin atoms complete to Sn^{4+} which is in contrast to Figure 49. Moreover the XPS and

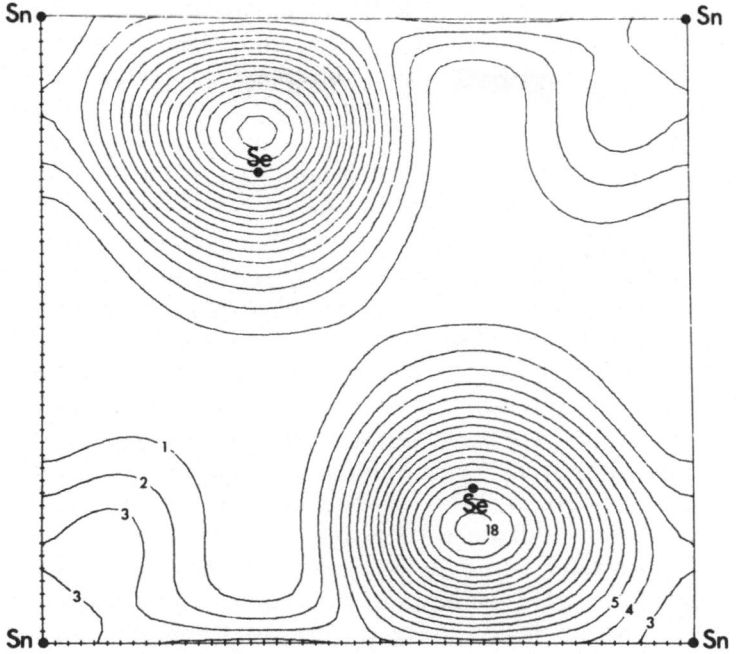

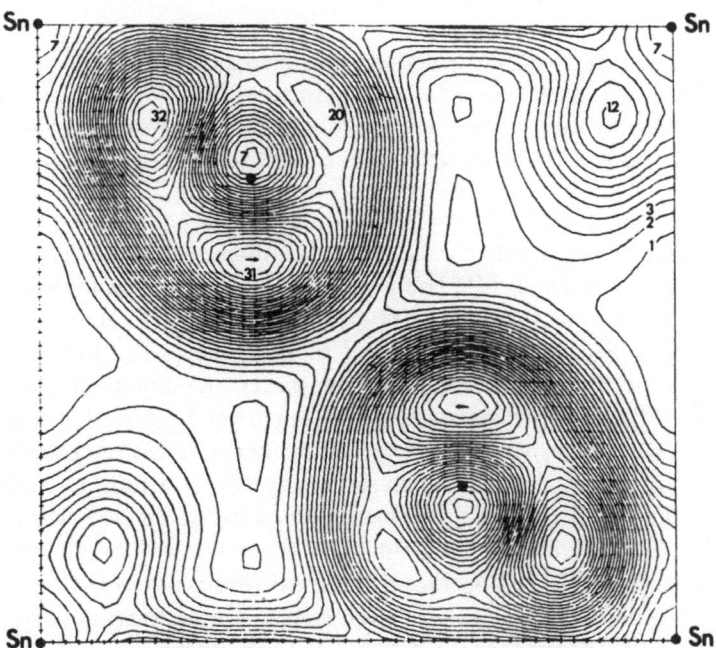

Fig. 50. Partial band charge density for bands 1 and 2 of SnSe₂ corresponding mainly to the anion *s*-states (upper plot) and partial band charge density for bands 3 to 8 of SnSe₂ including cation *s*- and anion *p*-states (lower plot). Plane and units are given as in Figure 45 (after [70]).

calculated density of states results displayed in Figure 48 indicate the presence of somewhat filled cation states. A detailed analysis of the valence band density of states shows that the upper three peaks, labelled A, B and C correspond essentially to anion p-states whereas the experimentally well resolved structure around $-8\,\mathrm{eV}$ (D) originates predominantly from cation s-states. Some cation character, including p- and d-like character is in fact found spread over the whole upper part of the spectrum. These facts may be illustrated by charge density plots for isolated groups of bands. Thus in Figure 50 (upper plot) we display the charge for the lowest two valence bands and in Figure 50 (lower plot) the charge for bands 3 to 8. The results show that no clear cut separation into anion s and p bands is possible. Indeed, the spheroidal shape of the charge cloud around Se in Figure 50 (upper plot) is misleading; the displacement of the density maximum with respect to the Se site indicates an appreciable p-admixture to bands 1 and 2. Similarly, the charge cloud corresponding to bands 3–8 (lower plot) is not centered on the anion site. Its central minimum which for pure p-bands should in fact be a node, is shifted away from the Se position, thus indicating $s - p$ hybridization. The total charge shown in Figure 49 is somewhat more centered. There are three equivalent planes $(1\bar{1}0)$ (shown), $(1\bar{2}0)$ and $(\bar{2}10)$ at angles of $120°$ with respect to each other which explain the symmetry of the shown charge distribution. The charge seems to arise to large extent from sp^3 hybridization. Three of the four tetrahedral legs point to cations in the layer whereas the fourth leg (pointing to the neighboring layers) constitutes lone pair electrons. The three charges constituting the Sn–Se bonds formed by each Se are not completely balanced by the charge of the lone pair. Each Se therefore carries a dipole compatible with the hexagonal symmetry of the crystal and pointing into the three-fold SnSe_2 layer.

The total charge at the cation site may be described as arising from a hybridization of the type sp^3d^2 which is appropriate for octahedral coordination [57]. On the other hand ionic contributions to the bonding are not negligible. The cation p-states are mainly found as empty conduction band states in the band structure calculation.

2.1.2. *The Five-Layer Compound Bi_2Te_3*

A number of layer compounds crystallize in the so-called $\mathrm{Bi}_2\mathrm{Te}_3$ structure which is derived from the CdI_2 structure. A perspective view of the structure is shown in Figure 51. Instead of three atomic sheets as in the CdI_2 structure, each sandwich is built by five atomic sheets. For $\mathrm{Bi}_2\mathrm{Te}_3$ the sequence of these sheets is Te(2)–Bi–Te(1)–Bi–Te(2). As indicated by the numbers 1 and 2 in parentheses, the anions in this structure are present in two different chemical states. As seen from Figure 51, the anions Te(1) forming the middle atomic sheet are octahedrally coordinated to six cations whereas the outermost anions Te(2) are strongly bound to three cations as in the CdI_2-type structure. In addition these Te(2) anions are responsible for the weaker inter-layer coupling. An interlayer distance of about 3.6 Å which is considerably smaller than the van der Waals distance of 4.4 Å suggests the existence of a small amount of direct chemical bonding between the five layer sandwiches. The cations occupying two atomic sheets are chemically identical and octahedrally coordinated.

A schematic bonding picture has been proposed for $\mathrm{Bi}_2\mathrm{Te}_3$ neglecting the weak inter-layer interaction [71]. In this picture which predicts a semiconducting behavior bonding has been proposed involving sp^3d^2-type orbitals on Bi and the middle layer Te(1)

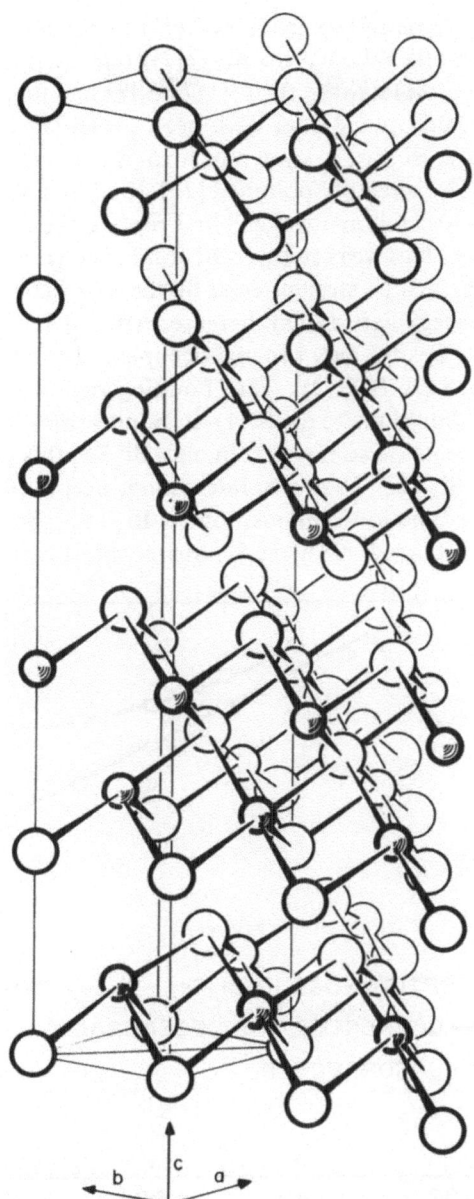

Fig. 51.   Perspective view of the Bi_2Te_3 structure. The small circles mark the cations, whereas the large open circles indicate the anions of the five-fold layers (after [57]).

anions. Only p-type orbitals of the outer Te(2) anions should contribute to the bonding. All s- and p-electrons of Bi and Te(1) and only p-electrons of Te(2) are assumed to contribute to the bonding, whereas the two s-electrons of Te(2) form a lone pair. A fully saturated bonding structure with two electrons per bond is obtained. The total number of bonding valence electrons per unit cell is twenty-four (four from each of the Te(2) atoms,

five from each of the Bi atoms and six from the Te(1) atom) and they are shared between
the twelve bonds per unit cell. Whereas the Bi–Te(2) bond is assumed to be strongly covalent with a fair ionic component, the Bi–Te(1) bond is less ionic and thus weaker.

Interesting experimental optical data have been presented [72] which seem to confirm this bonding picture. Bi_2Te_3 is known to form mixed crystals with Se. Moreover,
an ordered form Bi_2SeTe_2 is readily obtained [72]. If tellurium is replaced by selenium,
the more electronegative selenium atom will first replace the more weakly bound Te(1)
atoms in the middle layer. This weaker Te(1)–Bi bond which is assumed to be responsible
for the optical gaps will thus be strengthened by the increased ionicity of the Se(1)–Bi
bond. As a result the energy gap should increase. After all Te(1) sites are occupied by
Se atoms, additional Te(2) sites may become occupied at random. This net effect is to
attract more charge into the $BiTe(2)_{1-x}Se_x$ bonding region (i.e. the outer regions of
the sandwiches) which should make the Se(1)–Bi bond somewhat less ionic. Therefore,
the energy gap should decrease somewhat. In pure Bi_2Se_3 this gap should however still
be larger than in pure Bi_2Te_3. This very interesting, non-monotonic behavior of the
lowest energy gap as a function composition in $Bi_2Te_{3-x}Se_x$ has been observed in
absorption [72] and is displayed in Figure 52. Moreover, it is shown in [72] that a large

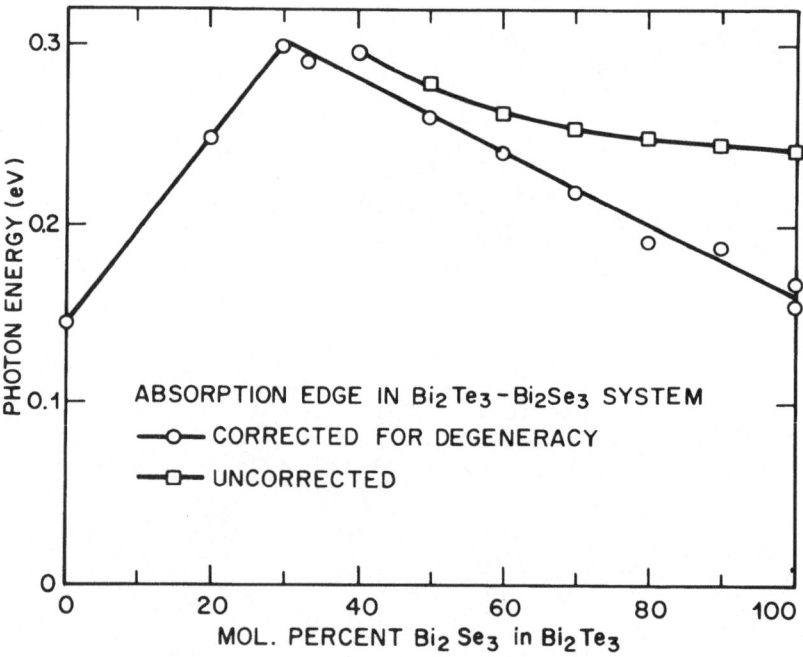

Fig. 52. Experimental absorption edge in the $Bi_2Te_{3-x}Se_x$ mixed crystal system as a function of
composition x (after [72]).

number of peaks at higher energy (up to 10 eV) observed in reflectivity exhibit similar
two-slope behavior. Recent X-ray crystallographic studies confirmed the existence of
similar ordering effects in the $Sb_2Te_{3-x}Se_x$ mixed crystal series [73].

In addition to these very interesting bonding properties met in poly-anionic compounds, Bi_2Te_3 has been widely studied for its galvanomagnetic and thermoelectric

properties. Experimental results published before 1963 are reviewed in [74]. These experiments indicate a small gap with multivalley structure in both valence and conduction bands.

Several theoretical studies on Bi_2Te_3 exist so far. The bandstructure was first calculated in 1963 by use of the APW method based on ad-hoc potentials [75]. No relativistic corrections were taken into account, except for the spin-orbit coupling, whose effect was estimated semi-empirically. This approximation, which is frequently used in semi-empirical methods seems to be insufficient if used in connection with an a-priori method. It is thus not surprising that the APW results of [75] are not sufficient for an understanding of the various physical properties of Bi_2Te_3. Three more band structure calculations, all based on the pseudopotential scheme have been reported:

(i) The semiempirical approach (EPM) is used in [76]. The Fourier coefficients of the pseudopotential are fitted such as to obtain an energy gap throughout the Brillouin zone without any band overlap. This semiconducting band structure is then perturbed by a spin-orbit interaction which to a first order is evaluated in the tight-binding scheme. The resulting band structure is claimed to give a reasonable interpretation of the optical spectrum and the galvanomagnetic properties of the material.

(ii) The pseudopotential used in [77] is a non-local model potential of the Heine-Animalu type [78]. These ionic potentials are screened by a free-electron type dielectric function. Spin-orbit terms are included fully in the Hamiltonian matrix. The spin-orbit interaction was found necessary to avoid band overlap and to create a semiconductor. Slight corrections to the model potentials are introduced to distinguish between the two types of Te-atoms. The resulting band structure is claimed to be consistent with galvanomagnetic experiments. No interpretation of optical data is attempted.

(iii) Finally, in the calculations of [79] a non-local pseudopotential of the Lin-Kleinman type [80] is used. Similarly to (ii) spin-orbit corrections are fully taken into account. Screening of the ionic potentials is performed by use of a dielectric function such that the potential converges for $q \to 0$. The empirical parameters defining the potential are adjusted to reproduce galvanomagnetic results.

Even though the calculations of (ii) and (iii) give similar results for bands in the vicinity of the Fermi level, the three calculations cannot be used to establish a consistent picture of the valence electron structure in Bi_2Te_3. More theoretical work is thus desirable. Let us now present some particular results of these pseudopotential band-structure calculations. All band structures are calculated assuming a rhombohedral unit cell of Bi_2Te_3 which contains one formular unit. This unit cell may be chosen in place of a hexagonal unit cell extending over three sandwiches. The space group of the structure is D_{3d}^5. The basic translation vectors of the rhombohedral cell are

$$\mathbf{a}_1 = \frac{a}{\sqrt{3}}\mathbf{i} + \frac{c}{3}\mathbf{k},$$

$$\mathbf{a}_2 = \frac{a}{2\sqrt{3}}\mathbf{i} + \frac{a}{2}\mathbf{j} + \frac{c}{3}\mathbf{k},$$

$$\mathbf{a}_3 = \frac{a}{2\sqrt{3}}\mathbf{i} - \frac{a}{2}\mathbf{j} + \frac{c}{3}\mathbf{k}, \tag{C28}$$

where $a \approx 4.4$ Å and $c \approx 30.5$ Å as shown in Table 15. The positions of the five atoms in the unit cell are given with respect to these lattice vectors by

$$\text{Te}(1) \quad (0, 0, 0)$$

$$\text{Te}(2) \pm (x, x, x), \qquad x = 0.212 \tag{C29}$$

$$\text{Bi} \quad \pm (y, y, y), \qquad y = 0.400.$$

The Brillouin zone of the rhombohedral lattice is presented in Figure 53. The zone is highly compressed along the Γ-Z direction because of the large c/a ratio. High symmetry points and lines are indicated.

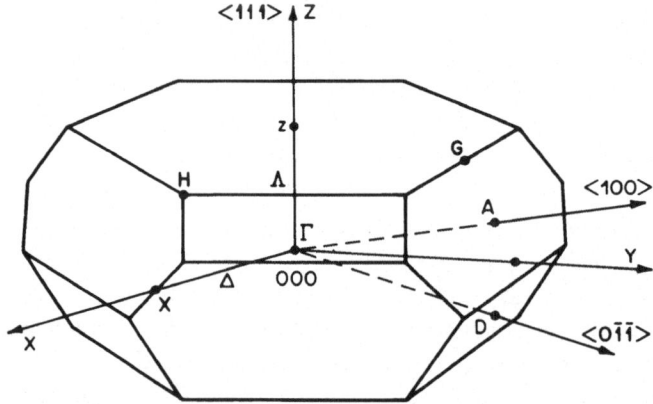

Fig. 53.   Perspective view of the rhombohedral Brillouin zone of Bi_2Te_3. Some high symmetry points are indicated (after [76]).

We may note with respect to the number and location of band extrema reported by the galvanomagnetic measurements that:

(i) the number of extrema is twelve for a general **k**-point;

(ii) the number of extrema is six, if located in the Γ-A-D reflection plane; and

(iii) the number of extrema reduces to three, if located at A or D.

No multivalley structure is found for extrema located at Γ or Z.

In Figures 54(a) and (b) we display the bandstructures as calculated by the EPM of [76] without and with spin-orbit interaction. It can be seen from these figures, that the semiconducting properties are not altered by the inclusion of spin-orbit effects. The pseudopotential form factors are adjusted to yield a semiconductor even in the absence of spin-orbit interaction. This procedure seems to be more satisfying than to invoke spin-orbit effects to stabilize the crystal structure. On the other hand, an unusual set of Bi and Te form factors has to be used (see [76]) to achieve the stability. A group of at least ten valence bands (in two groups of five each) are shown as the uppermost valence bands. This result seems to indicate that the anion p-levels (nine bands) overlap with the cation s-levels. This situation has some support from recent UPS data [81] which shall be shown later.

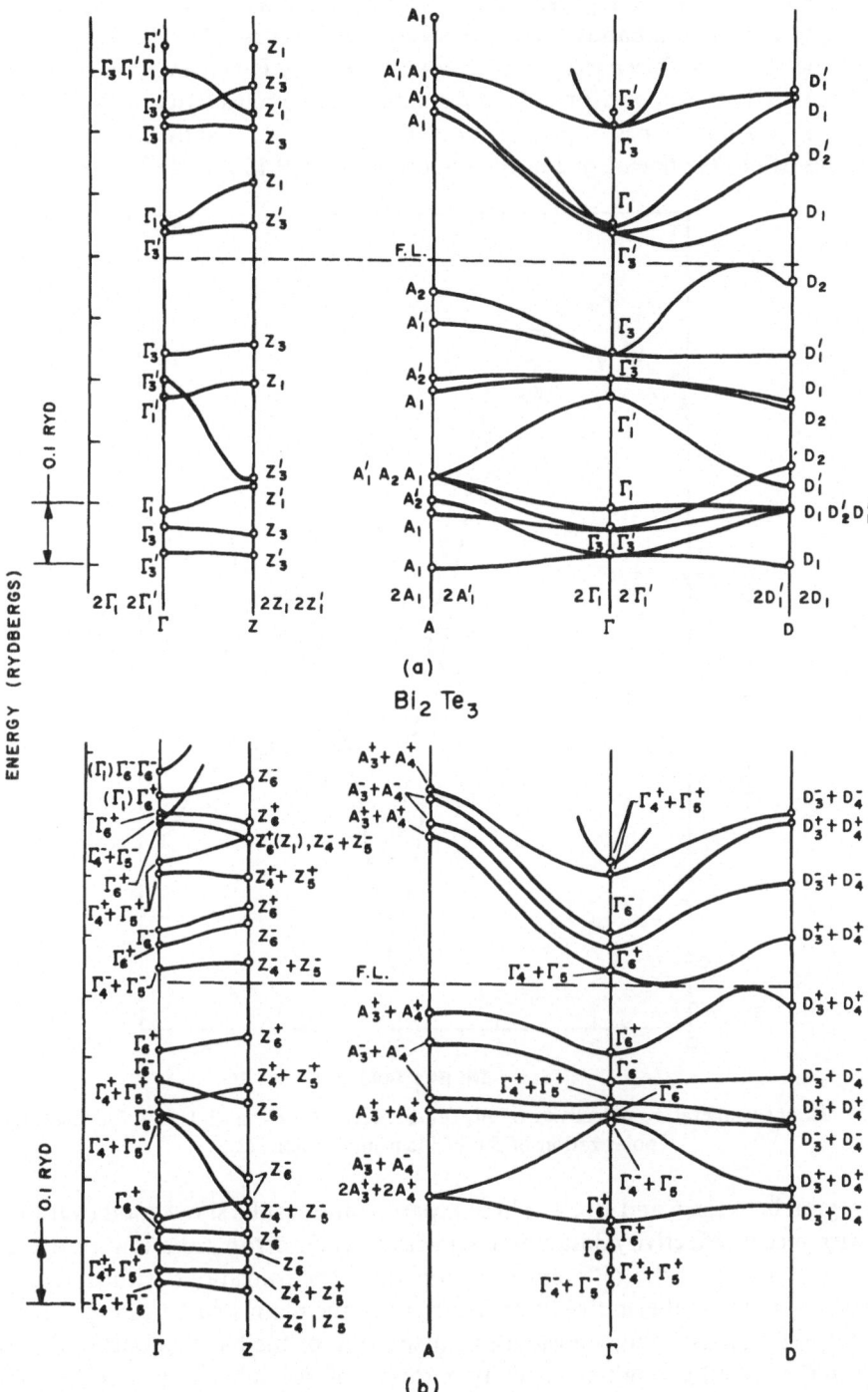

Fig. 54. (a) EPM band structure of [76] for Bi_2Te_3 neglecting spin-orbit interaction, and (b) with the inclusion of spin-orbit coupling.

The band structures in Figures 54(a) and (b) predict a six-valley structure for both valence and conduction bands occurring along the direction Γ–D. To test the validity of the band structure for energies further away from the Fermi level, reflectivity results of [72] are used for comparison. The room temperature reflectivity of Bi_2Te_3 between 0 and 12 eV for both types of polarization ($E \perp \hat{c}$ and $E \| \hat{c}$) is shown in Figure 55 [72]. Two series of peaks appear, one in the region between 0.15 and 4 eV and another one

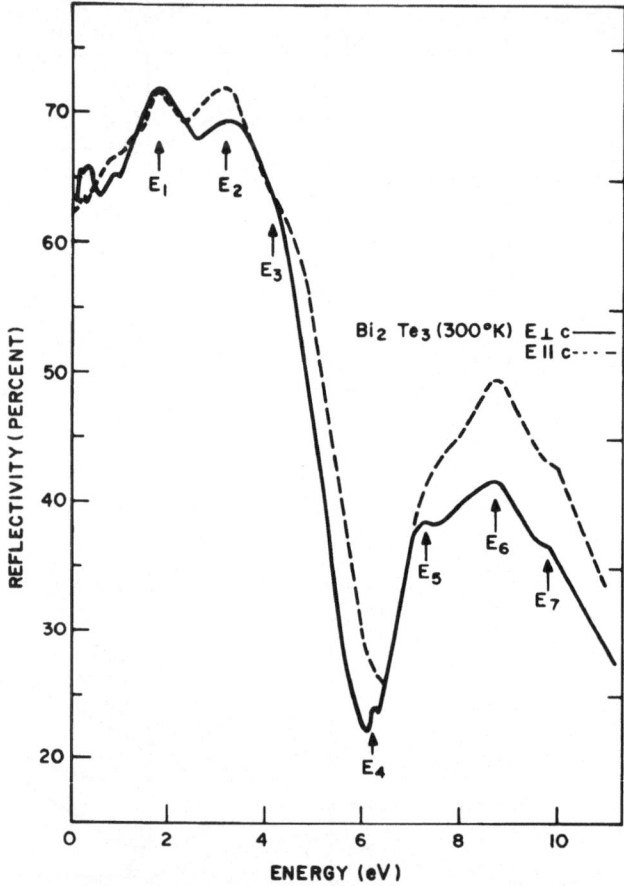

Fig. 55. Room temperature reflectivity of Bi_2Te_3 between 0.1 eV and 11 eV for $E \| \hat{c}$ and $E \perp \hat{c}$ polarization of the incident light (after [72]).

in the region between 6 and 10 eV. These experimental results show good overall agreement with recent reflectivity data obtained with a synchrotron radiation source [82]. In [72] use was made of the different selection rules for transitions with different light polarization and two alternative interpretations of the main structures were given. The first interpretation assuming considerable dispersion of the bands along Γ–Z (perpendicular to the layers) giving rise to the two groups of peaks has clearly to be dismissed on the basis of the band structure calculations. The second alternative which involves two sets of flat (along Γ–Z) bands which are separated by about 4 eV is in better agreement with the calculations. The fine structure in the reflectivity labelled by E_1 to E_7 in

Figure 55 is assigned to particular transitions at the Γ-point. It is pointed out however, in [76] that there are a large number of closely spaced bands which lead to this group of structure. Structures between 1.5 eV and 4 eV are likely to result fi om one spin-orbit split group of bands and correspond to the basic transitions $\Gamma_3^+ \to \Gamma_3^-$ ($E\|\hat{c}$ and $E\perp\hat{c}$), $\Gamma_3^- \to \Gamma_1^+$ ($E\perp\hat{c}$). Similarly, the basic transitions $\Gamma_1^+ \to \Gamma_3^-$ ($E\perp\hat{c}$), $\Gamma_3^+ \to \Gamma_3^-$ ($E\|\hat{c}$, $E\perp\hat{c}$) $\Gamma_3^- \to \Gamma_1^+$ ($E\perp\hat{c}$) are likely to yield the structures in the reflectivity between 6 and 9 eV. In [72] the triplet structures (E_1, E_2, E_3) and (E_5, E_6, E_7) visible in $E\perp\hat{c}$ light are interpreted as arising from spin-orbit split $\Gamma_3^\pm \to \Gamma_3^\mp$ transitions exclusively (for $E\|\hat{c}$ there should be singlet structure, a trend to which has been observed). The presence of more bands obtained in the calculations (see Figure 54) certainly allows more freedom in assignment. The lowest indirect and direct transitions, at 0.13 and 0.18 eV respectively can be attributed to $\Omega_{3,4} \to \Omega_{3,4}$ transitions along $\Gamma-D$.

A large number of additional experimental optical results has been reported since the fundamental studies of [72]. These include reflectivity [83], electroreflectance [84, 85] and wavelength modulated reflectivity [86] experiments. In spite of the tentative assignments to the various band structure calculations an unequivocal, clear picture of the electronic structure and the related optical transitions in Bi_2Te_3 is still missing.

As already mentioned, the other two pseudopotential calculations [77, 79] concentrate on the explanation of the galvanomagnetic properties of Bi_2Te_3. Since the details of the applied methods and the various results differ from those of [76] we shall present some results of the most recent calculation [79] which is based on pseudopotentials of the Lin-Kleinman type [80]. In this formulation of the pseudopotential a non-local potential acting on wave functions of s-like character only is added to a local unscreened model pseudopotential. In [79] this potential is screened by a dielectric function such that the potential Fourier transform converges to $\frac{2}{3}E_F$ as q goes to zero. Spin-orbit effects are taken into account directly in terms of a non-local pseudopotential which effectively describes the influence of the core spin-orbit splitting (given by a parameter) onto the valence pseudo-wavefunctions.

The various parameters of this model potential are chosen and slightly varied in agreement with atomic data and other crystal band calculations involving Bi and Te atoms as constituents. The resulting band structure (excluding the lowest Γ_6^+ level) is presented in Figure 56. As can be seen by comparing these results to those of Figure 54, substantial differences exist in the two band structures. The third independent band structure calculation of [77] yields results whose gross features correspond to those given in Figure 56. In spite of various different details the band structure of Figure 56 may thus be regarded as representative for the two model-potential calculations.

As an important result, the nine topmost valence bands (most likely corresponding to the anion p-states) are well separated from the cation (around -1 eV) and anion (below -2 eV) s-states. It may be argued that the lowest band (not shown) corresponds to the Te(1) s-states and is thus separated from the two Te(2) s-bands. This situation is in contrast to the EPM calculation of [76] in which the anion p-bands and cation s-bands seem to overlap. No information about the lower bands is given in [76].

Recent UPS measurements [82] are quantitatively in disagreement with all band structure calculations. In Figure 57, we display the energy distribution curves of electrons emitted by photons of various energies [86]. These results, though not conclusively interpreted yet suggest a width of the upper valence bands (probably including the

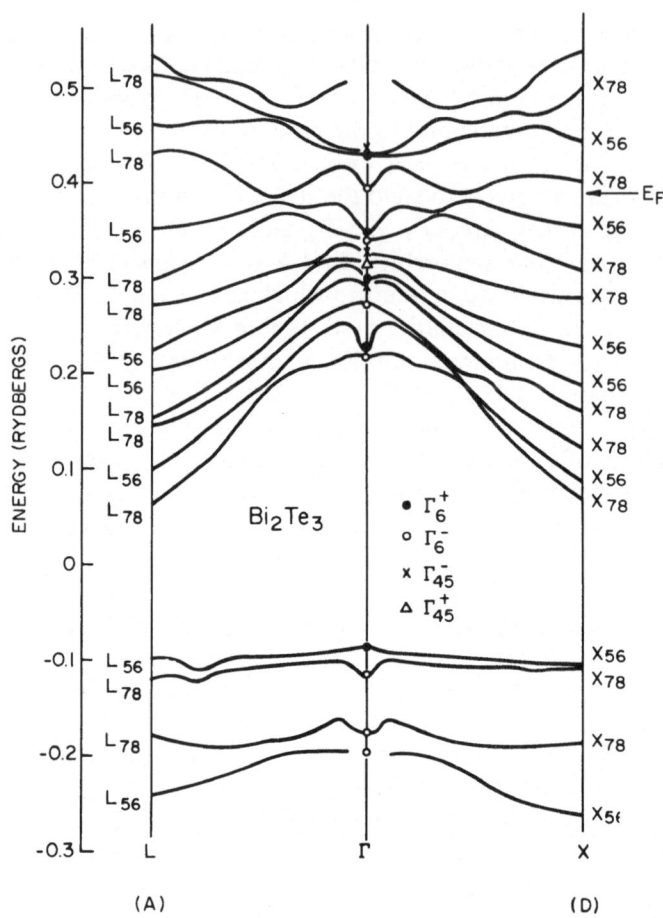

Fig. 56. Pseudopotential band structure of Bi_2Te_3 obtained in [79] using Lin-Kleinman type pseudopotentials.

cation s-bands) of about 4 eV followed by a wide window and the split anion s-bands around -10 to -12 eV. The uppermost group of valence bands in Figure 56 is about 4 eV broad but does not include the cation s-bands. This same group of bands is at least 6 to 7 eV wide in the calculations of [76, 77]. It is, however, conceivable that the position of the s-bands in Figure 56 is easily shifted upwards in energy by modifying the non-local s-potential, without changing the important features of the band structure at the Fermi level.

The hole band edge is located along Γ–X (X corresponds to D and L to A in the notation of [76]). The minimum of the conduction band is located along Γ–L. The results thus support six-valley models for both conduction and valence bands. The indirect gap obtained from the calculation is about 0.14 eV in good agreement with experiment. The band structure of [77] is very similar in the vicinity of the Fermi level, with the exception of the valence band edge at L, which supports the three-valley model for holes. In spite of a large number of experimental data available [74, 87, 88] no clear experimental evidence has been presented yet to our knowledge which unambiguously

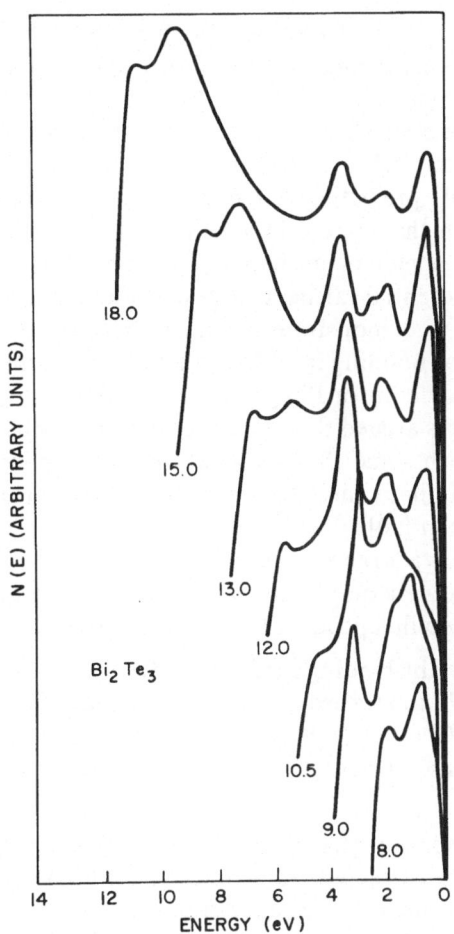

Fig. 57.   UPS spectra of Bi_2Te_3 for varying incident photon energy (after [81]).

determines the number of valence and conduction band extrema in Bi_2Te_3. On the other hand, the existing theoretical studies show too many inconsistencies among themselves to answer this question either.

2.1.3. *Lead Iodide* PbI_2

A large number of metal di-halides including PbI_2 crystallize in the CdI_2-type structure which has been discussed in detail before. The structure is identical to the 1T-phases of the group IVB and VB transition metal dichalcogenides and is also found in SnS_2 and $SnSe_2$. A perspective view of the atomic arrangement within the unit cell is presented in Figure 1. The lattice parameters are listed together with the various theoretical approaches in Table 15. Aside the fundamental CdI_2 stacking sequence, a large number of polytypes has been reported for PbI_2. Moreover, a first detailed experimental and theoretical comparative analysis of the two lowest order polytypes of PbI_2 has been reported recently. This study, which we shall report later in some detail gives direct quantitative information on the effect of interlayer coupling. PbI_2 has been thoroughly investigated experimentally. Results reported before 1972 are compiled in a review article [89]. Most of the experiments have the optical spectrum as subject which is very rich of sharp excitonic structures.

Before any quantitative band structure calculations were available, several schematic models, based on experimental information were presented. Thus a bonding model has been proposed for PbI_2 predicting its semiconducting properties on the basis of the general '8-N rule' [57]. In this picture in which the anions complete their outermost valence shell, lead keeps a pair of unshared (non-bonding) s-electrons. Quantitative results, based on pseudopotential valence charge densities essentially confirm this simple chemical picture, but also demonstrate its inherent oversimplifications. A detailed discussion on the chemical bonding in PbI_2 shall be given at the end of this section.

An independent band scheme for PbI_2 has been proposed in [89] based on a variety of experimental results. It is argued that the upper valence bands should be constructed from iodine $5p$ states (with some Pb $6s$ admixture) and the lowest conduction band should be cation p-like. Similar conclusions have been drawn from the interpretation of reflectivity measurements in [90].

During the last few years a large number of highly resolved optical measurements, in particular of the excitonic structure at the band edge have become available [91]. The experiments stimulated two independent band structure calculations:

(i) The semiempirical tight-binding method (ETB) is applied in [92, 93]. As a particular feature of this calculation the exchange crystal potential is screened by a dielectric function as discussed in Chapter B. Spin-orbit coupling and other relativistic effects are added as perturbations. The resulting energy band structure is fitted to the optical absorption data.

(ii) The empirical pseudopotential method (EPM) is applied in [94]. The pseudopotential form-factors are extracted from model potentials and slightly adjusted to yield the correct bandstructure in the vicinity of the optical gap. The effect of spin-orbit coupling is discussed in detail in a perturbation approach and also taken into account directly in the Hamiltonian matrix. Using the pseudo-wavefunctions, electronic charge densities are calculated and discussed in view of the chemical bonding in PbI_2 [70, 94].

We now present some detailed results of the two calculations.

In the ETB calculations of [92, 93] atomic eigenvalues and eigenfunctions of the $6s$ and $6p$ levels of lead and the $5s$ and $5p$ levels of iodine are used as input data. The crystal potential is evaluated with the usual $\rho^{1/3}$ approximation for the exchange potential which is screened by a dielectric function to reduce its range. Only two-center integrals are retained, the relevant ones between centers within the same layer are semiempirically adjusted in order to improve the energetic positions of the conduction bands [93]. All non-vanishing integrals between neighboring layers are fully considered, thus allowing a study of different polytypes of PbI_2 [93]. Relativistic effects are accounted for by a perturbative approach. Darwin and mass-velocity corrections are directly included in the atomic eigenvalues. In Figure 58 we show the tight-binding bandstructure of [92] without the inclusion of spin-orbit effects. The direct gap of about 2.5 eV appears between the states A_1^+ and A_3^- (the notations g and u are used in [92] for the $\pm$ superscripts). As seen from the positions of the atomic levels, also indicated in Figure 58, the lowest two bands correspond to the iodine $5s$ states. These bands are

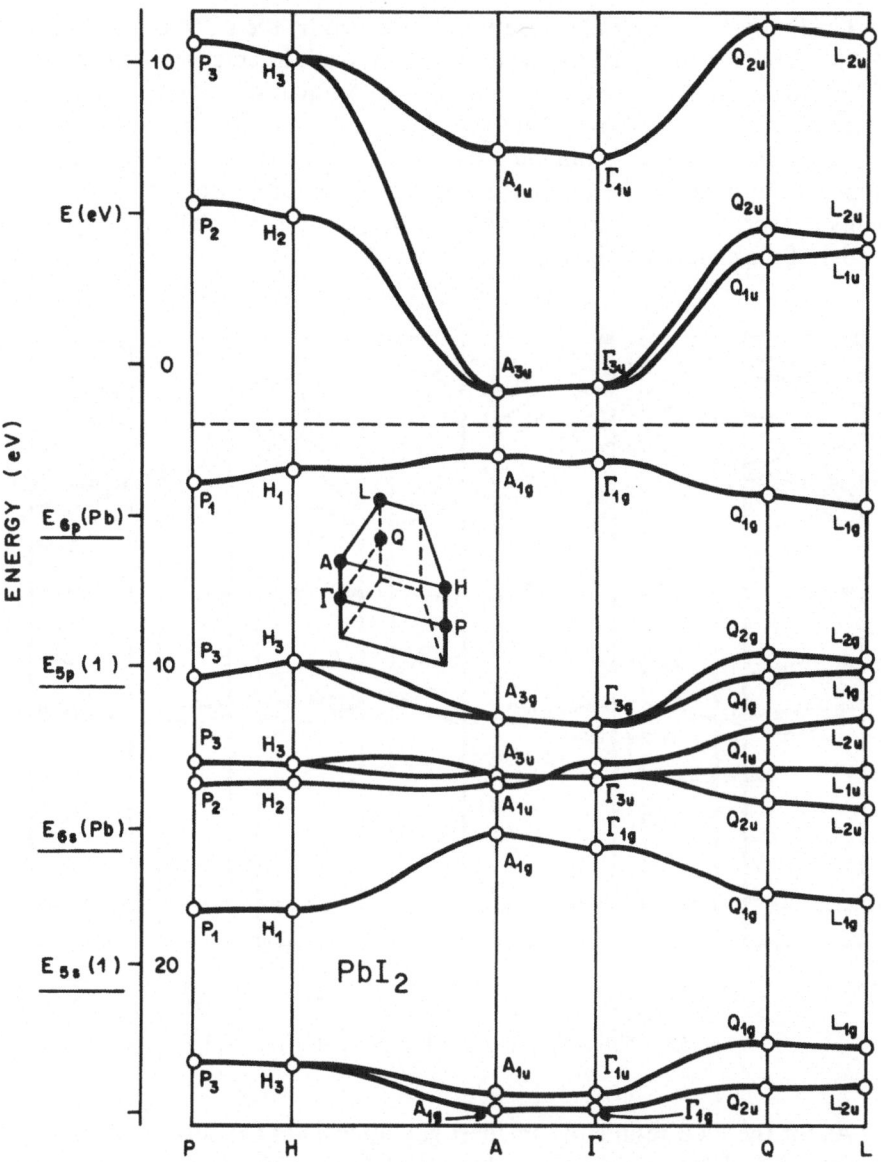

Fig. 58.   Two-dimensional ETB band structure of PbI₂ (after [92]).

followed by six iodine p-bands. The widely split-off topmost valence band corresponds to the lead $6s$-levels but mixes strongly with the third band (iodine p-like). A detailed picture of this situation which is confirmed by the pseudopotential results of [94] shall be given later in connection with charge density plots. The lowest conduction bands correspond to the crystal-field split lead p-levels. The splitting in Figure 58 is too large and has been corrected in [93] by empirically adjusting several interaction parameters.

Qualitatively, very similar results are obtained by the EPM calculation of [94]. The EPM band structure of PbI_2 neglecting spin-orbit interaction is shown in Figure 59. The lead potential used in this calculation is an Animalu-Heine type model potential [78], appropriately screened by a Penn-type [65] semiconductor dielectric function. A first-order iodine potential is obtained by extrapolating the renormalized Animalu-Heine type [78] form factors of In, Sn, Sb and Te. These form factors are then varied empirically to reproduce the numerical value of the gap, $Eg \approx 2.5$ eV and to yield the correct order of the conduction bands.

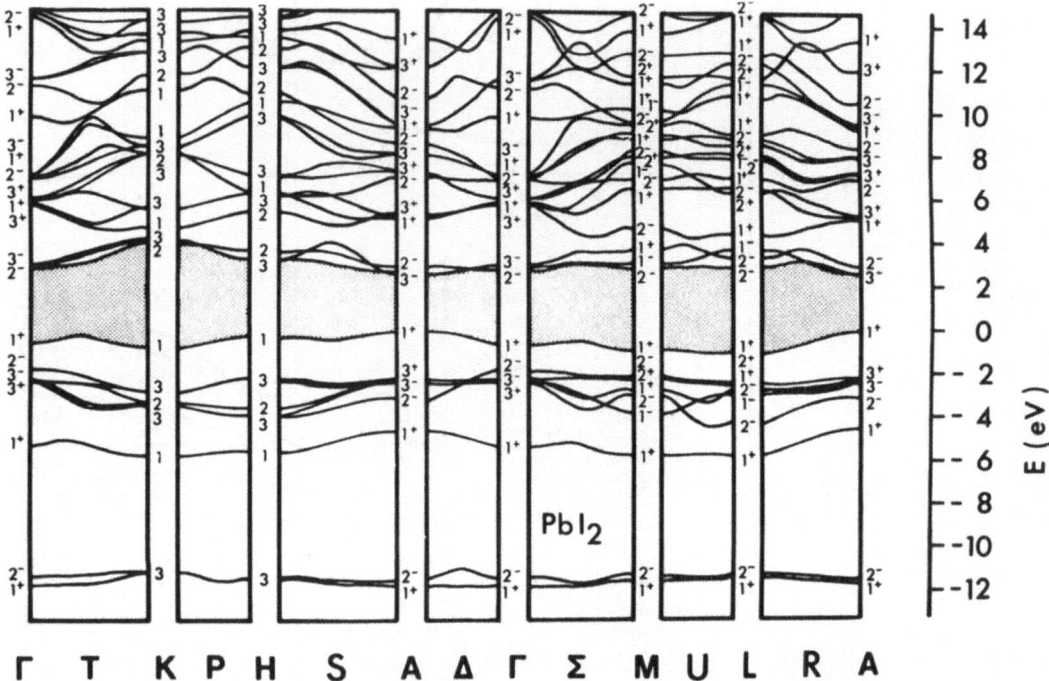

Fig. 59.   Three-dimensional EPM band structure of PbI_2 (after [94]).

As for the ETB results the smallest gap appears at the surface of Brillouin zone at point A (which is group theoretically equivalent to point Γ) between the non-degenerate state A_1^+ and the two-fold degenerate level A_3^-. The third p-like level A_2^- appears about 0.6 eV higher in energy. Optical transitions from A_1^+ to A_3^- are allowed for light polarized $\mathbf{E} \perp \hat{c}$ only. This is in contrast to experiment [91] which shows an excitonic transition at 2.5 eV for both polarizations, though the transition for $\mathbf{E} \perp \hat{c}$ is about four times stronger. The experimental facts can be explained by invoking spin-orbit coupling. In fact, double group selection rules for the group D_{3d} show that the band edge transition should be visible for both polarizations: A_1^+ goes into A_4^+ and A_3^- splits into A_4^- and two one-dimensional states A_5^-, A_6^- which are degenerate by time-reversal symmetry; A_2^- goes into A_4^- too. Transitions from A_4^+ into A_4^- are allowed for both polarizations suggesting that the band edge should appear between these states.

The ordering of the three cation p-like conduction bands may best be described in an atomic-like semiempirical model [94], which treats crystal-field and spin-orbit effects as perturbations.

Let the crystal field split the state with $m_l = 0$ angular momentum from the states with $m_l = \pm 1$ by the amount ΔE_C. The spin-orbit interaction which couples states with the same l and the same total angular momentum component $m_j = m_l + m_s$ may be parametrized by λ. The three eigenvalues, each two-fold (Kramers) degenerate are then given by

$$E_1 = \lambda$$

$$E_{2,3} = \tfrac{1}{2}(\Delta E_c - \lambda) \pm (\tfrac{1}{4}\Delta E_c^2 + \tfrac{1}{4}\lambda^2 + \tfrac{1}{2}\Delta E_c\lambda)^{1/2} \tag{C30}$$

while the eigenfunctions of E_1 are pure p_{xy}-like and transform like A_5^-, A_6^-, the eigenfunctions of $E_{2,3}$ are mixtures of p_z and p_{xy} transforming like A_4^-. One finds in particular

$$|E_2\rangle = [(\lambda + E_2)^2 + 2\lambda^2]^{-1/2} [\sqrt{2}\,\lambda\,|p_{xy}\rangle + (\lambda + E_2)\,|p_z\rangle],$$

$$|E_3\rangle = [(\lambda + E_3)^2 + 2\lambda^2]^{-1/2} [\sqrt{2}\,\lambda\,|p_{xy}\rangle + (\lambda + E_3)\,|p_z\rangle]. \tag{C31}$$

For $\lambda \to 0$, $|E_2\rangle$ becomes pure p_z-like and $|E_3\rangle$ becomes pure p_{xy}-like, corresponding to the pure crystal-field case. The oscillator strengths of optical dipole transitions into the (lowest) conduction state $|E_3\rangle$ are proportional to

$$|M_{xy}|^2 = \frac{2\lambda^2}{(\lambda + E_3)^2 + 2\lambda^2} \quad \text{for} \quad \mathbf{E} \perp \hat{c}$$

and to

$$|M_z|^2 = \frac{(\lambda + E_3)^2}{(\lambda + E_3)^2 + 2\lambda^2} \quad \text{for} \quad \mathbf{E} \parallel \hat{c} \tag{C32}$$

For $\lambda \to 0$ one correctly finds the single group selection rules $|M_{xy}|^2 = 1$ and $|M_z|^2 = 0$. Figure 60 illustrates these results: the lower plot shows the energies of the three states and the upper plot gives the relative oscillator strength of transitions into the lowest conduction band as a function of $\lambda/\Delta E_c$. To compare these results to experiment we show in Figure 61 experimental reflectivity data for both polarizations $\mathbf{E} \perp \hat{c}$ and $\mathbf{E} \parallel \hat{c}$ [91]. A strong excitonic resonance appears at the band edge at about 2.5 eV. The slight shift between the two polarizations is due to excitonic fine structure effects, like exchange and to polariton effects [91]. The three structures at 2.5, 3.3, and 3.95 eV can be attributed to transitions from the A_4^+ valence band into the A_4^-, (A_5^-, A_6^-) and A_4^- conduction bands respectively. All three transitions may have excitonic character and appear at M_0-type critical points. Three shoulders, corresponding to M_1-type saddle point transitions with similar polarization behavior are seen around 3.0, 3.9, and 4.3 eV. Due to spin-orbit coupling the transition at 2.5 eV is allowed for both polarizations of light, though the transition is about four times stronger for $\mathbf{E} \perp \hat{c}$ than for $\mathbf{E} \parallel \hat{c}$. The intermediate exciton at 3.3 eV is visible only for $\mathbf{E} \perp \hat{c}$ as the selection rules for a $A_4^+ \to (A_5^-, A_6^-)$ transition predict. The third exciton at 3.9 eV again is visible for both polarizations corresponding to the second $A_4^+ \to A_4^-$ transition. The energies of these three transitions together with the observed $4:1$ ratio of oscillator strengths of the lowest exciton yield the values $\Delta E_c = +0.8$ eV and $\lambda = 0.3$ eV of the empirical bandstructure. This situation is indicated in Figure 60 by the broken line. The crystal-field splitting ΔE_c has to be positive (p_z higher than p_{xy}) to obtain agreement with experiment. The p-state spin-orbit splitting $3\lambda = 0.9$ eV is somewhat smaller than that of atomic lead (1.3 eV) [23].

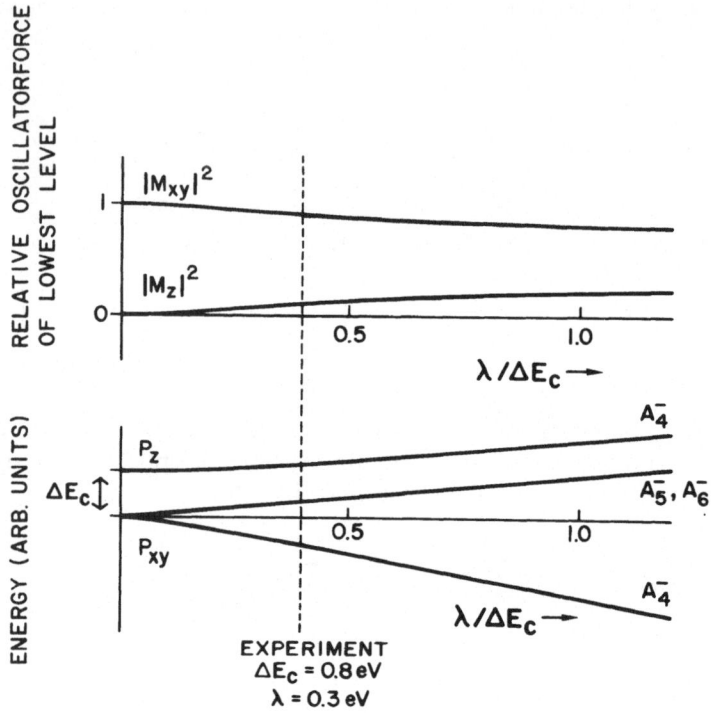

Fig. 60. Band model for the PbI$_2$ conduction bands at point A in the Brillouin zone. The lower plot shows the energy as a function of the normalized spin-orbit interaction $\lambda/\Delta E_c$. The upper plot indicates the variation of oscillator strength for transitions into the lowest conduction band as a function of $\lambda/\Delta E_c$. The dotted lines approximately indicates the experimental situations (after [94]).

The 2.5 eV band edge exciton series of PbI$_2$ has found several contradictory interpretations. Optical absorption measurements revealed the existence of a hydrogenic exciton series, which could be explained in terms of an isotropic Wannier exciton series $E_n = E_{\text{gap}} - R/n^2$ with $E_{\text{gap}} = 2.552$ eV and $R = 0.127$ eV except for the $n = 1$ line which was shifted to higher energies by about 0.070 eV [91]. An alternative interpretation was given in terms of two overlapping series, one allowed for $\mathbf{E} \perp \hat{c}$ and one for $\mathbf{E} \| \hat{c}$ [97]. More recent polarization dependent measurements disproved this interpretation [98]. The results displayed in Figure 62 show clearly that all lines belong to the same Wannier series and that the intensity ratio $I_\perp / I_\|$ is about constant for all lines. In addition to the imaginary part $\varepsilon_2(\omega)$ of the dielectric function, the imaginary part of $1/\varepsilon$ is displayed as obtained by Kramers-Kronig transformation. The two functions allow to distinguish between transverse ($\varepsilon_2(\omega)$) and longitudinal (Im $1/\varepsilon$) excitations. The energetic splitting

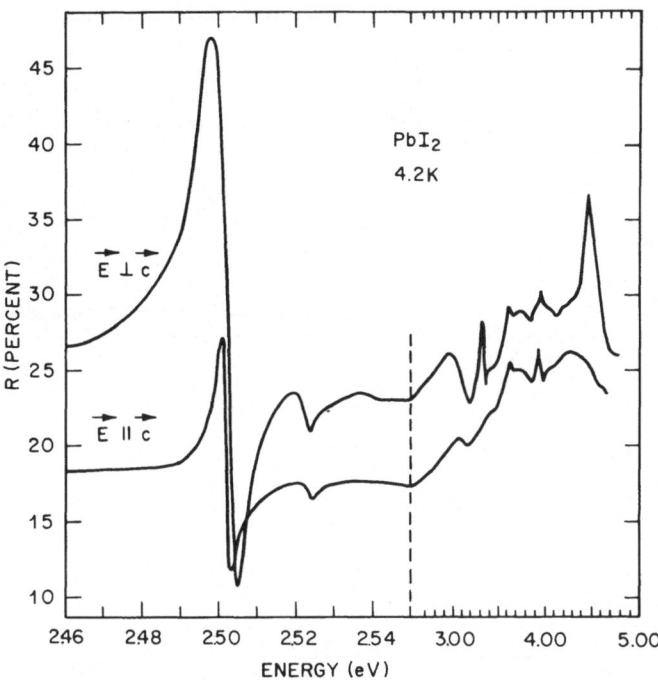

Fig. 61.   Reflectivity spectrum of PbI$_2$ at 4.2 K for $\mathbf{E}\|\hat{c}$ and $\mathbf{E}\perp\hat{c}$ polarziations on as grown surface. Note the change of scale on the abscissa of 2.55 eV (after [92]).

between the two resonances is a consequence of a combination of spin-orbit coupling and exciton exchange interaction [99]. Excitonic fine-structure effects of this kind cannot be obtained in the effective mass approximation but may be considered as perturbations on the Wannier-type exciton series. Qualitatively the effect may be understood as follows:

(i) The spins of electron and hole in an exciton may in the absence of any spin-orbit interaction be classified in terms of two-particle spin-functions i.e. singlet and triplet. In PbI$_2$ a transition into the singlet is allowed for $\mathbf{E}\perp\hat{c}$ whereas a transition into the triplet should be possible for $\mathbf{E}\|\hat{c}$. Further selection rules, however, allow only the singlet to be visible in the absence of spin-orbit interaction.

(ii) Only singlet states or states with singlet components of the $n = 1$ level are affected by the exchange interaction.

(iii) The exciton exchange interaction may be divided into one constant part and one part which depends on the angle between the directions of polarization and propagation of the exciton. This second part gives rise to the longitudinal-transverse splitting. Only transverse components are created by light.

(iv) The presence of spin-orbit interaction couples the singlet-triplet components to each other and thus relaxes the strict selection rules. In particular, in PbI$_2$ it is responsible for the $\mathbf{E}\|\hat{c}$ (triplet) absorption. Since spin-orbit coupling can be regarded as a perturbation, the $\mathbf{E}\|\hat{c}$ (triplet) absorption is weak compared to the $\mathbf{E}\perp\hat{c}$ (singlet) absorption. The amount of longitudinal-transverse splitting, carried over from the singlet to the triplet by the spin-orbit coupling should follow the same 4:1 ratio as the

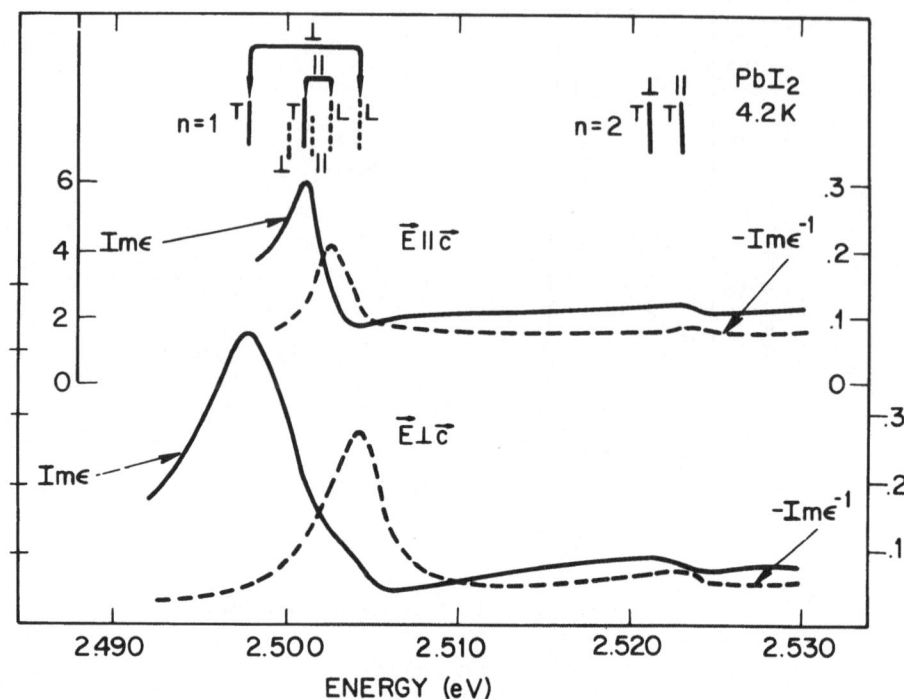

Fig. 62.   Imaginary part Im ε of the dielectric function and $- \mathrm{Im}(\varepsilon^{-1})$ for 2H–PbI$_2$ in the band edge exciton region for $\mathbf{E}\|\hat{c}$ and $\mathbf{E}\perp\hat{c}$. The longitudinal (L) transverse (T) splittings are indicated (after [91]).

intensity ratio in the absorption. This is approximately found experimentally, as seen from the lines marked in Figure 62.

The explanation of the excitonic fine structure still leaves the problem of the large $n = 1$ anomaly to be solved. It has been shown [91] that this anomaly cannot result from any anisotropy effects nor from simple dielectric screening effects which tend to produce opposite ground state anomalies of too large binding energies. Recent optical data [100] revealed a new line which was tentatively attributed to the $n = 3$ exciton line. The position of this line indicates a normal Wannier-type exciton series without ground state anomaly and a small exciton ionization energy of 30 meV. These findings could not be reproduced consistently [91] and may be related to extrinsic effects.

An alternative explanation involves 'central cell corrections' to the effective mass approximation of the exciton which are necessary, if valence and conduction states become rather localized in space [98]. This 'central cell correction' may be used to account for a smooth transition found in nature between Wannier-type excitons in small gap materials and Frenkel-type excitons in large gap materials. The necessity of such a correction is evident if one looks at the exciton binding energy which in the effective mass approximation (Wannier case) is given by

$$R = -\frac{e^4 m^*}{2\hbar^2 \varepsilon^2}. \tag{C33}$$

In the limiting case of extremely tight binding (i.e. flat bands, $m^* = \infty$, $\varepsilon \approx 1$) this theory produces an unphysical infinite binding. For intermediate cases Equation (C33) may still be used in connection with 'central cell corrections' which act on the most localized (i.e. $n = 1$) state.

In [98] an estimate of the amount of this $n = 1$ ground state anomaly is made on the assumption that both electron and hole are essentially localized at the lead sites. This 'cationic' picture of the exciton is confirmed by charge density calculations which we shall present later. It is shown in [91, 98] that while the exciton radius is still fairly large (about 15 to 20 Å), cationicity and layer geometry cause the anomaly to be particularly large. It is assumed that due to these circumstances electron and hole can effectively interact only in a small volume around the individual lead atoms, characterized by the Pb atomic radius r_0. Given an exciton envelope function, roughly characterized by a radius a_0, the electron-hole Coulomb interaction is ineffective over a fraction of order r_0/a_0 of the total exciton volume. Since the electron-hole potential energy is on the average $2R$ (from the virial theorem), this predicts a decrease of the binding energy of order $2R \, r_0/a_0$ which can be a large correction to the ground state. These simple qualitative arguments given in [91, 98] explain reasonably the observed ground state anomaly of about 0.07 eV. Nevertheless, more experiments seem to be desirable in particular in view of the new results of [100] to clear the puzzling question about the band edge exciton in PbI_2.

The small changes of physical properties generated by polytypism are subject of a recent experimental and theoretical study [93]. In this work the consequences of layer compound polytypism on optical properties and electronic band structures are investigated for the cases of 2H and 4H PbI_2. We follow here the polytype notation of [93] which enumerates the anion sheets (2H, 4H) and not the layers. Thus the 2H-phase corresponds to the standard one layer CdI_2-type structure with D_{3d}^3 symmetry, whereas the 4H-phase is characterized by two layers (four anion sheets) per unit cell. The corresponding space group is C_{6v}^4. In the basic 2H structure, the stacking sequence is AcB, where the capital letters denote anions and c represents the cations. The anion stacking sequence corresponds to hexagonal close packing. In the next higher polytype, 4H–PbI_2 the stacking sequence is AcB, CaB, corresponding to a rotation of the upper layer by 180° around the $\hat{c}$-axis going through the site B. The anion stacking sequence shows mixed cubic, hexagonal close-packing character (see Chapter A). The three dimensional unit cells of 2H and 4H–PbI_2 contain one and two molecules respectively. The lattice constants of both cells in the x, y cleavage plane are the same ($a = 4.56$ Å) while in the $\hat{c}$-direction the 4H unit cell is doubled with respect to 2H ($c_{2H} = 6.98$ Å, $c_{4H} = 13.96$ Å).

General, symmetry related rules giving a correspondence between microscopic properties of the two polytypes cannot be found since D_{3d}^3 (2H) and C_{6v}^4 (4H) do not form a group-subgroup relationship. Not even the point group C_{6v} is a subgroup of D_{3d}. This implies that the electronic bandstructure of the 4H phase cannot just be derived from that of the 2H phase by a perturbational approach. The properties of the 4H phase have rather to be considered in a separate calculation. This is attempted in [93] in a tight-binding calculation. The resulting band structure along Γ–A' ($\hat{c}$-direction) is shown in Figure 63. For comparison the results for 2H–PbI_2 [92] are indicated. The 2H-bands are improved over those shown in Figure 58 by including spin-orbit interaction and by fitting the uppermost conduction band to experiment. Only the Γ–A'

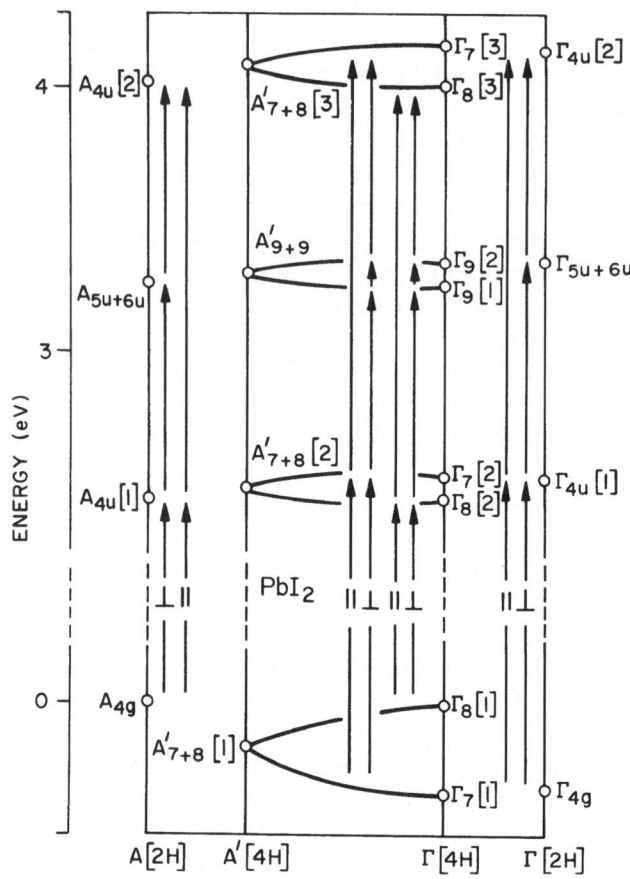

Fig. 63. ETB band structure of 4H–PbI$_2$ along the vertical Γ–A' line of the three-dimensional hexagonal Brillouin zone. Energy levels of 2H–PbI$_2$ at the points Γ and A are also shown for comparison. The arrows indicate allowed transitions for specific polarization (see Table 19) (after [93]).

section of the Brillouin zone is shown, since no essential changes occur in the x–y plane by going from the 2H to the 4H phase. Since the real space unit cell is doubled in length (along $\hat{c}$) the extension of the Brillouin zone Γ–A' is one half of the Γ–A distance in 2H–PbI$_2$. It follows from time reversal symmetry in connection with the symmetry properties of C_{6V}^4, which contains a two-fold screw axis perpendicular to the x–y plane, that all states on the top-face (A') of the 4H Brillouin zone are at least two-fold degenerate (see also discussion of D_{6h}^4 in Chapter A). Thus the bands can intersect at A' with non-vanishing slope not creating any critical points. The interesting features of the optical spectrum of 4H–PbI$_2$ should therefore be explainable in terms of transitions at Γ.

In Figure 64 the reflectivity spectra are shown for 2H– and 4H–PbI$_2$ measured with $E \perp \hat{c}$ polarized light [93]. A comparison of the main features of the two spectra yields these results:

(i) No splitting is found for the band edge exciton at 2.5 eV. There is only a small 10 meV shift.

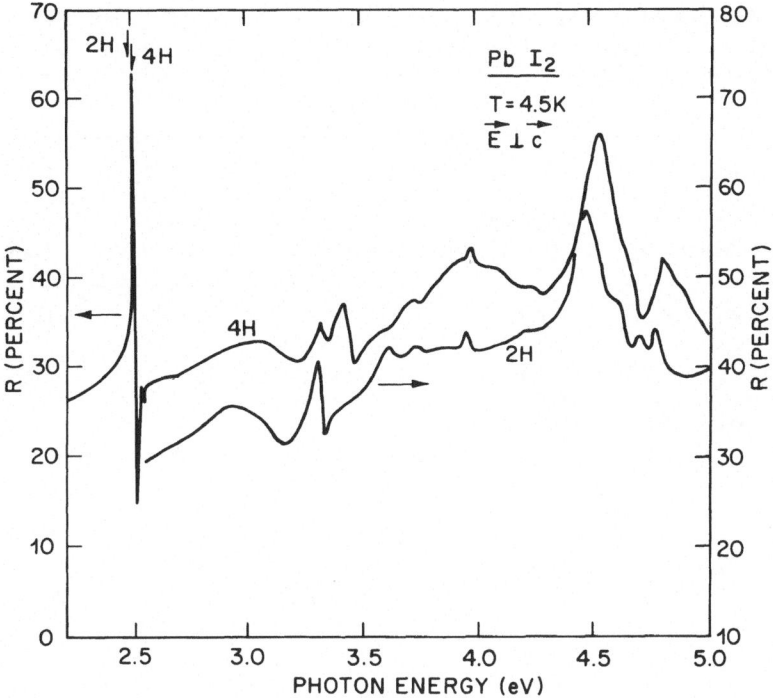

Fig. 64.   Reflectivity spectra of 4H–PbI$_2$ and 2H–PbI$_2$ for E$\perp\hat{c}$ polarization at 4.5 K (after [93]).

(ii) The 3.3 eV exciton is split into a doublet in 4H–PbI$_2$ with 0.1 eV separation. The oscillator strength is transferred mainly to the higher energy component.

(iii) No splitting is found for the 3.9 eV exciton.

Examining the bandstructure in Figure 62 one may easily explain these features.

(i) The band edge exciton at 2.5 eV (connected to a M_0 critical point) is related to the $\Gamma_8(1) \to \Gamma_8(2)$ transition. No peak is expected for the $\Gamma_8(1) \to \Gamma_7(2)$ transition which is electric dipole forbidden.

(ii) The 3.3 eV exciton of the 2H-phase is split in 4H–PbI$_2$ since two (M_0) critical points appear corresponding to the two allowed transitions $\Gamma_8(1) \to \Gamma_9(1)$ and $\Gamma_8(1) \to \Gamma_9(2)$. The small splitting of 0.1 eV illustrates how narrow the conduction bands are along Γ-A in both polytypes.

(iii) The $\Gamma_8(1) \to \Gamma_8(3)$ (M_0-type) transition is associated with the 3.9 eV exciton.

As for (i) selection rules forbid the second $\Gamma_8(1) \to \Gamma_7(3)$ transition. Thus only one structure is observed as for 2H–PbI$_2$.

More detailed assignments are compiled in Table 19. The results show that the change of polytype and thus of interlayer interaction preserves the main features of the spectrum. The energy bands of the 4H- phase look as if they are just folded back at A'. Although this is not required by group theory, the effect occurs since band separations are considerably larger than interlayer interaction energies for the conduction bands of PbI$_2$. The reason

TABLE 19

Measured and calculated energies of the relevant features in the optical $\vec{E} \perp \hat{c}$ reflectivity spectrum of 2H and 4H PbI_2 and assigned transitions. For $\vec{E} \| \hat{c}$ in 4H PbI_2 (no experimental data available) the predicted transitions are also given. The asterisks denote exciton transitions (see Figures 63, 64) (after [93]).

2H PbI_2 $\vec{E} \perp \hat{c}$		4H PbI_2 $\vec{E} \perp \hat{c}$		4H PbI_2 $\vec{E} \| \hat{c}$	
Energy (meas.) (eV)	Assignment	Energy (meas.) (eV)	Assignment	Energy (calc.) (eV)	Transition
2.498*	$A_4^+ \to A_4^-(1)$	2.508*	$\Gamma_8(1) \to \Gamma_8(2)$	2.508*	$\Gamma_8(1) \to \Gamma_8(2)$
		2.63	$\Gamma_8(1) \to \Gamma_7(2)$?		
2.95	$\Gamma_4^+ \to \Gamma_4^-(1)$	3.05	$\Gamma_7(1) \to \Gamma_7(2)$	3.05	$\Gamma_7(1) \to \Gamma_7(2)$
3.31*	$A_4^+ \to A_{5+6}^-$	$\begin{cases} 3.330* \\ 3.425* \end{cases}$	$\begin{matrix} \Gamma_8(1) \to \Gamma_9(1) \\ \Gamma_8(1) \to \Gamma_9(2) \end{matrix}$	– –	– –
3.62*		3.6*			
3.74		3.73			
3.94*	$A_4^+ \to A_4^-(2)$	3.975*	$\Gamma_8(1) \to \Gamma_8(3)$	3.97	$\Gamma_8(1) \to \Gamma_8(3)$
3.95	$\Gamma_4^+ \to \Gamma_{5+6}^-$	3.95	$\Gamma_7(1) \to \Gamma_9(1)$		
		4.05	$\Gamma_7(1) \to \Gamma_9(2)$		
				≈ 4.5	$\Gamma_7(1) \to \Gamma_7(3)$
4.480*		4.54*			
4.60		4.67			
4.71		4.805			
4.78		4.86			

for this is found in the character of the conduction band wave functions which are localized in the middle of each sandwich around the lead atoms and do not appreciably interact with neighbouring layers. This particular situation is not commonly met in layer-compounds [70].

Concluding the discussion of the electronic properties of PbI_2 we shall present some information on the nature of chemical bonding in this compound.

The electronegativity difference ($\Delta x = 0.9$) between divalent lead and monovalent iodine is appreciable and PbI_2 should therefore be considerably more ionic than $SnSe_2$ which crystallizes in the same structure. This general result is confirmed by the total valence charge density, which is displayed in Figure 65. The charge distribution is calculated from the EPM results of [94] using three 'special k-points' as discussed in the section on $SnSe_2$. Like for $SnSe_2$ the charge contour plot is displayed in a (110) plane, indicated as shaded area in Figure 1. It can be seen from Figure 65 that there is only a small covalent charge built up between neighbouring lead and iodine atoms, both of them being surrounded by nearly spherical charge clouds. Since iodine has seven valence electrons two p-like electrons are transferred from lead to the two iodine atoms to complete their valence shells. The experimental bond length of 3.15 Å compares somewhat more favorably to the sum of ionic radii (3.36 Å) of the configuration Pb^{2+}, I^{1-} than to the sum of covalent radii (2.80 Å). The transfer of two lead electrons leaves the lead $6s$ states occupied and one would expect them to be accommodated in non-bonding

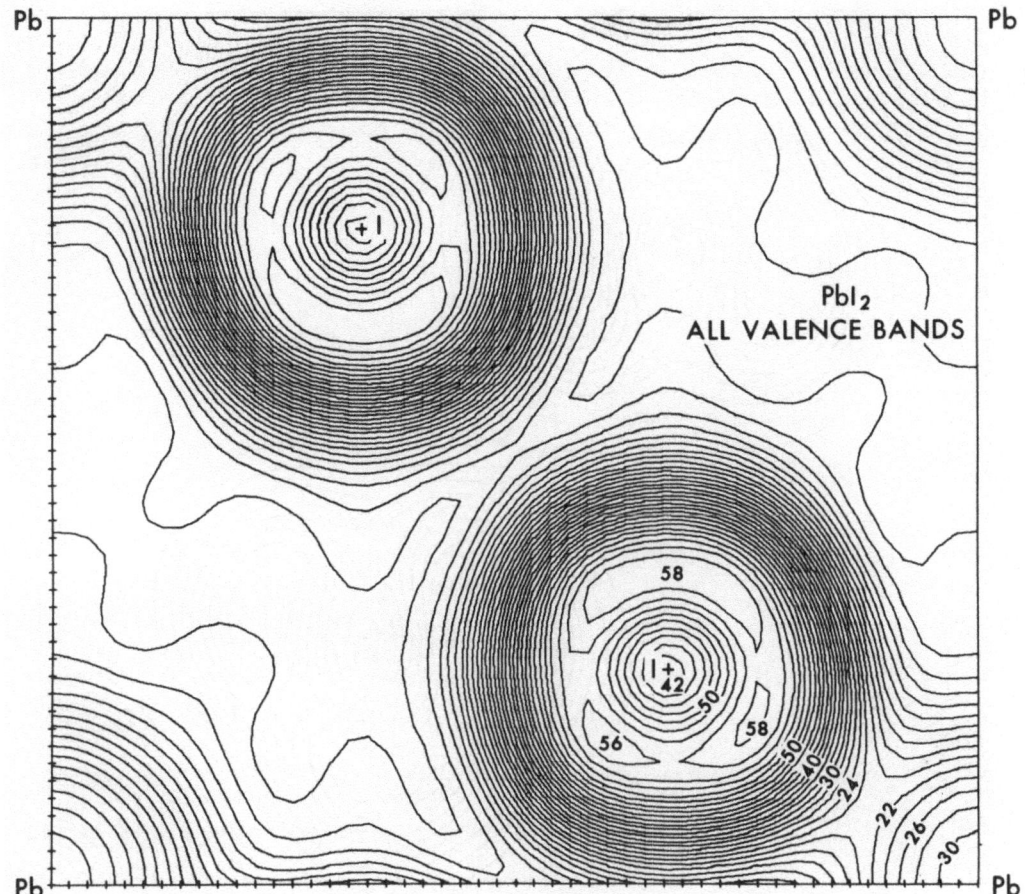

Fig. 65. Total valence charge density of the PbI_2 as obtained from the EPM calculation of [94]. As for $SnSe_2$ the contours are displayed in a (110) plane in units of electrons per unit cell.

or lone-pair levels. This tendency of the heavier group IV elements to retain a non-bonding pair of s-electrons has already been pointed out in early qualitative considerations [57]. It is of interest to determine the exact energetic position of these s-like states. It has been argued in the bandstructure calculations of [92, 93, 94] that the lead $6s$ states are likely to correspond to the split-off, topmost valence band. From Figure 66 it indeed follows that the charge corresponding to the highest valence band (band 9) does have appreciable s-character near Pb. The charge on iodine in Figure 66 clearly results from an sp^3 hybridization. This sp^3 hybrid, however, is inverted such that the charge is distributed among the *anti-bonding* sites predicted for exited states by molecular orbital theory [24]. The antibonding character is confirmed by the nodes appearing between neighboring Pb and I atoms.

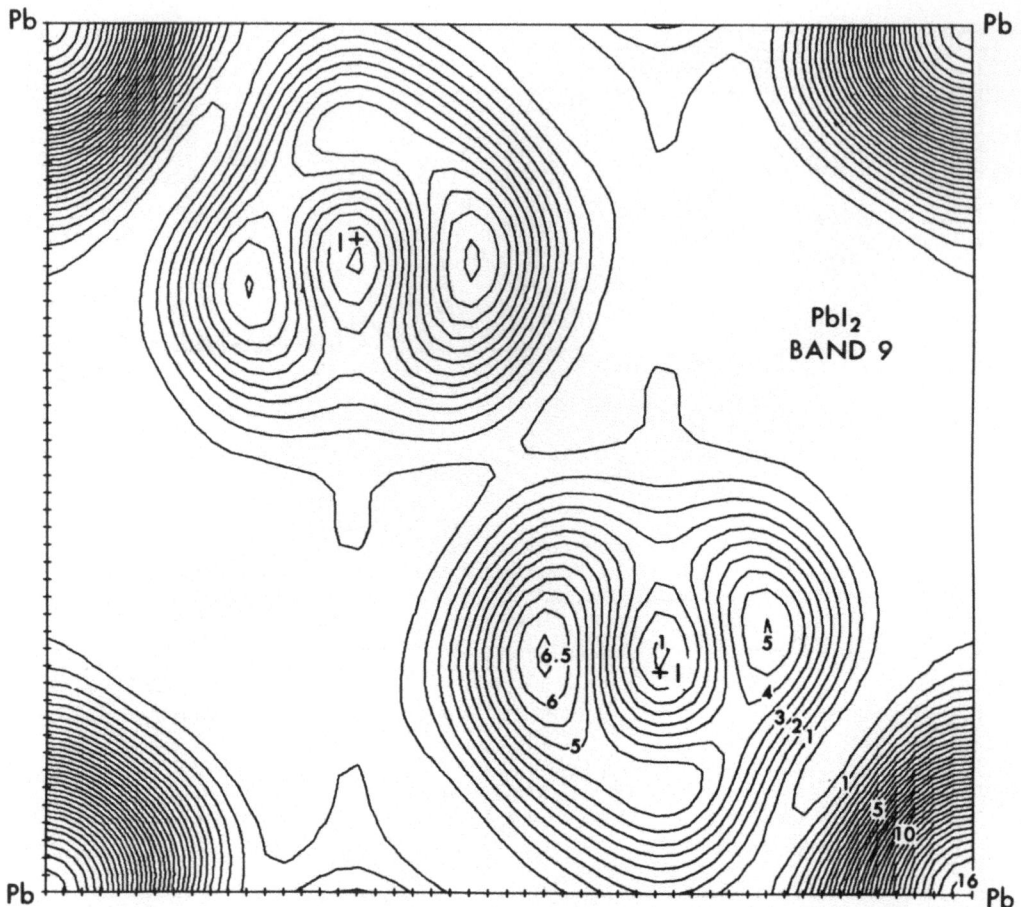

Fig. 66.   Partial band charge density of the uppermost valence band (band 9) in PbI_2 (after [94]).

The *bonding* partner of band 9 is band 3 which lies some 5 eV lower in energy and whose density contours are shown in Figure 67. It is obvious from this figure that the charge maxima of the sp^3 hybrid around I now fall upon the bonding sites and that the nodes between neighbouring Pb and I have disappeared. A simple chemical argument would have predicted the valence s-electrons of Pb to form lone pairs by occupying the non-bonding $6s$ states of Pb. It now becomes evident that these states combine with the valence orbitals of iodine to produce a bonding and an antibonding *pair* of bands. Since both bands are occupied their net effect on the cohesive energy in PbI_2 is negligible, and in this sense these electrons can be viewed as non-bonding.

The group of closely spaced bands (4–8) lying between the bonding and antibonding bands discussed above has predominantly p character on iodine with little s-admixture. (See Figure 68.) A small covalent charge appears between neighbouring Pb and I atoms

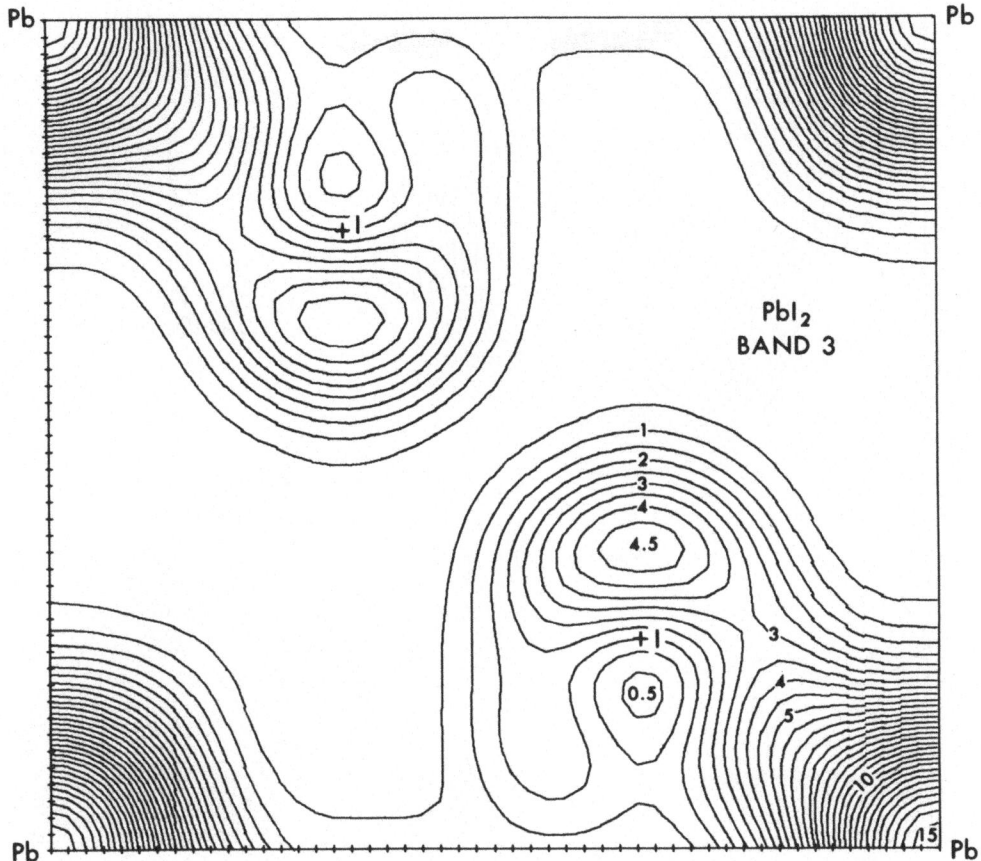

Fig. 67. Partial band charge density of the third valence band in PbI_2 (after [94]).

showing that PbI_2 has some covalent character. The lowest two valence bands (1–2) are essentially iodine s bands with a small p-contribution accounting for the deviations from spherical symmetry (see Figure 69).

The effect of hybridizing 'non-bonding' states with orbitals of the other atom in a bonding-antibonding fashion has also been found in another layer halide BiI_3 and for the Ga–Ga cation pair in GaSe [70]. These results shall be presented in later sections. No effect of this kind has been found for Se or Te which also retain lone-pair electrons [70].

A valence band density of states, calculated from the EPM bandstructure of [94] is compared to an experimental XPS spectrum [101] in Figure 70 and in Table 20. Another XPS spectrum of PbI_2 is reported in [102]. The experimental spectrum differs from the calculated curve in its quantitative aspects. In fact, the XPS results seem rather to favor the existence of a non-bonding band of Pb s-electrons at -8 eV (structure D)

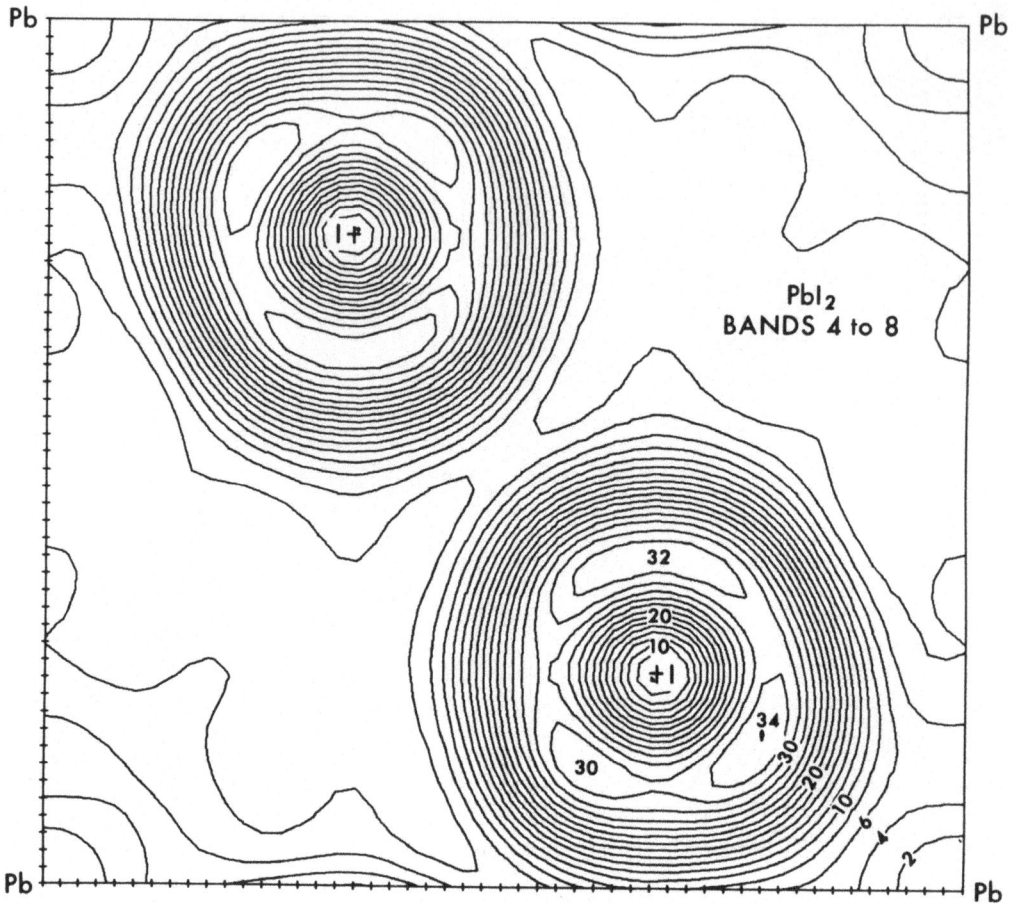

Fig. 68.   Partial band charge density of bands 4 to 8 in PbI_2 (after [94]).

without any of the aforementioned hybridization effects. The conclusions regarding the hybridization, however, may be misleading as the results on BiI_3 show (discussed in the next section), a theoretically very similar hybridization effects as in PbI_2 which are coupled to a very asymmetric density of states just like the one experimentally observed for PbI_2. Thus the quantitative disagreement between theory and XPS-data for PbI_2 (Figure 70 and Table 20) does not rule out the predicted hybridization effects. One reason for the occurrence of these effects may be the unusually large size of the ions in PbI_2 and BiI_3. In fact, the sum of the respective ionic radii exceeds the experimental bond lengths by some 5–10% whereas e.g. for the tin-dichalcogenides the two distances nearly agree. More theoretical work, however, is necessary to explain these effects and the discrepancies between existing calculations and photoemission spectra.

Coming back to the problem of the cationic exciton [98] we note that the lowest three

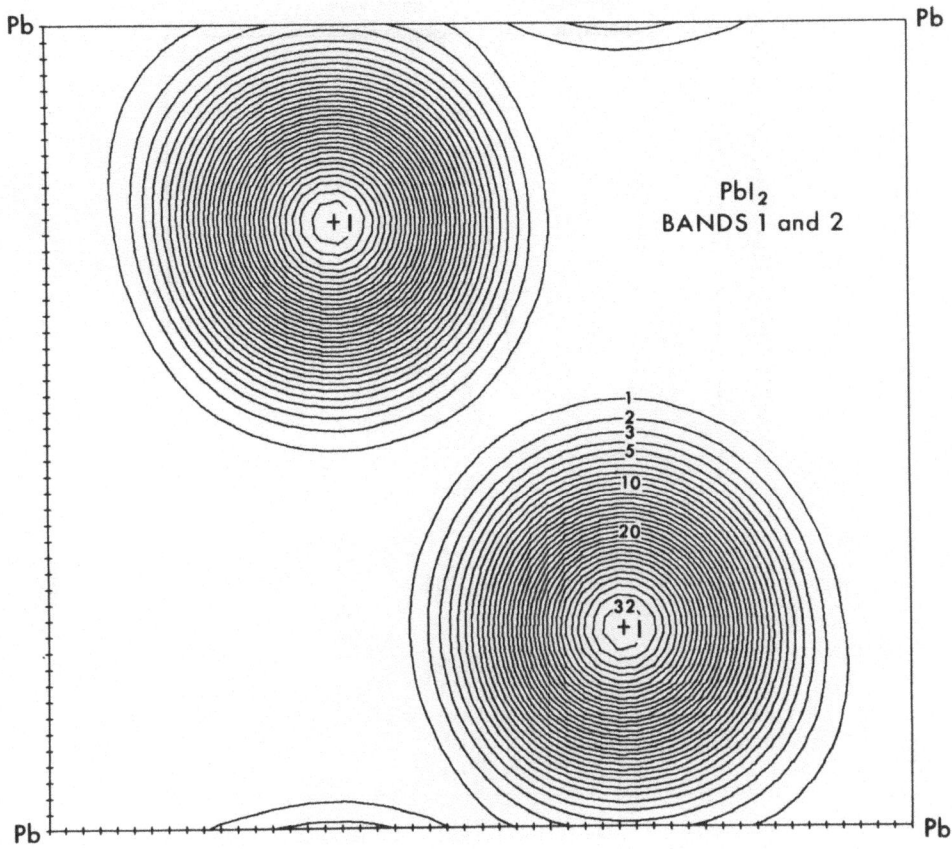

Fig. 69. Partial band charge density of bands 1 and 2 in PbI$_2$ (after [94]).

conduction bands, whose integrated (hypothetical) charge density is shown in Figure 71, are indeed formed predominantly by cation p-states. As seen from Figure 71 the wavefunctions are highly concentrated around the lead atoms and these states show very small interlayer coupling. Together with the charge distribution of the topmost valence band (Figure 66) these results strongly support the hypothesis of cationic band-edge excitons in PbI$_2$.

2.1.4. *Bismuth Iodide BiI$_3$*

BiI$_3$ crystallizes in a structure which derives from the octahedral CdI$_2$-type structure. Each sandwich consists of three atomic sheets, the cations forming the middle sheet. In BiI$_3$ these sheets exhibit a honeycomb-like arrangement of Bi atoms. In contrast to CdI$_2$, the honeycomb centers are empty. Thus only $\frac{2}{3}$ of the regular CdI$_2$ type lattice sites are occupied creating cation voids. Every Bi atom and Bi void is octahedrally

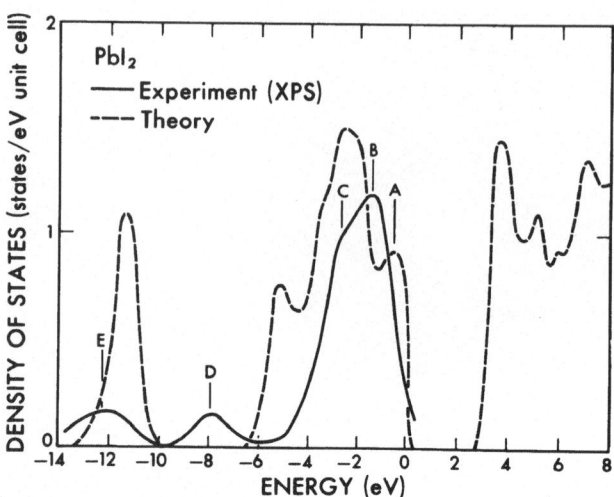

Fig. 70.   Calculated density of states for PbI$_2$ based on the EPM calculation of [54] compared to
experimental photoemission data of [101] (see also Table 20) (after [70]).

TABLE 20

Listing of energies for the main peaks in the experimental photoemission
spectrum of PbI$_2$ compared to theoretical density of states calculations
(see Figure 70) (after [70]).

	PbI$_2$			
Structure	Experiment [101]	EPM [70], [94]	ETB[a] [92]	Predominant nature of states
A	–	− 0.5	–	Pb(s) and I(sp^3)
B	− 1.5	− 2.5	− 1.5	I(p)
C	− 3.0	− 3.5		
D	− 8.0	− 5.5	− 5.5	Pb(s) and I(sp)^3
E	− 12.2	− 11.5	− 12.5	I(s)

[a] The band structure had to be shifted up with respect to the topmost valence band by 8.5 eV.

surrounded by iodine atoms. Because of the Bi voids the anions are only twofold coordinated.

In the standard polytype of BiI$_3$, the unit cell extends over three layers of two molecules each. This corresponds to an ABC stacking sequence of Bi-voids. The space group
is D_{3d}^5. To facilitate calculations one may 'synthesize' a one-layer polytype according
to a simple A over A stacking of Bi voids. The changes in the electronic system, induced

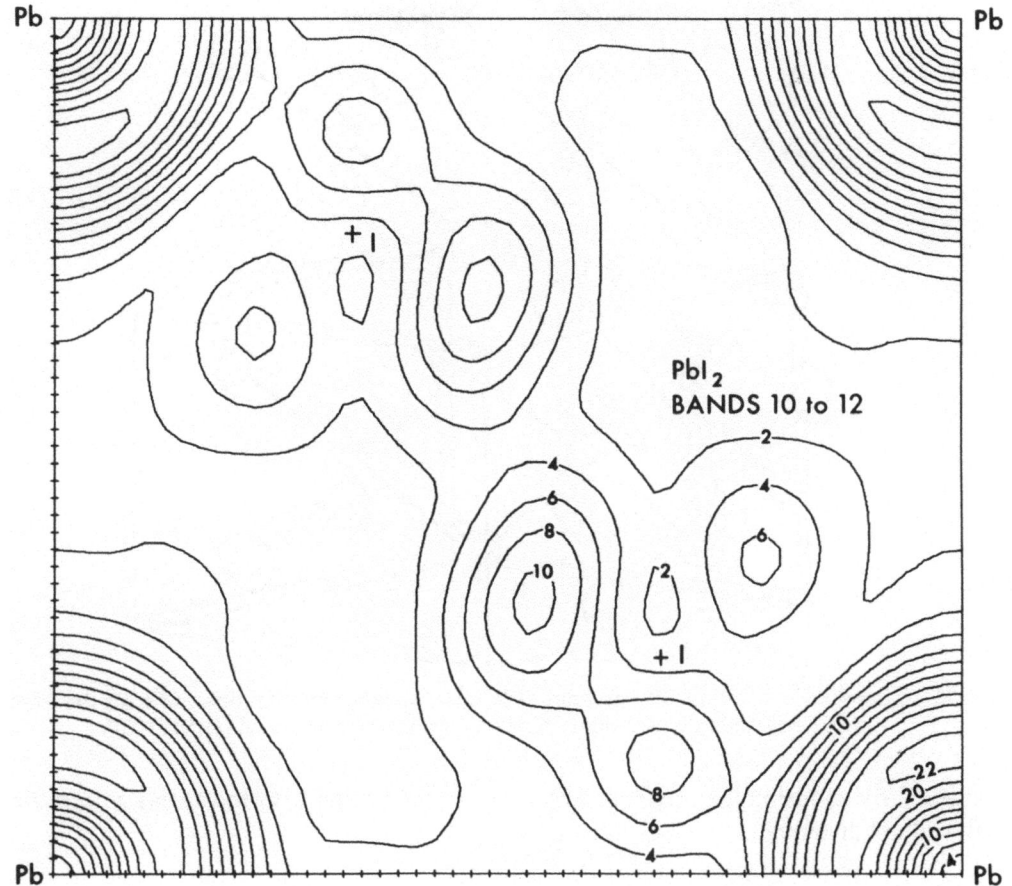

Fig. 71.   Hypothetical charge density of the first three conduction bands in PbI$_2$ (after [94]).

by this artificial stacking sequence are minor, since the relative arrangement of the outermost anion sheets is unaltered. The changes may safely be disregarded in first order. A perspective view of the one-layer unit cell and the atomic arrangement is shown in Figure 72. The six atoms per unit cell are shaded. The crystal space group is D_{3d}^1, which differs only slightly from the space group D_{3d}^3 of SnSe$_2$ and PbI$_2$ and has been discussed in detail in Chapter A. The difference consists of an interchange of the 3 two-fold axes of rotation (C_2) with the three vertical planes of reflection (σ_d). This interchange results from the orientation of the original PbI$_2$ lattice with respect to the new BiI$_3$ lattice. The new BiI$_3$ lattice can be viewed as a $\sqrt{3} \times \sqrt{3}$ superlattice of the PbI$_2$ lattice. The interchange of group operations has no effect on the crystal point group and only minor consequences for several 'small groups' of **k**-vectors on the boundary of the Brillouin zone (see Chapter A). The dimensions of the hexagonal unit cell are listed in

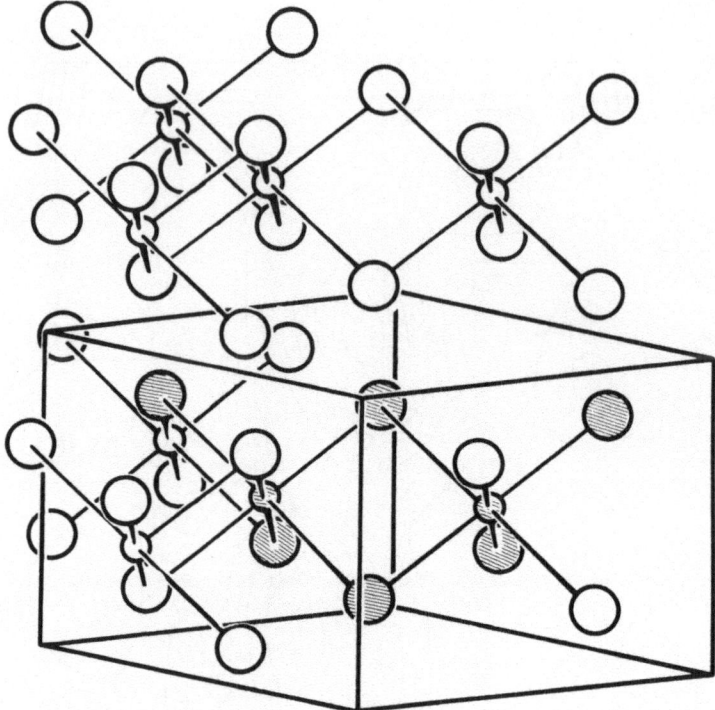

Fig. 72.   Perspective view of the BiI_3 structure. The small circles mark the cations, whereas the large
open circles indicate the anions. Note the existence of cation voids (after [103]).

Table 15. By choosing the origin at a Bi void site the positions of the atoms within the
unit cell are given by

$$Bi: (\tfrac{2}{3}, \tfrac{1}{3}, 0) \qquad (\tfrac{1}{3}, \tfrac{2}{3}, 0),$$

$$I: (\tfrac{2}{3}, 0, -u) \qquad (\tfrac{1}{3}, 0, u),$$

$$(0, \tfrac{1}{3}, u) \qquad (0, \tfrac{2}{3}, -u),$$

$$(\tfrac{1}{3}, \tfrac{1}{3}, -u) \qquad (\tfrac{2}{3}, \tfrac{2}{3}, u), \qquad u = \tfrac{1}{4}. \tag{C34}$$

A schematic bandstructure model for BiI_3 similar to that of PbI_2 has been proposed in
[89]. The model is based on various optical experiments and the comparison to other
layer halides. In this model the topmost valence bands are assumed to be iodine p-like
with some Bi s-admixture while the conduction bands are assumed to be Bi p-like.

To our knowledge only one recent EPM calculation [103] is available so far. In this
calculation which does not account for spin-orbit effects the atomic pseudopotentials
are derived from Animalu-Heine type [78] model potentials for bismuth and from
earlier calculations on SbSI [104] for iodine. The resulting form factors are slightly
readjusted by requesting the minimum gap to be about 2 eV and the maximum of the
valence band density of states to appear about 2 eV below the top of the valence bands.
This latter information was derived from XPS results [105] which shall be shown later.
No reflectivity data are used to adjust the potentials.

The resulting band structure of BiI_3 along various symmetry directions in the hexagonal Brillouin zone is shown in Figure 73. The shaded region indicates the forbidden gap between the twenty-sixth and twenty-seventh bands. The results suggest that BiI_3 is a direct gap material with the fundamental gap of about 2.2 eV appearing at the point A between the states A_1^+ and A_3^-. Without spin-orbit interaction, band edge transitions are allowed for light polarized $E \perp \hat{c}$. Spin-orbit coupling splits the A_3^-

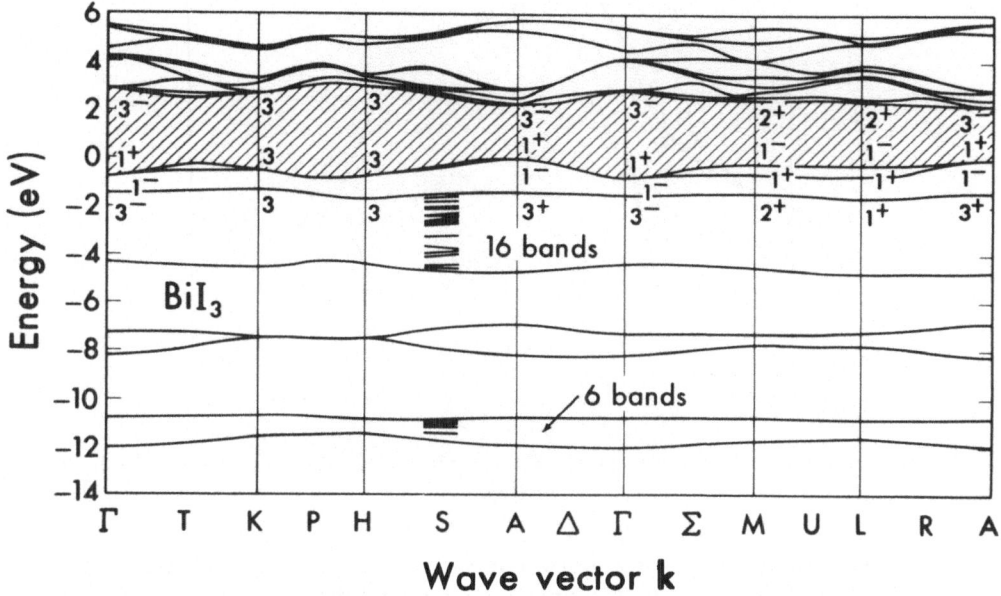

Fig. 73. EPM band structure of BiI_3. Only the outer bands of several dense groups of valence bands are explicitly indicated (after [103]).

conduction band and couples it to the slightly higher A_2^- conduction band. The situation is analogous to that in PbI_2 and need not to be discussed here. For details we refer to the PbI_2 discussion.

Because of very dense clustering of some bands only the outermost bands of these clusters are shown in Figure 73 for clarity. For bands close to the fundamental gap the detailed configuration is shown.

In contrast to most EPM calculations, experimental optical reflectivity data did not serve as an input into the process of determining the atomic form factors of BiI_3. Nevertheless, a comparison of the calculated and experimental reflectivities favorably supports the empirical calculation. In Figure 74 the calculated and measured reflectivities [103] of BiI_3 are compared with each other. The experimental spectrum is obtained in the usual $E \perp \hat{c}$ geometry and is compared to an unpolarized theoretical spectrum. Individual structure in the calculated spectrum is assigned to band to band transitions located in specific regions in k-space (see Figure 74). For transitions between 2 and 4 eV, the highest pair of valence bands (bands 25, 26) and the ensemble of the lowest six conduction bands (bands 27 to 32) are involved. These (2 + 6) bands correspond mainly to the s- and p-states of the two bismuth atoms in the unit cell. This shall be

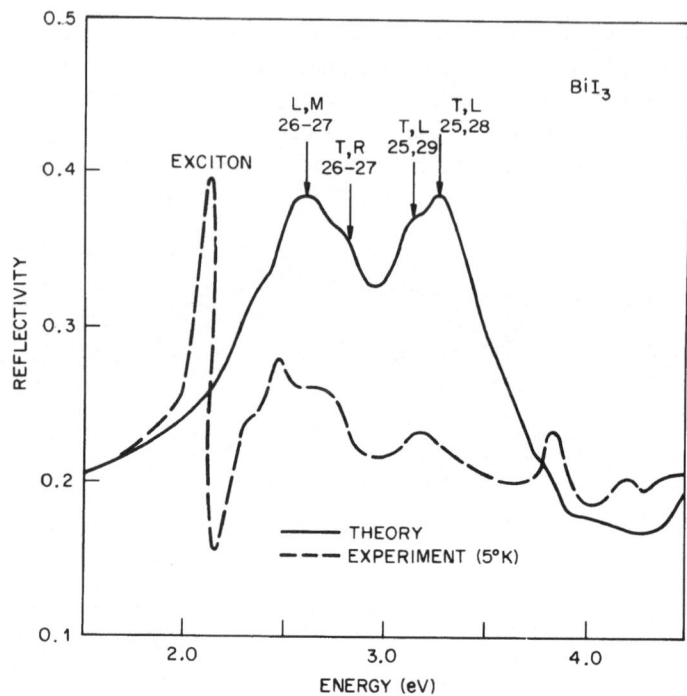

Fig. 74. Comparison of the experimental (5 K) and calculated reflectivity spectra of BiI₃ (after [103]).

confirmed later by charge density plots. The situation thus very much resembles the PbI$_2$ 'cationic' absorption edge. The anisotropic crystal field splits the cation p-states in the conduction band according to p_{xy}- and p_z-like characters with the p_{xy}-like states at lower energy. The prominent two-peak structure 2.5 eV and 3.5 eV in the calculated spectrum is a result of this crystal field effect. Spin-orbit interaction which is not included in the calculation can cause additional splittings combined with mixings between states of equal total angular momentum, which is likely to superimpose additional structure onto the simple theoretical two-peak structure.

The overall structure of the calculated reflectivity spectrum of BiI$_3$ resembles much the spectrum of PbI$_2$ [94]. As mentioned before, the transitions below 4 eV are mainly cationic in character, i.e., the transitions take place between cation s- and p-like states. Structure above 4 eV which is not represented in Figure 74 corresponds mainly to interionic, charge transfer like transitions. Thus the 'ionic gap' in BiI$_3$ is similarly to PbI$_2$ of the order of 5–6 eV. In the low energy region (below 4 eV) the experimental curve is somewhat more spread out than the calculated reflectivity. This difference is mainly due to spin-orbit effects, which are not considered in the calculation. No detailed comparison between experimental data and calculations including spin-orbit effects are reported so far. The strong resonance around 2.1 eV is due to excitonic effects and is not reproduced in the calculated spectrum.

The absorption and emission spectra of BiI$_3$ in the vicinity of the band edge are subject to numerous discussions. In addition to a regular, strong band edge exciton series [106], a hydrogen-like series of lines is reported [107] which converges toward the *long*-wavelength side. This reversed hydrogen series is attributed to the optical excitation of electrons into discrete levels of a state with *negative* mass. This system of two electrons with a negative reduced mass is named 'bielectron'. An alternative system is a 'bihole' which cannot easily be distinguished from a 'bielectron' by optical experiments. The results presented in [107] are reported to fit the inverse serial dependence of a hydrogenic Wannier exciton, i.e.,

$$E_n = E_g + \frac{R}{n^2} = 15\,978 + \frac{1995}{n^2}\ \text{cm}^{-1} \qquad (C35)$$

with $n = 3, 4, 5, 6, 7$. The absorption results of [107] are shown in Figure 75. We shall briefly outline the model of the 'bielectron' promoted in [107] to interpret these spectra.

A regular Wannier type exciton may be described by an effective Schrödinger equation for its envelope function $F(r)$ as follows [99]:

$$\left[\frac{p^2}{2\mu} - \frac{e^2}{\varepsilon r}\right] F(r) = E_{ex} F(r). \qquad (C36)$$

In Equation (C36) $\mathbf{r} = \mathbf{r}_e - \mathbf{r}_h$ denotes the relative coordinate between electron and hole. The center of mass motion has been separated away. $\mathbf{p}$ is the conjugate momentum to $\mathbf{r}$ and the system's reduced mass μ is defined by

$$\frac{1}{\mu} = \frac{1}{m_e} + \frac{1}{m_h}. \qquad (C37)$$

If μ becomes negative, no bound states are obtained from Equation (C36). If, however, μ is negative and Coulomb repulsion $(+(e^2/\varepsilon r))$ instead of attraction $(-(e^2/\varepsilon r))$ is assumed, Equation (C36) restores itself with $E'_{ex} = -E_{ex}$. In other words, for $\mu < 0$ and Coulomb repulsion ('bielectron' or 'bihole'), discrete eigenvalues as given in Equation (C35) can result from Equation (C36). The situation of a regular exciton and a 'bielectron' are schematically demonstrated in Figure 76. The regular band edge exciton appears for electrons excited into the first conduction band (1st CB) and holes left in the valence band (VB). The 'bielectron' is expected to occur if electrons are simultaneously excited into the first (1st CB) and into second (2nd CB) conduction bands. The excitation energy at which the lines in Figure 75 are observed would correspond to the energy difference between the two conduction bands plus the hydrogenic energy E'_{ex}. The formation of a 'bielectron' thus requires a very particular conduction band structure. Inspection of Figure 73 shows that the existence of such a particular structure cannot be ruled out. In fact, an isolated conduction band with negative mass and appropriate geometry appears at about the correct energy.

Serious objections to the 'bielectron' model, however, are presented on experimental basis [108]. Repeated experiments on a large number of samples did not support the 'bielectron' model. Not all samples show the additional absorption lines and if they are seen, their intensities do not follow the predicted dependence on the main quantum number n. Further arguments against the 'bielectron' model are based on the observed line widths of exciton and 'bielectron': the observed 'bielectron' lines are significantly narrower than the exciton lines. In [108] an alternative model of singly charged donor-acceptor pairs located at distinct lattice sites is proposed to explain the rich absorption spectrum of BiI_3. The existence of 'bielectrons' in BiI_3 is definitely not confirmed by the

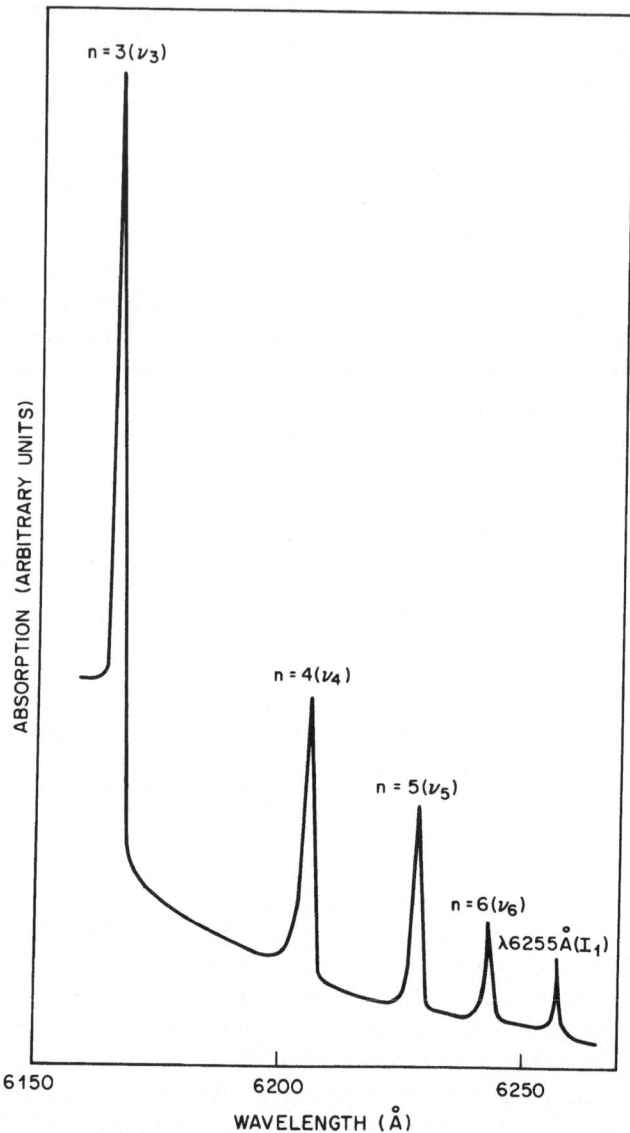

Fig. 75. Absorption spectrum of the reverse hydrogen-like 'bielectron' series in BiI$_3$ at 4.2 K (after [107]).

experiments reported in [108]. The very interesting suggestion of the 'bielectron' model, however, deserves further theoretical and experimental examination.

In this final part we present calculations of the valence band density of states combined with charge density distributions to illustrate the nature of chemical bonding of BiI$_3$. The calculated density of states [103] is compared in Figure 77 to the experimental XPS spectrum of [105]. Except for the lower two peaks (D and E) whose position deviates by about 1–2 eV from the experimental data (where they merge to one peak) all other theoretical structures agree reasonably with the XPS results. Since the lowest group of

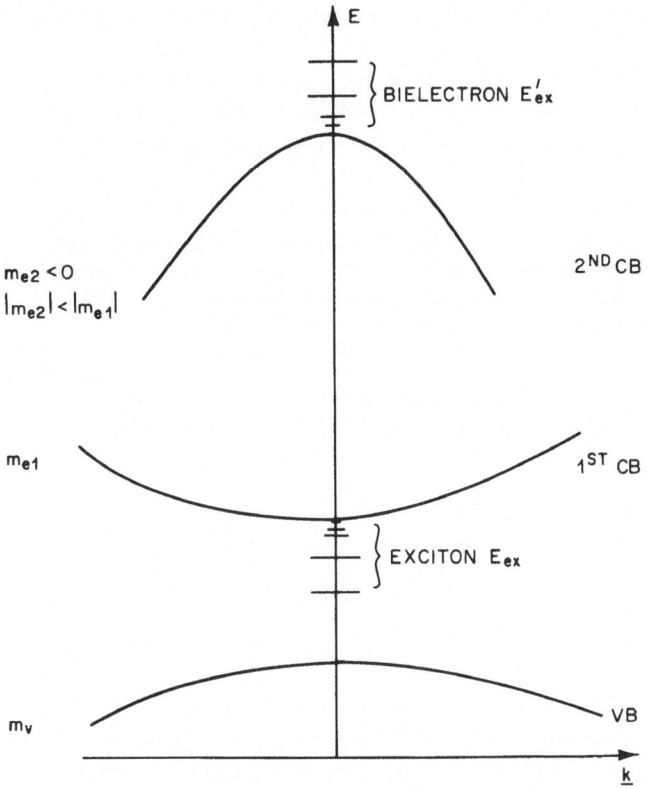

Fig. 76.   Proposed bandstructure scheme leading to the 'bielectron' model of [107].

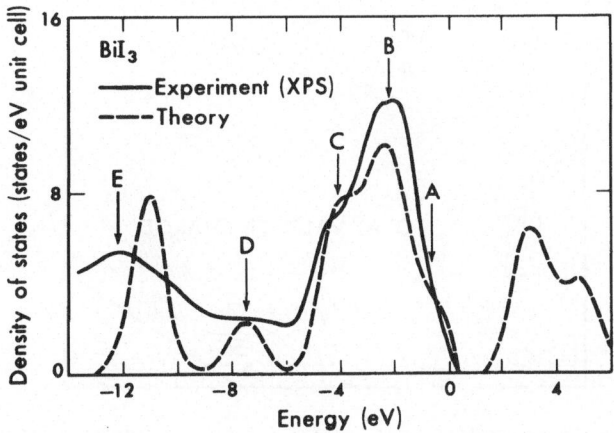

Fig. 77.   Calculated density of states for BiI_3, based on the EPM calculation of [103] compared to experimental photoemission data of [105] (see Table 21).

valence bands (peak E) corresponds to essentially non-bonding anion s-like states, the 1 eV deviation of peak E is of no importance in the discussion of chemical bonding in BiI_3. The exact theoretical and experimental positions of valence band structure are compiled in Table 21.

The various peaks in the density of states are identified by the corresponding valence electron charge distributions. The charge density plots are displayed in a (100) plane, extending over two half-layers of BiI_3 and containing the anion-cation bonds.

TABLE 21

Listing of energies for the main peaks in the experimental photoemission spectrum of BiI_3 compared to a theoretical density of states calculation (see Figure 77) (after [103]).

	BiI_3		
Structure	Experiment [105]	EPM [103]	Predominant nature of states
A	–	-0.5	$Bi(s)$ and $I(sp^3)$
B	-2.2	-2.4	$I(p)$
C	-4.2	-4.0	
D	-10.0	-7.5	$Bi(s)$ and $I(sp^3)$
E	-12.2	-11.0	$I(s)$

The filling of the anion valence shells, required by simple crystal chemical arguments, can be achieved by transferring the three p-electrons of the cations. Similar to Pb^{2+} in PbI_2, Bi^{3+} in BiI_3 would thus retain only its s-electrons. The calculated total valence charge which is shown in Figure 78 supports this ionic picture to some extent. It also

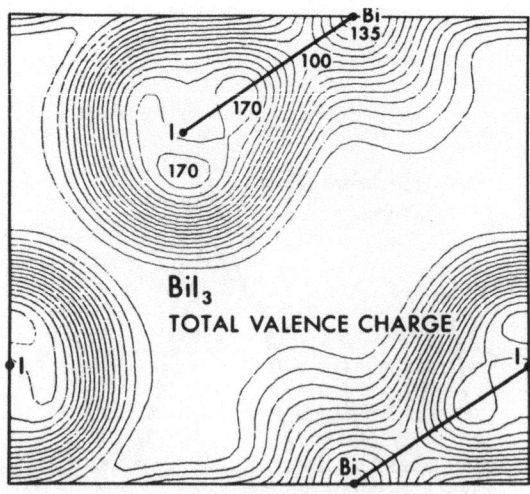

Fig. 78.   Total valence charge density for BiI_3 as obtained from the EPM calculations of [103]. The contours are displayed in a (100) plane of the BiI_3 unit cell and are normalized to two electrons per band.

indicates the presence of some covalent character in the bonding by the non-sphericity of the charge arrangement around the anions. In comparison with other layer-compounds one may thus place BiI_3 on an ionicity scale between the less ionic $SnSe_2$ and SnS_2 and the slightly more ionic PbI_2. The relationship between the experimental bond length (3.04 Å), the sum of ionic radii (3.20 Å) and the sum of covalent radii (2.79 Å) resembles closely that of PbI_2. The ionic configuration Bi^{3+}, I^{1-} leaves the bismuth $6s$ states occupied which can be seen in the total valence charge distribution of Figure 78. The details of the electron distributions corresponding to various groups of bands are shown in Figure 79. The first six valence bands (peak E in the density of

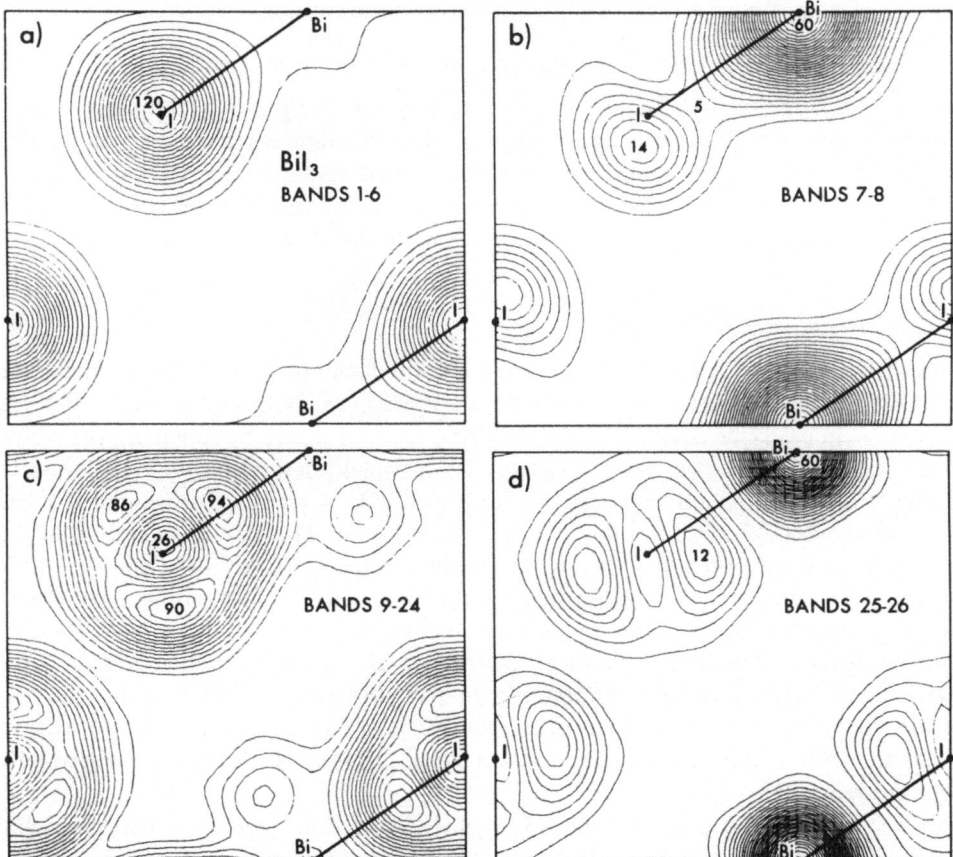

Fig. 79.   Partial band charge densities for individual groups of valence bands in BiI_3 (see Figures 73 and 77 to identify the various bands) (after [103]).

states) correspond to non-bonding anion s-like states (Figure 79(a)). A very small p-hybridization is indicated by the slight off-center location of the charge maximum. The charge distribution of the next higher group of bands (peak D) originating from bands 7 and 8 is shown in Figure 79(b). Together with the topmost pair of valence bands (shoulder A), displayed in Figure 79(d), these states show strong cation s-character with the admixture of some anion sp^3-type hybrids. Very similar to PbI_2, a hybridization of the 'non-bonding' cation s-states with anion sp^3 hybrids also exists in BiI_3. Both, bonding (bands 7, 8) and anti-bonding (bands 25, 26) configurations are occupied, thus cancelling any net bonding effect. Peaks C and D in the density of states which originate from the group of bands 9 to 24 correspond to anion p-like states, as seen from Figure 79(c). The comparison with XPS data shows that similar to PbI_2, the cation s-states are too weakly bound by about 2 eV in the calculated spectra (see Tables 20, 21).

In Figure 80, finally, charge distributions of the states forming the smallest gap at point A are displayed. Both valence and conduction band show strong cation admixtures. An optical transition at the band edge thus corresponds to the excitation of cation s-electrons into cation p-states, similarly to the case of PbI_2.

2.2. COMPOUNDS WITH TRIGONAL PRISMATIC ARRANGEMENT OF THE CATIONS

Strictly speaking, only the heavier ($n = 4, 5$) group VB and group VIB transition metal ions are found in trigonal prismatic coordination. For lighter transition metal cations and for all non-transition metal cations the distorted octahedral coordination is more favorable. In the following section, however, we shall discuss the electronic properties of a series of non-transition metal compounds with slightly modified structure but with overall trigonal prismatic symmetry.

2.2.1. The Gallium Chalcogenides

There exists a class of *poly*-cationic compounds which exhibit a quasi-trigonal prismatic coordination. Well known members of this class are GaS, GaSe and InSe. Their crystal structure can easily be derived from the MoS_2-type structure by replacing the middle layer transition metal atoms by *two* sheets of non-transition metal atoms. A similar, but slightly more complicated structure is found for GaTe. A perspective view of the four-fold layers in GaS, GaSe and InSe is shown in Figure 81. The diatomic cation molecule replacing the transition metal cation can readily be recognized. Each cation is almost tetrahedrally surrounded by three anions and one cation. The anions forming the outermost atomic sheets of the layers are three-fold coordinated if one neglects interlayer interaction. The arrangement of having top and bottom anion sheets on top of each other is characteristic for trigonal prismatic symmetry. It is evident that the most important feature of this structure is the occurrence of like-atom bonds (cation-cation). We shall see that this feature dominates a large number of electronic properties of these compounds.

In Table 22 we list several materials with the four-fold layer structure for which theoretical studies have been undertaken. The trigonal prismatic symmetry requires a hexagonal unit cell extending over at least two-layers if one wants to consider interlayer effects. In the case of the $3R$, three-layer modification (γ–GaSe) a rhombohedral unit cell extending over one layer can be defined. Moreover, two different low-order, two-layer polytypes exist, as discussed for the transition metal compounds TaS_2 and MoS_2 in the first part of this chapter. These polytypes, known as C27 and C7 structure types are labelled as ε- and β-modifications in the case of the gallium chalcogenides. The β-modification (C7 structure in Figure 27) is currently found in GaS [109] and much less frequently in transport reacted flakes of GaSe [110]. Its space group is D_{6h}^4 and its hexagonal unit cell with the dimensions indicated in Table 22 contains four molecular

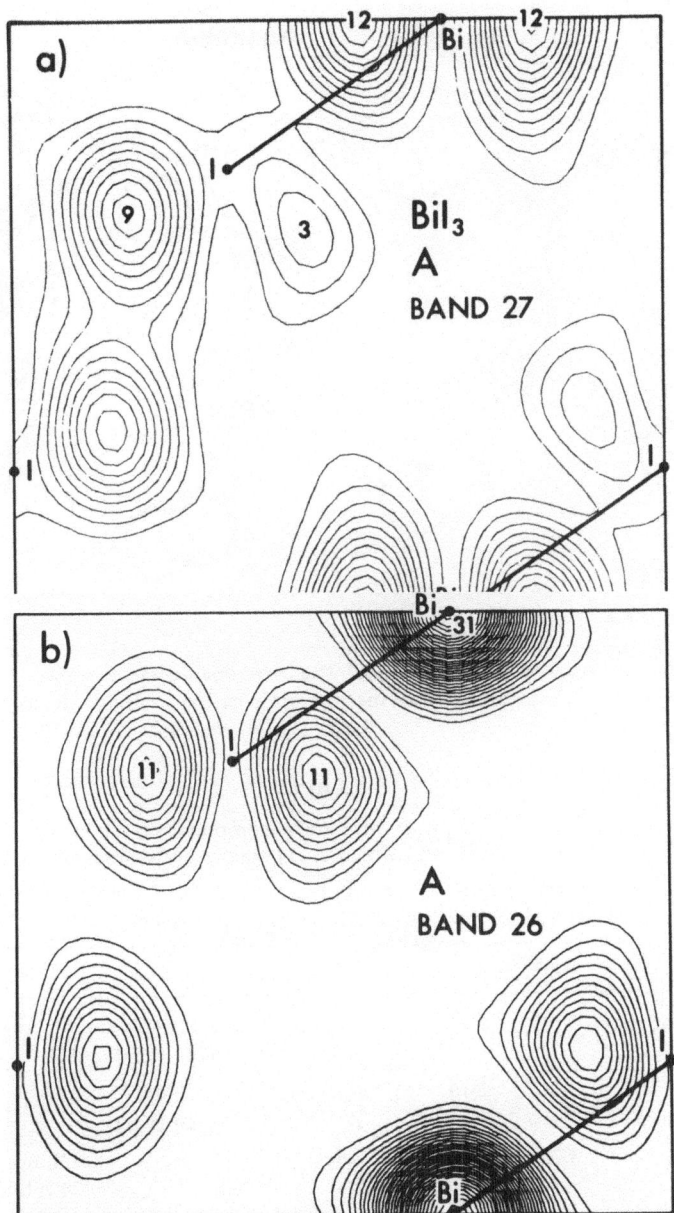

Fig. 80. Partial band charge densities for (a) the lowest conduction band state at A (band 27) and (b) the topmost valence band state at A (band 26) for BiI$_3$ (after [103]).

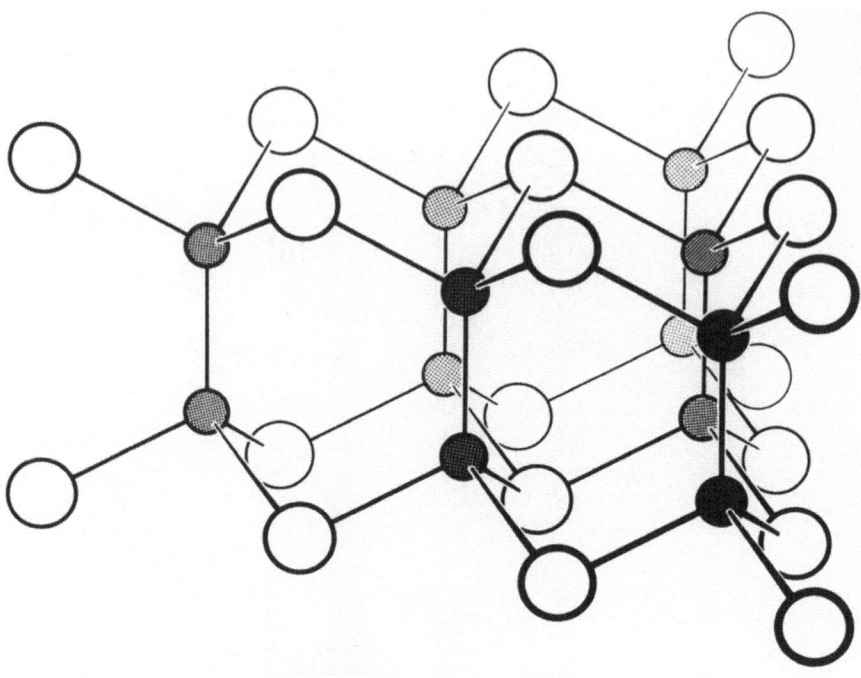

Fig. 81.   Perspective view of a layer of GaSe. The large open circles represent the anions, the small shaded circles indicate the cations. Note the di-atomic cation molecule in the middle of the layer (after [112]).

TABLE 22

Table of the GaS, GaSe family of non-transition metal compounds crystallizing in a trigonal prismatic structure for which theoretical studies are undertaken.

	Lattice constants		Theoretical approaches	
	a(Å)	c/a	ETB	EPM
Single layer GaSe D_{3h}^1	3.75	2.12	Bassani, Pastori [120] Kamimura, Nakao [121]	
Single layer GaS D_{3h}^1	3.58	2.16	Bassani, Pastori [120] Kamimura, Nakao [121]	
β-GaSe D_{6h}^4	3.75	4.24	Nagel [122]	Bourdon [123] Schlüter [124] Schlüter et al. [125]

units of two layers. Each layer of GaS or β-type GaSe can be brought into coincidence with one of its neighboring layers by a translation $\tau = \pm c/2$ and a rotation in the plane of the layers through 180° (screw axis of D_{6h}^4). This rotation is responsible for the relatively high stacking fault energy observed in β–GaSe [111] which prevents release of residual strain and thus probably is responsible for the poor resolution of the exciton spectra in β-samples [112].

All Bridgeman and most transport grown samples of GaSe contain intimate mixtures of ε-(C7 structure in Figure 27) and γ-type stacking. While the stacking sequence of the outermost anion (double) sheets of the γ-modification is reminiscent of cubic close-packing, that of the ε-type derives from hexagonal close packing. The corresponding space groups are C_{3v}^5 and D_{3h}^1. In both modifications a layer can be transformed into one of its neighbors by a translation only. Unlike the β-crystals, glide stacking faults therefore have a very small energy [111]. Moreover (hypothetical) introduction of ordered stacking faults transform the γ- into the ε-modification and vice versa. For this reason transport reacted flakes normally contain statistical mixtures of the two stacking types. The three different stacking types are schematically shown in Figure 82. Recently a fourth, more complicated 4H-modification of GaSe has been reported [113] (see also Chapter A) (δ-modification).

All theoretical studies are carried out either on a two-dimensional one-layer model (space group D_{3h}^1) or on the β-modification which exhibits the highest degree of symmetry (space group D_{6h}^4). The atomic positions of the four molecules of β–GaS/GaSe are given by:

$$\text{Ga:} \left(\tfrac{1}{3}, \tfrac{2}{3}, u_{\text{Ga}}\right)$$

$$\left(\tfrac{1}{3}, \tfrac{2}{3}, \tfrac{1}{2} - u_{\text{Ga}}\right)$$

$$\left(\tfrac{2}{3}, \tfrac{1}{3}, \tfrac{1}{2} + u_{\text{Ga}}\right)$$

$$\left(\tfrac{2}{3}, \tfrac{1}{3}, 1 - u_{\text{Ga}}\right)$$

$$\text{S, Se:} \left(\tfrac{2}{3}, \tfrac{1}{3}, u_s\right)$$

$$\left(\tfrac{2}{3}, \tfrac{1}{3}, \tfrac{1}{2} - u_s\right)$$

$$\left(\tfrac{1}{3}, \tfrac{2}{3}, \tfrac{1}{2} + u_s\right)$$

$$\left(\tfrac{1}{3}, \tfrac{2}{3}, 1 - u_s\right) \tag{C38}$$

where $u_s = 0.09 \pm 0.005$ and $u_{\text{Ga}} = 0.18 \pm 0.05$ [114]. The best X-ray data available on the u-parameters of β-type GaSe do not permit the interatomic distances to be determined better than $\pm 4\%$. This amounts to variations of the Ga–Ga bond length from 2.23 to 2.39 Å. These distances seem to be too short if compared to twice the covalent radius of Ga (2.52 Å) and to the shortest observed metallic bond length (2.45 Å) [114]. In fact, new Ga–Ga bond length values reported for the δ-modification are reasonably larger (2.46 Å) [113]. Bandstructure calculations show that the Ga–Ga bond length critically influences the valence electron spectrum of GaSe.

A number of schematic band pictures based on various experimental results and on crystal-chemical arguments have been proposed. The first speculations on the electronic structure of GaS and GaSe are given in [115], based on a two-dimensional approximation of a gas of 'nearly free electrons'. The semiconducting behavior of these poly-cationic compounds is emphasized in [57] and [116] to be due to the occurrence of small (diatomic) cation molecules. First reflectivity data are correlated to atomic energy levels to establish an 'experimental' band picture of GaS and GaSe in [117]. Reflectivity data and bond length values are discussed in [118]. It is shown that only small ionic contributions to the bonding may exist in these polycationic compounds. Similar conclusions on

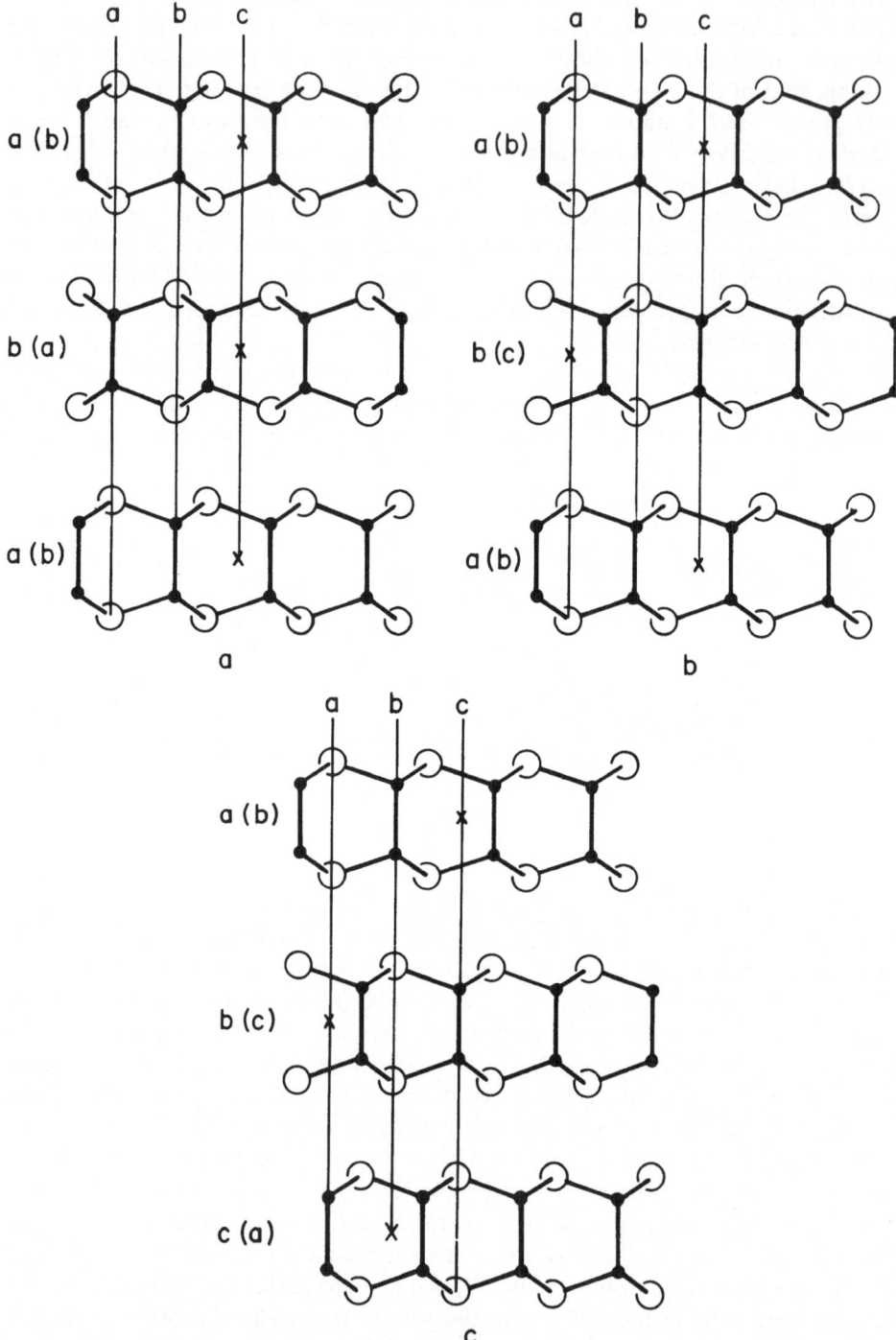

Fig. 82. Schematic view of different stacking modification of GaS and GaSe: (a) GaS and β-type
GaSe, (b) ε-type GaSe, and (c) γ-type GaSe (after [111]).

the small amount of ionicity in GaSe and GaTe are drawn from chemical shifts of X-ray K absorption edges [119].

The standard crystal chemical argument of filling of the anion valence subshells would result in an ionic configuration of $Ga^{2+}Se^{2-}$. Whereas the Ga and Se ions would attract each other and form an ionic bond, the Ga–Ga molecules in the middle of each layer would be highly unstable in this ionic configuration. This ionic picture is contrasted by a purely covalent picture in which electrons are transferred to satisfy covalent bonding requirements. In this picture one cation s-electron is promoted to the p-orbital and one anion p-electron is transferred into a cation p-level. This Ga^{1-}, Se^{1+} configuration allows Ga to form four covalent bonds with sp^3-type orbitals, one to its neighboring Ga atom and three to Se atoms. Because of different bond strengths no perfect sp^3-hybridization will take place and the Ga–Ga bond is likely to be formed predominantly by s-electrons whereas the Ga–Se bonds involve more p-electrons. The anion s-electrons would occupy non-bonding lone-pair orbitals.

It is clear, that none of these extreme pictures is absolutely correct, and more elaborate, quantitative band structure calculations are necessary to understand the chemical bonding in GaS and GaSe.

There are three independent tight binding calculations (ETB) reported for GaSe and three calculations applying the empirical pseudopotential scheme (EPM) (see Table 22). The ETB calculations of [120, 121] are based on two-dimensional one-layer models one of which [121] is empirically extended into three dimensions whereas the ETB calculation of [122] considers three-dimensional β-type GaSe. All EPM calculations [123, 124, 125] are based on β-type GaSe. Only in the ETB calculations of [120] and [121] the electronic structure of GaS is empirically considered. Let us first summarize the important features of the various ETB calculations and present some of their results:

(i) The ETB calculations of [120] are performed in the two-dimensional limit which assumes complete decoupling of individual sandwiches. The crystal wavefunctions are expanded in Bloch functions which are formed from occupied, atomic Ga and Se $4s$ and $4p$ orbitals. Only two-center integrals and nearest neighbor interactions are taken into account. The first estimates of these interactions obtained from using the orbitals of the free atoms are reduced by semiempirically considering the distortion of orbitals in the bonding directions. This empirical variation is used to adjust the value of the fundamental gap. If one pictures the (molecular) orbitals of GaSe in terms of σ and π like orbitals, the calculation allows for full mixing of σ and π states away from symmetry points. The resulting bandstructure is shown in Figure 83 (left plot). The lowest two bands correspond to the anion s-states; they are followed by two π-bands formed mostly from the p_z-functions of Se and the s functions of Ga. A group of four σ-bands corresponding mostly to the p_{xy}-like orbitals of Se falls below the uppermost π-band of predominantly Ga p_z-like orbitals. The Γ_5 (or Γ_3^+) conduction band edge (the $\pm$ superscripts [notation used in Chapter A] correspond to the symmetry behavior with respect to a horizontal reflection plane in D_{3h}^1) corresponds to a two-fold degenerate σ-band. Optical transitions from the Γ_1 (or Γ_1^+) valence band into the lowest conduction band would therefore be forbidden for $E\|\hat{c}$ which is in disagreement with experiment [126]. This disagreement is not surprising since tight-binding calculations based on occupied atomic orbitals only are known to give an insufficient description for the conduction bands. On the other hand, the valence bands in Figure 83 (left plot) may well describe

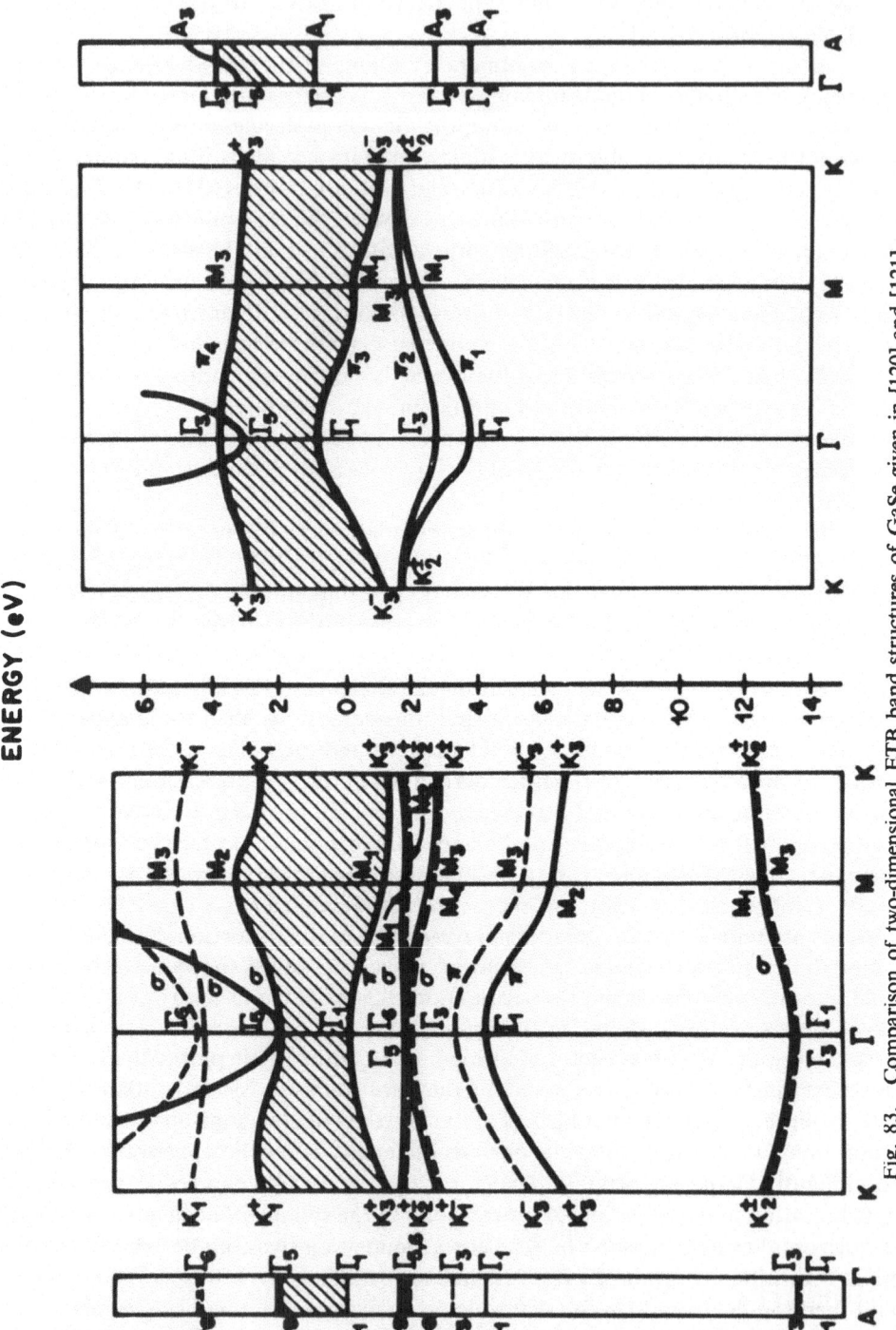

Fig. 83.   Comparison of two-dimensional ETB band structures of GaSe given in [120] and [121].

the electronic structure of GaSe. Density of states maxima estimated from this band structure are listed in Table 23.

The results for GaS are qualitatively similar and not explicitly shown. The Γ_1^+ conduction band is expected to drop below the Γ_5 (or Γ_3^+) state to explain the strong indirect band edge of GaS.

(ii) The ETB calculations of [121] are performed in a similar way. In contrast to the preceding discussion, however, it is assumed that the electrons in the σ-bands form strong co-planar bonds similar to graphite and are thus fully occupied and expected to lie well below the π-bands. This assumption is *not* supported by the results of [120]. The π-band structure including only p_z-like orbitals (s–p_z mixing is neglected!) is calculated and presented in Figure 83 (right plot). Three of the four π-bands are occupied, the fourth Γ_3 (or Γ_2^-) π-band is empty. An optical transition from the Γ_1 (or Γ_1^+) valence band into the Γ_3 (or Γ_2^-) conduction band would be forbidden (remember the significance of $\pm$ superscripts in the group D_{3h}^1). It is thus proposed in [121] that an antibonding σ-band of Γ_5 (or Γ_3^+) symmetry takes the role of the lowest conduction band. The situation at the fundamental gap is then similar to that of [120] (Figure 83, left plot) but still contradicts experiments. Interlayer interactions are introduced semiempirically in [121] and effective masses for electrons and holes are calculated. They are presented in Table 24 and compared to experimental [127] values and other calculations [123, 124]. Considering the most recent experimental results on photoemission, optical absorption, reflectivity and electrical transport, the ETB calculations of [121] disagree very strongly with experiment. This may be attributed to the fact that the number of experimental data available at the time of the empirical calculation was very limited.

(iii) A fully three-dimensional tight-binding calculation of β-type GaSe is reported in [122]. The Ga–Ga bond length is chosen in agreement with the calculations of [120] to be 2.54 Å. In addition to the occupied $4s$ and $4p$ states of the Ga and Se atoms *unoccupied 5s states* are included in the tight-binding basis set. All wavefunctions given in terms of Slater functions in [128] are subsequently expanded in terms of Gaussians to allow analytical integration of the matrix elements. Only two-center integrals are considered with the exception of crystal-field integrals which are explicitly calculated and scaled by an empirical parameter. Interactions between atoms separated up to three lattice constants are included. It is argued in [122] that this large number of interaction integrals is necessary to produce a positive definite overlap matrix of the non-orthogonal orbitals. The used atomic-like basis functions are finally 'contracted' by a Gaussian factor $e^{-\gamma r^2}$ with γ varying between 0.01 for the $5s$ functions to 0.09 for the $4s$ functions to adjust the direct gap at Γ to the experimental value. The indirect gap at M is adjusted by varying the anisotropy ratio of the p_{xy}- and p_z-like crystal-field integrals. The band structure of β–GaSe plotted along the high-symmetry lines of the three-dimensional hexagonal Brillouin zone is displayed in Figure 84. The number of bands is doubled with respect to the one-layer results shown in Figure 83. The splitting between 'pairs' of bands is a measure of the varying interlayer interaction which can reach values up to about 1.5 eV. No splitting occurs on the entire top-face of the Brillouin zone due to the time-reversal symmetry induced degeneracy in D_{6h}^4 (see discussion in Chapter A). The valence bands of Figure 83 (left plot) and Figure 84 qualitatively resemble each other. The lowest four selenium s-like states are followed by two 'pairs' of π-states and a group of eight σ-states. The topmost 'pair' of valence

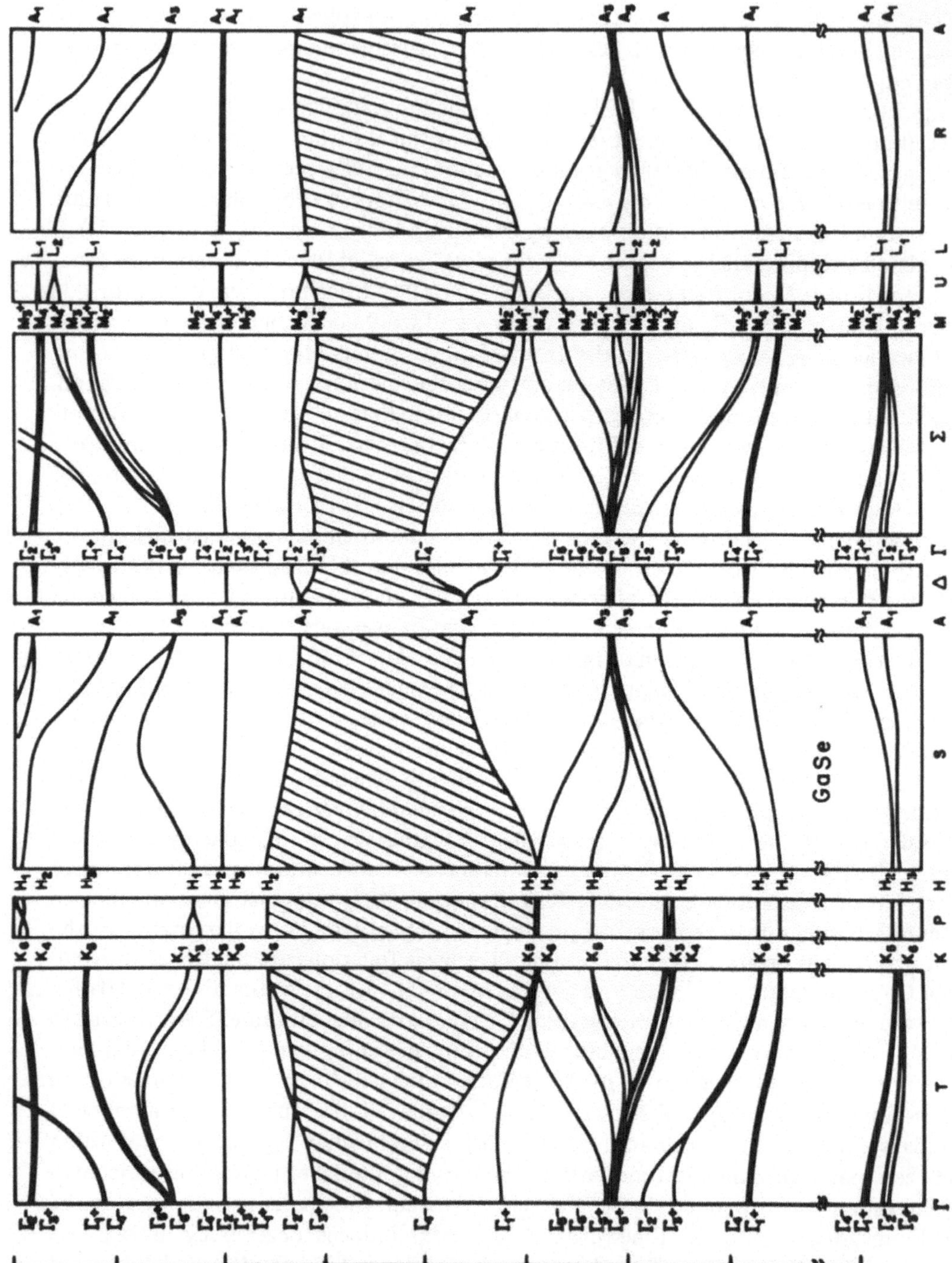

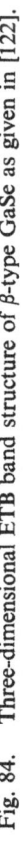

Fig. 84.   Three-dimensional ETB band structure of β-type GaSe as given in [122].

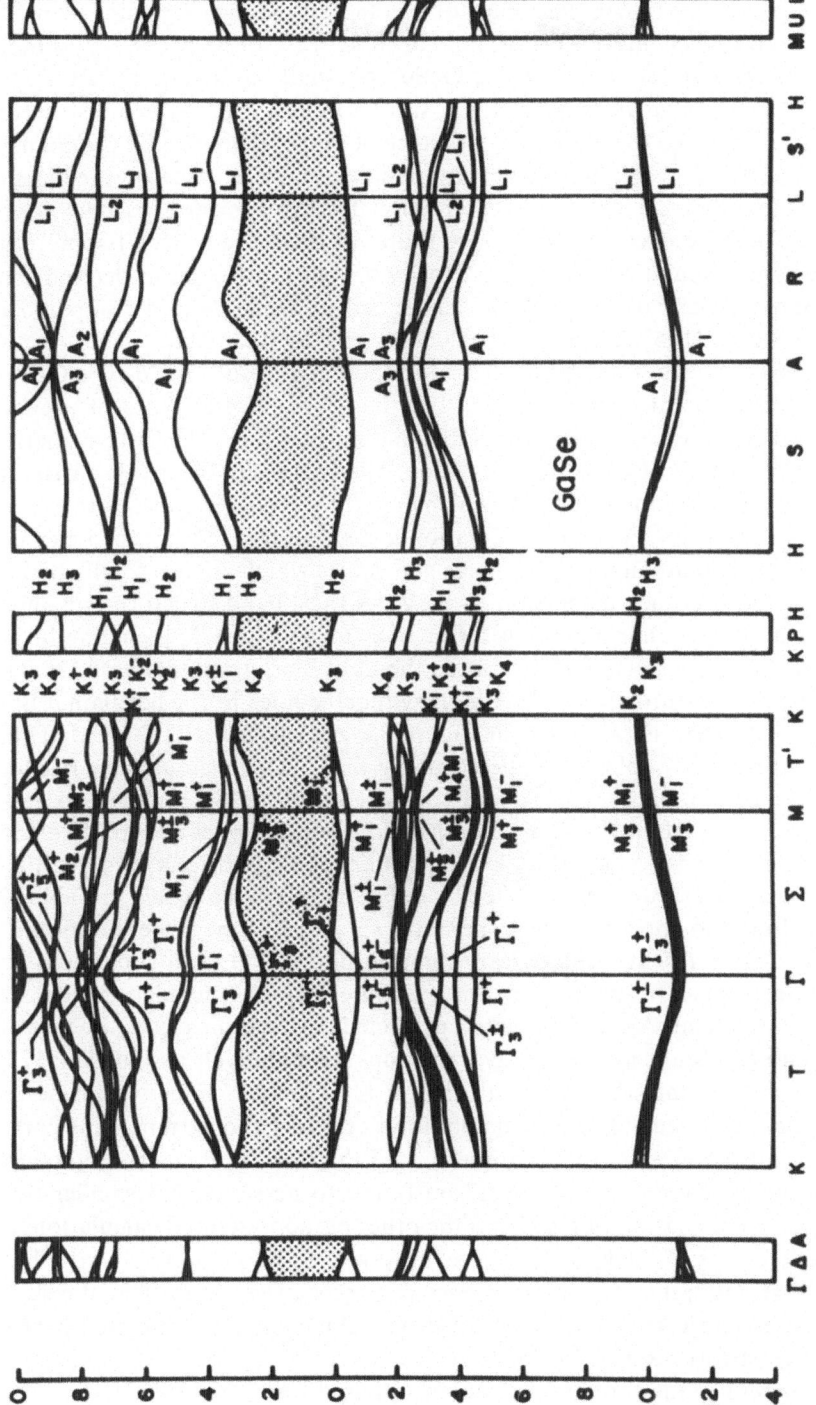

Fig. 85. Three-dimensional pseudopotential band structure of β-type GaSe obtained in [123] using Lin-Kleinman type pseudopotentials.

bands is π-like in character. Its antibonding partners form the lowest 'pair' of conduction bands. The second extremely flat 'pair' of conduction bands originates from the Se $5s$ orbitals. Even higher unoccupied atomic orbitals seem to be necessary to more correctly describe the conduction bands in GaSe. We shall see by comparing to the pseudopotential results, that in spite of these shortcomings, the ETB band structure of [122] represents very well the electronic structure of GaSe. Estimates of the changes in the band structure due to different stacking modifications are made on the basis of these results. It is found that the interlayer interaction decreases as one goes from β-type to ε-type GaSe. The differences, however, are found to be in the order of some meV only. From the band calculations of [122] a density of states histogram is derived. The energetic positions of the various peaks are indicated in Table 24 and compared to experimental XPS data [105] and to other calculations [120, 123, 125].

We now proceed to the description of the pseudopotential results for β-type GaSe.

(i) Several calculations are performed in [123] based on several types of pseudopotentials. In each case the Ga–Ga bond length is chosen as 2.30 Å. The first calculation based on model potentials of the Animalu-Heine type [78] yields relatively satisfying valence bands but gives conduction bands which are off by 2–4 eV compared to experiment and other calculations. Another calculation using non-local unscreened pseudopotentials of the Lin-Kleinman type [80] yields gap values which are too small and low lying bands showing too much dispersion (~ 3 eV). A third band structure using non-local Lin-Kleinman type pseudopotentials screened by a Penn-type [65] semiconductor dielectric function yields results, comparable to the tight binding results of [122]. The corresponding band structure along some high symmetry lines in the hexagonal Brillouin zone of β-type GaSe is presented in Figure 85. A characteristic feature of this band structure is the widely split-off highest valence band, which should give rise to a gap or dip in the corresponding density of states. No such dip has however been observed experimentally [105]. Furthermore, the width of the upper group of valence bands of about 5 eV seems to be about 3 eV too small compared to XPS data [105]. Maxima in the density of states are estimated from Figure 85 and indicated in Table 23. The direct gap is found to appear between the states Γ_1^- and Γ_3^+ (the representation Γ_1^- of [123] corresponds to Γ_4^- in the standard notation of D_{6h}^4). Optical transitions are allowed for polarizations $\mathbf{E}\|\hat{c}$ and $\mathbf{E}\perp\hat{c}$ if spin-orbit interaction is included, which is in agreement with experiment. An indirect gap is predicted between the top of the valence band at $K(K_3)$ and Γ_3^+. Some higher energy transitions are tentatively assigned in [123], but no joint density of states or $\varepsilon_2(\omega)$ calculation is presented. Effective masses of electrons and holes are calculated by fitting parabolas to the band extrema; the corresponding values are compared to experiment in Table 24. Interlayer-interactions estimated from the splitting of band 'pairs' are found to be somewhat smaller than reported in the ETB calculation of [122] and the other pseudopotential calculations of [124, 125]. Most features of the band structure, however, agree qualitatively with these other EPM calculations and no separate discussion is necessary to be presented here. A detailed analysis of the electronic structure of β-type GaSe shall be presented next in connection with the EPM results of [124, 125].

(ii) An independent pseudopotential calculation based on empirically varied form factors is reported in [124]. A Ga–Ga bond length of 2.39 Å is chosen in this calculation. Later, the calculation is repeated with similar pseudopotential form factors and an

TABLE 23

Listing of energies for the main peaks in the experimental photoemission spectrum of GaSe compared to theoretical density of states calculations (see Figure 105) (after [70]).

Structure	Experiment XPS [105]	ETB [120]	ETB [122]	EPM [123]	EPM [125]	Predominant nature of states
A	− 0.8[a]	− 0.8[b]	− 1.9	− 0.5[c]	− 0.9	Ga–Ga (p_z) bonding, Se (p_z)
B	− 2.0	− 2.0	− 3.9	− 2.1	− 1.9	Se (p_{xy})
C	− 3.6	− 2.9	− 4.5	–	− 2.9	
D	− 5.6	− 5.5	− 5.8	− 5.0	− 5.6	Ga–Ga (s) antibonding
E	− 6.5	− 6.2	− 6.8		− 6.5	Ga–Ga (s) bonding
F	− 12.8	− 12.5	− 14.7	− 10.5	− 12.1	Se (s)

[a] The valence band edge is placed to obtain best agreement with the EPM results of [125]. If the spectrum is shifted such that the Ga 3d emission coincides with consistent values obtained for other Ga-compounds (GaP, GaAs, GaSb) i.e. − 18.6 eV, all indicated experimentalv alues should be shifted by − 0.6 eV. The values listed in [105] seem to contain some errors.
[b] Estimated from the band structure in Figure 83 (left plot).
[c] Estimated from the band structure in Figure 85.

TABLE 24

Table of calculated and experimental effective masses for the direct and indirect gap in GaSe (after [127]).

	Experiment [126], [127]		ETB [121]		EPM [124]		EPM [123]	
	⊥	‖	⊥	‖	⊥	‖[b]	⊥	‖
m_h	0.8	0.2	1.67	∞	1.05	∼0.2	>0.6	0.4
m_e direct	0.17	0.3	0.15[a]	0.36	0.58	∼0.2	0.12	0.51
m_e indirect	0.5	1.6	2.64	–	0.6	–	–	–
μ direct	0.14	0.12	–	–	0.37	∼0.1	∼0.1	0.22

[a] Calculated assuming a reduced mass of 0.14.
[b] These values are calculated without repulsive barrier between layers (see [124]).

increased Ga–Ga bond length of 2.52 Å [125]. This increase in bond length improves agreement with XPS [105] and reflectivity [125] data considerably. First order pseudo-potential form factors are obtained by interpolating between scaled atomic form factors used in various III–V and II–VI compounds [129]. Then the form factors are adjusted to approximately reproduce the experimental direct and indirect gaps. Adjustments are mainly restricted to the Ga form factors since it is believed that the Ga potential may be altered due to the direct Ga–Ga contact. The resulting band structure (using a Ga–Ga bond length of 2.39 Å) is able to explain most of the chemical bonding properties [124], the optical properties in the vicinity of the fundamental gap [126] and the electrical properties [127]. Its accuracy for energies further away from the fundamental gap, however, is unclear. In fact, preliminary joint density of states calculations based on this band structure could not be matched well with existing reflectivity data [117]. A density of states histogram derived from the band structure also show some disagreement with photoemission data [130]. A simple and obvious modification of the band structure of [124] is obtained by increasing the Ga–Ga bond length to twice the covalent radius of Ga (2.52 Å) [125]. This new band structure which accounts much better for photoemission and reflectivity data is shown in Figure 86 (the lowest four Se s-bands, located around −12 eV are not shown). With respect to the original calculation, an increase in the Ga–Ga bond length decreases the gap between bands 5, 6 and 7, 8 (the first two groups of bands in Figure 86). As we shall see later, these bands have predominantly bonding and antibonding Ga s-character. The same deformation also raises the topmost valence band (18) and its 'partner' (17) completely above the group of σ-bands. Similar situations are found in [122] and [123] as seen from Figures 84 and 85. The splitting of band 'pairs' varies between 0.1 eV and 1 eV depending on the overlap of the respective wave functions of two adjacent layers.

In this paragraph we shall analyse the absorption and reflection spectra of GaSe near the fundamental gap and interpret them in terms of the band structure results of [124, 125]. The excitonic spectrum of GaSe associated with the band edge shows many interesting details and shall also be discussed. Since no basic differences between the electronic structures of different stacking polytypes are expected, the β-type calculations may safely be used to interpret the experimental data obtained on samples containing mixed λ, ε stackings. In particular, the selection rules for optical transitions near the fundamental edge are believed to be independent of stacking order. From Figure 86 it is found that the direct gap occurs at Γ between the states Γ_4^- and Γ_3^+ (D_{6h}^4 notation). Selection rules show that this transition is allowed for light polarized $\mathbf{E} \| \hat{c}$. Experiment confirms [131] that light of this polarization is strongly absorbed in GaSe at energies near the band gap. However, there is in this region also an absorption for $\mathbf{E} \perp \hat{c}$, which is only about one to two orders of magnitude weaker than that for $\mathbf{E} \| \hat{c}$ [131]. This can be explained if in addition to orbital symmetry one also considers spin symmetry. This is best done by going into the double group $\mathbf{D}_{6h}^4 = D_{6h}^4 \times D^{1/2}$, where $D^{1/2}$ is the two-dimensional representation of the Pauli spinors (see discussion in Chapter A). In fact one readily finds that Γ_4^- goes into Γ_8^- and Γ_3^+ into Γ_8^+ of $\mathbf{D}_{6h}^4$ respectively. The corresponding double group selection rules admit dipole transitions for both polarizations $\mathbf{E} \| \hat{c}$ and $\mathbf{E} \perp \hat{c}$. However, since the $\mathbf{E} \perp \hat{c}$ transition is only spin-allowed and since the light polarization operator does not operate on spin functions, the experimentally observed absorption for $\mathbf{E} \perp \hat{c}$ results from the presence in GaSe of

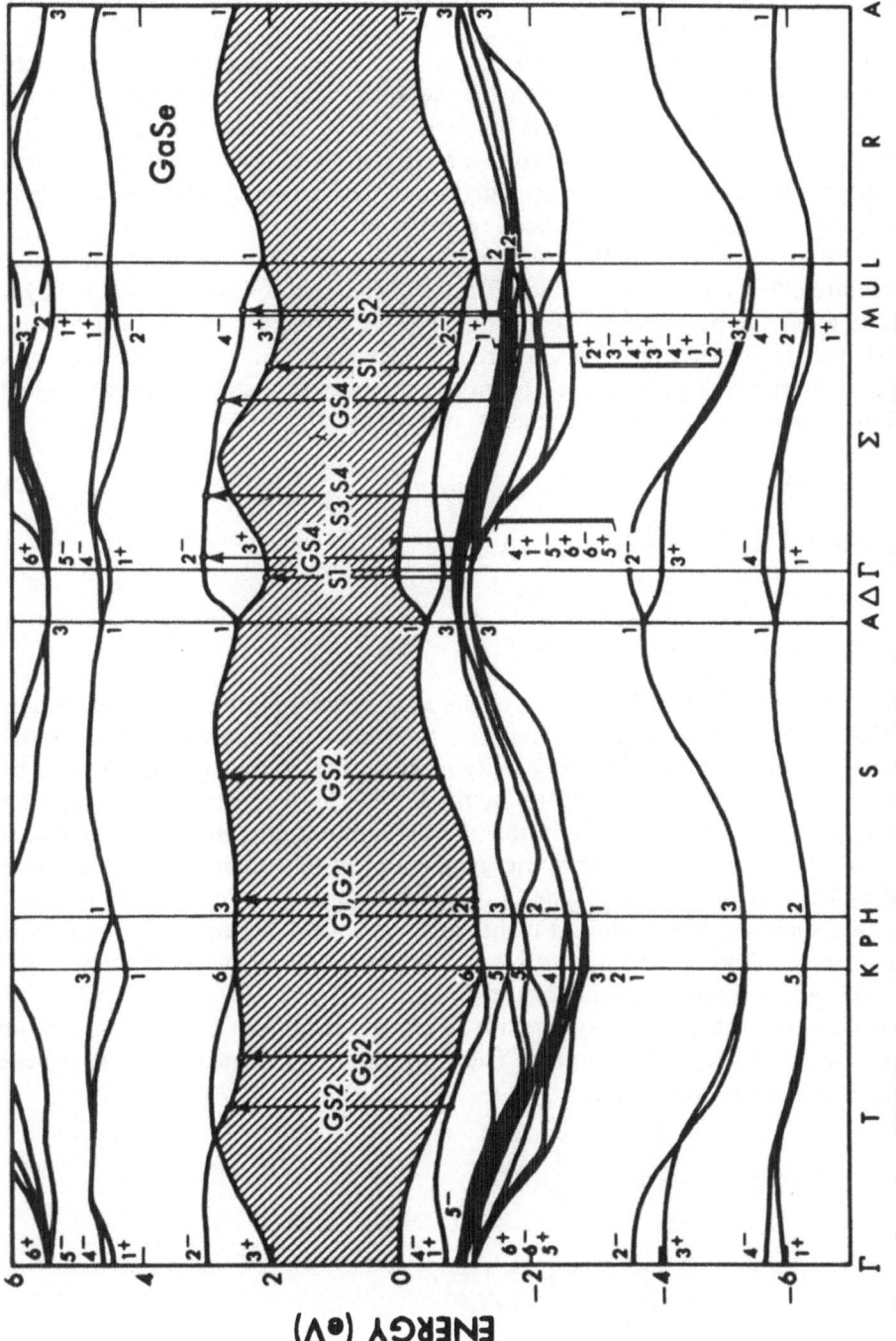

Fig. 86. Three-dimensional EPM band structure of β-type GaSe after [125]. The shown band structure differs slightly from that presented in [124] due to a change in the Ga–Ga bond length. The low lying anion s-bands are not shown.

spin-orbit coupling. According to the band structure shown in Figure 86 valence and conduction bands of GaSe are *not* orbitally degenerate at Γ. Spin-orbit coupling therefore results from interband mixing and thus is relatively weak. The mixing occurs between the valence band Γ_4^- and one (Γ_6^-) of the four closely spaced σ-bands $\Gamma_5^\pm$, $\Gamma_6^\pm$ which are estimated to lie about 1 eV below Γ_4^-. Interband mixing of the conduction band can be neglected in first order because the separation between this band and any other band of appropriate symmetry is too large. The p_{xy}-like character of the $\Gamma_5^\pm$, $\Gamma_6^\pm$ σ-bands readily explains why the proposed mixing gives rise to a finite transition probability for $\mathbf{E} \perp \hat{c}$. No quantitative calculation of the absorption coefficients based on bandstructure results, however, has been reported yet.

The direct gap exciton in GaSe is well resolved and allows the study of a number of fine structure effects. To discuss some of these effects, theoretical corrections beyond the effective mass approximation (EMA) have to be invoked [126]. Let us, however, start the discussion with the EMA approach. The standard EMA equation for an isotropic exciton envelope function $F(\mathbf{r})$ in the presence of a static external magnetic field $\mathbf{H}$ (as in magneto-optical experiments) defined by

$$\mathbf{A}(\mathbf{r}) = \tfrac{1}{2}(\mathbf{H} \times \mathbf{r}) \tag{C39}$$

has the form

$$\left[\frac{p^2}{2\mu} - \frac{e^2}{\varepsilon|\mathbf{r}|} + \frac{e}{\mu'c}\mathbf{A} \cdot \mathbf{P} + \frac{e^2}{2\mu c^2}A^2 - \frac{2eh}{Mc}\mathbf{K} \cdot \mathbf{A} \right] F(\mathbf{r}) = E_{ex}F(\mathbf{r}) \tag{C40}$$

$\mathbf{p}$ and $\mathbf{r}$ are momentum and coordinate of the relative internal motion of electron and hole. The presence of an external magnetic field does not allow any more a strict separation of internal and centre of gravity motion. The coupling between these two motions is expressed in Equation (C40) by the $\mathbf{K} \cdot \mathbf{A}$ term (magneto-Stark term) which may be regarded as a perturbation on the internal motion. $M = m_e + m_h$ describes the total mass of the exciton. The kinetic energy operator $p^2/2\mu$ contains the reduced mass, defined by $1/\mu = (1/m_e) + (1/m_h)$. The Coulomb interaction $-e^2/\varepsilon|\mathbf{r}|$ is screened by an appropriate dielectric function ε. Further magnetic field dependent terms are the Zeeman-term $(e/\mu'c)\mathbf{A} \cdot \mathbf{p}$ which involves a modified reduced mass given by $1/\mu' = (1/m_e) - (1/m_h)$, and the diamagnetic term $(e^2/2\mu c^2)A^2$.

To account for the uniaxial symmetry of GaSe anisotropic effective masses and dielectric functions may be introduced. The various terms in Equation (C40) labelled as H_1 to H_5 then have the form:

$$H_1 = \frac{1}{2\mu_x}(p_x^2 + p_y^2) + \frac{1}{2\mu_z}p_z^2,$$

$$H_2 = -\frac{e^2}{\sqrt{\varepsilon_x\varepsilon_z}}\frac{1}{\sqrt{x^2 + y^2 + \varepsilon_x/\varepsilon_z z^2}},$$

$$H_3 = \frac{e}{\mu_x'c}(A_xp_x + A_yp_y) + \frac{e}{\mu_z'c}A_zp_z, \tag{C41}$$

$$H_4 = \frac{e^2}{2\mu_x c^2} (A_x^2 + A_y^2) + \frac{c^2}{2\mu_z c^2} A_z^2,$$

$$H_5 = -\frac{2eh}{M_x c} (K_x A_x + K_y A_y) - \frac{2eh}{M_z c} K_z A_z.$$

The effect of anisotropy in the field free case is studied in [132] and [133]. By introducing eliptical coordinates the anisotropy may be treated as a perturbation $P(\alpha)$ scaled by an anisotropy parameter $\alpha = \mu_x \varepsilon_x / \mu_z \varepsilon_z$. The zero-order (unperturbed) solutions are hydrogenic functions with eigenvalues

$$E_{ex}^n = -\frac{R}{n^2} \frac{\mu_x}{\varepsilon_x \varepsilon_z} \qquad (C42)$$

and are localized within the radii

$$a_{ex}^n = a_B n^2 \frac{\sqrt{\varepsilon_x \varepsilon_z}}{\mu_x}, \qquad (C43)$$

where R stands for the Rydberg constant and a_B is the first Bohr radius of the hydrogen atom. For not too strong anisotropies, the exciton eigenstates may still be classified according to the hydrogen quantum numbers n, l, m. The perturbation $P(\alpha)$ removes some of the accidental $(l-)$ degeneracies of these levels and introduces mixing between states which transform identically in the group of rotations, D^∞. This mixing occurs e.g. between s and d_{z^2} states which is appreciable for states with $n \geq 3$. For $n = 1, 2$ levels the mixing may be neglected. In Figure 87 the dependence of $n = 1, 2$ s- and p-levels on the anisotropy parameter α is displayed [132]. It follows from this figure that if in addition to an s-level at least one of the p-levels of the same principal quantum number can be observed, then α can be determined experimentally. As we shall see, generally only excitonic s-states are visible. However, magnetic field perturbations can mix some s- and p-states and thus permit by extrapolation to determine the anisotropy parameter α [126]. Before we analyse experimental magneto optical spectra done for GaSe, let us consider the effect of spin on the exciton states.

Since spin-orbit coupling is weak (but not negligible) in GaSe it is reasonable to analyse the exciton states in terms of eigenstates of definite spin multiplicity and consider spin-orbit coupling as perturbation. Thus one singlet state and three triplet states of different symmetries exist. Due to the symmetry of the bandstructure no first-order singlet triplet mixing occurs. Furthermore, symmetry analysis shows that the singlet-state is observable for light with $E \| \hat{c}$. Transitions into the triplet states, even though formally allowed for $E \perp \hat{c}$ have zero oscillator strength, since they involve a change of the spin of the excited electron which cannot be induced by an electric dipole operator. The observation of some excitonic structure for $E \perp \hat{c}$ is thus due to interband spin-orbit coupling which mixes triplet and singlet states associated with different band gaps (no first-order mixing between the original triplet-singlet states occurs!).

A combination of the following selection rules determines whether an exciton state can be created by the absorption of a photon:

(i) the probability of finding electron and hole in the same position has to be non-

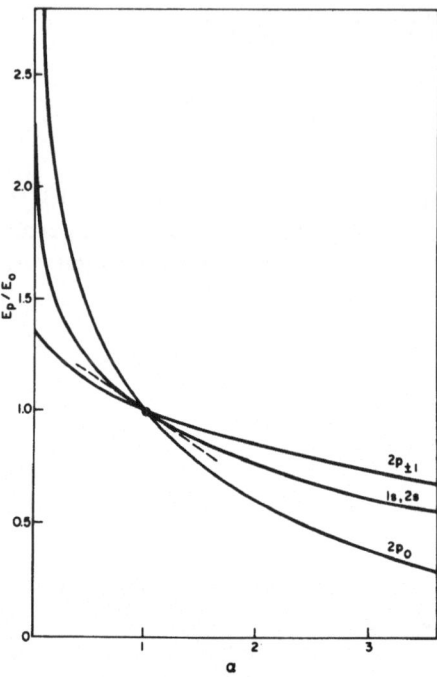

Fig. 87. The dependence on the anisotrophy parameter $\alpha = \mu_x \varepsilon_x / \mu_z \varepsilon_z$ of the $n = 1, 2$ s and p exciton states (after [132]).

zero, i.e. $|F(0)|^2 \neq 0$ which is satisfied for s-states or states with s-admixture (higher order transitions with $|\nabla F(0)|^2 \neq 0$ neglected);

(ii) only singlet states or states with singlet admixture can be created;

(iii) the total symmetry of the exciton state has to match with the polarization of the incident light, i.e. the representation of the exciton state

$$\Gamma_{ex} = \Gamma_{electron} \times \Gamma_{hole} \times \Gamma_{envelope} \times \Gamma_{spin}$$

has to contain the representation of the polarization vector of the photon which is to be absorbed.

In this scheme only singlet s-excitons originating from bands of appropriate orbital symmetry can be seen in first order. Several perturbations like anisotropy, spin-orbit coupling or external magnetic fields cause relaxation of these strict rules.

The $n = 1$ ground state of the exciton in a sample of ε-γ-GaSe is shown in Figure 88 as observed in reflectivity with light incident at $45°$ to the layer normal [126]. For the polarization vector perpendicular to the plane of incidence ($\mathbf{E} \perp PI$), which corresponds to $\mathbf{E} \perp \hat{c}$, a group of relatively weak lines t_i is observed. This group of lines represents the various *triplet* fine structure components. The fine structure splitting is believed to be due to ε-γ stacking faults which are sensed by the exciton wavefunctions [126]. As the plane of polarization is turned into the plane of incidence ($\mathbf{E} \| PI$) the polarization vector acquires a component along $\hat{c}$ and the *singlet* components, labelled s_i, become visible. Indeed for $\mathbf{E} \| \hat{c}$, the transitions into the singlet states are allowed, whereas those into the triplet states remain allowed by spin-orbit coupling only. The lines s_i therefore are much

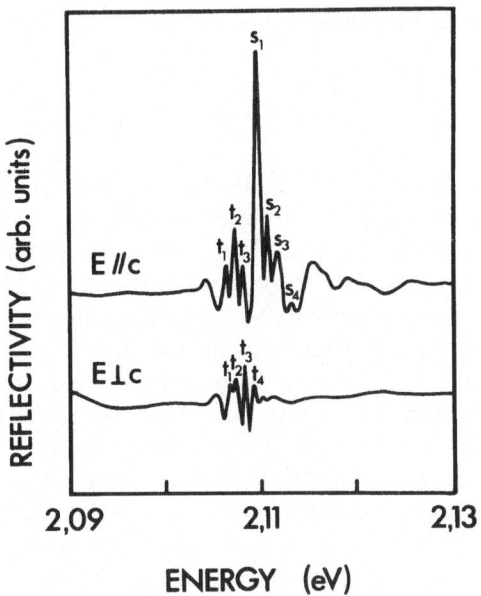

Fig. 88. Fine structure of the exciton ground state in γ-, ε-, GaSe as seen in reflectivity at 4.2 K. The measurements are taken with light incident at an angle of 45° with respect to the crystal $\hat{c}$-axis and with the polarization vector parallel ($\mathbf{E}\|PI$) and perpendicular ($\mathbf{E}\perp PI$) to the plane of incidence (after [126]).

stronger than the corresponding lines t_i in spite of the fact that, because of the high index of refraction $n = 3$ of GaSe, the $\mathbf{E}\perp\hat{c}$ component is still considerably larger than the $\mathbf{E}\|\hat{c}$ component. The relative intensities of the lines s_i and t_i readily permit corresponding singlet and triplet components to be identified and thus determine the singlet-triplet splitting for this geometry to be $\Delta E_{s-t}^{1s} = 1.9$ meV. This splitting in energy between singlet and triplet is due to exchange interactions which are not included in the standard EMA Hamiltonian. The effect of the exchange may be added as a perturbation to the EMA Hamiltonian [99] and can be written as

$$H_{ex} = J(\mathbf{K}) \cdot \delta_M \cdot \delta(\mathbf{r}), \tag{C44}$$

where the symbols δ_M and $\delta(\mathbf{r})$ indicate that H_{ex} is non-zero only for s-like singlet states. H_{ex} may be split into two terms:

(i) The short range exchange J_1 which is independent of the centre of mass momentum $\mathbf{K}$ of the exciton and which may be written in terms of valence and conduction band Wannier functions a_v and a_c centered at the origin

$$J_1 \sim 2 \left\langle a_c(\mathbf{r}_1) a_v(\mathbf{r}_2) \left| \frac{e^2}{r_{1,2}} \right| a_v(\mathbf{r}_1) a_c(\mathbf{r}_2) \right\rangle. \tag{C45}$$

(ii) The long range exchange $J_2(\mathbf{K})$ involving functions not centered at the origin which has to be evaluated from a sum over all lattice vectors $\mathbf{R}_l$

$$J_2(\mathbf{K}) \sim 2 \sum_{l=0} e^{i\mathbf{K}\cdot\mathbf{R}_l} \left\langle a_c(\mathbf{r}_1 - \mathbf{R}_l) a_v(\mathbf{r}_2) \left| \frac{e^2}{r_{12}} \right| a_v(\mathbf{r}_1 - \mathbf{R}_l) a_c(\mathbf{r}_2) \right\rangle \tag{C46}$$

$J_2(\mathbf{K})$ is non-zero whenever the transition has non-zero dipole moments

$$\mathbf{d} = \langle a_v(\mathbf{r})| \, \mathbf{r} \, |a_c(\mathbf{r})\rangle \neq 0. \qquad (C47)$$

The actual value of $J_2(\mathbf{K})$ depends on the relative orientation of $\mathbf{d}$ and $\mathbf{K}$. In GaSe $\mathbf{d}$ is predominantly parallel to $\hat{c}$ with some x, y admixture due to spin-orbit coupling. The direction of exciton propagation ($\mathbf{K}$) inside the crystal depends on the angle of the incident light. The $\mathbf{K}$-dependence of J_2 thus gives rise to spatial dispersion, i.e. to the so-called longitudinal-transversal shift of the singlet exciton states. It follows from Equation (C44) that the exchange energy is proportional to $1/n^3$ and therefore only appreciable in the ground state. In Figure 88 the triplet state observed for $\mathbf{E} \perp \hat{c}$ is purely transverse whereas the singlet observed in the $\mathbf{E} \| PI$ configuration has mixed longitudinal-transverse character. Its position is expected to shift due to $J_2(\mathbf{K})$ with varying angle of incident light. Assuming no energy correction to the triplet state the exciton binding energy may be calculated from the positions of the observed $n = 1$, 2 and 3 levels (only $n = 1$ is shown in Figure 88). An energy of $E_{\mathrm{ex}} = 19.8 \pm 0.1$ meV is obtained. The interpretation of recent electro-reflectance data yields $E_{\mathrm{ex}} = 21$ meV [134]. Furthermore, the observed series may be fitted to the theoretical expression (Equation (C42) for anisotropic excitons [132] by taking the dielectric constant values $\varepsilon_x = 10.2$ and $\varepsilon_z = 7.6$ in accordance with [135] and by assuming a perpendicular reduced mass $\mu_x = 0.14 \pm 0.05$. These values correspond to a zero-order $1s$ envelope function which extends over about 32 Å. A value for the exciton anisotropic parameter α cannot be obtained from these measurements, as the observed $n = 1$, 2 s-like excitons are equally affected by α (see Figure 87).

More information on the excitonic series is obtained from magneto-optical spectra [126]. Absorption results for which the magnetic field $\mathbf{H}$ is applied parallel to the incident light beam (Faraday geometry) and parallel to the crystal $\hat{c}$-axis ($\mathbf{E} \perp \hat{c}$) are presented in Figure 89. The light is circularly polarized with the topmost spectrum being taken with left-hand circular polarization (lcp) and the others being taken with right-hand circular polarization (rcp). Since the observed ground- and first excited states do not carry any orbital angular momentum, the shifts between corresponding rcp and lcp lines are entirely due to the electron-hole spin which is non-zero for the triplet states ($S_z = \pm 1$). The measured splitting yields an effective exciton g-factor of $g_z = 2.7 \pm 0.2$. The various lines may be fitted with this value of g_z and a value of $\mu_x = 0.13$ in the diamagnetic term H_4 of Equation (C41) which is the only magnetic field dependent term effective on s-states in this geometry (see Figure 90). The singlet nature of the state whose zero-field ionization energy is 17.9 meV is clearly recognizable. This good characterization of the multiplet nature of the exciton states is strong evidence for the weak spin-orbit coupling in GaSe.

It is clear that the perturbative approach becomes invalid once the magnetic field energy H_4 becomes comparable with the binding energy of the state under consideration. For very high fields, $H_4 \gg E_{\mathrm{ex}}$ adiabatic approaches may be used in which the motion perpendicular to $\mathbf{H}$ is determined solely by the magnetic field whereas along the field axis an effective electron-hole Coulomb attraction is still active [136, 137]. It can be shown that in this limiting case of high fields the excitonic levels are described by Landau-type levels which are subject to an additional (field-dependent) quantization in an effective quasi-one-dimensional Coulomb-like potential. This additional quantiza-

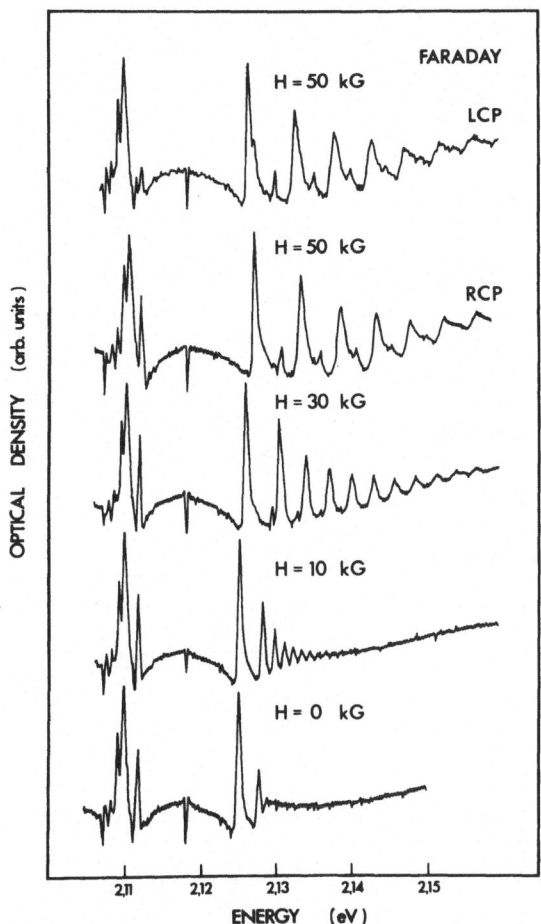

Fig. 89.   Magneto-absorption spectrum of γ-, ε-GaSe in Faraday geometry. The uppermost spectrum is obtained in left circularly polarized light (lcp), all others in right circularly polarized light (rcp) (after [126]).

tion is responsible for the appearance of so-called satellite-lines as they are visible in Figure 89 for fields above 30 kG for higher excited states. Thus the quadratic field-dependence for low fields of lines in the spectrum of Figure 90 becomes linear (Landau) for very high fields with the simultaneous appearance of satellite lines. In an alternative picture these satellite lines may be viewed as mixtures of higher angular momentum states of the field-free exciton. The question of connecting the high-field satellite lines to the low-field hydrogenic exciton lines is very difficult and is not completely solved yet. Connection-rules proposed in [136] seem to be verified by numerical solutions of the general magnetic field dependent hydrogenic Schrödinger equation [138]. The connection is based on the rule of non-crossing of levels of the same symmetry. This rule [136] suggests that *all* exciton states connect with the sublevels of the *lowest* Landau-type level. Levels corresponding to higher Landau-type states go into the continuum as the magnetic field decreases. In addition the non-crossing rule of levels may be accompanied by transfer of oscillator strength. These very general problems of magneto-optics whose

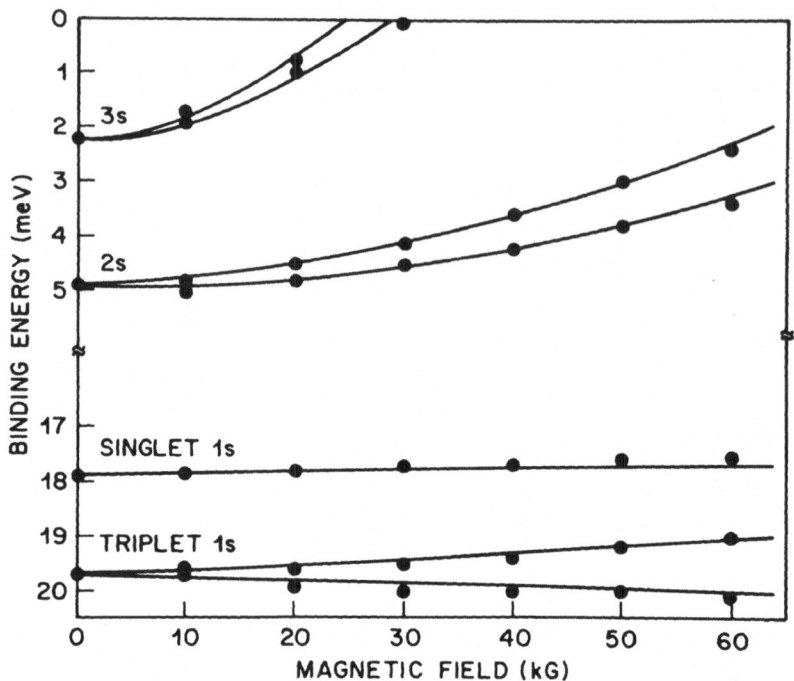

Fig. 90.   Magnetic field dependence of the exciton states $n = 1$, 2, and 3 of GaSe taken from Figure
89. The curves are calculated and fitted according to the text (after [126]).

details go beyond the scope of this review have been investigated to a great extent on the
basis of the exciton spectra of GaSe.

Another series of absorption spectra taken in the Voigt geometry ($H \perp$ incident light)
for different fields is shown in Figure 91. The light is linearly polarized with $E \perp H$. The
discussion of this geometry is complicated by a complete breakdown of angular momentum quantum numbers. Assuming the exciton states to be quantized along the crystal
$\hat{c}$-axis all magnetic field dependent perturbation terms H_3, H_4, H_5 have non-zero
matrix elements:

(i) The Zeeman term H_3 has only off-diagonal elements and mixes the p-states whose
axes are perpendicular to the magnetic field.

(ii) The diamagnetic term H_4 has only diagonal elements which are scaled by the
anisotropy ratio of the reduced effective masses μ_x/μ_z.

(iii) The magneto-Stark term H_5, whose effect is equivalent to an electric field applied
perpendicular to both H and the direction of exciton propagation K, gives rise to the
only off-diagonal term between s- and p-states. This matrix element scales with $K_z = 2\pi n/\lambda$, where n is the index of refraction at the excitation wavelength λ.

As mentioned earlier the field induced s–p mixing is of considerable importance since
it permits the relative positions of the involved s and p levels to be determined by
extrapolating to $H = 0$. An experimental value for the anisotropy parameter α may
thus be obtained from Figure 91. A fit of the $n = 2$ levels to the observed spectra of
Figure 91 using the above mentioned non-zero matrix elements is presented in Figure

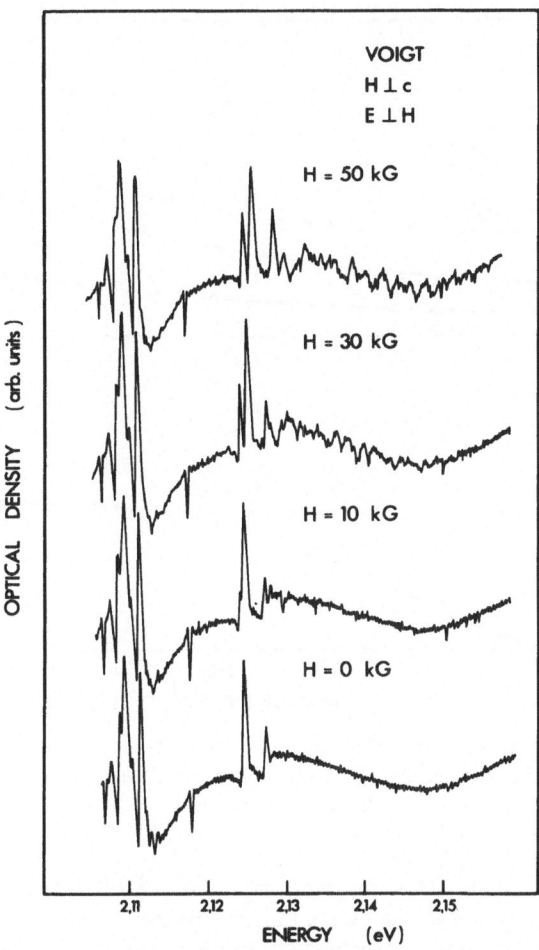

Fig. 91. Magneto-absorption spectrum of γ-, ε-GaSe in Voight geometry for linearly polarized light $\mathbf{E} \perp \mathbf{H}$, $\mathbf{H} \perp \hat{c}$ (after [126]).

92. The strong line appearing at the low energy side of the $n = 2$ s-state in Figure 91 is interpreted to originate from the $2p_y$ state which via the $\mathbf{K} \cdot \mathbf{A}$ magneto-Stark term acquires rapidly increasing s-character and thus becomes visible [126]. Extrapolating to $\mathbf{H} = 0$ the energy difference between this $2p_y$ state and the slightly higher s-state allows to determine the anisotropy parameter to be

$$\alpha = \frac{\mu_x \varepsilon_x}{\mu_z \varepsilon_z} = 1.8.$$

With the dielectric constants ε_x and ε_z of [134] one thus obtains the remarkable value for the anisotropy ratio of the reduced masses

$$\frac{\mu_x}{\mu_z} = 1.3 \pm 0.2$$

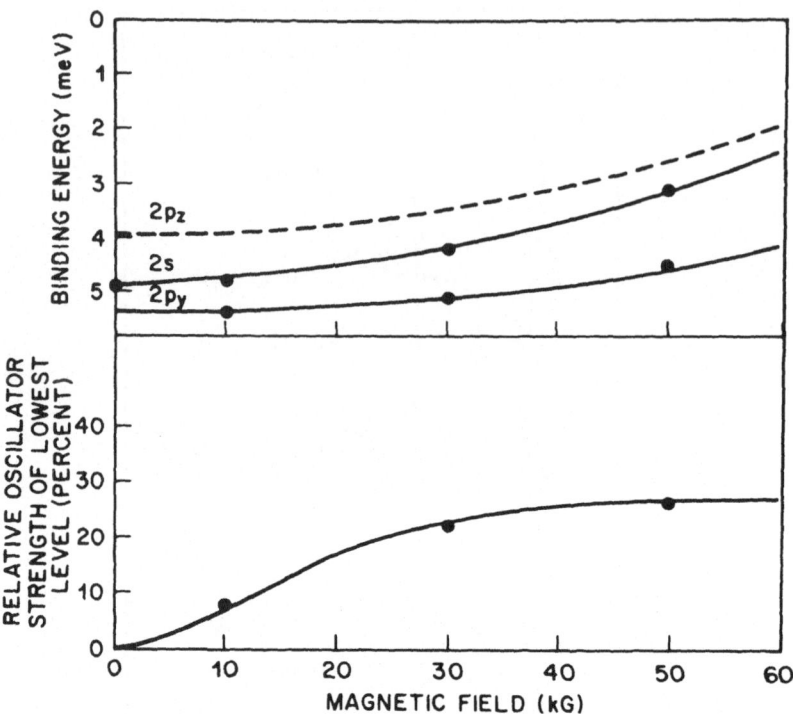

Fig. 92. Magnetic field dependence of the mixed 2s, 2p levels and of the relative oscillator strength of the lowest $n = 2$ level as derived from Figure 91. The curves are calculated and fitted according to the text (after [126]).

which compares well with the value $\mu_x/\mu_z = 1.0 \pm 0.2$ obtained from comparing the high-field diamagnetic shifts of the 1s line in Faraday and Voigt geometries. An intensity fit (Figure 92) of the $2p_y$-line determines all involved parameters. In particular, the K_z component of the exciton wave vector can be evaluated to be $K_z = 7.3 \times 10^{-2} M_z \, \text{Å}^{-1}$. Comparing this value with that $K_z = 2\pi n/\lambda = 3.2 \times 10^{-3} \, \text{Å}^{-1}$ derived from the wavelength $\lambda = 5835$ Å of the absorbed light and from a refractive index $n = 3$ one finds a rather low value $M_z = 5 \times 10^{-2}$ for the total exciton mass, which is about one order of magnitude lower than the value obtained from independent electrical measurements [127]. It should be noted here that exciton lines are in the regions of anomalous dispersion. Thus the appropriate refractive index may well exceed the one used above by a factor of ten due to polariton effects and thus may yield a correct total exciton mass.

This detailed analysis of the excitonic spectra of GaSe under the influence of external magnetic fields yields complete information on the direct gap band parameters. The remarkable result is the relatively small mass of carriers perpendicular to the plane of layers. This is in contradiction to earlier interpretations of the electronic structure of GaSe [121] but agrees well with electrical measurements [127] and is also indicated by some theoretical bandstructure calculations (see Table 24). From all this evidence one is led to conclude that if some of the bands in GaSe have the expected two-dimensional character, those determining the optical and electrical properties at the fundamental band gap are definitely three-dimensional.

   In the preceding discussions only the direct band gap and the associated exciton
series were considered. It is, however, argued that GaSe exhibits an indirect conduction
band minimum (at M) several tens of meV below the Γ-minimum [139]. The bandstructures of GaS_xSe_{1-x} ($0 \leq x \leq 1$) mixed crystals are similar but the direct and
indirect gaps as well as the difference between them increase with increasing x. Due to
this situation the direct band gap excitons are even for GaSe in resonance with the
indirect band edge. The line shape of the exciton resonating with indirect band edge
continuum can well be described by a Fano-type theory [139]. This resonance behavior
of the free exciton is made evident by a study of the recombination kinetics describing
the exciton luminescence [139]. The proposed model is schematically shown in Figure
93. The deduced energy difference E_2 between direct and indirect band edge is about

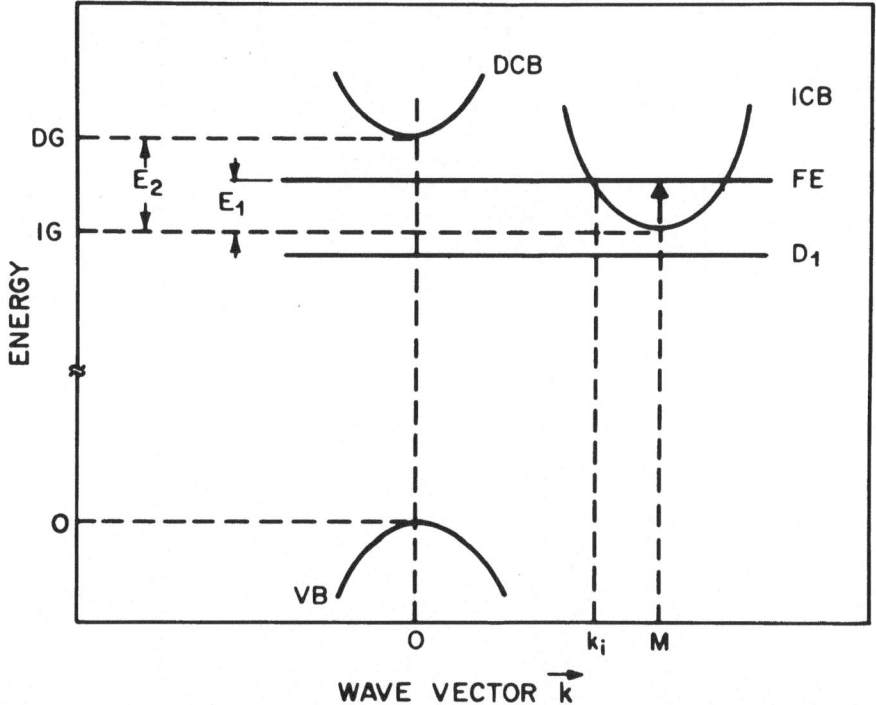

Fig. 93.   Model of the direct-indirect gap in GaSe. VB is the valence band, DCB and ICB are the
direct and indirect conduction bands, respectively, FE is a free exciton $n = 1$ level associated with
DCB and D_1 is a representative donor level associated with ICB (after [139]).

40 meV for GaSe which leaves about 20 meV for E the difference between the direct
exciton ground state and the indirect band edge. This difference is small enough in
GaSe to allow the exciton lines to be well resolved. As the difference increases with
increasing x in GaS_xSe_{1-x} the exciton gradually disappears due to the increasing
resonance behavior (density of states).

   The difference E_1 between exciton and indirect band edge may also be increased by
applying hydrostatic pressure to GaSe [140]. It is found that both exciton and indirect

band edge decrease with pressure but with different rates i.e. $\partial E_{\mathrm{ex}}/\partial P = -4.2 \pm 0.3 \times 10^{-6}$ eV bar^{-1} and $\partial E_{\mathrm{tail}}/\partial P = -6.5 \pm 0.3 \times 10^{-6}$ eV bar^{-1} [140]. The behavior of the absorption edge for various values of hydrostatic pressure is shown in Figure 94. The disappearance of the exciton peak under pressure may thus qualitatively be related to the increased resonating behavior of the exciton with the indirect band edge.

The negative pressure coefficient of the band gap in GaSe, which is also found in other layer compounds [141–143] may be understood on the basis of the three-dimensional band- and bond-picture in these materials. From Figures 84–86 it follows

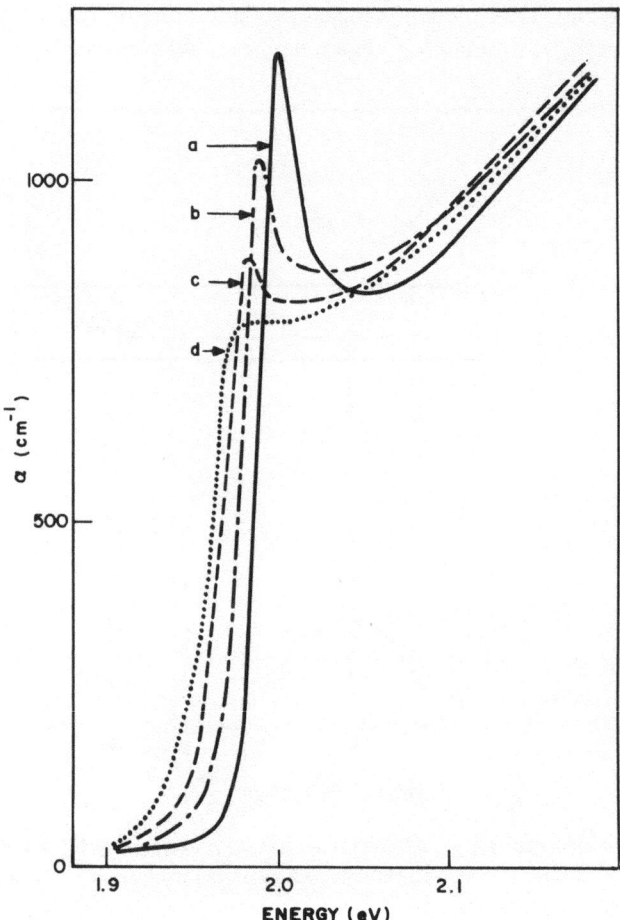

Fig. 94. Absorption coefficient for GaSe for $\mathbf{E} \perp \hat{c}$ light polarization with hydrostatic pressure as parameter. Curve a (full line) 1 bar, curve b (dashed-dotted line) 2.3 kbar, curve c (dashed line) 4.4 kbar, curve d (dotted line) 6.7 kbar (after [140]).

that the optical gap appears between states whose energy is determined by two types of interactions:

(i) strong *intra*-layer interactions (in the case of GaSe, the strong Ga–Ga bond) which give rise to a mean gap (~ 2–3 eV);

(ii) weaker *inter*-layer interactions which present some (0–1 eV) modulation on the *intra*-layer band gap.

Figure 95 indicates schematically the change in the band structure occurring if hydrostatic pressure is applied. Pressure first decreases the (soft) interlayer spacing without much affecting the (stiff) intralayer bonding. This increases the inter-layer interaction and via the bandstructure 'modulation' (ii) decreases the fundamental gap, thus resulting in a negative pressure coefficient $\partial E_g/\partial P$. Moreover, one may extrapolate that this situation will change, once the interlayer interaction becomes under pressure comparable in strength to the intralayer interaction. Then the pressure coefficient should

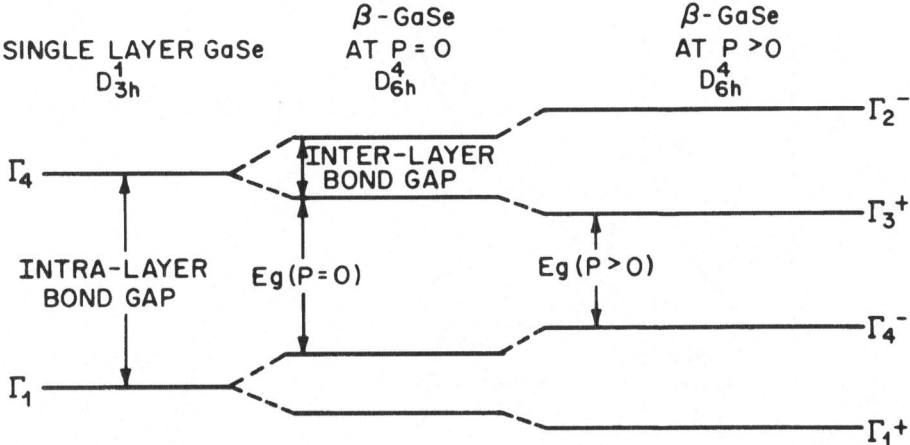

Fig. 95. Schematic band model at the fundamental gap for layer crystals like GaSe under hydrostatic pressure. The single-layer levels (left hand side) are split by inter-layer interaction (middle). Under moderate pressure (see [140]) the inter-layer interaction increases (right hand side) reducing thus the optical gap.

become small and eventually change sign. First experimental indications of this behavior seem to be observed in GaSe for pressures above 40 kbar [144].

A number of experimental papers consider the electronic structure of GaSe farther away from the fundamental gap. First reflectivity data are reported in [117]. Particular interest is centred on a structure around 3.4 eV. This structure is extensively studied in absorption, reflectivity and modulated reflectivity [145] and subsequently assigned to various types of excitonic transitions. Recent reflectivity, wavelength modulated reflectivity and electro-reflectivity measurements [125] seem to indicate the existence of an M_0-type critical point exciton at this energy. The measured binding energy (the $n = 2$ state is resolved) of 33 meV is compatible with the reduced masses $\mu_\| = 0.3$ and $\mu_\perp = 0.17$. The corresponding band to band transition, labelled S1 in Figure 86 is supposed to take place at Γ between the σ-like state Γ_5^- and the first π-like conduction state Γ_3^-. It is of predominantly anion (Se) character, i.e. its initial state contains almost pure Se p_{xy} character and its final state shows appreciable Se s-admixture.

High resolution reflectivity curves between 3 eV and 6 eV have recently been reported in [125] and [146]. The results of [125] are shown in Figure 96 together with a calculated curve derived from the band structure of Figure 86. The calculated reflectivity curve had to be shifted by 0.4 eV to higher energies to achieve satisfactory agreement with experiment. Since all important transitions below 5.5 eV terminate at the first isolated conduction band pair (bands 19, 20) the shift of the reflectivity curve may easily be obtained by rigidly shifting upwards the main portions of bands 19 and 20. Although

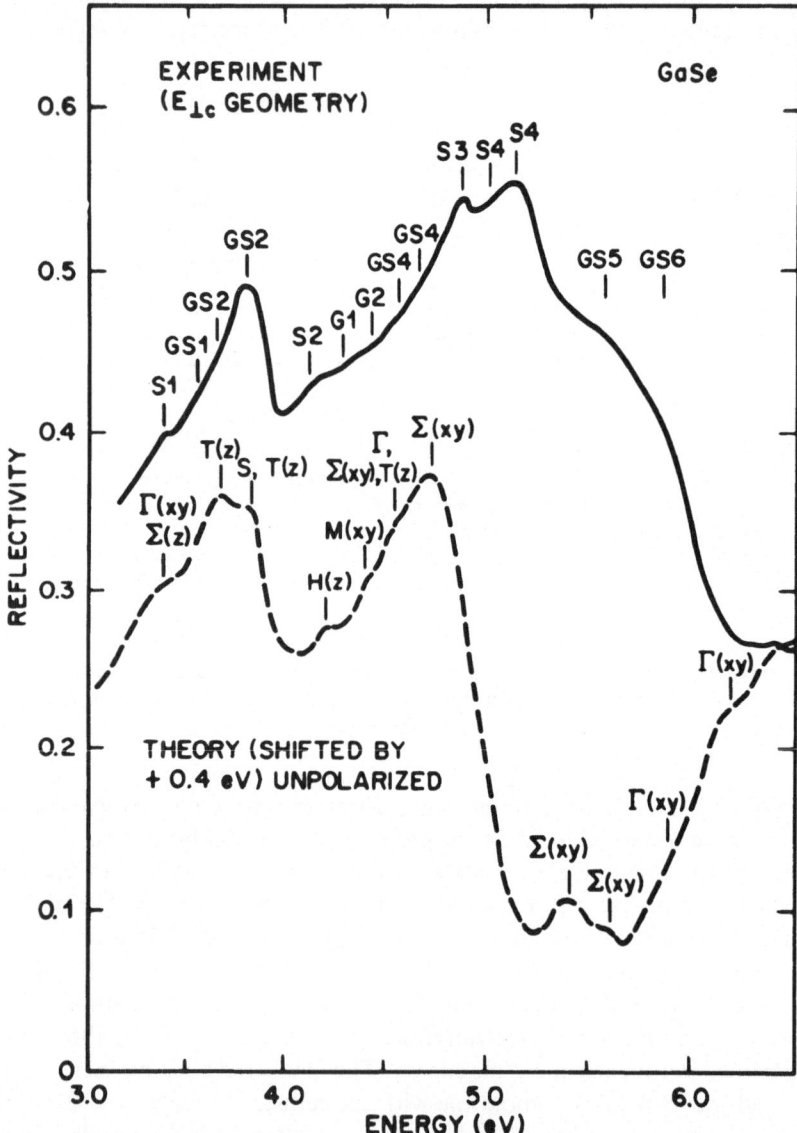

Fig. 96. Experimental (full line) at 5 K and calculated (broken line) reflectivity of GaSe. The theoretical curve has been shifted by 0.4 eV to higher energies as described in the text. Structure in the experiment spectrum are labelled according to Table 25. The corresponding labeling of the theoretical curve refers to k-space location and polarization dependence of the transitions involved (after [125]).

the experiment is performed in the $E \perp \hat{c}$ geometry a small finite angle of incidence ($\sim 5°$) and the finite aperture of the light beam ($\sim 5°$) allow the presence of an $E \| \hat{c}$ component. Because of this $E \| \hat{c}$ component which is combined with a strong matrix element anisotropy (see e.g. the discussion of the band edge exciton) the experimental spectrum is most conveniently compared to an unpolarized calculated spectrum. The precise positions of structure obtained from the respective derivative spectra are listed together with bandstructure assignments in Table 25. The main transitions are also indicated

TABLE 25

Listing of experimental (5 K) energies for the main structures in the reflectivity of GaSe below 6 eV. Theoretical assignments are given for GaSe and the experimental (5 K) energies of the corresponding structures in GaS are indicated (see Figures 86, 96, 98) (after [125]).

GaSe			GaS
Experiment [125]		Theory [125]	Correspondence
Labelling	Energy 5 K	Assignment	Energy 5 K
Sl(excit.)	3.37		4.55
S1	3.40	$\Gamma(xy)16$–19	4.61
GS1	3.58	$\Sigma(z)18$–19	
	3.70	$T(z)18$–19	
GS2	3.80	$T, S(z)17$–19	4.03
GS3	–		4.20
S2	4.07	$M(x, y)15$–20	4.82
G1	4.28	$H(z)18$–19	4.30
G2	4.41		4.42
GS4	$\begin{cases} 4.54 \\ 4.67 \end{cases}$	$\begin{cases} \Sigma(xy)16\text{–}20 \\ \Gamma(xy)16\text{–}20 \\ T(z)17\text{–}20 \end{cases}$	4.82
S3	4.86	$\Sigma(xy)16$–20	5.57
G3	5.03	–	5.03
S4	5.13	$\Sigma(xy)14$–20	5.78
G4	–		5.23
GS5	5.56	$\begin{cases} \Sigma(x, y)18\text{–}20 \end{cases}$	5.90
GS6	5.83		6.20

directly in the bandstructure diagram (Figure 86). The labelling of transitions is chosen to indicate whether cation gallium (G) or anion selenium (S) or both atomic characters are present in the initial and final states defining the transition. The first main peak (GS2) around 3.8 eV originates from transitions between the uppermost valence band pair (17, 18) and the first conduction band pair (18, 20) around T and S and corresponds to an excitation of the Ga–Ga bond and of non-bonding Se p_z to s transitions. The second main peak (S3, S4) around 5 eV originates mainly from transitions starting from the lower lying Ga–Se σ-bands (14, 16) around Σ and involves exclusively Se p_{xy} to s transitions.

It is of interest to note that these two groups of transitions which roughly determine the lower part ($E < 6$ eV) of the reflectivity spectrum of GaSe take place at locations in the hexagonal Brillouin zone which are close to symmetry determined 'special k-points' [47]. The significance and definition of these points has been discussed in the sub-sections on SnS$_2$, SnSe$_2$ in this chapter. The results seem to indicate that the low-energy part of the spectrum of GaSe can be adequately approximated by evaluating the bandstructure at just two k-points. It should be emphasized, however, that important structure in the dielectric response of GaSe appears at energies higher than 6 eV. This can be seen from Figure 97 where a complete reflectivity spectrum of GaSe for energies between 3 eV and 25 eV is shown [147]. The spectra are taken with a synchrotron radiation source at various angles of incidence (20°, 45° and 60°). The photons are polarized linearly in the plane of incidence thus allowing a variation of the $\mathbf{E} \| \hat{c}$ component. The numerous structures indicated in the spectrum are obtained from an analysis of the second derivative spectrum. As seen from the figure two big main peaks appear above 7 eV followed by some structure between 20 and 25 eV which originates from transitions from the localized Ga $3d$ core levels into the lowest groups of conduction bands. Some attempts of assigning the structure above 7 eV to band to band transitions are reported in [147]. The results in Figure 97 seem to indicate the presence of two strong structures in the GaSe conduction band density of states. Indeed the high energy transitions above 20 eV reveal in addition to the 0.45 eV spin-orbit splitting of the Ga $3d$-levels the presence of a gap between a first and second group of conduction bands. This finding is in good accord with the bandstructures of [122], [124] and [125] (see Figures 84, 86). Further independent reflectivity data in a wide energy range between 2 eV and 80 eV are reported in [148]. The results though not showing the resolution of those in Figure 97 essentially agree with [147]. By Kramers-Kronig analysis of these reflectivity data an energy loss function $-\text{Im}\,(1/\varepsilon)$ is calculated and compared to direct measurements obtained by electron spectroscopy [149]. A strong plasmon peak is found at 15.8 ev in both approaches which corresponds to the plasma oscillation of all valence electrons. Several subsidiary peaks or shoulders are found around 7 eV, 10 eV and 21–25 eV corresponding to the various fall-off regions in reflectivity shown in Figure 97.

Interesting investigations about the exact location of the Ga $3d$-core levels obtained by XPS, reflectivity and energy loss experiments are under way [150] and should give information on the existence of 'core-excitons' associated with the various excitations [151].

A first complete experimental and theoretical discussion of the optical spectra of the GaS$_x$Se$_{1-x}$ mixed crystal series for energies up to 6 eV is presented in [125]. The spectral changes observed in the mixed crystal series are interpreted on the basis of the theoretical understanding of the GaSe spectrum (Figure 96). In Figure 98 an illustration of the general variation of the experimental reflectivity spectrum is given. Most of the assigned structures of GaSe can be followed through the series to GaS (Table 25). Analysis reveals linear shifts for almost all important structures. Moreover, according to their shifts (Δ) all structures (including direct and indirect gap) can be grouped into three different classes with zero ($\Delta \approx 0$), intermediate ($\Delta \approx 300$–400 meV) and strong

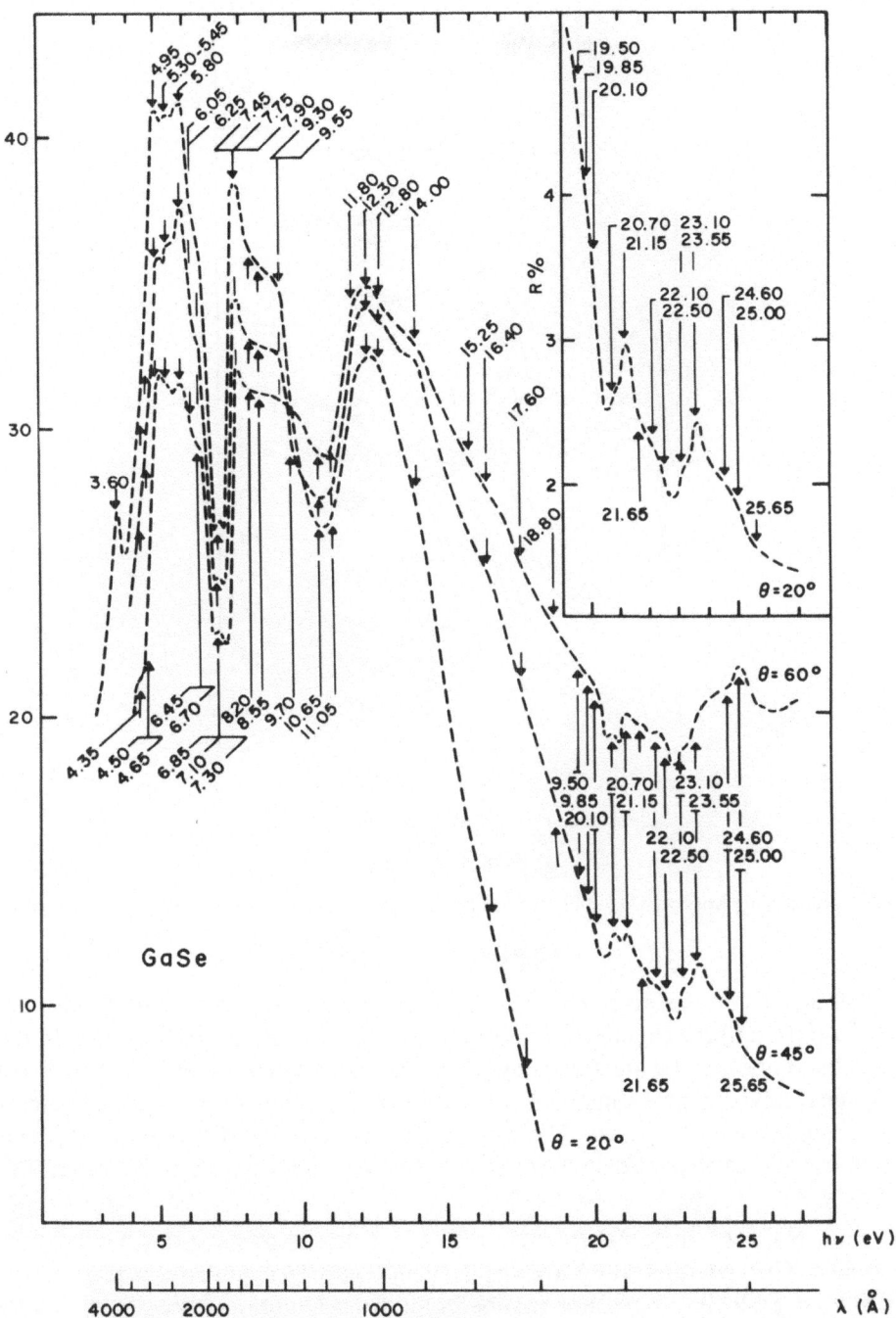

Fig. 97. Experimental reflectivity at room temperature of GaSe obtained using synchrotron radiation. Spectra for various angles of incidence are given. The exact locations of structures are inferred from second derivative spectra (after [147]).

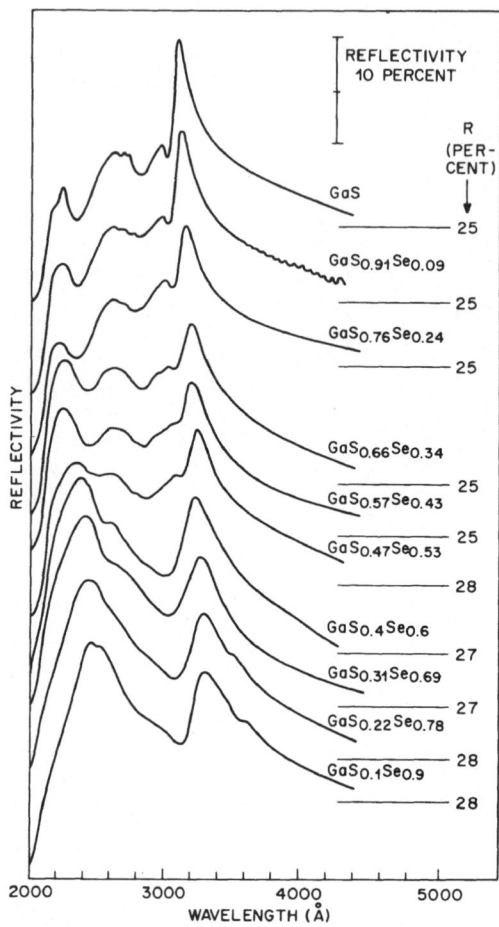

Fig. 98.   Reflectivity spectra $\mathbf{E} \perp \hat{c}$ of ten GaS_xSe_{1-x} mixed crystals measured at 5 K between 3 eV and 6 eV (after [125]).

($\Delta \approx 780$ meV) shifts respectively. A direct correspondence can be found between the observed shifts and the calculated amount of atomic-like charge on the anion in GaSe. The relationships may be illustrated by calculated charge density contour maps and by real-space resolved contour maps of the transition functions $\mathbf{F}_{if}(\mathbf{r}) = \psi_i(\mathbf{r})\mathbf{P}\psi_f(\mathbf{r})$ connecting initial and final states. These quantities are already discussed in the section describing $SnSe_2$, SnS_2. The three classes of observed shifts are described in (i), (ii) and (iii) below.

(i) Four main structures labelled G1 to G4 in Table 25 and Figure 96 belong to the $\Delta = 0$ class. Their energies which are displayed in Figure 99 as a function of composition x show no overall shift if sulphur is substituted for selenium. As seen from the corresponding charge density plots (shown in a (110) plane of GaSe) little charge sits around the anions and the transitions take place in the Ga–Ga bond (Figure 100). This obviously explains the negligible shift in transition energy if sulphur replaces selenium.

(ii) Several transitions labelled GS1 to GS6 belong to the class of intermediate shifts

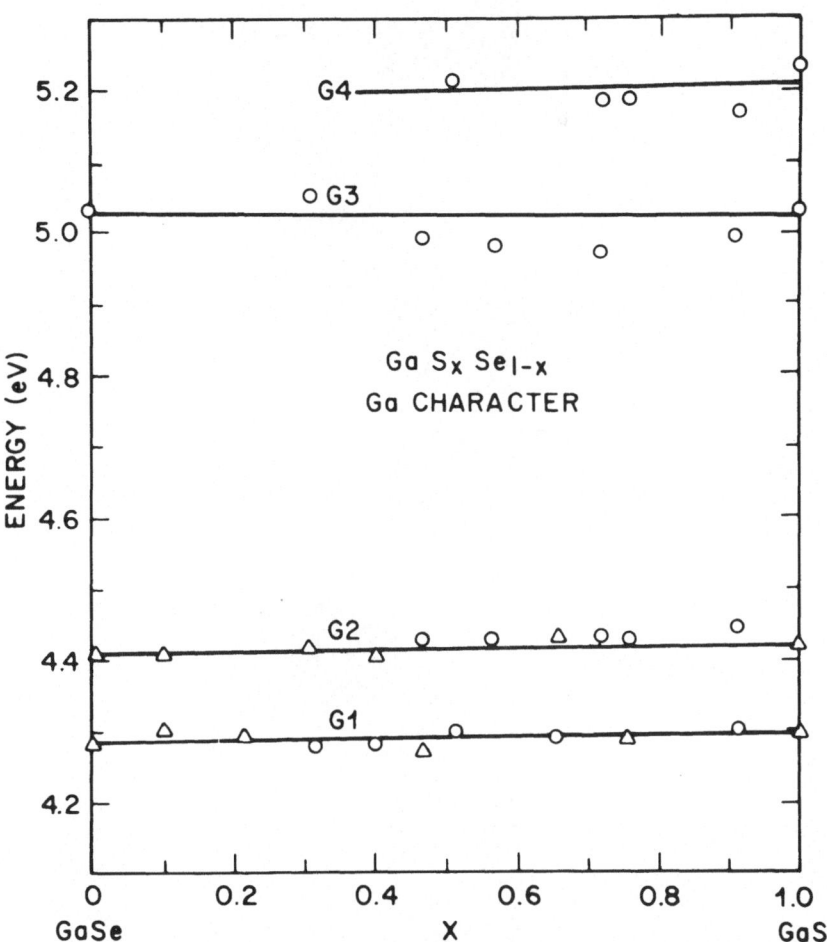

Fig. 99.   Energy diagram of structures observed in the GaS_xSe_{1-x} mixed crystal reflectivities which exhibit $\Delta = 0$ energy shift. The structures are labeled G1 to G4 indicating the predominant anion (gallium) character in the wavefunctions of initial and final states. Well resolved structures are indicated by circles, weak structures by triangles (after [125]).

(Figure 101). Both initial and final states of the representative GS2 transition show charge distributions of mixed cation and anion character (Figure 102). It should be mentioned that the indirect gap belongs to this class of intermediate shifts.

(iii) Four main transitions S1 to S4 exhibit a large energy shift of $\Delta \approx 780$ meV as shown in Figure 103. In Figure 104 charge densities and the transition function are displayed for the S1 transition. The final state (band 19 at Γ) has partially antibonding Ga–Ga character combined with strong Se s, p_z-type characters. This explains the sensitivity of the conduction band edge to anion alloying. Moreover, the initial state (band 16 at Γ) exhibits exclusively Se p_{xy}-character and therefore also is strongly affected by alloying. The corresponding transition is localized on the anion sites as seen from the transition function map (Figure 104 bottom) and corresponds mainly to Se p_{xy} to s atomic-like transitions.

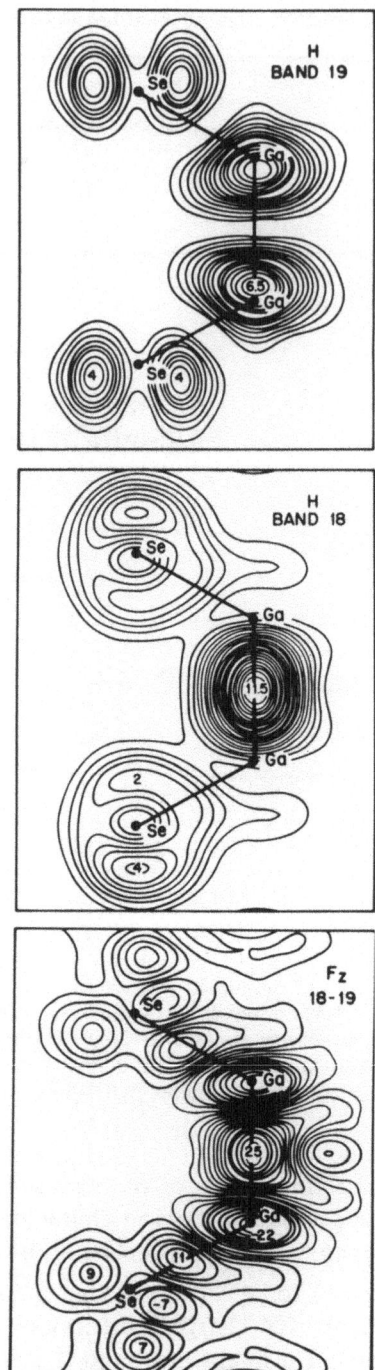

Fig. 100. Charge density contour plots of valence band state (middle figure) and conduction band state (top figure) at point H in the Brillouin zone of GaSe. The transition between these states which gives rise to the G1 (G2) structures indicated in Figure 96 and in Table 25 is chosen as representative for gallium (G)-like transitions of Figure 99. The transition function F_z (bottom figure) which indicates the real space location (around the gallium atoms) of the transition is defined in the text. All contour plots are displayed in a (110) plane extending over one layer of GaSe. The charge density contours are given in units of electrons per (two-layer) unit cell; the transition function contours are given in atomic units (after [125]).

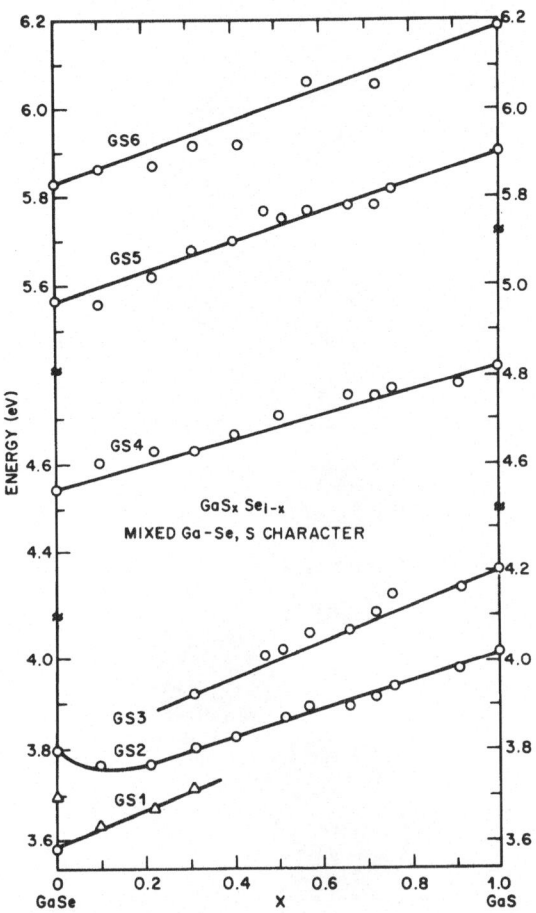

Fig. 101. Energy diagram of reflectivity structures, labelled GS1 to GS6 exhibiting intermediate ($\Delta = 400$ meV) energy shift. The labelling refers to mixed cation (Ga) and anion (Se, S) characters of the involved wavefunctions (after [125]).

The direct gap at Γ also shows the strong experimental shift $\Delta \approx 780$ meV if selenium is substituted by sulphur. As claimed in other publications [124] the direct gap transition corresponds partially to the excitation of the Ga–Ga bond. However, considerable contributions from anion p_z to s atomic-like transitions are also found which are responsible for the large energy shift of the direct gap. The spectral analysis presented in [125] of the GaS$_x$Se$_{1-x}$ mixed crystal series therefore is directly related to pure atomic-like properties of the constituents rather than to changes in ionicity or bond length.

In this final part of the discussion we come back to the nature of chemical bonding in GaSe. The various band structure models, the related density of states spectra combined with photoemission data and charge density plots present a sufficient amount of information to gain some insight into the bonding properties. A large amount of experimental input is available in forms of UPS [152, 153], XPS [153, 105] and synchrotron radiation data [147, 155]. In Figure 105 a theoretical density of states as obtained from the band structure calculation of [125] is compared to the XPS spectrum of [105]. The

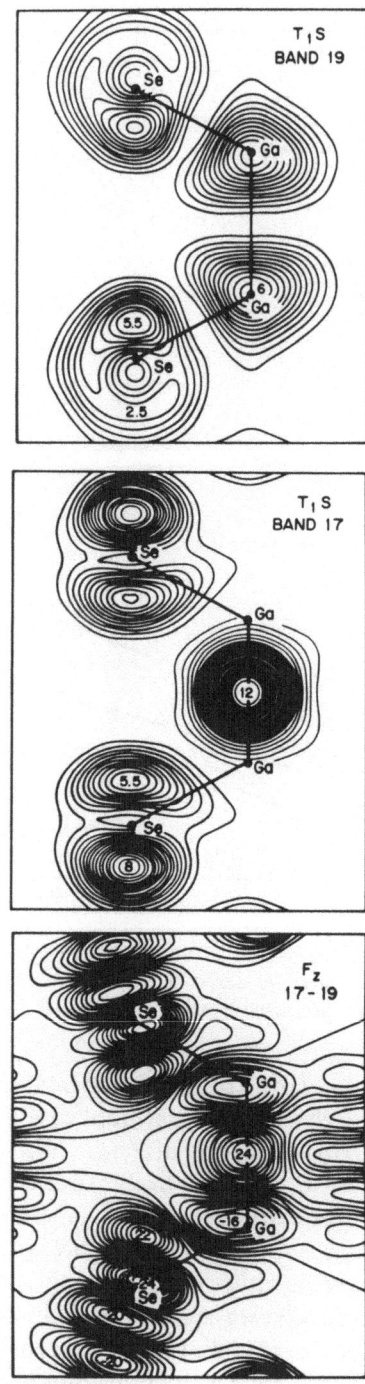

Fig. 102.   Charge density and transition function contour plots for one representative transition
(GS2) of the group of transitions exhibiting intermediate (Δ = 400 meV) energy shift in GaSe as
displayed in Figure 101. For explanations see caption of Figure 100 (after [125]).

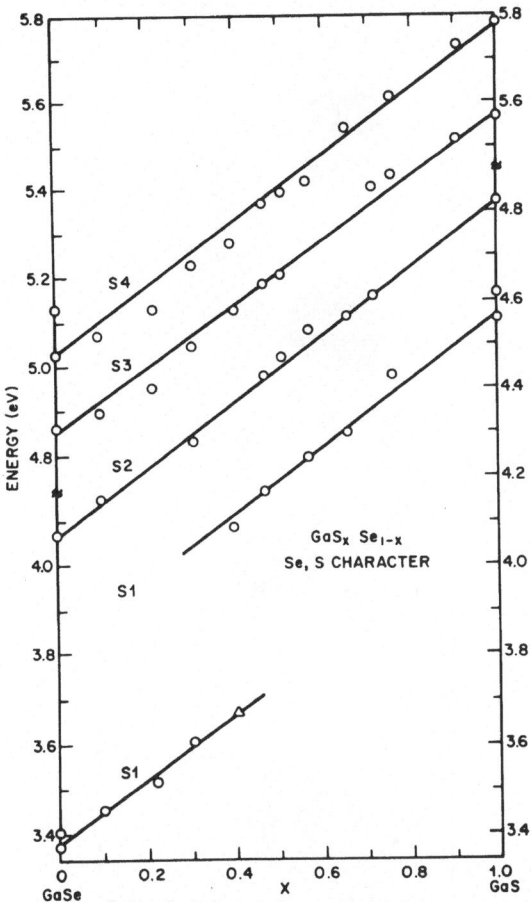

Fig. 103. Energy diagram of reflectivity structures, labelled S1 to S4 exhibiting strong ($\Delta = 780$ meV) energy shift. The labelling refers to predominant anion (Se, S) character of the involved wavefunctions (after [125]).

energy values of prominent structure are listed in Table 23 and compared to other band structure results. Results published for the various experiments mainly differ in the location of the valence band edge which results in relative shifts of the entire spectra. In fact, no unambiguous method is known yet to determine absolute XPS energy values better than several tenths of eV. The experimental spectrum shown in Figure 105 and the peak energy values quoted in Table 23 are shifted to obtain best agreement with the theoretical EPM results. This alignment, however, places the Ga $3d$-emission (not shown in Figure 105) at -18.0 ± 0.3 eV which seems to be too high in energy compared to the value of -18.6 ± 0.3 eV of other compounds (GaP, GaAs, GaSb) containing Ga as constituent.

Similar to PbI_2 and BiI_3, in GaSe the cations possess more electrons than necessary for the saturation of the anion valences. These excess electrons in GaSe also occupy an anti-bonding band rather than a non-bonding band. The situation in GaSe, however, differs from the previous examples of PbI_2 and BiI_3 by the presence of a di-atomic

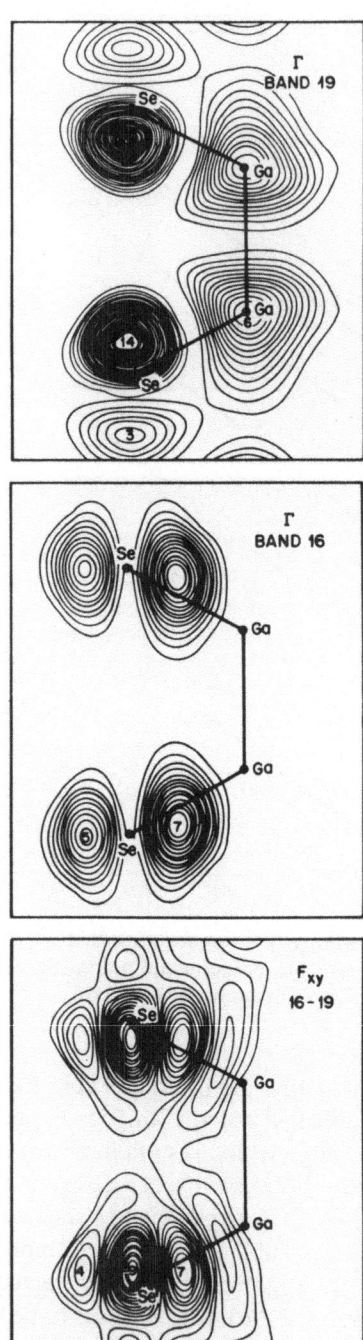

Fig. 104.   Charge density and transition function contour plots for one representative transition (S1) in GaSe of the group of transitions exhibiting strong ($\varDelta = 780$ meV) energy shift as displayed in Figure 103. For explanations see caption of Figure 100 (after [125]).

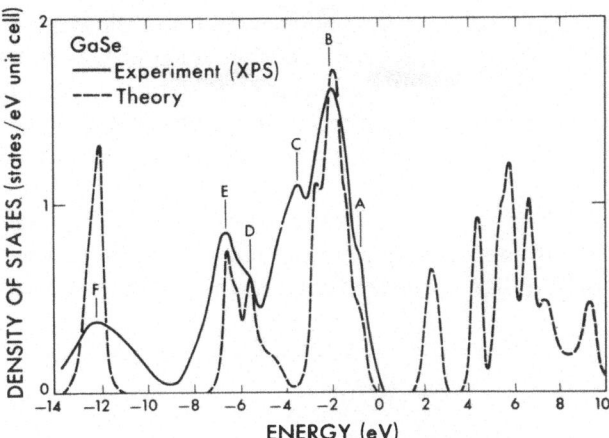

Fig. 105. Calculated density of states for GaSe, based on the EPM calculation of [125] compared to experimental photoemission data from [105] (see Table 23).

molecule of cations in the middle of each sandwich. The corresponding covalent bond can be seen in the total valence charge distribution shown in Figure 106. The charge density map is evaluated in a (110) plane extending over one layer (one half unit cell of β-GaSe), the charge values are given in units of electrons per unit cell. Figure 106 also shows the strongly asymmetric charge distribution around the anions indicating the presence of covalent character in the anion-cation bond. As already argued at the

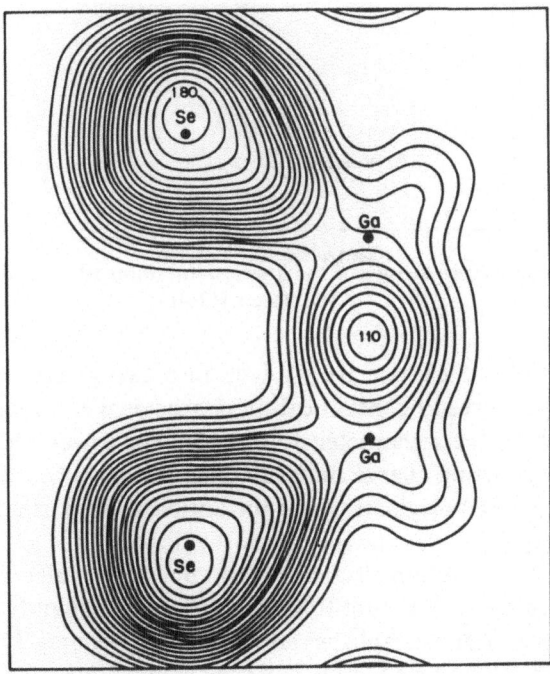

Fig. 106. Total valence charge density for GaSe displayed in a (110) plane as derived from the EPM calculations of [124].

beginning of this section, too strong ionicity would not allow a stable structure with direct cation-cation contact. Moreover, the experimental Ga–Se bond-length corresponds well to the sum of covalent radii whereas it is about 15 % smaller than the sum of ionic radii. The total charge may be decomposed into charge contributions resulting from individual groups of bands according to the density of states diagram in Figure 105.

The lowest peak (F) at about -12.1 eV corresponds to the anion s-states. They are obviously non-bonding and a charge contour map (Figure 107) shows that the electrons occupying them behave like core electrons. The next higher structure in Figure 105

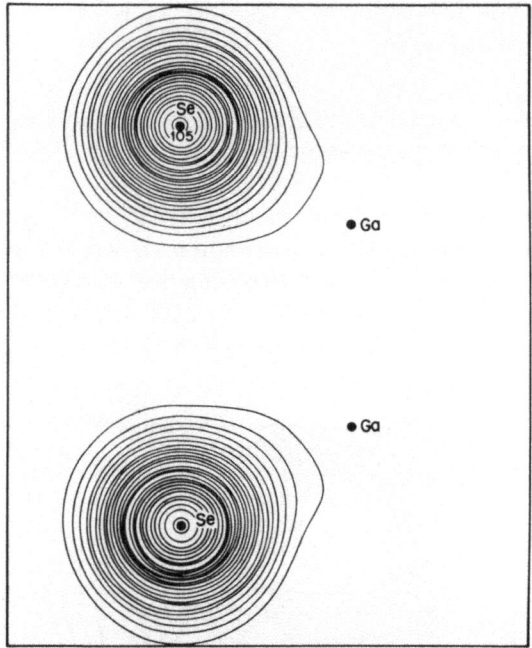

Fig. 107. Partial band charge density of the lowest four anion s-bands in GaSe (peak F in Figure 106) (after [124]).

between -8 eV and -5 eV can be resolved into two peaks labelled E and D. The corresponding states already contain cation character, i.e. a mixture of s- and p_z-like states of Ga in which the s character dominates. The lower peak (E) corresponds to states which show a strong charge build-up constituting a Ga–Ga bond (Figure 108). The next higher peak (D) originates from the antibonding partner of this bond (Figure 109). These antibonding states like the topmost valence bands in PbI_2 and BiI_3 seem to accommodate the excess cation electrons. The uppermost valence band 'pair' in GaSe, giving rise to a shoulder A in Figure 105 is formed predominantly by the p_z-like orbitals of the Ga and Se atoms and establishes a second Ga–Ga bond (Figure 110). With what was said above one is therefore tempted, in the chemist's language, to describe the total Ga–Ga interaction as the result of a superposition of σ $4s$ bonding, σ^* $4s$ antibonding and σ $4p_z$ bonding states. The paradox of this description illustrates the difficulties

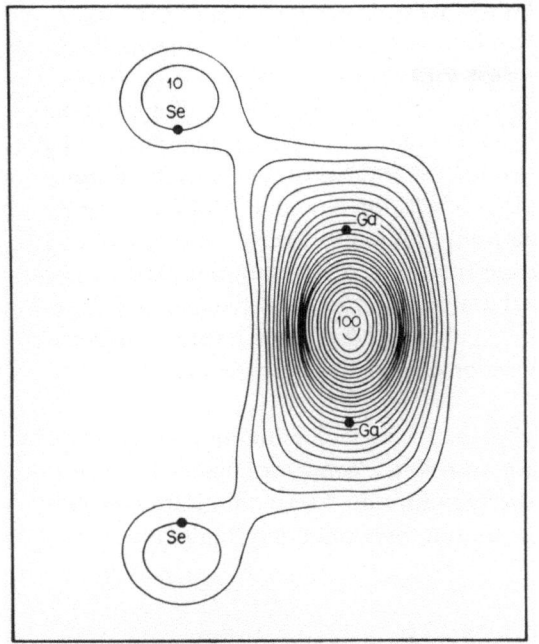

Fig. 108. Partial band charge density for bands 5 and 6 in GaSe (peak *E* in Figure 106) (after [124]).

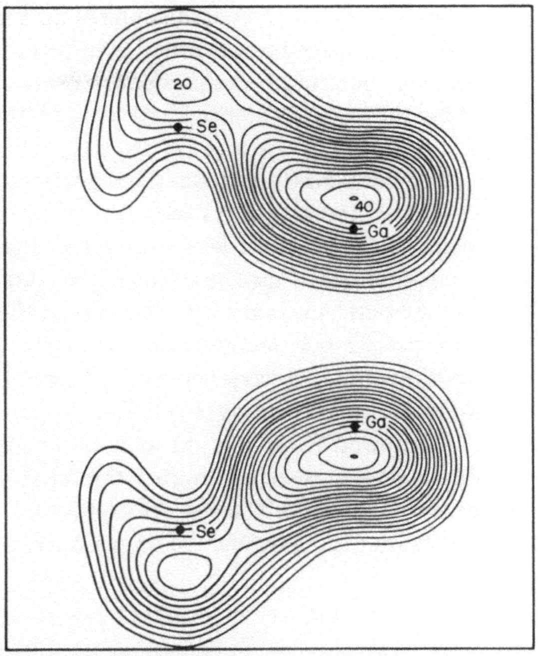

Fig. 109. Partial band charge density for bands 7 and 8 in GaSe (peak *D* in Figure 106) (after [124]).

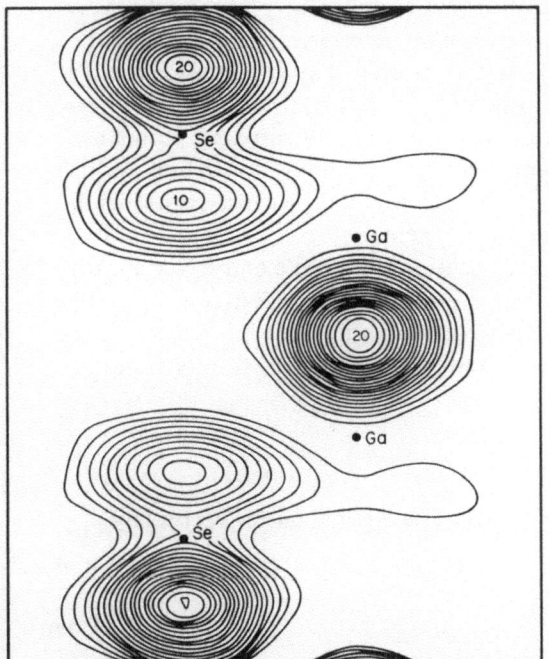

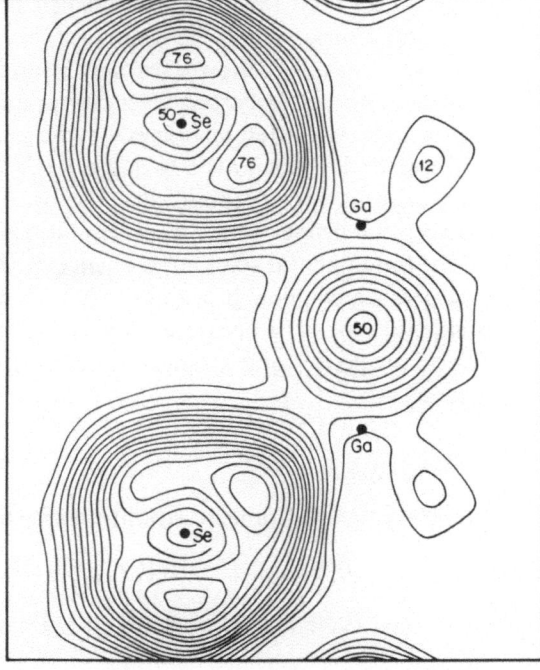

Fig. 110. Partial band charge density for the upper-most pair of valence bands (bands 17, 18) in GaSe (shoulder *A* in Figure 106) (after [124]).

Fig. 111. Partial band charge density for bands 9 to 16 in GaSe (peaks *B* and *C* in Figure 106) (after [124]).

encountered by a too narrow interpretation of the bond picture: in spite of the concen-
tration of charge between the Ga atoms the wave functions extend over the whole unit
cell and the mutual orthogonality between the various bonding bands is warranted.
Below the highest valence band 'pair' lie the eight σ-bands mainly corresponding to the
p_{xy}-orbitals of the Se and Ga atoms (peaks B and C in Figure 105). As compared to the
experimental XPS spectrum the theoretical bandwidth of these states seems to be about
0.5–1.0 eV too small. As seen from Figure 111 in which the charge is plotted, the Se
orbitals predominate. Moreover, the accumulation of charge between the Ga and Se
atoms indicates that these bands are responsible for the covalent component of the
Ga–Se bond. A quantitative measure of the total ionic component is calculated in [124]
in terms of a 0.4 electron effective charge on Se. This value has since been confirmed
experimentally to vary between 0.4 and 0.5 electrons by analyzing chemical shifts of
X-ray K absorption edges [119].

It followed from this discussion that the chemist's views of bonding in GaSe as
presented at the beginning of this chapter are somewhat oversimplified, that their
general traits, nevertheless, are confirmed by the calculations. The main discrepancies
arise from the inexistence of an unequivocal correlation between bonds and bands.

2.3. HEXAGONAL NETWORK STRUCTURES

Among the most studied layer compounds are the hexagonal network structure graphite
and boron-nitride (BN). Their structures are made up of two-dimensional arrays of
hexagons. The distance between nearest atoms in a given plane is much smaller (1.42 Å
in graphite) than the distance between nearest atoms in different planes (3.37 Å in
graphite). Consequently the planes are very loosely bound between each other, while the
atoms in the same plane are strongly bound. To a good degree of approximation (better
than in the other discussed layer compounds) one may neglect to first order the interaction between different planes. In the periodic table boron and nitrogen atoms are
adjacent to carbon, thus BN is the III–V compound corresponding to carbon and is
iso-electronic with graphite. Because of this overall resemblance, a close relationship
between the properties of graphite and BN is to be expected.

The fundamental stacking sequence of three-dimensional graphite and BN is shown
in a perspective view in Figure 112. In the following section we shall first discuss the
bandstructure of graphite and its relationship to galvanomagnetic properties. The
optical properties of graphite are then considered and compared to the results of boron
nitride.

2.3.1. *Graphite*

For graphite the atomic positions are given in terms of hexagonal lattice vectors by

$$(0, 0, u) \qquad (\tfrac{1}{3}, \tfrac{2}{3}, v),$$
$$(0, 0, u + \tfrac{1}{2}) \qquad (\tfrac{2}{3}, \tfrac{1}{3}, v + \tfrac{1}{2}), \tag{C48}$$

where u can be taken as zero and v is practically zero and cannot exceed 0.05. The exact
space group associated with this *buckled* (not planar) graphite structure is C_{6v}^4. If both
u and v are taken as zero, the space group of the planar graphite is $D_{6h}^4 = C_{6v}^4 \times I$. Since

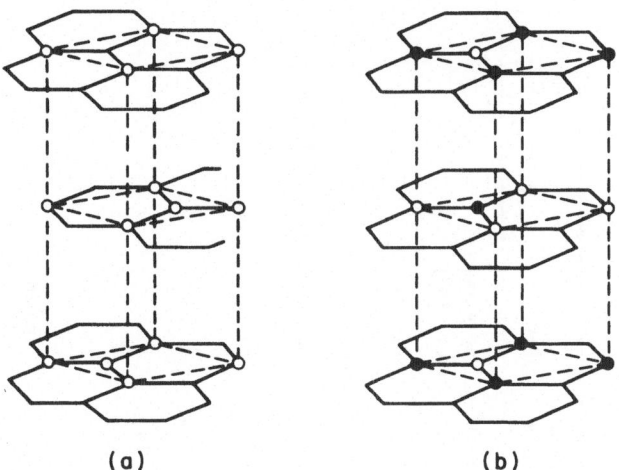

(a)   (b)

Fig. 112. Perspective view of the three-dimensional crystal structures of (a) graphite and (b) boron nitride (after [167]).

C_{6v}^4 is a subgroup of D_{6h}^4 it is easy to establish the symmetry induced differences in band structures for buckled and planar graphite. The symmetry reduction leads only in the double group case (spin included) to band splittings. In particular the spin-degeneracy is lifted on the lines Σ, T and at the point K as well as in the general point. For the points A, L, H, the lines S and R (i.e. on the surface of the Brillouin zone, Figure 3, Chapter A) the states remain spin-degenerate because of time reversal symmetry. Because spin-orbit coupling is negligible for the first row elements and because the interesting parts of the band structure are located around H, as we shall see, full D_{6h}^4 symmetry may safely be assumed in all calculations. For a thorough discussion of D_{6h}^4 see Chapter A.

Note from Figure 112 that out of the four carbon atoms in the unit cell two have neighbors directly above and below in adjacent planes whereas two atoms do not. Thus the inclusion of interlayer interaction will allow to distinguish between those atoms.

Whereas the gross features of the graphite band structure, in particular, optical properties at higher energies can well be understood in the two-dimensional approximation, the exact evaluation of e.g. Fermi surface properties calls for the inclusion of interlayer interaction.

A very large number of theoretical papers has been published on the band structures of graphite and boron nitride [156–170]. The various studies having different objectives are carried out on two-dimensional as well as three-dimensional structural models. The most important studies are listed and grouped according to the type of approach in Table 26. Basically, one may divide the various calculations into:

(i) studies of the galvanomagnetic properties of graphite, i.e. of the very detailed band structure in the vicinity of the Fermi level; and into

(ii) studies of the bonding and higher energy optical properties.

This grouping is characteristic of the existence of two different types of bonds in graphite. The σ-bonds localized in the layers give rise to a strong (several eV wide) bonding-antibonding gap. These σ-bonds involve three electrons per atom. The fourth

TABLE 26

Table of the hexagonal network structures graphite and boron nitride and related theoretical studies

Crystals	a Å	c/a	Theoretical approaches					
			ETB	$k \cdot p$ + Fourier	ab-initio LCAO	Self-consistent LCAO cluster	OPW	EPM
Graphite 2-dim D_{6h}	2.46	—	Wallace (π) [156] Corbato (π) [162] Lomer (σ) [159] Bassani et al. (π, σ) [120] Coulson et al. (π) [157]		Painter, Ellis (π, σ) [165] Zupan (π, σ) [167]	Zunger (π, σ) [170]	Nagayoshi et al. (π, σ) [168]	
Graphite 3-dim D_{6h}^4	2.46	2.81	Wallace (π) [156] Coulson et al. (π) [157] Doni et al. (π) [163] Greenaway et al. (π) [175]	McClure (π) [161] Slonczewski et al. (π) [160] Johnson et al. (π) [169]	Painter, Ellis (π) [165] Zupan (π) [167]		Nagayoshi et al. (π) [168]	Van Haeringen et al. (π, σ) [164]
BN 2-dim D_{3h}	2.50	—	Doni et al. (π, σ) [163]		Zupan (π, σ) [167]	Zunger (π, σ) [170]		
BN 3-dim D_{6h}^4	2.50	2.66	Doni et al. (π) [163] Taylor et al. (π) [158]		Zupan (π) [167]		Nakhmanson, Smirnov (π, σ) [166]	

electron occupies a p_z-like orbital which forms weak π-bonds, and is influenced by interlayer coupling. It is this latter case, which gives graphite its characteristic properties. Thus, calculations involving σ-bonds are in general restricted to one-layer approximations whereas three-dimensional calculations concentrate on the π-band structure (see Table 26). A number of review articles appeared on the subject of the electronic structure of graphite [171, 172]. The reader is referred to these articles for complete discussions and very extensive listings of experimental studies. Let us first discuss the group of calculations, concentrating on the galvanomagnetic properties of graphite.

The two-dimensional π-band structure was first discussed in an empirical tight binding scheme in [156]. The results indicated that the highest valence band and the lowest conduction band are degenerate in energy at the six corners of the two-dimensional hexagonal Brillouin zone. Later calculations of the single-layer band-structure [157, 159, 162] agree with these findings. Furthermore, there are no other bands with energies near the degenerate energy, so that the only part of the band-structure important for transport phenomena is the region near the zone corners.

Interaction between layers lifts some of the degeneracy. The splitting is small compared to the band-width but is appreciable compared to the kinetic energies of the carriers. In the three-dimensional case there are two conduction and two valence bands (not counting spin degeneracy) two of which are required by symmetry to be degenerate along the vertical zone edges (H–K–H in Figure 3 of Chapter A). The three-dimensional calculation of [156] shows one conduction band degenerate with one valence band along HH but no band overlap. More complicated band structures involving band overlap in addition to the symmetry induced degeneracy have subsequently been proposed [160, 161]. Two of these studies led to the so-called Slonczewski-Weiss-McClure model (SWM) whose basic ideas shall be presented.

All calculations show that the interesting part of the Brillouin zone is located quite near the vertical zone edges. Thus it is sensible to make a Taylor ($\mathbf{k} \cdot \mathbf{p}$) expansion of the Hamiltonian in terms of k_x, k_y (distances from the zone edge in the x and y directions). However, in the z-direction, a Fourier expansion of the Hamiltonian is made as the layer planes are widely separated and the series is rapidly convergent (a case of ideal tight-binding). In this scheme the band structure of the four π-bands ($u_1 - u_4$) may be described by a 4×4 matrix:

$$
H = \begin{bmatrix}
E_1 & 0 & H_{13} & H_{13}^* \\
0 & E_2 & H_{23} & -H_{23}^* \\
H_{13}^* & H_{23}^* & E_3 & H_{33} \\
H_{13} & -H_{23} & H_{33}^* & E_3
\end{bmatrix}.
\tag{C49}
$$

The order of the wavefunctions u_1 to u_4 is as follows: u_1 is the symmetric (antibonding) combination of p_z-orbitals centered at the two atoms which are on top of each other (see Figure 112). This band usually has the highest energy. u_2 is the antisymmetric (bonding) combination of these orbitals giving rise to the lowest band. The bands corresponding to u_1 and u_2 are degenerate at H. u_3 and u_4 are symmetric and antisymmetric combinations of p_z-orbitals centered at the other two atoms. The corresponding bands are degenerate along the whole zone edge (HH) and have intermediate energy. The various interaction energies in Equation (C49) may be parametrized as follows [161]:

$$E_1 = \gamma_1 \Gamma + \Delta + \tfrac{1}{2}\gamma_5 \Gamma^2$$

$$E_2 = -\gamma_1 \Gamma + \Delta + \tfrac{1}{2}\gamma_5 \Gamma^2$$

$$E_3 = \gamma_2(1 + \cos \xi) = \tfrac{1}{2}\gamma_2 \Gamma^2$$

$$H_{13} = \frac{1}{\sqrt{2}}(-\gamma_0 + \gamma_4 \Gamma)\sigma e^{i\alpha}$$

$$H_{23} = \frac{1}{\sqrt{2}}(\gamma_0 + \gamma_4 \Gamma)\sigma e^{i\alpha}$$

$$H_{33} = \gamma_3 \Gamma \sigma e^{i\alpha} \qquad\qquad\qquad\qquad (C50)$$

$\alpha = 1/[\tan(-k_x/k_y)]$ and $\sigma = \tfrac{1}{2}\sqrt{3a}\,|k_{xy}|$ describe the $x - y$ dispersion around the zone edge (HH) whereas $\xi = k_z c$ describes the z-dispersion along (HH); $\Gamma = 2\cos(\xi/2)$. Spin-orbit coupling is neglected in the equations given above. A discussion of spin-orbit sub-Hamiltonians is given in [173]. The first two non-vanishing Fourier terms in the k_z-expansion are included in Equation (C50). There are seven parameters $\gamma_0 - \gamma_5$ and Δ which determine the band structure. The quantity γ_0 determines mainly the $x - y$ dependence of the energy. It is the only parameter in the single layer approximation. γ_1 represents the main splitting caused by interlayer interaction. The quantity Δ reflects the fact that the atoms sit at inequivalent sites due to the crystal structure. The parameter γ_2 is responsible for the band overlap due to interlayer interaction between the atoms which sit on top of each other (Figure 112). It is this parameter which controls the warping of the two-fold degenerate band and thus the location of the carriers. γ_5 describes the same effect for the other two atoms. The quantity γ_3 gives rise to the anisotropy in the xy plane, it causes the Fermi surface to be distorted trigonally. The parameter γ_4 has no qualitative effect on the band structure but influences the z-dispersion in a quantitative way. In certain limits, like e.g. for $\gamma_3 = \gamma_4 = 0$ the matrix (C49) may be factorized and analytic solutions can easily be written down. A typical band structure scheme along the line H–K–H (edge of the hexagonal Brillouin zone) is shown in Figure 113 [171]. The Fermi level ζ is indicated, its position giving rise to the existence of electrons and holes. A perspective view of the electron-hole Fermi-surface located at the edges of the Brillouin zone is shown in Figure 114 [161]. The surface is very anisotropic, with the length in the $\hat{c}$-direction about thirteen times the width perpendicular to the $\hat{c}$-direction. As already mentioned, the sign of γ_2 controls the location of the free carriers (a sign change in γ_2 interchanges the electron with the hole pockets). The attempt to consistently interpret a large number of experimental de Haas-van Alphen data, Shubnikov-de Haas data, cyclotron resonance data, magnetoreflection data, infrared laser absorption data and galvanomagnetic data results in a negative sign for γ_2 [171, 174], thus placing the electron pocket at K and the hole pockets around H (see Figure 114). The question of the carrier assignment has provoked many discussions [171] and may still be considered not to be solved completely. Assuming the

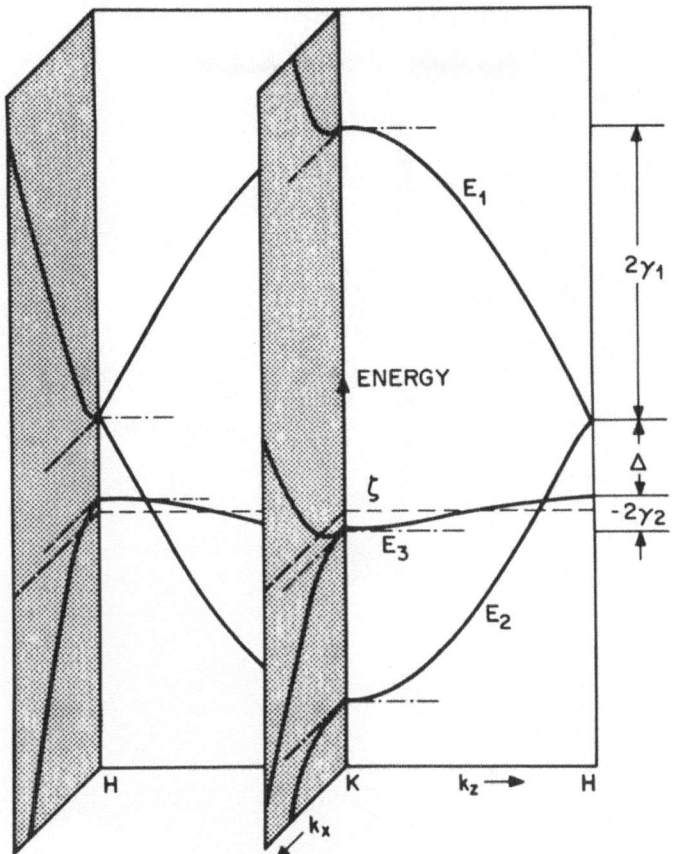

Fig. 113.   Energy band structure of graphite along the Brillouin zone edge (HH). The energy is plotted vertically and the $k_x k_z$ plane is horizontal. The dashed line represents the Fermi level ζ for pure graphite. Some of the SWM-model parameters are indicated (after [171]).

popular negative sign for γ_2 the range of estimated values of the band parameters is listed in Table 27.

In a recent study the optical properties of the graphite π-bands between 0.5 eV and 6 eV are computed using a three-dimensional full zone Fourier expansion [169]. This expansion can be viewed as a general form of the aforementioned SWM method and allows a description of the π-band structure throughout the Brillouin zone. The coefficients in the Fourier expansion are evaluated by a fit to the SWM parameters as well as to various optical data [175–177].

The study of the optical response of graphite is the motivation for a second group of band structure calculations. As already mentioned, most of these calculations are based on a two-dimensional model including σ- and π-bands [120, 165, 167, 168, 170]. The calculations are extended to three-dimensions for the π-bands only in some studies [163, 165, 167, 168, 175]. Solely the EPM calculation of [164] treats σ- and π-bands in three dimensions.

It is first argued in [120] and later confirmed by other calculations and by comparison

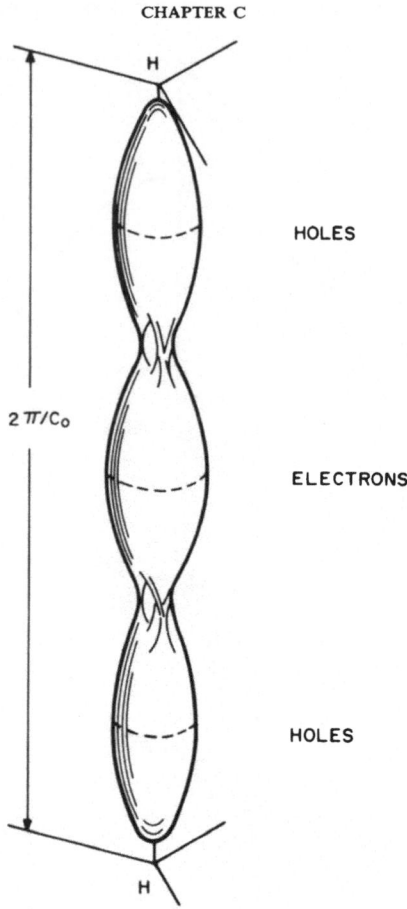

Fig. 114.   Sketch of the proposed Fermi surface of pure graphite. For negative values of the SWM-parameter γ_2 the central surface contains electrons whereas the outer surfaces contain holes (see Figure 113). The length to width ratio of each surface is about 13. The trigonal anisotropy is exaggerated for clarity (after [161]).

TABLE 27

Table of the Slonczewski–Weiss–McClure (SWM) parameters which describe the Fermi-surface of graphite in accordance with experiments (after [171]).

SWM parameter	Value (eV) [171]
γ_0	2.8 to 3.2
γ_1	0.27 to 0.40
γ_2	-0.018 to -0.02
γ_3	0.14 to 0.29
γ_4	0.2 to 0.3
γ_5	-0.018 to -0.02
Δ	0.005 to 0.1

to experiment [175] that with the exception of some fine structure in low energy infrared $\pi-\pi$ transitions [177, 178] the optical spectra of graphite may well be explained by the two-dimensional approximation.

A representative band structure of two-dimensional graphite is given by the ETB calculations of [120]. The procedure of the calculation is identical to that of GaSe discussed in the preceding section and shall not be repeated here again. The resulting bandstructure for the two-dimensional σ (full lines) and π (broken lines) states is shown in Figure 115. The group notations are those of [120], the superscripts $\pm$ referring to a reflection in the $x-y$ plane and the subscripts g, u referring to inversion. Note, that Q corresponds to M and P to K in the notation used in Chapter A. Some transitions, giving rise to structure in the optical spectrum are indicated in the figure. The imaginary part of the dielectric function, $\varepsilon_2(\omega)$, is calculated in the approximation of constant matrix elements [120] and shown in Figure 116. No polarization effects can be seen due to the constant matrix element approximation. The contribution due to interband transitions between π-bands and between σ-bands are indicated separately (full curves) in Figure 116. The broken line represents the experimental data of [179]. The peak at 4.5 eV is attributed to the saddle point transition $Q_{2g}^- \rightarrow Q_{2u}^-$ in Figure 115. Interlayer interaction gives rise to fine structure around this peak [175, 178]. The peak at 14.5 eV is due to pure $\sigma \rightarrow \sigma^*$ transitions at $Q_{2u}^+ \rightarrow Q_{2g}^+$. A second peak due to the transition $Q_{2u}^+ \rightarrow Q_{1g}^+$ can be seen at 16 eV. The third peak which is shown at 20 eV, is due to transitions $Q_{1g}^+ \rightarrow Q_{2g}^+$ but has no relevance since this transition is forbidden by parity. The inclusion of transition matrix elements would thus suppress it.

Three-dimensional extensions of the π-bands of [120] are reported in [163] and [175]. The results are presented in Figure 117. Allowed interband transitions are indicated with the respective polarization dependence. The continuous arrows indicate transitions at the symmetry points that are allowed in the three-dimensional as well as in the two-dimensional approximation. The broken arrows indicate transitions that are allowed only in the three-dimensional approximation and that are therefore much weaker. No doublet structure is observed in [175] at the saddle point transition at 4.6 eV and it is argued that this may be due to the degeneracy at the point L and to the fact that the order of even and odd states is opposite in the valence and conduction bands at the point Q. More recent thermoreflectance data [178], however, seem to indicate some doublet structure around 4.6 eV. The same experiments [178] show a doublet structure in the infrared range at 0.74 eV and 0.88 eV. It is attributed to transitions starting from K_1^- into levels above the Fermi energy around K_3^- and to transitions starting from filled levels near K_3^- into the upper band K_2^-. Additional experimental evidence for the interlayer splitting of π-bands in graphite is obtained from UPS-data [177]. A splitting of 0.8 ± 0.1 eV for the occupied valence bands at both Q and $K(P)$ is reported which agrees well with the thermoreflectance data [178] at K. This splitting at K yields a value of 0.4 eV for the γ_1 parameter of the SWM model which is in good agreement with other experiments.

While the various band structure calculations agree as to the origin of the 4.6 eV structure in the optical spectrum of graphite, the assignment of higher energy transitions is somewhat less unequivocal. The first principle LCAO calculations of [165], though qualitatively resulting in a similar band structure as that of Figure 116, show the Γ_{3u}^+ conduction band some 6 eV higher in energy. The onset of $\sigma \rightarrow \sigma^*$ transitions which in

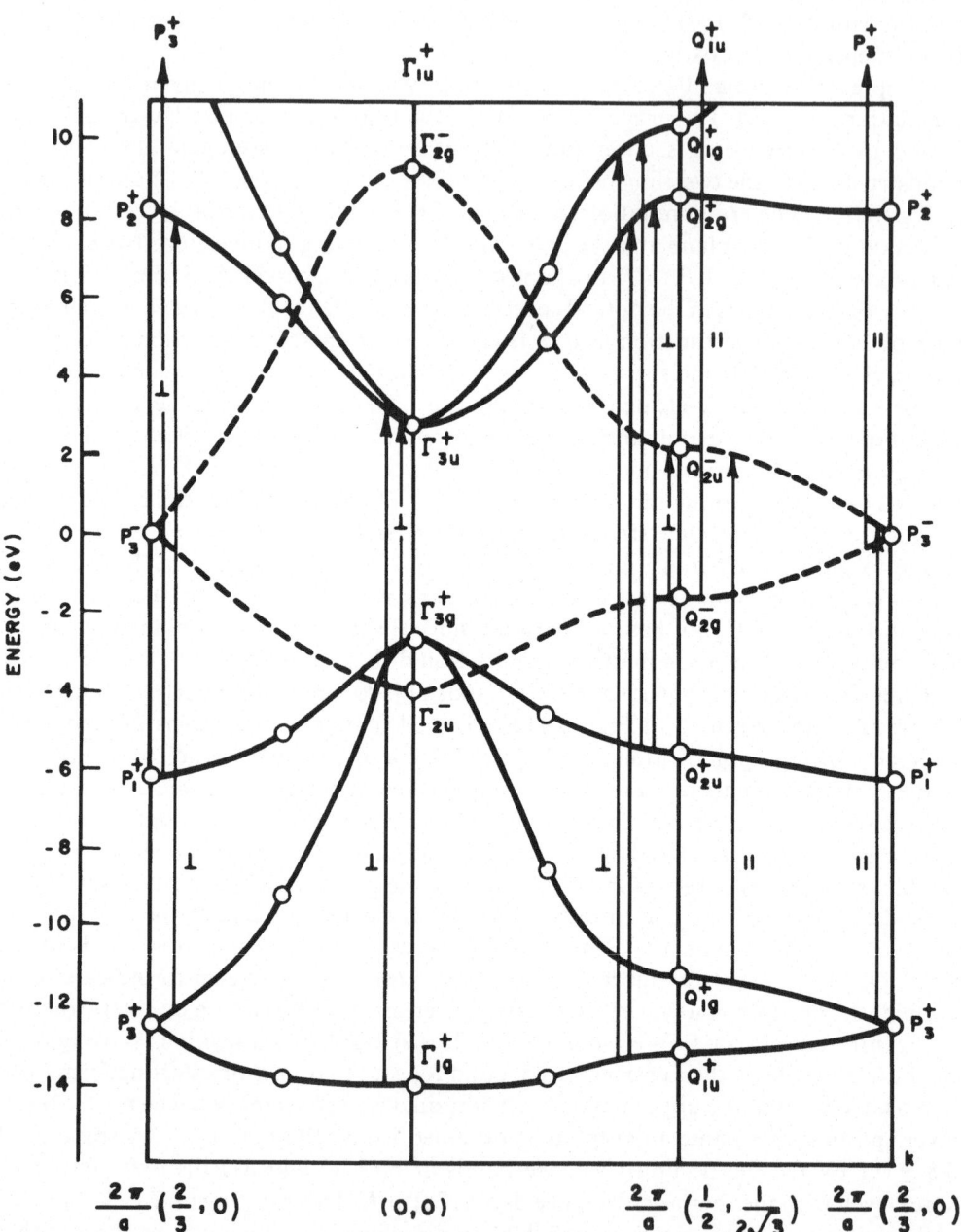

Fig. 115. Two-dimensional ETB band structure of graphite after [120] and [175]. σ and π bands are indicated by continuous and broken lines respectively. The allowed transitions at critical points are indicated for both polarizations.

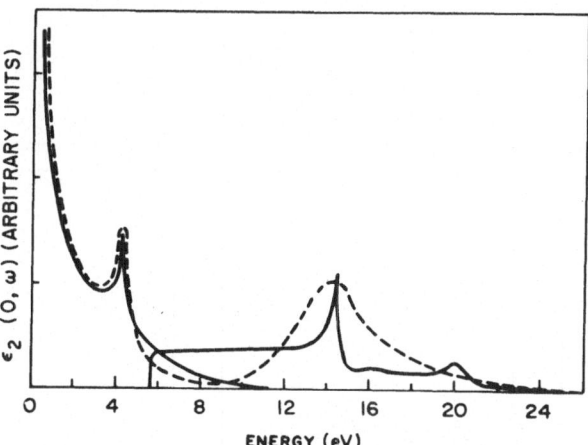

Fig. 116. Imaginary part of the dielectric constant (assuming constant matrix elements) in two-dimensional graphite as calculated from the ETB band structure of [120] shown in Figure 115 (full line). The contribution due to interband transitions between π bands and between σ bands are indicated separately. The broken line indicates experimental data of [179].

[120] appears around 6 eV would thus be shifted to about 12.2 eV immediately followed by the reflectivity peak at 14.5 eV. This band structure, on the other hand, does not give any explanation for the 6.0 eV structure in the $\mathbf{E}\|\hat{c}$ reflectivity [175]. Transitions along the line T' connecting $Q(M)$ and $P(K)$ between the topmost σ-valence band and the lowest π^* conduction band which according to [120] occur around 6 eV for $\mathbf{E}\|\hat{c}$ appear only above 10 eV in the LCAO results of [165].

A compilation of various theoretically and experimentally obtained band structure features of graphite including optical transitions is listed in Table 28. The notation of the various transitions is chosen in accordance with Figures 115 and 117.

2.3.2. Boron Nitride

The two-dimensional crystal structure of hexagonal boron nitride consists of hexagons with vertices alternatingly occupied by boron and nitrogen atoms. The value of the lattice parameter is very close to the value for graphite. The group of two-dimensional boron nitride is D_{3h} in contrast to D_{6h} for graphite because of the lack of all symmetry operations which interchange the two sublattices among themselves. In particular inversion is absent in D_{3h} ($D_{6h} = D_{3h} \times I$). The stacking sequence of the two-dimensional layers in three-dimensional boron nitride is shown in Figure 112. The atomic positions are given in terms of hexagonal lattice vectors by

$$B(0, 0, 0), \qquad (\tfrac{1}{3}, \tfrac{2}{3}, \tfrac{1}{2})$$

$$N(\tfrac{2}{3}, \tfrac{1}{3}, 0), \qquad (0, 0, \tfrac{1}{2}). \tag{C51}$$

The corresponding space group is D_{6h}^4, half of its operations interchange the two layers. The stacking sequence of boron nitride may also be obtained by repeating layers and by shifting them by half a hexagon diameter ($\tfrac{1}{3}(1\bar{1}0)$). Since the essential properties of the electronic band structure of boron nitride may be understood in the two-dimensional approximation we shall discuss the single-layer model first. The band structures of

TABLE 28

Listing of various relevant bandstructure energies of graphite as obtained by theoretical approaches and by experiments.

Feature	π-valence band width (eV)	Total σ–π valence band width (eV)	Optical transitions[a] involving only π-bands $K_1^- \to K_3^-$ / $K_3^- \to K_2^-$	$Q_{2g}^- \to Q_{2u}^-$	Optical transitions[b] involving σ- and π-bands $\Gamma_{3g}^+ \to \Gamma_{3u}^+ \perp$ ($\sigma \to \sigma^* \perp$)	$Q_{1g}^+ \to Q_{2u}^+$ ($\sigma \to \pi^* \parallel$)	$Q_{2u}^+ \to Q_{2g}^+$ ($\sigma \to \sigma^* \perp$)	Type
Theory								
Johnson, Dresselhaus [169]	17.5	–	~0.8	~4.5	–	–	–	Fourier
Bassani, Pastori [120]	5.0	14.0	–	4.6	~6	~11	~14	ETB 2-dim
Doni, Pastori [163]	4.5	–	~0.5	4.4	~6	–	–	ETB 3-dim
Painter, Ellis [165]	7.3	~20	~0.4	4.6	~12	~16.5	~16.3	LCAO ab initio
van Haeringen, Junginger [164]	~8	–	0.3 / 1.0	~4.5	~10	–	–	EPM
Zupan [167]	5.3	~16	–	5.1	~13	15–17	–	SCF CNDO TB
Nagayoshi et al. [168]	~6	~23	~1.2	5.1	~6	–	14.6	OPW TB
Zunger [170]	5.9	21–24	–	4.6–4.8	–	–	–	molecular cluster 2-dim
Experiment								
Greenaway et al. [175]	–	–	–	4.8	shoulder? ~6	–	–	reflectivity
Taft, Philipp [179]	–	–	~0.8	~4.5	–	~14	~14	reflectivity
Guizetti et al. [178]	–	–	0.74 / 0.88	4.6 / 4.85	–	–	–	thermo-reflectance
Hamrin et al. [180]	–	~30	–	–	–	–	–	ESCA
Chalkin [182]	~5.5	~20	–	–	–	–	–	X-ray emission
Willis et al. [176]	–	~20	–	4.76 / 4.82	11.5	–	15.0	UPS

[a] Transitions shown in Figure 117.
[b] Transitions shown in Figure 115.

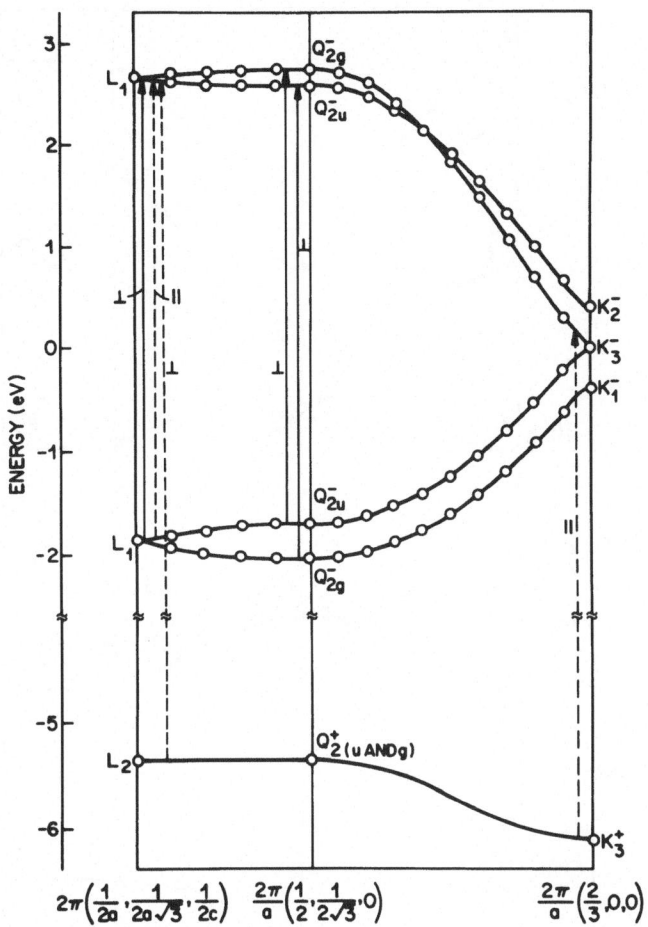

Fig. 117.   ETB π-band structure of three-dimensional graphite. The position of the lowest σ-band is taken from [120]. The continuous arrows indicate transitions at the symmetry points that are allowed in the three-dimensional as well as in the two-dimensional approximation. The broken arrows indicate transitions that are allowed only in the three-dimensional approximation (weaker). Polarization dependencies are also indicated (after [163] and [175]).

two-dimensional graphite and boron nitride may be correlated by considering a perturbation scheme [163]. For this purpose one may consider the effect on the band-structure of graphite with a perturbative potential given by the difference between the potentials of boron nitride and graphite:

$$V_{\mathrm{pert}}(\mathbf{r}) = V^{\mathrm{BN}}(\mathbf{r}) - V^{\mathrm{graphite}}(\mathbf{r}). \tag{C52}$$

The potential $V_{\mathrm{pert}}(\mathbf{r})$ may in turn be split into two parts which are symmetric and antisymmetric with respect to the exchange of B and N atoms, i.e. with respect to inversion. Since boron and nitrogen are adjacent to carbon in the periodic table, the symmetric part of $V_{\mathrm{pert}}(\mathbf{r})$ may be taken as zero. The effect of the antisymmetric potential $V_a(\mathbf{r})$ can be evaluated qualitatively by means of perturbation theory and symmetry considerations. $V_a(\mathbf{r})$ is invariant under the operations of D_{3h} of boron nitride and changes its

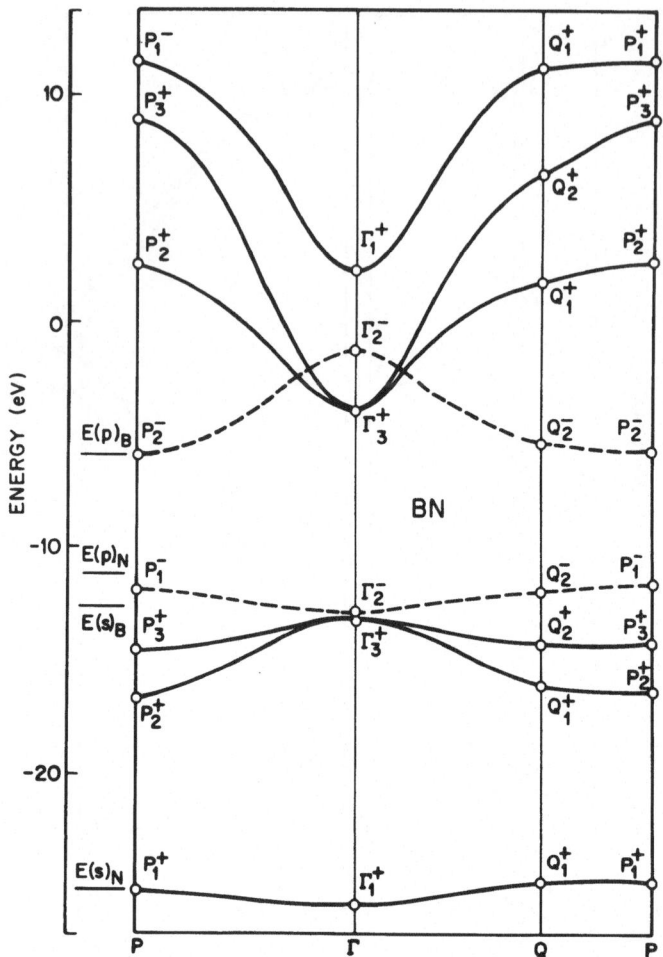

Fig. 118. ETB band structure of two-dimensional boron nitride. The σ and π bands are indicated by continuous and broken lines respectively (after [163]).

sign under the inversion. The representations to which $V_a(\mathbf{r})$ belongs may now be determined at high symmetry points of graphite. The general matrix element theorem (see Chapter A) may then be used to determine which states of graphite can be coupled and which degeneracies can be lifted by the perturbation $V_a(\mathbf{r})$. It follows from this study that first order corrections are zero at Γ, Q and P except for the doubly degenerate states P_3^+ and P_3^- which split in D_{3h}. To second order the valence state Q_{2g}^- interacts with the conduction state Q_{2u}^- thus increasing the gap between these states. A similar but smaller effect is expected for the states Γ_{2u}^- and Γ_{2g}^- because of the relatively large energy difference between them. Thus the perturbation scheme predicts a splitting of the degenerate graphite π-band P_3^- and an opening of the gap at Q. The resulting band structure is that of a true semiconductor in contrast to the zero gap semiconductor graphite. Similar considerations apply to the lowest σ-bands in Figure 115.

A number of first-principles [166, 167] and empirical [158, 162, 170] calculations of the band structure of boron nitride are reported (see Table 26). In Figure 118 we present

as representative two-dimensional band structure the ETB results of [163]. The calculations are performed in the same fashion as the graphite, GaSe and GaS calculations of
[120] which have been described. The opening of a gap between the π-bands at P and
the split-off of the nitrogen s-band is evident from the figure. A joint density of states
and a dielectric function considering constant matrix elements are calculated from this
bandstructure. After a gap of 5.8 eV the dielectric function rises fast and has a peak at
6.6 eV due to the saddle point transitions at Q. The dielectric function, $\varepsilon_2(\omega)$, obtained
from a Kramers-Kronig transformation of experimental reflectivity measurements [183]
on the other hand exhibits two peaks at 6.2 and 7.0 eV. It is argued in [163] that interlayer interactions are responsible for this splitting. A three-dimensional extension [163]
of the band structure of Figure 118 is presented in Figure 119. While for the corresponding case of graphite transport properties (i.e. the SWM-parameters) were used to
empirically determine interlayer coupling strengths for boron nitride optical data [183]
are consulted. The values of the corrections turn out to be approximately equal in both
cases. It is important, however, to note that the order of the valence π-bands at Q
(Q_{2u}^-, Q_{2g}^-) in boron nitride is reversed with respect to graphite (Q_{2g}^-, Q_{2u}^-) (see Figures
117 and 119). Since the conduction band ordering is unchanged, allowed optical

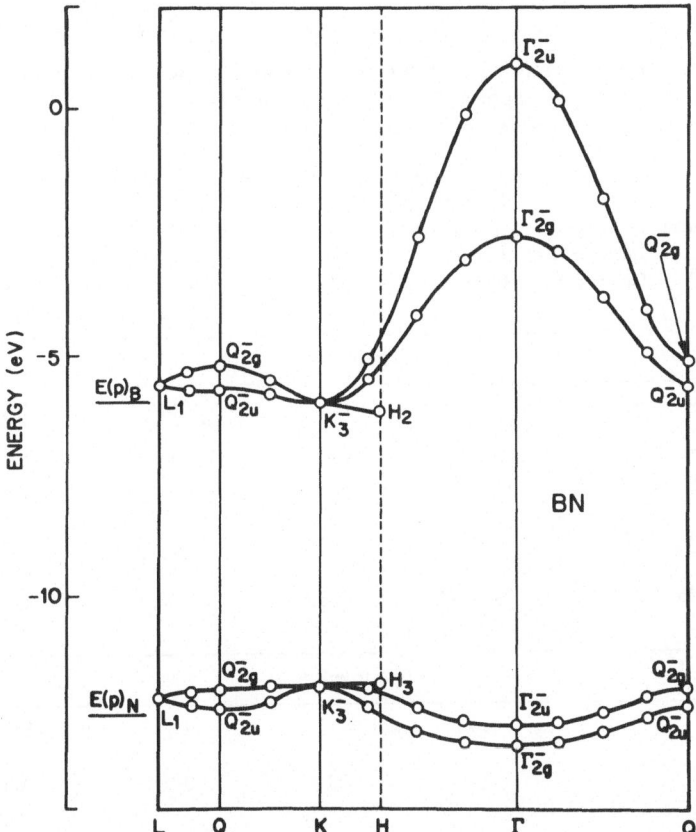

Fig. 119. ETB π-band structure of three-dimensional boron nitride. Note the reversed valence band
order at Q with respect to the band structure of graphite in Figure 117 (after [163]).

transitions for $\mathbf{E}\perp\hat{c}$ show different behaviour in boron nitride than in graphite. In fact, the two allowed transitions $Q_{2g}^- \rightarrow Q_{2u}^-$ and $Q_{2u}^- \rightarrow Q_{2g}^-$ have about equal energy in graphite (see Figure 117) whereas in boron nitride they are split by twice the interlayer interaction (~ 0.8 eV). The reversed band order thus explains the observed double structure in boron nitride at 6.2 eV and 7.0 eV and the existence of a single peak at 4.6 eV in graphite. The band inversion is related to the difference in the lattice structure for graphite and boron nitride. The results of various independent band calculations using different methods (see Table 26) agree qualitatively with the above assertion. The quantitative features, like e.g. magnitude of band gap and width of the valence bands or optical transition energies are compiled in Table 29 and compared to various experimental results like e.g. ESCA-data [180], relectivity data [183, 185] or X-ray emission data [184]. Considerable quantitative inconsistencies exist for some features. More experimental and quantitative theoretical studies should resolve these inconsistencies. The general qualitative aspects of the electronic structure of boron nitride, however, seem to be well understood on the basis of the existing experimental and theoretical results. In general, the properties of boron nitride are considerably simpler to explain than those of graphite because of existence of a large semiconducting gap.

TABLE 29

Listing of various relevant bandstructure energies of boron nitride as obtained by theoretical approaches and by experiments.

				BN		
			Position of nitrogen s-band (eV)	Optical transitions[a] involving π-bands		
Feature	Band gap (eV)	p-valence band width (π, σ) (eV)		$Q_{2u}^- \rightarrow Q_{2g}^-$ $\perp$	$Q_{2g}^- \rightarrow Q_{2u}^-$ $\perp$	Type
Theory						
Doni, Pastori [163]	5.4	~ 5	~ -14	6.2	7.0	ETB
Zupan [167]	4.9	6.8	-15	6.4	11.0	SCF CNDO TB
Taylor, Coulson [158]	4.6	–	–	~ 5.0	–	ETB (2-dim)
Zunger [170]	3.7	3.8	-18.8	~ 6.0	–	molecular cluster (2-dim)
Nakhmanson, Smirnov [166]	2.45 3.8	15.5	~ -25	~ 6.5	~ 7.0	OPW
Experiment						
Hamrin et al. [180]	–	–	-19.5	–	–	ESCA
Choyke [183]	–	–	–	6.2	7.0	reflectivity
Fomichev [184]	3.6	14.0	~ -19			X-ray emission
Vilanove [185]	< 5.0	–	–	6.5	none	reflectivity

[a] Transitions shown in Figure 119.

References

[1] J. A. Wilson and A. D. Yoffe: *Adv. Phys.* **18** (1969), 193.

[2] F. R. Gamble and T. H. Geballe: in *Solid State Chemistry*, ed. by B. Hannay, Plenum Press, in press and A. D. Yoffe, XII Int. conf. Phys. Semicond. ed. by M. H. Pilkuhn, Teubner, Stuttgart, 1974, p. 611.

[3] J. A. Wilson, F. J. Disalvo, and S. Mahajan: *Phys. Rev. Letters* **32** (1974), 882; and *Adv. Phys.* **24** (1975), 117.

[4] A. H. Thompson, K. R. Pisharody, and R. F. Koehler: *Phys. Rev. Letters* **29** (1972), 163.

[5] D. W. Fischer: *Phys. Rev.* **B8** (1973), 3576.

[6] D. L. Greenaway and R. Nitsche: *J. Phys. Chem. Solid* **26** (1965), 1445.

[7] A. R. Beal, J. C. Knight, and W. Y. Liang: *J. Phys.* **C5** (1972), 3531.

[8] L. E. Conroy and K. C. Park: *Inorg. Chem.* **7** (1968), 459.

[9] J. B. Goodenough: *Phys. Rev.* **171** (1968), 466.

[10] R. Huisman, R. Dejonge, C. Hass, and F. Jellinek: *J. Solid State Chem.* **3** (1971), 56.

[11] J. Bernard and Y. Jeanin: *Adv. Chem.* **39** (1963), 191.

[12] R. M. White and G. Lucovsky: *Solid State Comm.* **11** (1972), 1369.

[13] P. Krusius, J. von Boehm, and H. Isomaki: *J. Phys.* **C8** (1975), 3788.

[14] H. W. Myron and A. J. Freeman: *Phys. Letters* **44A** (1973), 167; and A. Zunger and A. J. Freeman: *Phys. Rev.* **B16** (1977), 906.

[15] D. E. Ellis and A. Seth: *Int. Quantum Chem. Symp.* **7** (1973), 223.

[16] R. B. Murray and A. D. Yoffe: *J. Phys.* **C5** (1972), 3038.

[17] P. A. Lee, G. Said, R. Davis, and T. H. Lim: *J. Phys. Chem. Solid* **30** (1969), 2719.

[18] F. Hulliger: in *Structure and Bonding*, ed. by C. Jorgensen, J. Neilands, R. Nyholm, D. Reimen, and R. Williams, Springer, New York 1968, Vol. 4, p. 83: and B. Koiller and L. Falicov: *J. Phys.* **C5** (1972), 63.

[19] R. A. Bromley and R. B. Murray: *J. Phys.* **C5** (1972), 738.

[20] R. B. Murray, R. A. Bromley, and A. D. Yoffe: *J. Phys.* **C5** (1972), 746.

[21] C. Y. Fong, J. Camassel, S. Kohn, and Y. R. Shen: *Phys. Rev.* **B13** (1976), 5442.

[22] L. F. Mattheiss: *Phys. Rev.* **B8** (1973), 3719; and L. F. Mattheiss: *Phys. Rev. Letters* **30** (1973), 748.

[23] F. Herman and S. Skillman: *Atomic Structure Calculations*, Prentice Hall, Englewood Cliffs, New Jersey 1963.

[24] L. Pauling: *The Nature of the Chemical Bond*, 3rd edition, Cornell Univ. Press, New York 1960, p. 518.

[25] F. R. Gamble: *J. Solid State Chem.* **9** (1974), 358.

[26] A. H. Thompson, F. R. Gamble, and J. F. Revelli: *Solid State Comm.* **9** (1971), 981.

[27] P. M. Williams and F. R. Shepherd: *J. Phys.* **C6** (1973), L36.

[28] H. W. Myron and A. J. Freeman: *Phys. Rev.* **B11** (1975), 2735.

[29] N. V. Smith, M. M. Traum, and F. J. Disalvo: *Solid State Comm.* **15** (1974), 211.

[30] M. M. Traum, N. V. Smith, and F. J. Disalvo: *Phys. Rev. Letters* **32** (1974), 1241; and N. V. Smith and M. M. Traum: *Phys. Rev.* **B11** (1975), 2087.

[31] N. V. Smith and M. M. Traum: *Surface Sci.* **45** (1974), 745.

[32] H. N. S. Lee, H. McKinzie, D. S. Tannhauser, and A. Wold: *J. Appl. Phys.* **40** (1969), 602.

[33] H. N. S. Lee, M. Garcia, H. McKinzie, and A. Wold: *J. Solid State Chem.* **1** (1970), 190.

[34] J. E. Graebner: *Bull. Am. Phys. Soc.* **18** (1973), 386.

[35] P. M. Williams and F. R. Shepherd: *J. Phys.* **C6** (1973), L32.

[36] J. C. McMenamin and W. E. Spicer: *Phys. Rev. Letters* **29** (1972), 150.

[37] R. H. Williams and A. J. McEvoy: *Phys. Status Solidi* **B47** (1971), 217.

[38] R. H. Williams, J. M. Thomas, M. Barber, and N. Alford: *Chem. Phys. Letters* **17** (1972), 142.

[39] R. Bachman, H. C. Kirsch, and T. H. Geballe: *Solid State Comm.* **9** (1971), 57.

[40] W. Y. Liang: *J. Phys.* **C6** (1973), 551.

[41] A. R. Beal, J. C. Knight, and W. Y. Liang: *J. Phys.* **C5** (1972), 3540.

[42] R. V. Kasowski: *Phys. Rev. Letters* **30** (1973), 1175.

[43] C. Y. Fong and M. L. Cohen: *Phys. Rev. Letters* **32** (1974), 720.

[44] M. H. van Maaren and H. B. Harland: *Phys. Letters* A29 (1969), 571.
[45] R. A. Bromley: *Phys. Rev. Letters* 29 (1972), 357.
[46] H. P. Hughes and W. Y. Liang: *J. Phys.* C7 (1974), L162.
[47] A. Baldereschi: *Phys. Rev.* B7 (1973), 5312; and D. J. Chadi and M. L. Cohen: *Phys. Rev.* B7 (1973), 692.
[48] J. P. Walter, C. Y. Fong, and M. L. Cohen: *Solid State Comm.* 12 (1972), 3031.
[49] C. Y. Fong, J. P. Walter, and M. L. Cohen: *Phys. Rev.* B11 (1975), 2759.
[50] R. S. Tittle and M. W. Schafer: *Phys. Rev. Letters* 28 (1972), 808.
[51] R. A. Bromley: *Phys. Letters* 33A (1970), 242; and R. A. Bromley, R. B. Murray, and A. D. Yoffe: *J. Phys.* C5 (1972), 759.
[52] D. R. Edmondson: *Solid State Comm.* 10 (1972), 1085.
[53] P. G. Harper and D. R. Edmondson: *Phys. Status Solidi* B44 (1971), 59.
[54] K. Wood and J. B. Pendry: *Phys. Rev. Letters* 31 (1973), 1400.
[55] B. L. Evans and P. A. Young: *Phys. Status Solidi* 25 (1968), 417.
[56] E. Ehrenfreund, A. C. Gossard, F. R. Gamble, and T. H. Geballe: *J. Appl. Phys.* 42 (1971),1491.
[57] E. Mooser and W. B. Pearson: in *Progress in Semiconductors*, ed. Gibson, Kröger, Burgess, John Wiley, New York 1960, Vol. 5, p. 103; and E. Mooser and W. B. Pearson: *Phys. Rev.* 101 (1956), 492.
[58] M. Y. Au-Yang and M. L. Cohen: *Phys. Rev.* 178 (1969), 1279.
[59] C. Y. Fong and M. L. Cohen: *Phys. Rev.* B5 (1972), 3095.
[60] I. Ch. Schlüter and M. Schlüter: *Phys. Status Solidi* B57 (1973), 145.
[61] G. Mula and F. Aymerich: *Phys. Status Solidi* B51 (1972), K35; and F. Aymerich, F. Meloni and G. Mula: *Solid State Comm.* 12 (1973), 139.
[62] R. B. Murray and R. H. Williams: *J. Phys.* C6 (1973), 3643.
[63] J. Camassel, M. Schlüter, S. Kohn, J. P. Voitchovsky, Y. R. Shen, and M. L. Cohen: *Phys. Status Solidi* B75 (1976), 303.
[64] G. Domingo, R. S. Itoga, and C. R. Kannewurf: *Phys. Rev.* 143 (1966), 538.
[65] D. Penn: *Phys. Rev.* 128 (1962), 2093.
[66] R. H. Williams, R. B. Murray, D. W. Govan, J. M. Thomas, and E. L. Evans: *J. Phys.* C6 (1973), 3631.
[67] L. van Hove: *Phys. Rev.* 89 (1953), 1189.
[68] G. Gilat and L. J. Raubenheimer: *Phys. Rev.* 144 (1966), 390.
[69] R. B. Murray and R. H. Williams: in *Proc. XII Int. Conf. Phys. Semicond.*, ed. by M. H. Pilkuhn, Teubner, Stuttgart 1974, p. 637.
[70] E. Mooser, I. Ch. Schlüter, and M. Schlüter: *J. Phys. Chem. Solid* 35 (1974), 1269; and M. Schlüter and M. L. Cohen: *Phys. Rev.* B14 (1976), 424.
[71] J. R. Drabble and C. H. I. Goodman: *J. Phys. Chem. Solid* 5 (1958), 142.
[72] D. L. Greenaway and G. Harbeke: *J. Phys. Chem. Solid* 26 (1965), 1585.
[73] T. L. Anderson and H. B. Krause: *Acta. Cryst.* B30 (1974), 1307.
[74] J. R. Drabble: *Progress in Semiconductors*, Heywood Co., London 1963, Vol. 7, p. 45.
[75] P. M. Lee and L. Pincherle: *Proc. Phys. Soc.* 81 (1963), 461.
[76] F. Borghese and E. Donato: *Nuovo Cimento* B53 (1968), 283.
[77] S. Katsuki: *J. Phys. Soc. Japan* 26 (1969), 58.
[78] A. O. E. Animalu and V. Heine: *Phil. Mag.* 12 (1965), 1249.
[79] R. Togei and G. R. Miller: in *The Physics of Semimetals and Narrow Gap Semiconductors*, ed. by D. L. Carter and R. P. Bate, Pergamon, New York 1971, p. 349.
[80] P. J. Lin and L. Kleinman: *Phys. Rev.* 142 (1966), 478.
[81] C. Olson, D. Lynch, and Z. Hurych: to be published.
[82] Z. Hurych and R. L. Benbow: *Phys. Rev.* B16 (1977) 3707.
[83] V. V. Sobolev: *Phys. Status Solidi* 30 (1968), 349.
[84] A. Balzarotti, E. Burattini, and P. Picozzi: *Phys. Rev.* B3 (1971), 1159.
[85] K. Taniguchi, A. Moritani, C. Hamaguchi, and J. Nakai: *Surface Sci.* 37 (1973), 212.
[86] V. Grasso, G. Mondio, and G. Saitta: *Phys. Letters* A42 (1973), 525.
[87] I. G. Austin: *Proc. Phys. Soc.* 76 (1960), 169.
[88] P. Drath: *Zeitschr. Naturforsch.* 21a (1968), 1146 and references therein.

[89] M. R. Tubbs: *Phys. Status Solidi* **B49** (1972), 11.
[90] D. L. Greenaway and G. Harbeke: *J. Phys. Soc. Japan Suppl.* **1** (1966), 151.
[91] G. Harbeke and E. Tosatti: *RCA Review* **36** (1975), 40 and references therein.
[92] E. Doni, G. Grosso, and G. Spavieri: *Solid State Comm.* **11** (1972), 493.
[93] E. Doni, G. Grosso, G. Harbeke, E. Meier, and E. Tosatti: *Phys. Status Solidi* **B68** (1975), 569.
[94] I. Ch. Schlüter and M. Schlüter: *Phys. Rev*, **B9** (1974), 1652.
[96] Ch. Gähwiller and G. Harbeke: *Phys. Rev.* **185** (1969), 1141.
[97] G. Baldini and S. Franchi: *Phys. Rev. Letters* **26** (1971), 503.
[98] G. Harbeke and E. Tossati: *Phys. Rev. Letters* **28** (1972), 1507.
[99] R. S. Knox: *Theory of Excitons*, Solid State Physics, Suppl. 3, Academic Press, New York 1963.
[100] F. Levy, C. Depeursinge, Le Chi Thanh, A. Mercier, E. Mooser, and J. P. Voitchovsky: in *Proc. XII Int. Conf. Phys. Semicond.*, ed. by M. H. Pilkuhn, Teubner, Stuttgart 1974, p. 1237.
[101] M. G. Mason and L. J. Gerenser: *Chem. Phys. Letters* **40** (1976), 476.
[102] T. Ishii, S. Kono, T. Matsukawa, T. Sagawa, and T. Kobayasi: *J. Electron Spect. and Rel. Phenom.* **5** (1974), 559.
[103] M. Schlüter, S. Kohn, M. L. Cohen, and C. Y. Fong: *Phys. Status Solidi* **B78** (1976), 737.
[104] C. Y. Fong, Y. Petroff, S. Kohn, and Y. R. Shen: *Solid State Comm.* **14** (1974), 681.
[105] S. P. Kowalczyk, L. Ley, F. R. McFeeley, and D. A. Shirley: *Solid State Comm.* **17** (1975), 463.
[106] B. L. Evans: *Proc. Roy. Soc*, **A289** (1966), 275.
[107] E. F. Gross, N. V. Starostin, M. P. Shepilov, and R. I. Shekhmametev: *Soviet Phys. Solid State* **14** (1973), 1681 and references therein.
[108] W. Czaja, G. Harbeke, L. Krausbauer, E. Meier, B. J. Curtis, H. Brunner, and E. Tosatti: *Solid State comm.* **13** (1973), 1445.
[109] H. Hahn and G. Frank: *Zeits. Anorg. Allg. Chem.* **278** (1955), 340.
[110] J. L. Brebner and E. Mooser: *Phys. Letters* **24A** (1967), 274.
[111] Z. S. Basinski, D. B. Dove, and E. Mooser: *Helv. Phys. Acta.* **34** (1961), 373.
[112] E. Mooser: private communication; see also [126].
[113] A. Kuhn, R. Chevalier, and A. Rimsky: *Acta Cryst.* **B32** (1976), 469.
[114] A. Kuhn, A. Chevy, and R. Chevalier: *Phys. Status Solidi* **A31** (1975), 469.
[115] G. Fischer: *Helv. Phys. Acta.* **36** (1963), 317.
[116] F. Hullinger and E. Mooser: *Prog. Solid State Chem.* **2** (1965), 330.
[117] F. Bassani, D. L. Greenaway, and G. Fischer: *Proc. VII Int. Conf. Phys. Semicond.*, Dunod, Paris 1961, p. 51.
[118] J. C. Phillips: *Phys. Rev.* **188** (1969), 1225.
[119] V. B. Sapre and C. Mande: *J. Phys. Chem. Solid* **34** (1973), 1351.
[120] F. Bassani and G. Pastori-Parravicini: *Nuovo Cimento* **50B** (1967), 95.
[121] H. Kaminura and N. Nakao: *Phys. Soc. Japan* **21**, Suppl. (1966), 27; and H. Kamimura and N. Nakao: *Phys. Soc.* **14** (1968), 1313.
[122] S. Nagel: Diploma Thesis, University Marburg, Germany, unpublished.
[123] A. Bourdon: *J. Phys. (Paris)* **35** (C3) (1974), 261.
[124] M. Schlüter: *Nuovo Cimento* **13B** (1973), 313.
[125] M. Schlüter, J. Camassel, S. Kohn, J. P. Voitchovsky, Y. R. Shen, and M. L. Cohen: *Phys. Rev.* **B13** (1976), 3534.
[126] E. Mooser and M. Schlüter: *Nuovo Cimento* **18B** (193), 164.
[127] G. Ottaviani, C. Canali, F. Nava, Ph. Schmid, E. Mooser, R. Minder, and I. Zschokke: *Solid State Comm.* **14** (1974), 933.
[128] R. E. Watson and A. J. Freeman: *Phys. Rev.* **124** (1961), 1117.
[129] M. L. Cohen and T. K. Bergstresser: *Phys. Rev.* **141** (1966), 789.
[130] A. Baldereschi, K. Maschke, and M. Schlüter: *Helv. Phys. Acta.* **47** (1974), 434.
[131] J. L. Brebner: *Phys. Chem. Solid* **25** (1964), 1427.
[132] J. A. Deverin: *Helv. Phys. Acta* **42** (1969), 397; and *Nuovo Cimento* **63B** (1969), 1.
[133] A. Baldereschi and M. G. Diaz: *Nuovo Cimento* **68B** (1970), 217.
[134] Y. Ichii, Y. Sasaki, C. Hamaguchi, and J. Nakai: *Solid State Comm.* **17** (1975), 451.
[135] P. C. Leung, G. Andermann, W. G. Spitzer, and C. A. Mead: *Phys. Chem. Solid* **27** (1966), 849.
[136] F. Bassani and A. Baldereschi: *Surface Sci.* **37** (1973), 304.

[137] L. Fritsche: *Phys. Status Solidi* **34** (1969), 195; and L. Fritsche and F. D. Heidt: *Phys. Status Solidi* **35** (1969), 987.

[138] N. Lee, D. M. Larsen, and B. Lax: *J. Phys. Chem. Solid* **34** (1973), 1059; and D. Cabib, E. Fabri, and G. Fiorio: *Nuovo Cimento* **10B** (1972), 185.

[139] A. Mercier, E. Mooser, and J. P. Voitchovsky: *Phys. Rev.* **B12** (1975), 4307.

[140] J. M. Besson, K. P. Jain, and A. Kuhn: *Phys. Rev. Letters* **32** (1974), 936.

[141] A. J. Grant and A. D. Yoffe: *Solid State Comm.* **8** (1970), 1919.

[142] A. J. Niilisk and J. J. Kirs: *Phys. Status Solidi* **31** (1969), K91.

[143] R. Zallen and D. F. Blossey: *Optical and Electrical Properties of Compounds with Layered Structures*, Vol. 4, ed. by P. A. Lee, Reidel Publ., Dordrecht, The Netherlands 1975.

[144] J. M. Besson, R. Letoullec, J. P. Pinceaux, A. Chevy, and H. Fair: *V Int. Conf. High Pressure and Techn.*, Moscow 1975.

[145] Complete lists of experimental references are given in [125] and [142].

[146] S. Kohn, Y. Petroff, and Y. R. Shen: *Surface Sci.* **37** (1973), 205.

[147] P. Thiry: Thesis (Paris 1975) unpublished; and P. Thiry, R. Pincheaux, D. Dagneaux, and Y. Petroff: *Proc. XII Int. Conf. Phys. Semicond.*, ed. by M. H. Pilkuhn, Teubner, Stuttgart 1974, p. 1324.

[148] R. Mamy, L. Martin, G. Leveque, and C. Raisin: *Phys. Status Solidi* **62** (1974), 201.

[149] R. Vilanove: Thesis (Paris 1971) unpublished; and J. Perrin, J. Cazaux, and P. Soukiassian: *J. Phys. (Paris)*, to appear.

[150] Y. Petroff: *Proc. XIII Int. Conf. Phys. Semicond.*, ed. F. G. Fumi, Rome 1976, p. 975.

[151] G. Martinez, M. Schlüter, M. L. Cohen, R. Pinchaux, P. Thiry, D. Dagneaux, and Y. Petroff: *Solid State Comm.* **17** (1975), 5.

[152] A. J. McEvoy and R. H. Williams: *J. Phys.* **C5** (1972), L1222; and *Phys. Status Solidi* **A** (1971), K155.

[153] F. R. Shepherd and P. M. Williams: *Phys. Rev.* **B12** (1975), 5705.

[154] I. Adams, J. H. Thomas, M. Barber, and R. H. Williams: *Chem. Phys. Letters* **10** (1971), 297.

[155] R. H. Williams, G. P. Williams, C. Norris, M. R. Howells, and I. H. Munro: *J. Phys.* **C7** (1974), L29.

[156] P. R. Wallace: *Phys. Rev.* **71** (1974), 622.

[157] C. A. Coulson and R. Taylor: *Proc. Phys. Soc.* **65** (1952), 815.

[158] R. Taylor and C. A. Coulson: *Proc. Phys. Soc.* **65** (1952), 834.

[159] W. M. Lomer: *Proc. Roy. Soc. London* **A227** (1955), 330.

[160] J. C. Slonczewski and P. R. Weiss: *Phys. Rev.* **109** (1958), 272.

[161] J. W. McClure: *Phys Rev.* **108** (1057), 612; and ibid. **119** (1960), 606.

[162] F. J. Corbato: *Proc. III Conf. Carbon*, Pergamon, London 1959, p. 173.

[163] E. Doni and G. Pastori-Parravicini: *Nuovo Cimento* **64** (1969), 117.

[164] W. van Haeringen and H. G. Junginger: *Solid State Comm.* **7** (1969), 1723.

[165] G. S. Painter and D. E. Ellis: *Phys. Rev.* **B1** (1970), 4747.

[166] M. S. Nakhmanson and V. P. Smirnov: *Soviet Phys. Solid State* **13** (1971), 752 ; and ibid. **13** (1972), 2763.

[167] J. Zupan and D. Kolar: *J. Phys.* **C5** (1972), 3097; and J. Zupan: *Phys. Rev.* **B6** (1972), 2477.

[168] H. Nagayoshi, M. Tsukada, K. Nakao and Y. Uemura: *J. Phys. Soc. Japan* **35** (1973), 396.

[169] L. G. Johnson and G. Dresselhaus: *Phys. Rev.* **B7** (1973), 2275.

[170] A. Zunger: *J. Phys.* **C7** (1974), 76; and ibid. **7** (1974), 96.

[171] J. W. McClure: in *The Physics of Semimetals and Narrow Gap Semiconductors*, ed. by D. L. Carter and R. T. Bate, Pergamon, New York 1971, p. 127; and *IBM J. Res. Develop.* **8** (1964), 255.

[172] W. N. Reynolds: *Physical Properties of Graphite*, Elsevier Publ. London 1968.

[173] G. Dresselhaus and M. S. Dresselhaus: *Phys. Rev.* **140** (1965), A401.

[174] P. R. Schroeder, M. S. Dresselhaus, and A. Javan: *Phys. Rev. Letters* **20** (1968), 1292.

[175] D. L. Greenaway, G. Harbeke, F. Bassani, and E. Tosatti: *Phys. Rev.* **178** (1969), 1340 and references therein.

[176] R. F. Willis, B. Feuerbacher, and B. Fitton: *Phys. Rev.* **B4** (1971), 2441.

[177] B. Feuerbacher and B. Fitton: *Phys. Rev. Letters* **26** (1971), 840.

[178] G. Guizzetti, L. Nosenzo, E. Reguzzoni, and G. Samoggia: *Phys. Rev. Letters* **31** (1973), 154.
[179] E. A. Taft and H. R. Philipp: *Phys. Rev.* **138** (1965), A197.
[180] K. Hamrin, G. Johansson, V. Gelius, C. Nordling, and K. Siegbahn: *Phys. Scr.* **1** (1970), 277.
[181] J. M. Thomas, E. L. Evans, M. Barber, and P. Swift: *Trans. Faraday Soc.* **67** (1971), 1865.
[182] F. C. Chalkin: *Proc. Roy. Soc.* **A194** (1948), 42.
[183] W. J. Choyke: unpublished data (see [163]).
[184] V. A. Fomichev: *Soviet Phys. Solid State* **13** (1971), 754.
[185] R. Vilanove: *Compt. Rend. Acad. Sci. Paris* **B271** (1970), S136 and Thesis CNRS No. A05885 (1971) Unpublished.

Acknowledgment

The authors are indebted to Ines Schlüter for proofreading the manuscripts and for numerous editorial tasks.

R	group symmetry operation
$\mathbf{R}$	double group symmetry operation
ψ_i	basis function
$\Gamma(R)_{ij}^{(n)}$	group representation matrix element
$\rho_{ij}^{(n)}$	group projection operator
$X(R)$	trace of representation, character
φ_k	class of group symmetry operations
$f(\mathbf{r})$	function
E	energy
H	Hamiltonian
$\mathbf{t}, \tau$	real space vector
$\mathbf{a}_i$	lattice basis vector
α	rotational part of group symmetry operation
ε	identity
$\mathbf{L}$	lattice vector
$\mathbf{b}_i$	reciprocal basis vector
$\mathbf{G}, \mathbf{k}$	vector in reciprocal space
G	space group
$\mathbf{D}$	double space group
G^0	point group
T	translation group
G/T	factor group
$G_{\mathbf{k}}$	'small' space group
$G_{\mathbf{k}}^0$	'small' point group
$G_{\mathbf{k}}/T_{\mathbf{k}}$	'small' factor group
θ, ϕ, ψ	Eulerian angles
$D_{1/2}$	spin representation matrices
u_s	spin function
$\boldsymbol{\sigma}$	spin operator
T	time reversal operator
U	unitary operator
K	complex conjugation operator

LIST OF SYMBOLS – CHAPTER B

H	Hamiltonian
$\left.\begin{array}{l}\psi_{n\mathbf{k}}(\mathbf{r})\\ \phi_{n\mathbf{k}}(\mathbf{r})\\ \Phi_{n\mathbf{k}}(\mathbf{r})\end{array}\right\}$	Bloch functions
$\varphi(\mathbf{r})$	Atomic Löwdin function
$\hbar$	Planck's constant
m	effective electron mass
∇	gradient operator
$V(\mathbf{r})$	one electron potential
$\mathbf{r}, \mathbf{R}_l, \tau_\alpha, \mathbf{Q}, \mathbf{d}$	real space vectors
$\mathbf{k}, \mathbf{G}$	reciprocal space vectors
$X_j(\mathbf{r})$	atomic orbital
$R_{\alpha j}(r), u_{\alpha j}(r)$	radial wave function
$\rho(\mathbf{r})$	electron density
S, L	transformation matrices
$j_l(kr)$	spherical Bessel function
$Y_{lm}(\theta, \phi)$	spherical harmonic
θ, ϕ	polar angles
$|c\rangle$	core state
$|\mathbf{k}\rangle$	plane wave state
$S_\beta(\mathbf{G})$	structure factor
$V_{\beta L}(\mathbf{G})$	form factor
σ	spin Pauli matrices
$\mathbf{P}_l$	angular projection operator
P_l	Legendre polynomial of order l
P_c	core states projection operator
$G_\mathbf{k}(\mathbf{r}, \mathbf{r}')$	Green's function
$n_l(kr)$	spherical Neumann function
$U_j^{(I)}$	free electron wave function
Q	transfer matrix
$H_l^{(1)}$	Hankel function of first kind
ζ	complex variable
$J_{|M|}$	Bessel function

$V(q)$	potential Fourier transform form factor
$S(q)$	structure factor
$\varepsilon(q)$	static dielectric function (real part)
$\varepsilon_1(\omega)$	real part of dielectric function
$\varepsilon_2(\omega)$	imaginary part of dielectric function
$\hbar$	Planck's constant
E	energy
ω	light frequency
∇	gradient operator
$U_{\alpha\mathbf{k}}$	periodic part of Bloch function
$\rho(\mathbf{r})$	charge density
$\mathbf{R}_m$	lattice vector
$\mathbf{k}$	reciprocal space vector
$N(E)$	density of states
μ	reduced mass
$\mathbf{A}$	vector potential
$a(\mathbf{r})$	Wannier function
α	anisotropy parameter

PART II

PHONONS

INFRARED AND RAMAN INVESTIGATIONS OF
LONG-WAVELENGTH PHONONS IN LAYERED MATERIALS

T. J. WIETING and J. L. VERBLE

U.S. Naval Research Laboratory, Washington, D.C. 20375, U.S.A.

1. Introduction

Research interest in the phonon energies of layered crystal structures has grown rapidly during the last decade. The reasons for the interest vary from fundamental considerations concerning low-dimensional systems to the utility of these materials as superconductors, lubricants, and thermoelectric elements. Infrared and Raman studies of layered materials were initiated long before the current phase of research activity. The first infrared spectrum was obtained on brucite [Mg(OH)$_2$] in 1905 by Coblentz [1]; however, relatively few papers were reported in the literature before 1960. Krishnamurti's work [2] on Mg(OH)$_2$ and Ca(OH)$_2$ appears to have been the first extensive Raman study of layered compounds. Current activity has also greatly benefited from instrumental developments in spectrometers and laser light sources. Infrared and Raman techniques are now mature, and they have been used to advantage in far-infrared and high-resolution work on layered structures. (The former experimental difficulties can be appreciated by reading the survey of the Raman effect in solids by Menzies [3], which was written a little over twenty years ago.)

We note several reviews that include certain infrared and Raman investigations of layered materials. For the early period the articles of Bhagavantum [4], Menzies [3], and Mathieu [5] are particularly useful. More recent surveys are those of Veddar and Hornig [6], who give references to work on organic molecular crystals; Mitra [7], who evaluates the literature on brucite; and Yoffe [8], who summarizes the investigations of the transition-metal dichalcogenides. Apart from the micas [5] and the organic molecular crystals, this article reviews all of the reported work on long-wavelength lattice vibrations.

Section 2 introduces the seven families of layered materials that have been studied by infrared and Raman techniques. After presenting the crystal structures, we describe the correlation method for obtaining the symmetries of the normal vibrations; this method is then compared with the older analytical method of Bhagavantum and Venkatarayudu. As an aid in discussing the lattice vibrations of a layered material, we define a triperiodic pseudolattice for each family, which possesses the factor-group symmetry of an individual layer. Correlation diagrams and tables are compiled for all members of the seven families. Section 3 develops in brief the theory and practice of infrared absorption and Raman scattering. These two techniques are experimentally complementary, as indicated by the selection rules for the infrared- and Raman-active modes. This section also discusses the effective charge of a polar lattice vibration, determined by infrared measurements. Section 4 examines the general properties of the normal vibrations of layered crystals, especially the correlation splittings often observed in infrared and Raman spectra, and the relative absorption and scattering intensities of

321

T. J. Wieting and M. Schlüter (eds.), Electrons and Phonons in Layered Crystal Structures. 321–407.
All Rights Reserved.
Copyright © 1979 *by D. Reidel Publishing Company, Dordrecht, Holland.*

the conjugate modes. In addition, the linear-chain model for lattice vibrations is applied to layered structures. In Section 5 the experimental and theoretical investigations of individual materials are discussed and evaluated. The longest subsection treats in detail the many papers on GaSe, which has been the subject of some controversy. Several fundamental as well as unresolved problems in the transition-metal dichalcogenides are also mentioned. The section concludes with a discussion of chemical bonding and effective charges.

2. Structures and Symmetry

2.1. Families of layered materials

All of the layered crystal structures considered in this review are composed of identical layers bound together weakly, in most cases, by forces of the van der Waals type. In some layered polymorphs the coordination of the atoms has been found to vary from layer to layer; hence, the layers are not identical, even though they do retain similar features. Examples of these nonidentical structures are the 4Hb and 6R polytypes of TaS_2 and $TaSe_2$ [9, 10], which contain both octahedral and trigonal prismatic coordination. Their lattice vibrations have not as yet been investigated, and we therefore limit the following outline of crystal structures to those materials having identical layers. We shall discuss them as families of polytypes, since the fundamental structural and vibrational unit is the individual layer.

2.1.1. GaSe

The materials included in this family are GaS, GaSe, InSe, and their polytypes, denoted by β, ε, and γ [11–13]. Figure 2.1 shows the coordination of the Ga and Se atoms in an isolated layer, which is constructed of single planes (or sheets) of Se atoms lying on either side of a double plane of Ga atoms. The c-axis is normal to the basal plane of the layers and is the axis of highest symmetry. Polytypes of GaSe are generated by stacking the layers regularly in the three relative positions permitted by the layered structure [14]. In the case of β–GaS adjacent layers in the hexagonal unit cell are rotated by 60° with respect to each other and translated by $\sqrt{3}\,a/3$ in the hexagonal [210] direction (a is the

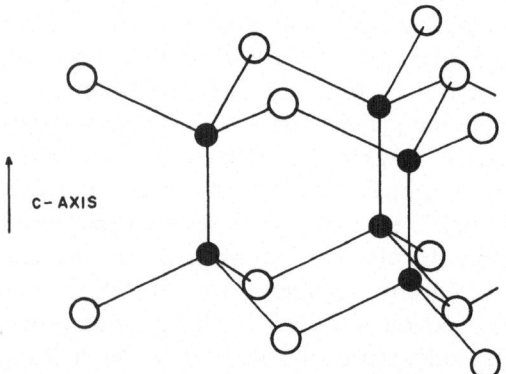

Fig. 2.1.   Coordination of the atoms in an isolated layer of GaSe. The black circles are the Ga atoms, and the open circles are the Se atoms.

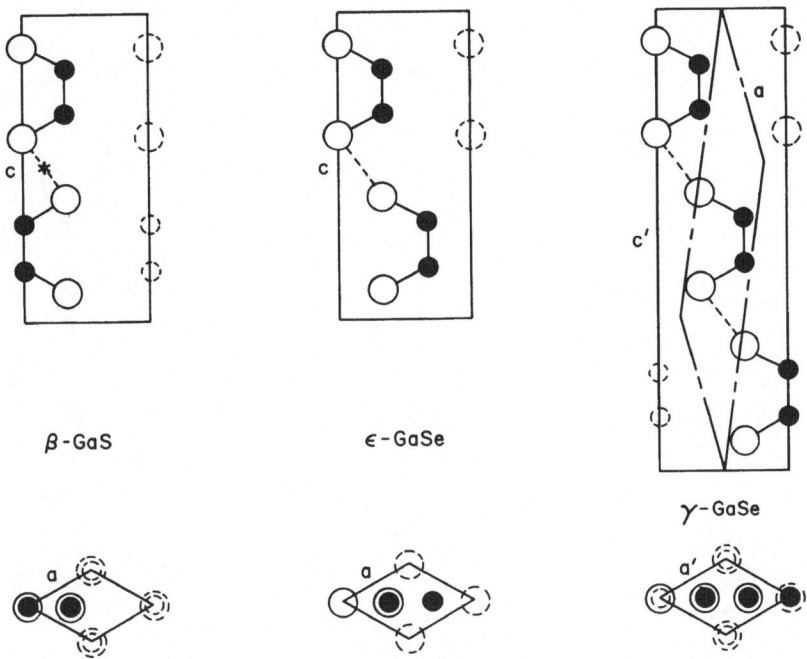

Fig. 2.2.   Atoms of the primitive unit cells of β-GaS, ε-GaSe, and γ-GaSe (the black circles are the Ga atoms). The upper half of the figure shows the hexagonal (110) sections of the cells, and the lower half shows the projections of the atoms onto the (001) plane. The rhombohedral cell is indicated by the long and short dashed line.

hexagonal unit-cell dimension). For the ε and γ polytypes no relative rotation of the layers is required, only a translation. The atoms of the primitive unit cells, which lie in the hexagonal (110) plane, are shown in Figure 2.2.

2.1.2. MoS_2

The materials included in the MoS_2 family [15, 16] are MoS_2, $MoSe_2$, $MoTe_2$, WS_2, WSe_2, and $NbSe_2$. Their polytypes are usually denoted by the prefixes 2H and 3R [17], which indicate the number of three-fold layers in the hexagonal (H) or rhombohedral (R), primitive unit cells.* The coordination of a single layer of MoS_2 is illustrated by the GaSe layer in Figure 2.1, if each pair of Ga atoms is combined into one atom at the center of the layer. The Mo atom and its six nearest-neighbor S atoms form a trigonal prism, whose base is perpendicular to the c-axis of the crystal. Figure 2.3 shows the atoms of the primitive unit cells of three polytypes of the MoS_2 family. 2H–$NbSe_2$ (or α–$NbSe_2$) has a different stacking sequence from that of 2H–MoS_2 (or β–MoS_2): the Nb atoms lie directly above or below each other in neighboring layers. However, the center of inversion between the layers is preserved, as well as the space group of the crystal D_{6h}^4 [14] and the site symmetries of the atoms.

* Another nomenclature for the polytypes [18, 11] uses the Greek letters α, β, γ, δ, and ε, which correspond to 2H–$NbSe_2$, 2H–MoS_2, 3R–MoS_2, 6R–MoS_2, and ε-MoS_2 (Section 4.1), respectively.

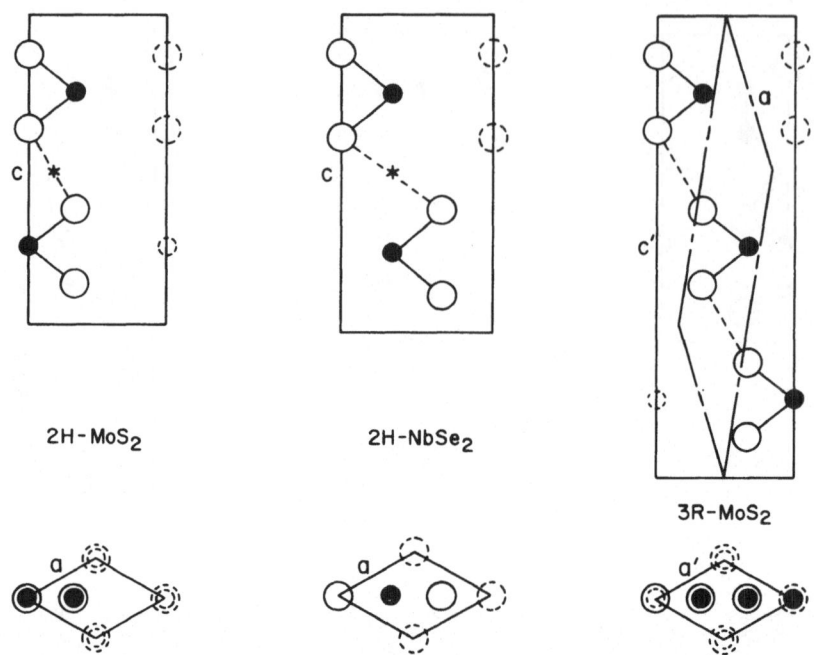

Fig. 2.3.   Atoms of the primitive unit cells of $2H-MoS_2$, $2H-NbSe_2$, and $3R-MoS_2$ (the black circles are the metal atoms). The upper half of the figure shows the hexagonal (110) sections of the cells, and the lower half shows the projections of the atoms onto the (001) plane. The rhombohedral cell is indicated by the long and short dashed line.

2.1.3. CdI_2

The materials included in the CdI_2 family are $CdCl_2$, $CdBr_2$, CdI_2, $MnCl_2$, $CoCl_2$, TiS_2, $TiSe_2$, ZrS_2, $ZrSe_2$, HfS_2, $HfSe_2$, SnS_2, PbI_2, $Mg(OH)_2$, and $Ca(OH)_2$. In this family the metal atoms are located at octahedral sites in the lattice, rather than at trigonal prismatic sites. The lattice vibrations of three polytypes have been studied: $CdCl_2$, $2H-CdI_2$, and $4H-CdI_2$ (Figure 2.4), where the prefixes 2H and 4H have a different meaning [19] from those of the MoS_2 family. To distinguish the polytypes of CdI_2, a particular iodine plane is chosen, and the relative orientations of the iodine planes immediately above and below it are determined. If the orientations of these two planes agree, the chosen layer is denoted by H (for hexagonal close packing of the layers); if the two layers are rotated by 60° with respect to each other, the layer is denoted by C (for cubic close packing of the layers). In $2H-CdI_2$ the three-fold layers are stacked directly above each other, and the sequence of the iodine planes is H, H, H, H . . . The primitive unit cell therefore contains one layer and two iodine planes.* In $4H-CdI_2$ the sequence of iodine planes is H, C, H, C, H, C . . ., and the primitive unit cell contains two layers and four iodine planes. Figure 2.4 shows the atoms of the unit cell, which lie in the hexagonal (110) plane. In the $CdCl_2$ structure, which includes $CdBr_2$, $MnCl_2$, and $CoCl_2$, the sequence of chlorine planes is C, C, C, C . . .; the

---

* The single layer polytype $2H-CdI_2$, is also designated as $1T-CdI_2$ (T = trigonal), following the nomenclature of the MoS_2 family.

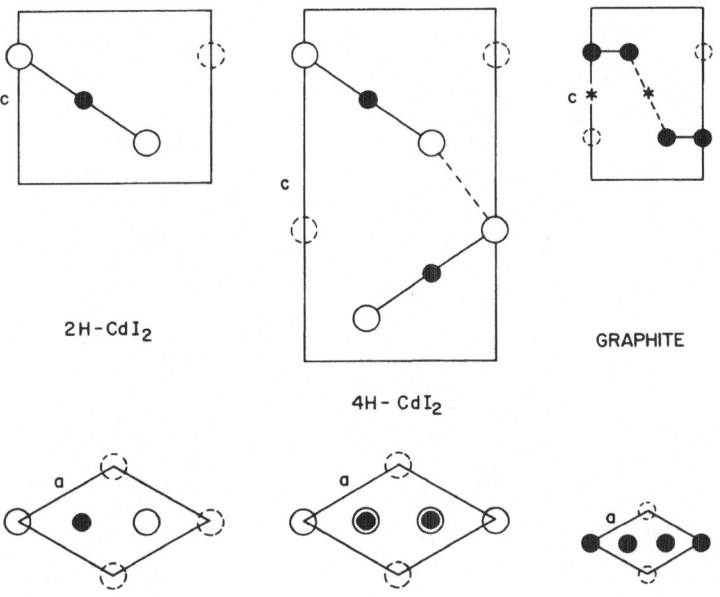

Fig. 2.4. Atoms of the primitive unit cells of 2H–CdI₂, 4H–CdI₂, and graphite. The upper half of the figure shows the hexagonal (110) sections of the cells, and the lower half shows the projections of the atoms onto the (001) plane.

rhombohedral primitive unit cell spans three layers but contains only one molecule. The symmetries and displacement directions of the normal modes of $CdCl_2$ are identical to those of $2H–CdI_2$.

2.1.4. *Graphite*

Although two polytypes of graphite [20, 21] are known, only the two-layer hexagonal form and its lattice vibrations have been investigated. The structural units of each plane (or layer) of the crystal are hexagons with carbon atoms at the corners, similar to the benzene ring. The packing of the carbon atoms is hexagonal, compared with cubic packing for the three-layer rhombohedral form. Figure 2.4 shows the atoms of the primitive unit cell, which coincide with the hexagonal (110) plane.

2.1.5. *As₂S₃*

The materials included in this family are As_2S_3 (orpiment) [22] and $As_2 Se_3$[23], which are monoclinic structures having two layers in the primitive unit cell. A single layer is composed of two molecules, and the layer symmetry is very nearly orthorhombic [24]. Each As atom has three nearest-neighbor S atoms, while each S atom has two nearest-neighbor As atoms. The atoms of the primitive unit cell are shown in Figure 2.5, where for clarity the view looking down the *b*-axis of the crystal (the axis of highest symmetry) is restricted to the atoms of a single layer. The atoms of this layer have *not* been shifted, as in the figure of Zallen *et al.* [24], in order to illustrate the close orthorhombic symmetry of the layer. In Sections 2.3.8 and 5.5 we shall discuss the physical meaning of the layer distortion and its effects upon the lattice vibrations.

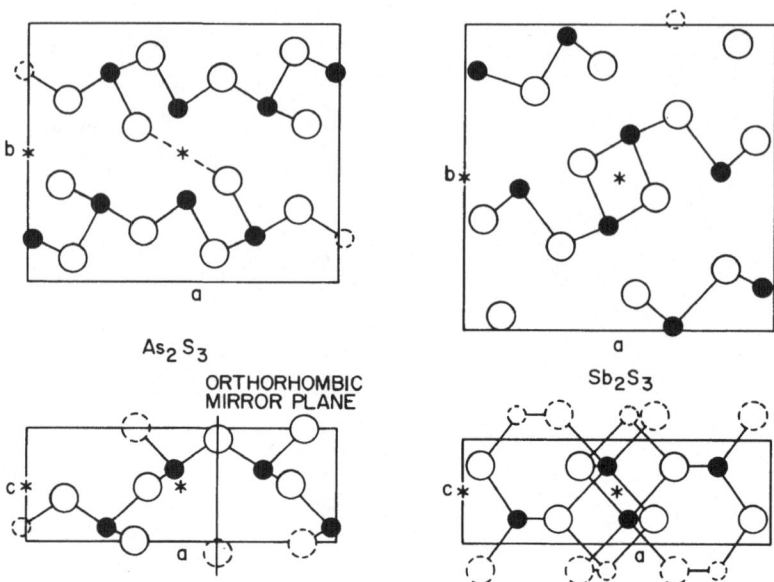

Fig. 2.5. Atoms of the primitive unit cells of As_2S_3 and Sb_2S_3. The projections shown in the lower half of the figure are restricted to the atoms of single layers of As_2S_3 and Sb_2S_3. The mirror plane in the orthorhombic pseudolattice of As_2S_3 is also indicated.

2.1.6. Sb_2S_3

The materials included in the orthorhombic stibnite family [25] are Sb_2S_3, Sb_2Se_3, and Bi_2S_3. Two layers lie within the primitive unit cell, and each layer is composed of two molecules or 10 atoms. The individual layers, which bear some resemblance to those of As_2S_3 but are more buckled, run parallel to the xz plane of the crystal, as shown in Figure 2.5. Half of the Sb atoms are linked to five nearest-neighbor S atoms, and the other half exhibit three-fold coordination with S atoms. Since all three members of this family have their a and b unit-cell dimensions nearly equal (within 0.01–0.15 Å), the lattices are almost tetragonal.

2.1.7. Bi_2Te_3

The materials included in this family are Bi_2Se_3 and Bi_2Te_3. Each layer of tellurobismuthite [26, 27] is composed of five planes of atoms in the sequence $Te^{(1)}$–Bi–$Te^{(2)}$–Bi–$Te^{(1)}$, where the $Te^{(2)}$ and Bi atoms occupy nearly octahedral sites. Although the primitive unit cell is rhombohedral and contains five atoms, the crystal structure is more conveniently referenced to the hexagonal unit cell (Figure 2.6), which contains 15 atoms. A section through the rhombohedral cell, which coincides with the hexagonal (110) plane, is also shown in the figure.

2.2. GROUP THEORY AND THE CORRELATION METHOD

Group theory is a powerful tool for analyzing the lattice vibrations of crystalline solids. The vibrational information that may be obtained from group theory includes: (i) the symmetry, number, and degeneracy of the allowed normal vibrations; (ii) the optical

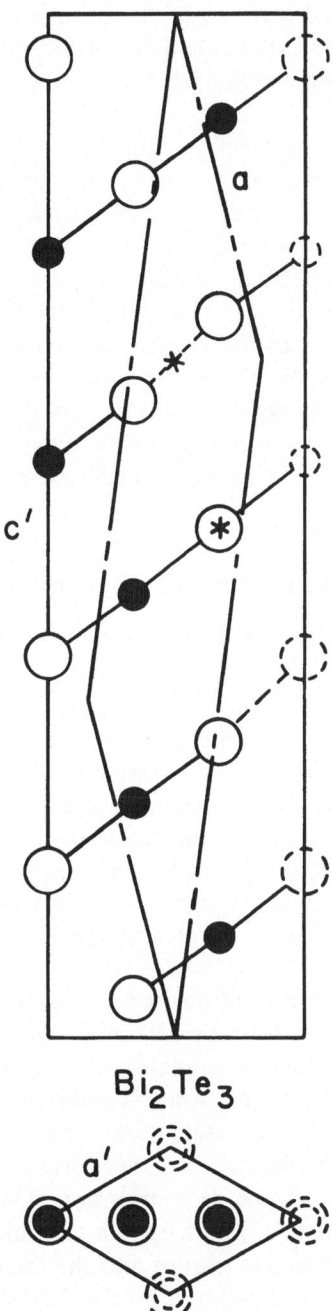

Bi_2Te_3

Fig. 2.6.   Atoms of the hexagonal unit cell of Bi_2Te_3. The upper half of the figure shows the hexagonal (110) section of the cell, and the lower half shows the projection of the atoms onto the (001) plane. The rhombohedral primitive unit cell is indicated by the long and short dashed line.

selection rules; and (iii) the directions of the atomic displacements for the normal modes. In this section and the next, we shall discuss the procedures and results of group theory, as applied to the long-wavelength lattice vibrations of a number of layered crystal structures.

There are several important symmetry groups that we should mention at the outset. The most general of these is the space group of the crystal, which contains all of the operations that transform the crystal into itself. For an arbitrary space group G, the set of symmetry operations can be written as $\{S \mid v(S) + x(m)\}$, where we have adopted the Seitz notation [28]. The set $\{S\}$ is the purely rotational part of the space group and contains the point-group operations of the crystal. The translations of the space group are of two kinds: $v(S)$ are the nonprimitive translations combined with certain rotational operations, and $x(m)$ are the primitive translation vectors of the lattice. When the translations $v(S)$ are identically zero for all rotational operations, the space group is said to be symmorphic; otherwise, the group is nonsymmorphic. If the equilibrium position vector of the κth atom in the lth unit cell is expressed as

$$\mathbf{X}(\kappa l) = \mathbf{X}(l) + \mathbf{X}(\kappa), \tag{2.1}$$

the effect of a space-group operation on the position vector is

$$\{S \mid v(S) + x(m)\}\mathbf{X}(\kappa l) = S\mathbf{X}(\kappa l) + v(S) + x(m),$$

$$= \mathbf{X}(\kappa' l'), \tag{2.2}$$

where $(\kappa' l')$ is an equivalent lattice point. Another useful group is the factor group of the space group of the crystal, which is isomorphic to a point group and consists of all operations of the space group with the exception of the primitive translations $x(m)$. In the case of a symmorphic space group, the factor group is identical to the set of purely rotational operations $\{S\}$. The site group for an atom of the crystal is the point group at the particular lattice site occupied by the atom. The site groups are subgroups of the factor group of the crystal; thus, correlations [29] exist between their irreducible representations.

Two equivalent methods have been devised for determining the irreducible representations and selection rules for lattice vibrations. The original method of Bhagavantum and Venkatarayudu (BV) [30, 31] makes use of the factor-group operations of the crystal to obtain a reducible representation of the factor group. This representation is then decomposed into irreducible representations according to well-established group-theoretical procedures. The symmetries of the normal vibrations and their optical selection rules are obtained from the decomposition. The correlation method, which was developed later by Hornig [32] and by Winston and Halford [33], utilizes the correlations between the various site groups and the factor group of the crystal. The main advantage of the correlation method over the BV method is its relative simplicity and ease of use. Although both methods will be discussed in this section, our principal interest is the correlation analysis of normal vibrations.

In applying the correlation method to layered materials, it is important to exploit the symmetry of the individual layers, which may differ from the symmetry of the crystal. As we shall see in Sections 4 and 5, the analysis of infrared and Raman data sometimes critically depends upon the predominance of the layer symmetry over the crystal symmetry. The layer symmetry can be described in two ways. If an individual layer of the crystal is removed and isolated, its symmetry is given by one of the eighty diperiodic groups in three dimensions, first listed by Wood [34]. These groups* admit symmetry operations in all three dimensions but restrict the periodicity of the lattice to two dimensions. Since an isolated layer extends indefinitely only in the two dimensions parallel to the plane of the layers, its lattice and symmetry group are diperiodic. Zallen *et al.* [24] have analyzed the lattice vibrations of As_2S_3 and As_2Se_3 by means of these diperiodic groups. In the present review we shall use only triperiodic groups to describe the symmetry of the layer. As triperiodic groups imply triperiodic structures, we shall define a pseudolattice that contains only one layer in the primitive unit cell; the real lattice of most crystals contains more than one layer. The pseudolattice (or pseudo-crystal) is composed of identical layers of a given material, stacked directly above or below the preceding layer without rotation or translation along the basal plane. In the direction normal to the layers, the periodicity of this lattice is equal to the width of a single layer plus an interlayer gap. Except for $2H–CdI_2$ (Section 2.1.3) the pseudolattice is obviously nonphysical. We shall call the factor group of the pseudolattice the layer factor group, in order to distinguish it from the factor group of the crystal. The pseudo-crystal provides the rationale for using triperiodic groups throughout a correlation analysis; diperiodic groups, in the opinion of the authors, are better suited to surface structures [38] and to the smectic state of liquid crystals [39]. The pseudocrystal also provides the basis for comparing the Brillouin zones and acoustical phonon dispersion curves in polymorphic materials (see Section 4.1). As these materials have one or more weakly interacting layers in the unit cell, their lattice vibrations can be interpreted in terms of extended and reduced zones. In the direction normal to the plane of the layers, the dispersion curves in the first or 'extended' zone of the pseudocrystal become the dispersion curves in the reduced zone of a given polytype.

The symmetry groups we employ in the correlation method for layered materials are the crystal factor group, the layer factor group, and the site groups of the individual atoms of the lattice, both pseudo and real. For a given factor group there are a limited number of sites having a particular point-group symmetry. These sites are listed for all space groups in the International Tables for X-Ray Crystallography [40]. Correlation tables for the point groups have been compiled by Wilson *et al.* [29] and by Fately *et al.* [41]. In applying the correlation method to the pseudolattice, the site groups of the atoms in this lattice are determined first. The irreducible representations of these site groups are then correlated with those of the layer factor group. The site groups of the atoms in the pseudocrystal may be higher, lower, or the same as the site groups in the real crystal. If the factor group of the crystal is identical to the layer factor group, the site symmetries will be the same; but if the layer factor group has a higher symmetry than the crystal factor group, the site symmetry of certain atoms may be higher in the pseudocrystal than in the real crystal.

As an example of the application of the correlation method, we consider the lattice vibrations of the 2H polytype of MoS_2 (Figure 2.3). In this material the primitive unit

---

* The analogous monoperiodic groups in three dimensions have been described by Tobin [35, 36] and applied to crystals of polymer chains [37]. Although Tobin calls them line groups or one-dimensional groups, the polymer chains are three-dimensional structures. It is the periodicity of the lattice that is restricted to one dimension.

cell contains two identical layers [15]. Each layer contributes three atoms, making a total of six atoms in the primitive unit cell. The space group of the crystal is D_{6h}^4 (P6$_3$/mmc), the layer factor group is D_{3h} ($\bar{6}$m2), and the site groups in both the real crystal and pseudolattice are D_{3h} for the Mo atoms and C_{3v} (3m) for the S atoms. We first look for the representations of the two site groups that transform as one of the translations T_x, T_y, or T_z. The reason for choosing representations with these transformation properties is that any normal mode of the lattice involves a linear combination of vibrational displacements along the x, y, and z axes of the crystal; hence, the representations that label the vibrational modes at a particular site must have the transformation properties of a translation. From Table 2.1 for the site group D_{3h}, we find that the A_2'' and E'

TABLE 2.1

Character tables for the point groups D_{3h} and C_{3v}. The irreducible representations are given in the first column; the characters of the twelve symmetry operations are given in the second (barred) column; the transformation properties of the coordinates x, y, and z and the infinitesimal rotations R_x, R_y, and R_z are given in the third column; and the transformation properties of the bilinear combinations of coordinates are given in the last column.

D_{3h}	E	$2C_3$	$3C_2$	σ_h	$2S_3$	$3\sigma_v$	Linear	Bilinear
A_1'	1	1	1	1	1	1		$x^2 + y^2, z^2$
A_2'	1	1	-1	1	1	-1	R_z	
E'	2	-1	0	2	-1	0	x, y	$x^2 - y^2, xy$
A_1''	1	1	1	-1	-1	-1		
A_2''	1	1	-1	-1	-1	1	z	
E''	2	-1	0	-2	1	0	R_x, R_y	yz, zx

C_{3v}	E	$2C_3$	$3\sigma_v$	Linear	Bilinear
A_1	1	1	1	z	$x^2 + y^2, z^2$
A_2	1	1	-1	R_z	
E	2	-1	0	$x, y; R_x, R_y$	$x^2 - y^2, yz; xy, zx$

representations transform as z and (x, y), respectively. Within the site group C_{3v}, the A_1 and E representations have the required transformation properties. In the correlation diagram of Figure 2.7, these representations are listed under their corresponding site groups in the left-hand column.

The next step is to obtain the correlations between the representations of the layer factor group D_{3h} and the site groups, D_{3h} and C_{3v}. Since the symmetry of the Mo atoms is the same as that of the layer, the A_2'' and E' representations in the two groups obviously correlate with themselves. This is indicated in the correlation diagram of Figure 2.7 by the connecting lines. To find the correlations between C_{3v} and D_{3h}, we must refer to a correlation table. Table 2.2 shows that A_1 in C_{3v} correlates with both A_1' and A_2'' in D_{3h}, and E in C_{3v} correlates with both E' and E'' in D_{3h}. The irreducible representations associated with the lattice vibrations of the pseudocrystal are therefore

$$2A_2'' + 2E' + A_1' + E''. \tag{2.3}$$

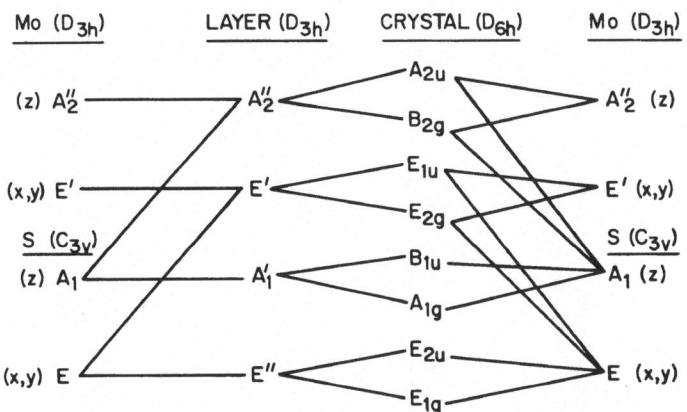

Fig. 2.7. Correlation diagram for 2H–MoS$_2$. The site symmetries of the atoms in the pseudolattice are on the left-hand side of the figure, and the site symmetries of the atoms in the real crystal are on the right-hand side.

TABLE 2.2

Correlation table for the point group D_{3h} and its subgroups.

D_{3h}	C_{3h}	D_3	C_{3v}	$\sigma_h \rightarrow \sigma_v(zy)$ C_{2v}	C_3	C_2	σ_h C_s	σ_v C_s
A_1'	A'	A_1	A_1	A_1	A	A	A_1	A'
A_2'	A'	A_2	A_2	B_2	A	B	A'	A''
E'	E'	E	E	$A_1 + B_2$	E	$A + B$	$2A'$	$A' + A''$
A_1''	A''	A_1	A_2	A_2	A	A	A''	A''
A_2''	A''	A_2	A_1	B_1	A	B	A''	A'
E''	E''	E	E	$A_2 + B_1$	E	$A + B$	$2A''$	$A' + A''$

The number of times an irreducible representation occurs in the decomposition of Equation (2.3) is directly related to the number of vibrational degrees of freedom transferred to the representation from the atomic sites. For example, in Figure 2.7 the A_2'' representation of the layer factor group correlates with the site representations A_2'' and A_1. According to the crystallographic tables [42], one Mo atom occupies the D_{3h} site and the two S atoms occupy the C_{3v} sites. The number of degrees of freedom transferred to A_2'' in the layer factor group therefore equals one from the Mo atom plus another one from the two S atoms (the two degrees of freedom in A_1 divide equally into A_2'' and A_1'). Similarly, the number of degrees of freedom transferred to E' in the layer factor group is two from the Mo atom plus another two from the two S atoms (E divides equally into E' and E''). The irreducible representations, A_2'' and E', therefore occur twice in Equation (2.3). The total number of modes, including possible degeneracies, must add up to $3r$, where r is the number of atoms in the primitive unit cell. In our example we find nine normal modes for the pseudocrystal corresponding to three atoms in the layer.

The correlations between the representations of the layer factor group and the crystal factor group are also readily obtained. Referring to Table 2.3, we find that A_2''

TABLE 2.3

Partial correlation table for the point group D_{6h} and its subgroups. Two distinct correlations exist for each of the subgroups D_{3h} and C_{3v}; however, C'_2 and σ_v are the appropriate operations for 2H-MoS$_2$, as these become the two-fold operation and the plane of symmetry in D_{3h} and C_{3v}, respectively.

D_{6h}	C'_2 / D_{3h}	C''_2 / D_{3h}	C_{6v}	C_{6h}	C'_2 / D_{3d}	C''_2 / D_{3d}	D_{2h}	$\sigma_h \to \sigma(xy)$, $\sigma_v \to \sigma(yz)$ / C_6	C_{3h}	C'_2 / D_3	C''_2 / D_3	σ_v / C_{3v}	σ_d / C_{3v}	S_6	D_2
A_{1g}	A'_1	A'_1	A_1	A_g	A_{1g}	A_{1g}	A_g	A	A'	A_1	A_1	A_1	A_1	A_g	A
A_{2g}	A'_2	A'_2	A_2	A_g	A_{2g}	A_{2g}	B_{1g}	A	A'	A_2	A_2	A_2	A_2	A_g	B_1
B_{1g}	A''_1	A''_2	B_2	B_g	A_{1g}	A_{2g}	B_{2g}	B	A''	A_1	A_2	A_2	A_1	A_g	B_2
B_{2g}	A''_2	A''_1	B_1	B_g	A_{2g}	A_{1g}	B_{3g}	B	A''	A_2	A_1	A_1	A_2	A_g	B_3
E_{1g}	E''	E''	E_1	E_{1g}	E_g	E_g	$B_{2g}+B_{3g}$	E_1	E''	E	E	E	E	E_g	B_2+B_3
E_{2g}	E'	E'	E_2	E_{2g}	E_g	E_g	A_g+B_{1g}	E_2	E'	E	E	E	E	E_g	$A+B_1$
A_{1u}	A''_1	A''_1	A_2	A_u	A_{1u}	A_{1u}	A_u	A	A''	A_1	A_1	A_2	A_2	A_u	A
A_{2u}	A''_2	A''_2	A_1	A_u	A_{2u}	A_{2u}	B_{1u}	A	A''	A_2	A_2	A_1	A_1	A_u	B_1
B_{1u}	A'_2	A'_2	B_1	B_u	A_{1u}	A_{2u}	B_{2u}	B	A'	A_1	A_2	A_1	A_2	A_u	B_2
B_{2u}	A'_1	A'_1	B_2	B_u	A_{2u}	A_{1u}	B_{3u}	B	A'	A_2	A_1	A_2	A_1	A_u	B_3
E_{1u}	E'	E'	E_1	E_{1u}	E_u	E_u	$B_{2u}+B_{3u}$	E_1	E'	E	E	E	E	E_u	B_2+B_3
E_{2u}	E''	E''	E_2	E_{2u}	E_u	E_u	A_u+B_{1u}	E_2	E''	E	E	E	E	E_u	$A+B_1$

correlates with A_{2u} and B_{2g}, E' correlates with E_{1u} and E_{2g}, A'_1 correlates with B_{1u} and A_{1g}, and E'' correlates with E_{2u} and E_{1g}. These correlations are indicated by the connecting lines in Figure 2.7. We note that each representation in the layer splits into a pair of modes in the crystal. Under the inversion operation, one member of the pair is odd and one is even. In Section 4 this property of paired (or conjugate) modes will prove to be important in specifying the displacement vectors of the atoms in a particular normal mode of the crystal. The final decomposition of the representation of the factor group for $2H\text{–}MoS_2$ is

$$2A_{2u} + 2B_{2g} + B_{1u} + A_{1g} + 2E_{1u} + 2E_{2g} + E_{2u} + E_{1g}. \tag{2.4}$$

These irreducible representations could have been obtained directly by correlating the representations of the crystal site group for each atom (Figure 2.7, right-hand column) with the representations of the crystal factor group.

From the character table for D_{6h} (Table 2.4), the optical selection rules for each of the representations in Equation (2.4) can be determined. We find that the A_{2u} and E_{1u}

TABLE 2.4

Character table for the point group D_{6h}.

D_{6h}	E	$2C_6$	$2C_3$	C_2	$3C'_2$	$3C''_2$	i	$2S_3$	$2S_6$	σ_h	$3\sigma_d$	$3\sigma_v$	Linear	Bilinear
A_{1g}	1	1	1	1	1	1	1	1	1	1	1	1		$x^2 + y^2, x^2$
A_{2g}	1	1	1	1	-1	-1	1	1	1	1	-1	-1	R_z	
B_{1g}	1	-1	1	-1	1	-1	1	-1	1	-1	1	-1		
B_{2g}	1	-1	1	-1	-1	1	1	-1	1	-1	-1	1		
E_{1g}	2	1	-1	-2	0	0	2	1	-1	-2	0	0	R_x, R_y	yz, zx
E_{2g}	2	-1	-1	2	0	0	2	-1	-1	2	0	0		$x^2 - y^2, xy$
A_{1u}	1	1	1	1	1	1	-1	-1	-1	-1	-1	-1		
A_{2u}	1	1	1	1	-1	-1	-1	-1	-1	-1	1	1	z	
B_{1u}	1	-1	1	-1	1	-1	-1	1	-1	1	-1	1		
B_{2u}	1	-1	1	-1	-1	1	-1	1	-1	1	1	-1		
E_{1u}	2	1	-1	-2	0	0	-2	-1	1	2	0	0	x, y	
E_{2u}	2	-1	-1	2	0	0	-2	1	1	-2	0	0		

representations transform as z and (x, y), respectively. The three long-wavelength acoustical modes correspond to rigid translations of the lattice; therefore, one of the A_{2u} and one of the E_{1u} representations are associated with these modes. In addition, since the infrared-active modes involve an electric dipole transition (Section 3.1), they too must transform in the same way as a translation. The remaining A_{2u} and E_{1u} representations in Equation (2.4) are associated with these infrared-active modes. For a mode to be Raman-active (Section 3.5), it must transform in the same way as bilinear combinations of the coordinates x, y, and z. Table 2.4 indicates that the E_{2g}, A_{1g}, and E_{1g} modes are all Raman-active.

The correlation method also yields information concerning the specific atoms involved in the vibrations and their displacement polarizations. If we consider the representations of the pseudolattice, we see that the A'_1 and E'' modes in the layer derive from representations of the sulfur site group. This implies that in the A'_1 and E''

modes only the sulfur atoms are in motion. The same argument carries over to the crystal: the B_{1u}, A_{1g}, E_{2u}, and E_{1g} modes involve the motion of sulfur atoms alone. Furthermore, in the pseudocrystal the Mo and S atoms vibrate together in the A_2'' and E' vibrations, because each of these representations originates from representations at the Mo and S sites. Applying the argument to the crystal again, we see that the A_{2u}, B_{2g}, E_{1u}, and E_{2g} modes involve the combined motion of the Mo and S atoms. (It should be mentioned that it is not always possible, as it was in our example, to identify modes that involve only the motion of particular atoms.) The nondegenerate A and B modes in the crystal originate from A representations at the atomic sites (Figure 2.7). Since these representations transform as the translation T_z, the non-degenerate modes in the crystal have their atomic displacements in the z direction, which is the c-axis of the crystal. The E modes, on the other hand, originate from E representations at the atomic sites, which transform as (x, y). Thus all of these doubly degenerate modes have displacements in the basal plane. The simple polarization breakup that occurs in the case of 2H–MoS$_2$ is due to the uniaxial symmetry of the crystal (see Table 2.5).

TABLE 2.5

Properties of the long-wavelength lattice vibrations of 2H–MoS$_2$.

Irreducible representation	Number of acoustical modes	Number of optical modes	Activity	Direction of vibration	Atoms involved	Transformation properties
A_{2u}	1	1	Infrared	c-axis	Mo + S	z
B_{2g}		2	Inactive	c-axis	Mo + S	
B_{1u}		1	Inactive	c-axis	S	
A_{1g}		1	Raman	c-axis	S	$x^2 + y^2$, z^2
E_{1u}	1	1	Infrared	Basal plane	Mo + S	x, y
E_{2g}		2	Raman	Basal plane	Mo + S	$x^2 - y^2$, xy
E_{2u}		1	Inactive	Basal plane	S	
E_{1g}		1	Raman	Basal plane	S	yz, zx

In order to underline the simplicity and elegance of the correlation method, we shall compare it with the BV method for 2H–MoS$_2$. The basic idea of this method is that certain atoms in the structure are left invariant under the factor-group operations. In order to determine which atoms remain invariant, all of the factor-group operations must be applied to each atom in the unit cell. After finding the number of atoms left invariant by each factor-group operation, a reducible representation of the group of the wave vector is formed. Since we are here concerned with long-wavelength lattice vibrations, we can set the wave vector equal to zero. The group of the wave vector is then isomorphic with the point group of the crystal, and the reducible representation can be decomposed according to the well-known formula of group theory [43]

$$a_j = h^{-1} \sum_R \chi^{(j)}(R)^* \, \chi(R), \qquad (2.5)$$

where a_j is the number of times the irreducible representation $\Gamma^{(j)}$ appears in the decomposition, h is the order of the group, $\chi^{(j)}(R)^*$ is the character of $\Gamma^{(j)}$, and $\chi(R)$ is the character of the reducible representation Γ. The summation R runs over all the elements or operations of the group.

Since the space group of 2H–MoS$_2$ is D_{6h}^4, nonprimitive translations are associated with half of the point-group operations; moreover, the order of the factor group is 24. The point-group operations for 2H–MoS$_2$ can be specified by the simple notation of Warren [44], in which an n-fold rotational axis is denoted by a coefficient, and an axis of rotation is denoted by a crystallographic direction. A positive sense of direction is defined by the right-hand rule. For example, a three-fold rotation around the hexagonal [001] axis would be written as 3[001]. Similarly, a rotation-inversion operation (rotation followed by inversion through the origin) would be written as $\bar{3}$[001]. The direction specified in these examples is the positive z axis. A three-fold rotation in the opposite sense would be written as 3[00$\bar{1}$], and the corresponding rotation-inversion operation would be $\bar{3}$[00$\bar{1}$]. For a general nonsymmorphic space group, nonprimitive translations are combined with one or more point-group operations. In 2H–MoS$_2$ there is only one nonprimitive vector, $\tau_0 = (0, 0, \frac{1}{2})$. The 24 factor-group operations for D_{6h} are listed in Table 2.6, and the effect of each operation on an arbitrary point

TABLE 2.6

Factor-group operations of 2H–MoS$_2$ and their effects upon an arbitrary point in the unit cell (hexagonal coordinate system [45]).

Factor-group operation	Schoenflies symbol	Transformed coordinates
1	E	x, y, z
6[001] + τ_0	C_6	$x - y, x, z + \frac{1}{2}$
6[00$\bar{1}$] + τ_0	C_6	$y, y - x, z + \frac{1}{2}$
3[001]	C_3	$\bar{y}, x - y, z$
3[00$\bar{1}$]	C_3	$y - x, \bar{x}, z$
2[001] + τ_0	C_2	$\bar{x}, \bar{y}, z + \frac{1}{2}$
2[210] + τ_0	C_2'	$x, x - y, \frac{1}{2} - z$
2[120] + τ_0	C_2'	$y - x, y, \frac{1}{2} - z$
2[$\bar{1}$20] + τ_0	C_2'	$\bar{y}, \bar{x}, \frac{1}{2} - z$
2[100]	C_2''	$x - y, \bar{y}, \bar{z}$
2[110]	C_2''	$y, x, \bar{z}$
2[010]	C_2''	$\bar{x}, y - x, \bar{z}$
$\bar{1}$	i	$\bar{x}, \bar{y}, \bar{z}$
$\bar{6}$[001] + τ_0	S_3	$y - x, \bar{x}, \frac{1}{2} - z$
$\bar{6}$[00$\bar{1}$] + τ_0	S_3	$\bar{y}, x - y, \frac{1}{2} - z$
$\bar{3}$[001]	S_6	$y, y - x, \bar{z}$
$\bar{3}$[00$\bar{1}$]	S_6	$x - y, x, \bar{z}$
$\bar{2}$[001] + τ_0	σ_h	$x, y, \frac{1}{2} - z$
$\bar{2}$[210] + τ_0	σ_d	$\bar{x}, y - x, z + \frac{1}{2}$
$\bar{2}$[120] + τ_0	σ_d	$x - y, \bar{y}, z + \frac{1}{2}$
$\bar{2}$[$\bar{1}$20] + τ_0	σ_d	$y, x, z + \frac{1}{2}$
$\bar{2}$[100]	σ_v	$y - x, y, z$
$\bar{2}$[110]	σ_v	$\bar{y}, \bar{x}, z$
$\bar{2}$[010]	σ_v	$x, x - y, z$

(x, y, z) in the unit cell is indicated in the right-hand column. The equilibrium position vectors of the six atoms in the unit cell are given in Table 2.7, referenced with respect to the origin of the unit cell [45]. The index κ runs from one to six and identifies the individual atoms.

TABLE 2.7

Equilibrium position vectors for the six atoms of the primitive unit cell of 2H–MoS$_2$. The origin is chosen at the inversion center $\bar{3}$m1, [45], the x-axis coincides with the hexagonal [100] direction in the crystal, and the components along x (or y) and z are given as fractions of the unit-cell dimensions, a and c, respectively.

Atom	κ	$\mathbf{X}(\kappa)$		
Mo	1	0,	$\sqrt{3}/3$,	1/4
Mo	2	1/2,	$\sqrt{3}/6$,	$-1/4$
S	3	1/2,	$\sqrt{3}/6$,	0.371
S	4	1/2,	$\sqrt{3}/6$,	0.129
S	5	0,	$\sqrt{3}/3$,	-0.129
S	6	0,	$\sqrt{3}/3$,	-0.371

The factor group operations are now applied in succession to each atomic position, in order to determine the interchanges that occur among the atoms. As the factor group does not involve primitive translations, we are at liberty to refer atoms transferred into a neighboring unit cell back to the original unit cell. We find, for example, that the factor-group operation 2[001] $+ \tau_0$ changes $1 \rightarrow 2$, $2 \rightarrow 1$, $3 \rightarrow 5$, $4 \rightarrow 6$, $5 \rightarrow 3$, and $6 \rightarrow 4$ (Table 2.8). Since no atom is left invariant, the character $\chi(R)$ of the representation of this operation is zero. In general the character of each operation is given by [31]

$$\chi(R) = \eta(R)[2 \cos \varphi \pm 1], \tag{2.6}$$

where $\eta(R)$ is the number of atoms left invariant by the operation R, φ is the angle of an n-fold axis of symmetry ($\varphi = 360°/n$), the plus sign applies to pure rotations, and the minus sign applies to rotation-reflection operations. Simple reflections or inversions correspond to $\varphi = 0°$ or $\varphi = 180°$, respectively. The 24 symmetry operations of the factor group and the values of $\eta(R)$ and $\chi(R)$ are given in Table 2.8. Using the characters of this table and the characters of the irreducible representations in Table 2.4, we obtain from Equation (2.5)

$$a(A_{1g}) = \tfrac{1}{24}[18 - 6 - 8 + 2 + 18] = 1,$$

$$a(A_{2g}) = \tfrac{1}{24}[18 + 6 - 8 + 2 - 18] = 0,$$

$$a(B_{1g}) = \tfrac{1}{24}[18 - 6 + 8 - 2 - 18] = 0,$$

$$a(B_{2g}) = \tfrac{1}{24}[18 + 6 + 8 - 2 + 18] = 2,$$

$$a(E_{1g}) = \tfrac{1}{24}[2(18) + 0 - 8 - 2(2) + 0] = 1,$$

$$a(E_{2g}) = \tfrac{1}{24}[2(18) + 0 + 8 + 2(2) + 0] = 2,$$

$$a(A_{1u}) = \tfrac{1}{24}[18 - 6 + 8 - 2 - 18] = 0,$$

$$a(A_{2u}) = \tfrac{1}{24}[18 + 6 + 8 - 2 + 18] = 2,$$

$$a(B_{1u}) = \tfrac{1}{24}[18 - 6 - 8 + 2 + 18] = 1,$$

$$a(B_{2u}) = \tfrac{1}{24}[18 + 6 - 8 + 2 - 18] = 0,$$

$$a(E_{1u}) = \tfrac{1}{24}[2(18) + 0 + 8 + 2(2) + 0] = 2,$$

and

$$a(E_{2u}) = \tfrac{1}{24}[2(18) + 0 - 8 - 2(2) + 0] = 1. \tag{2.7}$$

The irreducible representations of the factor group are therefore

$$2A_{2u} + 2B_{2g} + B_{1u} + A_{1g} + 2E_{1u} + 2E_{2g} + E_{2u} + E_{1g}, \tag{2.8}$$

which is the same set of irreducible representations obtained by the correlation method (Figure 2.7 and Equation (2.4)).

TABLE 2.8

Interchanges of the atoms in the primitive unit cell of $2H-MoS_2$ that are produced by the twenty-four factor-group operations.

Factor-group operation	$Mo(\kappa)$ 1	2	$S(\kappa)$ 3	4	5	6	$\eta(R)$	$\chi(R)$
1	1	2	3	4	5	6	6	18
$6[001] + \tau_0$	2	1	5	6	3	4	0	0
$6[00\bar{1}] + \tau_0$	2	1	5	6	3	4	0	0
$3[001]$	1	2	3	4	5	6	6	0
$3[00\bar{1}]$	1	2	3	4	5	6	6	0
$2[001] + \tau_0$	2	1	5	6	3	4	0	0
$2[210] + \tau_0$	1	2	4	3	6	5	2	-2
$2[120] + \tau_0$	1	2	4	3	6	5	2	-2
$2[\bar{1}20] + \tau_0$	1	2	4	3	6	5	2	-2
$2[100]$	2	1	6	5	4	3	0	0
$2[110]$	2	1	6	5	4	3	0	0
$2[010]$	2	1	6	5	4	3	0	0
$\bar{1}$	2	1	6	5	4	3	0	0
$\bar{6}[001] + \tau_0$	1	2	6	5	4	3	2	-4
$\bar{6}[00\bar{1}] + \tau_0$	1	2	6	5	4	3	2	-4
$\bar{3}[001]$	2	1	6	5	4	3	0	0
$\bar{3}[00\bar{1}]$	2	1	6	5	4	3	0	0
$\bar{2}[001] + \tau_0$	1	2	4	3	6	5	2	2
$\bar{2}[210] + \tau_0$	2	1	5	6	3	4	0	0
$\bar{2}[120] + \tau_0$	2	1	5	6	3	4	0	0
$\bar{2}[\bar{1}20] + \tau_0$	2	1	5	6	3	4	0	0
$\bar{2}[100]$	1	2	3	4	5	6	6	6
$\bar{2}[110]$	1	2	3	4	5	6	6	6
$\bar{2}[010]$	1	2	3	4	5	6	6	6

2.3. CORRELATION DIAGRAMS

In this section we shall complete the correlation diagrams and normal-mode decompositions for the layered materials reviewed as families in Section 2.1. The discussion of 2H–MoS$_2$ in Section 2.2 serves as an example of the vibrational information that can be obtained by the correlation method. Layered materials not specifically mentioned in this section but discussed in Section 5 have correlation diagrams and normal-mode decompositions identical to one of the following prototypes.

2.3.1. β–GaS

The space group for β–GaS is D_{6h}^4 (P6$_3$/mmc) [46], and the primitive unit cell contains two layers or eight atoms (Figure 2.2). The site group for the Ga and Se atoms in the pseudolattice and in the crystal is C_{3v} (3m). Correlations between the representations of the site group, layer factor group, and crystal factor group are shown in Figure 2.8.

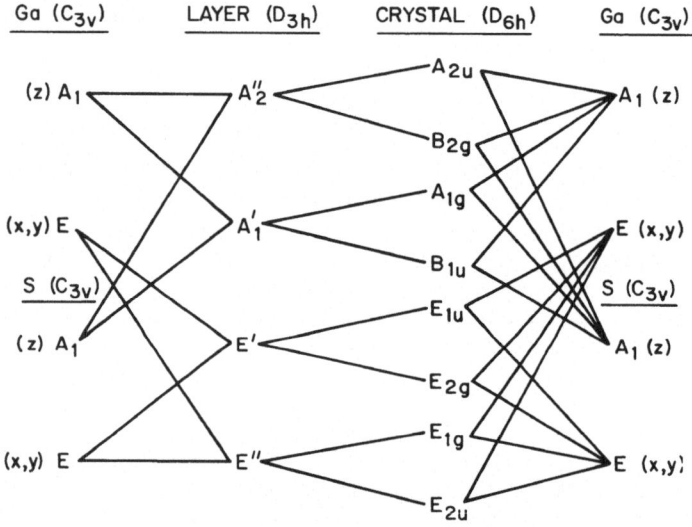

Fig. 2.8.   Correlation diagram for β-GaS.

The decomposition of the representation of the factor group into irreducible representations is

$$2A_{2u} + 2B_{2g} + 2B_{1u} + 2A_{1g} + 2E_{1u} + 2E_{2g} + 2E_{2u} + 2E_{1g}. \tag{2.9}$$

The modes are characterized in Table 2.9. Because of the identical site symmetries, the Ga and Se atoms are in motion in all 24 normal modes.

2.3.2. ε–GaSe

The space group for ε–GaSe is D_{3h}^1 (P6̄m2) [47, 11], and the primitive unit cell contains two layers or eight atoms (Figure 2.2). The symmetry of the pseudolattice is the same as that of the crystal, and the site group is C_{3v} (3m) for both Ga and Se. Correlations

TABLE 2.9

Properties of the long-wavelength lattice vibrations of β-GaS.

Irreducible representation	Number of acoustical modes	Number of optical modes	Activity	Direction of vibration	Atoms involved	Transformation properties
A_{2u}	1	1	Infrared	c-axis	Ga + S	z
B_{2g}		2	Inactive	c-axis	Ga + S	
B_{1u}		2	Inactive	c-axis	Ga + S	
A_{1g}		2	Raman	c-axis	Ga + S	$x^2 + y^2, z^2$
E_{1u}	1	1	Infrared	Basal plane	Ga + S	x, y
E_{2g}		2	Raman	Basal plane	Ga + S	$x^2 - y^2, xy$
E_{2u}		2	Inactive	Basal plane	Ga + S	
E_{1g}		2	Raman	Basal plane	Ga + S	yz, zx

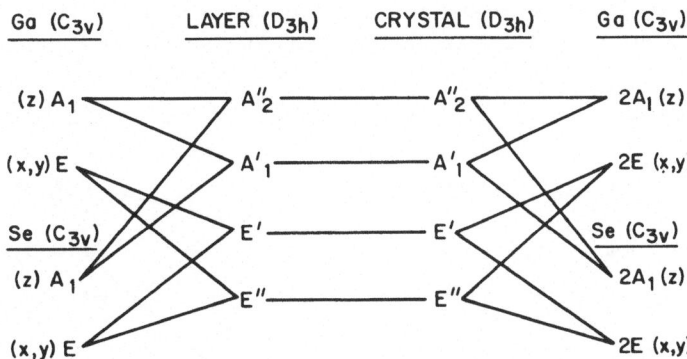

Fig. 2.9.   Correlation diagram for ε-GaSe.

to the layer and crystal factor groups are shown in Figure 2.9.* The decomposition into irreducible representations is

$$4A_2'' + 4A_1' + 4E' + 4E''. \tag{2.10}$$

The modes are characterized in Table 2.10. As in β–GaS, the Ga and Se atoms are in motion in all 24 normal modes. The E' modes are both infrared- and Raman-active, because of the lack of an inversion center in the ε polytype.

---

* In front of each site-group representation of the crystal (right-hand column of Figure 2.9) is the number 2, which means that 2 sets of atoms (Ge or Se) in the equivalent positions of C_{3v} are required to account for the eight atoms of the unit cell. These coefficients can be used to determine the number of times an irreducible representation occurs in a decomposition. For example, in ε-GaSe the two equivalent positions with C_{3v} symmetry in D_{3h}^1 [42] are multiplied by the coefficient 2, so that $2A_1$ has four degrees of freedom instead of two, and $2E$ has eight degrees of freedom instead of four. This explains the multiplicity of each irreducible representation in Equation (2.10).

TABLE 2.10

Properties of the long-wavelength lattice vibrations of ε-GaSe.

Irreducible representation	Number of acoustical modes	Number of optical modes	Activity	Direction of vibration	Atoms involved	Transformation properties
A_2''	1	3	Infrared	c-axis	Ga + Se	z
A_1'		4	Raman	c-axis	Ga + Se	$x^2 + y^2, z^2$
E'	1	3	Infrared + Raman	Basal plane	Ga + Se	$x, y; x^2 - y^2, xy$
E''		4	Raman	Basal plane	Ga + Se	yz, zx

2.3.3. γ–GaSe

The space group for γ–GaSe is C_{3v}^5 (R3m) [47], and the rhombohedral primitive unit cell contains three layers but only one molecule (Figure 2.2). In the pseudolattice and in the crystal, the Ga and Se atoms have C_{3v} (3m) site symmetry. The correlations to the layer factor group and crystal factor group are shown in Figure 2.10. When

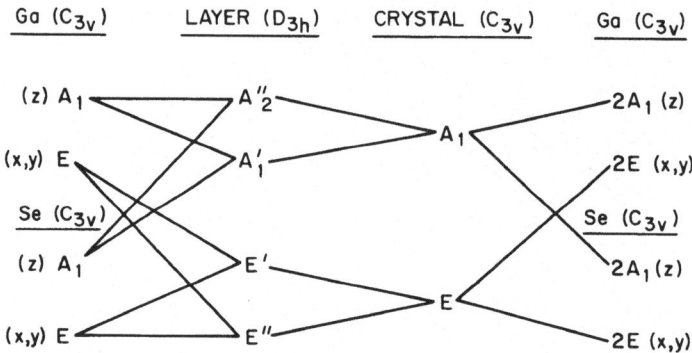

Fig. 2.10.   Correlation diagram for γ-GaSe.

discussing 2H–MoS$_2$, we found that the correlations between the layer and the crystal factor groups led to correlation splittings of the irreducible representations. In γ–GaSe the reverse is true: instead of correlation splittings, we find several confluences of the representations of the pseudolattice. Confluences will always occur when the symmetry of the pseudolattice is higher than that of the crystal. The decomposition into irreducible representations is

$$4A_1 + 4E. \tag{2.11}$$

Because of the low crystal symmetry, the normal modes parallel (or perpendicular) to the c-axis are indistinguishable by optical experiments (Table 2.11). However, in Section 4 we shall introduce physical arguments that will facilitate the selection-rule analysis of the data.

TABLE 2.11

Properties of the long-wavelength lattice vibrations of γ-GaSe.

Irreducible representation	Number of acoustical modes	Number of optical modes	Activity	Direction of vibration	Atoms involved	Transformation properties
A_1	1	3	Infrared + Raman	c-axis	Ga + Se	$z; x^2 + y^2, z^2$
E	1	3	Infrared + Raman	Basal plane	Ga + Se	$x, y; x^2 - y^2, yz; xy, zx$

2.3.4. 3R– and 3T–MoS₂

The space group of 3R–MoS$_2$ is C_{3v}^5 (R3m) [15], and as in γ–GaSe the primitive unit cell contains only one molecule (Figure 2.3). In the real crystal the Mo and S atoms have C_{3v} symmetry; in the pseudocrystal the site groups are D_{3h} ($\overline{6}$m2) for Mo and C_{3v} for S. Figure 2.11 shows the correlations to the layer and crystal factor groups,

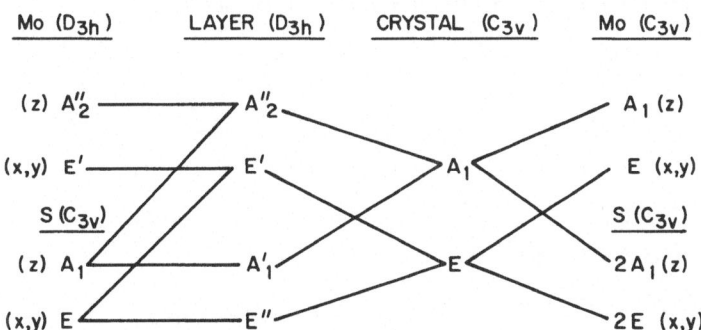

Fig. 2.11.   Correlation diagram for 3R–MoS₂.

and we note that confluences occur between the A_2'' and A_1' representations and between the E' and E'' representations. The decomposition into irreducible representations is

$$3A_1 + 3E. \tag{2.12}$$

The modes are characterized in Table 2.12. 3T–MoS$_2$, the only other three-layer polytype of MoS$_2$ [17], has the first two layers of the unit cell stacked as in 3R–MoS$_2$ (Figure 2.3); the third or lowest layer has the molybdenum atom positioned under the sulfur atoms of the second layer (2H–MoS$_2$ stacking). The space group of 3T–MoS$_2$ is C_{3v}^1 (P3m1), and the primitive unit cell contains three molecules. Equation (2.12), Figure 2.11, and Table 2.12 can all be applied to 3T–MoS$_2$, provided that we multiply the number of modes of each symmetry type by three. Although these two polytypes have not been studied experimentally, we include them in the correlation analyses because they illustrate certain physical principles discussed at length in Section 4.

TABLE 2.12

Properties of the long-wavelength lattice vibrations of 3R–MoS$_2$.

Irreducible representation	Number of acoustical modes	Number of optical modes	Activity	Direction of vibration	Atoms involved	Transformation properties
A_1	1	2	Infrared + Raman	c-axis	Mo + S	z; $x^2 + y^2$, z^2
E	1	2	Infrared + Raman	Basal plane	Mo + S	x, y; $x^2 - y^2$, yz; xy, zx

2.3.5. 2H–CdI$_2$

The space group of 2H–CdI$_2$ is D_{3d}^3 (P$\overline{3}$m1) [48], and the primitive unit cell contains one layer or three atoms (Figure 2.4). The site groups of the atoms are D_{3d} ($\overline{3}$m) for Cd and C_{3v} (3m) for I. Correlations to the layer factor group, which is the same as the crystal factor group (the pseudolattice and real crystal are identical), are shown in Figure 2.12. The decomposition into irreducible representations is

$$2A_{2u} + A_{1g} + 2E_u + E_g. \tag{2.13}$$

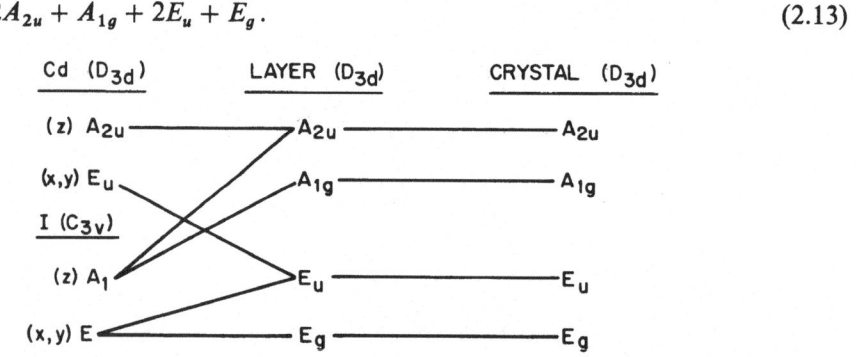

Fig. 2.12. Correlation diagram for 2H–CdI$_2$.

The modes are characterized in Table 2.13. Since the Cd atoms are located at centers of inversion, the even modes A_{1g} and E_g involve the motion of only I atoms. Equation (2.13) and Table 2.13 also apply to the CdCl$_2$ structure.

2.3.6. 4H–CdI$_2$

The space group of 4H–CdI$_2$ is C_{6v}^4 (P6$_3$mc) [49], and the primitive unit cell contains two layers or six atoms (Figure 2.4). The site group of the atoms in the crystal is C_{3v} (3m) for both Cd and I. Correlations to the layer and crystal factor groups are shown in Figure 2.13. A special problem arises in this correlation diagram, because the factor group D_{3d} is not a subgroup of C_{6v} (or vice versa); however, the correspondences between the representations in D_{3d} and C_{6v} can be determined from the transformation properties of the linear and bilinear combinations of coordinates, listed in Tables 2.13

TABLE 2.13

Properties of the long-wavelength lattice vibrations of 2H–CdI$_2$.

Irreducible representation	Number of acoustical modes	Number of optical modes	Activity	Direction of vibration	Atoms involved	Transformation properties
A_{2u}	1	1	Infrared	c-axis	Cd + I	z
A_{1g}		1	Raman	c-axis	I	$x^2 + y^2, z^2$
E_u	1	1	Infrared	Basal plane	Cd + I	x, y
E_g		1	Raman	Basal plane	I	$x^2 - y^2, yz;$ $xy, zx; R_x, R_y$

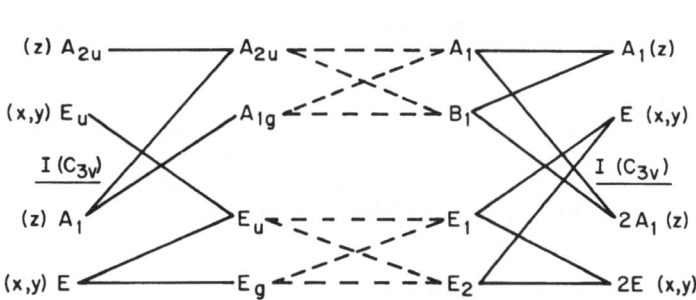

Fig. 2.13.   Correlation diagram for 4H–CdI$_2$.

and 2.14. Since the A_{2u} and A_{1g} modes in D_{3d} transform as z and $(x^2 + y^2, z^2)$, respectively, these two modes must become the A_1 modes in C_{6v}, which have the same transformation properties. To obtain the correct number of representations in C_{6v}, A_{2u} must also go over into B_1, and A_{1g} into B_1. Similar arguments apply to the basal-plane modes, E_u and E_g. Thus the decomposition into irreducible representations is

$$3A_1 + 3B_1 + 3E_1 + 3E_2. \tag{2.14}$$

TABLE 2.14

Properties of the long-wavelength lattice vibrations of 4H–CdI$_2$.

Irreducible representation	Number of acoustical modes	Number of optical modes	Activity	Direction of vibration	Atoms involved	Transformation properties
A_1	1	2	Infrared + Raman	c-axis	Cd + I	$z; x^2 + y^2, z^2$
B_1		3	Inactive	c-axis	Cd + I	
E_1	1	2	Infrared + Raman	Basal plane	Cd + 1	$x, y; yz, zx$
E_2		3	Raman	Basal plane	Cd + I	$x^2 - y^2, xy$

Since the Cd atoms are no longer located at centers of inversion, all of the normal modes involve the combined motion of Cd and I atoms.

2.3.7. *Graphite*

Although there has been some disagreement in the literature concerning the space group of graphite [21, 20], the accepted group appears to be D_{6h}^4 (P6$_3$/mmc), and not C_{6v}^4 (P6$_3$mc). The primitive unit cell contains two layers or four atoms (Figure 2.4). The symmetry of the pseudolattice is the same as that of the crystal, and the site group of the carbon atoms is D_{3h}. Correlations to the layer and crystal factor groups are shown in Figure 2.14, which indicates a simple doubling of the number of modes

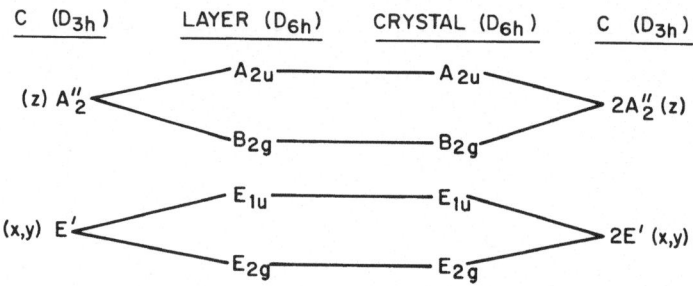

Fig. 2.14.   Correlation diagram for graphite.

in the crystal, but no change in the symmetry species. The decomposition into irreducible representations is

$$2A_{2u} + 2B_{2g} + 2E_{1u} + 2E_{2g}. \qquad (2.15)$$

The modes are characterized in Table 2.15.

TABLE 2.15

Properties of the long-wavelength lattice vibrations of hexagonal graphite.

Irreducible representation	Number of acoustical modes	Number of optical modes	Activity	Direction of vibration	Atoms involved	Transformation properties
A_{2u}	1	1	Infrared	c-axis	C	z
B_{2g}		2	Raman	c-axis	C	
E_{1u}	1	1	Infrared	Basal plane	C	x, y
E_{2g}		2	Raman	Basal plane	C	$x^2 - y^2, xy$

2.3.8. *As₂S₃*

The space group of the monoclinic crystal As_2S_3 is C_{2h}^5 (P2$_1$/n) [22], and the primitive unit cell contains two layers or 20 atoms (four molecular units; see Figure 2.5). The site group is C_1 (1) for the As and S atoms in both the pseudolattice and real crystal; the space group of the pseudolattice is C_s^2 (Pn). Correlations to the layer and crystal

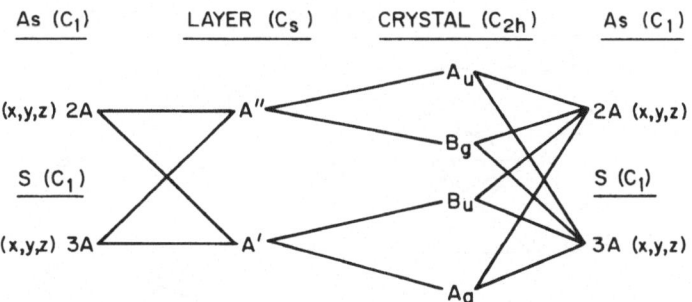

Fig. 2.15. First correlation diagram for As_2S_3. The site symmetries of the atoms in the monoclinic pseudolattice are on the left-hand side of the figure, and the site symmetries of the atoms in the real crystal are on the right-hand side.

factor groups are shown in Figure 2.15. The decomposition into irreducible representations is

$$15A_u + 15B_g + 15B_u + 15A_g. \tag{2.16}$$

Because of the inversion center located midway between the layers, the modes of the layer divide into pairs or even and odd modes in the crystal. Their optical activities are given in Table 2.16, where we have replaced the coordinates x, y, and z in the

TABLE 2.16

Properties of the long-wavelength lattice vibrations of As_2S_3.

Irreducible representation	Number of acoustical modes	Number of optical modes	Activity	Direction of vibration	Atoms involved	Transformation properties
A_u	1	14	Infrared		As + S	y
B_g		15	Raman		As + S	xy, yz
B_u	2	13	Infrared		As + S	x, z
A_g		15	Raman		As + S	x^2, y^2, z^2, zx

character table [50] by the coordinates z, x, and y, respectively, in order to conform to the 'second setting' for monoclinic As_2S_3 (b is the unique axis) [51].

Zallen *et al.* [24] followed a different approach to the correlation analysis of As_2S_3. Instead of assigning the space group C_s^2 to the pseudocrystal, they displaced the atoms of the layers slightly, by an amount no greater than the reported experimental error in the X-ray data, and obtained the orthorhombic group C_{2v}^7 ($Pmn2_1$). (Actually, Zallen and coworkers used the diperiodic group [34] DG32 ($Pnm2_1$) for the isolated As_2S_3 layer.) The site groups within the orthorhombic pseudocrystal are now C_1 (1) for four of the As atoms, C_1 (1) for four of the S atoms, and C_s (*m*) for two of the S atoms (the orthorhombic group has a mirror plane at two of the S atoms in

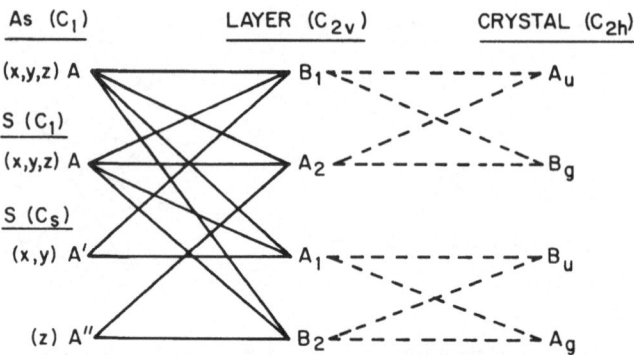

Fig. 2.16.   Second correlation diagram for As_2S_3. The site symmetries of the atoms in the orthorhombic pseudolattice are on the left-hand side of the figure.

the unit cell). Figure 2.16 shows the modified correlation diagram. The decomposition
of the representation of the *layer* factor group into irreducible representations is

$$8B_1 + 7A_2 + 8A_1 + 7B_2.$$ (2.17)

The modes are characterized in Table 2.17, where we have interchanged the x and y
coordinates in the character table [52] in order to conform to our orthorhombic axes
for As_2S_3. Since C_{2v} is not a subgroup of C_{2h} (or vice versa), we have a similar situation to that in 4H–CdI_2 (Section 2.3.6). The correspondences between the representations in C_{2v} and C_{2h} are again determined by the transformation properties of the

TABLE 2.17

Properties of the long-wavelength lattice vibrations of the pseudocrystal of As_2S_3.

Irreducible representation	Number of acoustical modes	Number of optical modes	Activity	Direction of vibration	Atoms involved	Transformation properties
B_1	1	7	Infrared + Raman		As + S	$y; yz$
A_2		7	Raman		As + S	xy
A_1	1	7	Infrared + Raman		As + S	$z; x^2, y^2, z^2$
B_2	1	6	Infrared + Raman		As + S	$x; zx$

coordinates x, y, and z (compare Tables 2.16 and 2.17). These correspondences are the
same as those obtained by Zallen *et al.* [24], who utilized the common subgroup C_s in
C_{2v} and C_{2h}.

2.3.9. Sb_2S_3

The space group of Sb_2S_3 is D_{2h}^{16} (Pnmb) [25, 53], and the primitive unit cell contains
two layers or 20 atoms (four molecular units, Figure 2.5). The layer factor group is

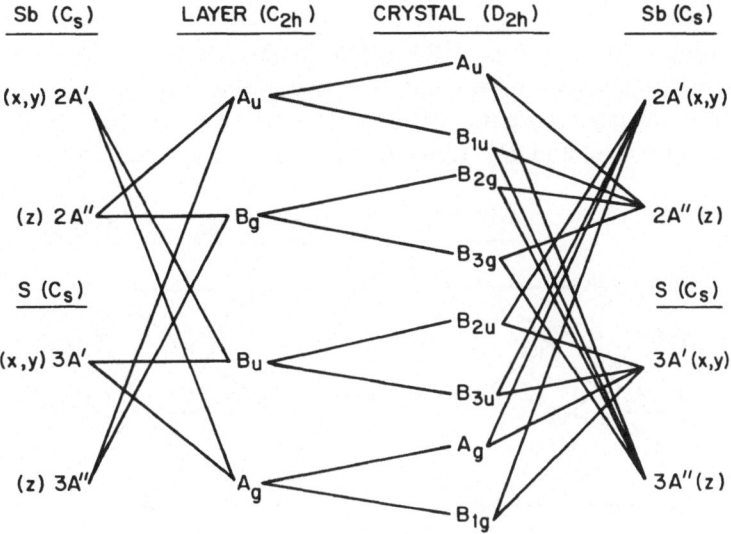

Fig. 2.17.   Correlation diagram for Sb$_2$S$_3$.

C_{2h}, and for all atoms in the pseudolattice and real crystal the site group is C_s. Correlations to the layer and crystal factor groups are shown in Figure 2.17. The decomposition into irreducible representations is

$$5A_u + 5B_{1u} + 5B_{2g} + 5B_{3g} + 10B_{2u} + 10B_{3u} + 10A_g + 10B_{1g}. \qquad (2.18)$$

The modes are characterized in Table 2.18. Because of the center of inversion within each layer, the modes in both the pseudolattice and real crystal are either even or odd.

TABLE 2.18
Properties of the long-wavelength lattice vibrations of Sb$_2$S$_3$.

Irreducible representation	Number of acoustical modes	Number of optical modes	Activity	Direction of vibration	Atoms involved	Transformation properties
A_u		5	Inactive	c-axis	Sb + S	
B_{1u}	1	4	Infrared	c-axis	Sb + S	z
B_{2g}		5	Raman	c-axis	Sb + S	yz
B_{3g}		5	Raman	c-axis	Sb + S	zx
B_{2u}	1	9	Infrared	ab plane	Sb + S	x
B_{3u}	1	9	Infrared	ab plane	Sb + S	y
A_g		10	Raman	ab plane	Sb + S	x^2, y^2, z^2
B_{1g}		10	Raman	ab plane	Sb + S	xy

2.3.10. Bi_2Te_3

The space group of Bi_2Te_3 is D_{3d}^5 (R$\bar{3}$m) [26]. Although the rhombohedral, primitive unit cell spans three layers, it contains only five atoms (Figure 2.6). The symmetries of the layer and the crystal are the same, and the site groups of the atoms are C_{3v} (3m) for two Bi and two Te, and D_{3d} ($\bar{3}$m) for the third Te. Correlations to the layer and

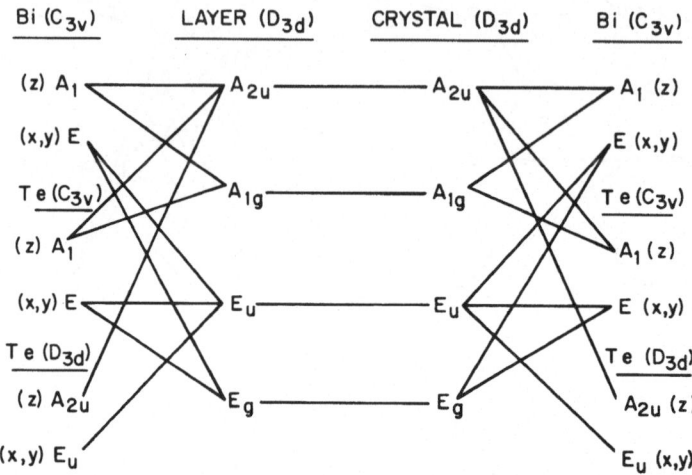

Fig. 2.18.   Correlation diagram for Bi_2Te_3.

crystal factor groups are shown in Figure 2.18. The decomposition into irreducible representations is

$$3A_{2u} + 2A_{1g} + 3E_u + 2E_g. \tag{2.19}$$

The modes are characterized in Table 2.19, and they reflect the odd and even pairing produced by the inversion center at one of the Te atoms (D_{3d}). The odd modes, A_{2u} and E_u, involve the motion of all three sets of equivalent atoms, but the even modes involve only two sets, $Bi(C_{3v})$ and $Te(C_{3v})$.

TABLE 2.19

Properties of the long-wavelength lattice vibrations of Bi_2Te_3.

Irreducible representation	Number of acoustical modes	Number of optical modes	Activity	Direction of vibration	Atoms involved	Transformation properties
A_{2u}	1	2	Infrared	c-axis	Bi + Te	z
A_{1g}		2	Raman	c-axis	Bi + Te(C_{3v})	$x^2 + y^2, z^2$
E_u	1	2	Infrared	Basal plane	Bi + Te	x, y
E_g		2	Raman	Basal plane	Bi + Te(C_{3v})	$x^2 - y^2, yz;$ xy, zx

3. Infrared Absorption and Raman Scattering

Infrared absorption (or reflection) and Raman scattering are the principal experimental techniques for studying long-wavelength phonons in crystalline materials. For a lattice vibration to be infrared-active, the atomic displacements must generate a net electric dipole moment in the primitive unit cell. This moment is associated with the lattice vibration itself and differs fundamentally from the electric moment in the Raman effect, which is induced in the crystal by an external field. If the lattice vibration has the proper symmetry, first-order infrared absorption will occur when the radiation frequency approaches the lattice frequency. The Raman effect, on the other hand, describes the inelastic scattering of photons by phonons (or other elementary excitations of the crystal). The shifts observed in the frequency of the scattered light are characteristic of the vibrational energy states of the material. If ω is the angular frequency of the incident light and ω_j is a vibrational angular frequency, the first-order Raman frequencies are given by $\omega \pm \omega_j$, where the minus sign indicates the so-called Stokes shifts toward lower energies and the plus sign indicates the 'anti-Stokes' shifts toward higher energies. The Raman effect is distinguished from luminescence or fluorescence by the fact that the scattered light tracks the incident light: the shifts are independent of the frequency of the incident light.

3.1. SELECTION RULES FOR INFRARED ABSORPTION

Infrared selection rules and oscillator strengths are governed by the matrix elements of the electric moment

$$p_{\rho,fi} = \int \Psi_f^* p_\rho \Psi_i \, \mathrm{d}V, \tag{3.1}$$

where p_ρ is the x, y, or z component of the moment, Ψ_i and Ψ_f are the time-dependent wave functions of the initial and final states, respectively, the asterisk denotes the complex conjugate of Ψ_f, and $\mathrm{d}V$ is an element of volume in real space. The transition rate for infrared absorption is proportional to the square of the time-independent part of $p_{\rho,fi}$. Infrared transitions are allowed if $\Psi_f^* \Psi_i$ transforms in the same way as p_ρ. The integral in Equation (3.1) is then symmetric and finite. This requirement defines the general selection rule for infrared absorption.

Additional information about selection rules can be obtained by expanding p_ρ in a Taylor series in the oscillator displacements:

$$
\begin{aligned}
p_\rho &= p_\rho^{(0)} + \sum_{n=1}^{3rN} \left(\frac{\partial p_\rho}{\partial u_n}\right)_0 u_n + \sum_{n=1}^{3rN} \sum_{n'=1}^{3rN} \left(\frac{\partial^2 p_\rho}{\partial u_n \partial u_{n'}}\right)_0 u_n u_{n'} + \cdots \\
&\equiv p_\rho^{(0)} + \sum_{\beta \kappa l} p_\rho^{(1)}(\beta\kappa) u_\beta(\kappa l) + \\
&\quad + \sum_{\beta \kappa l} \sum_{\beta' \kappa' l'} p_\rho^{(2)}(\beta\beta', \kappa\kappa', l - l') \, u_\beta(\kappa l) u_{\beta'}(\kappa' l') + \cdots,
\end{aligned}
\tag{3.2}
$$

where $p_\rho^{(0)}$ is the moment when $u_n = u_{n'} = 0$; rN is the number of atoms in the lattice (r is the number of atoms in a primitive unit cell and N is the number of unit cells); and β is the x, y, or z coordinate of the κth atom in the lth unit cell. The total number of values of the triple index $\beta\kappa l$ (or $\beta'\kappa'l'$) is $3rN$. The first partial derivative in Equation (3.2) defines the effective charge

$$e^*(\beta\kappa) = \left(\frac{\partial p_\rho}{\partial u_\beta(\kappa l)}\right)_0 = p_\rho^{(1)}(\beta\kappa). \tag{3.3}$$

We have eliminated the index l in the derivative, as the effective charge cannot depend on the location of the unit cell in the crystal.

Using the theory of small harmonic oscillations [54] and Bloch's theorem for a periodic lattice, we can rewrite Equation (3.2) in a form analogous to the polarizability in Born and Bradburn [55]:

$$p_\rho = p_\rho^{(0)} + \sum_{\beta\kappa, j} \frac{p_\rho^{(1)}(\beta\kappa)}{\sqrt{M_\kappa}} a_\beta(\kappa; j, \mathbf{q} = 0)\xi(j, \mathbf{q} = 0) + \tag{3.4}$$

$$+ \sum_{\substack{\beta\kappa l, \\ \mathbf{q}j}} \sum_{\substack{\beta'\kappa' \\ j'}} \frac{p_\rho^{(2)}(\beta\kappa l; \beta'\kappa')}{\sqrt{M_\kappa M_{\kappa'}}} e^{i\mathbf{q}\cdot\mathbf{x}} a_\beta(\kappa; \mathbf{q}j)a_{\beta'}^*(\kappa'; \mathbf{q}j')\xi(\mathbf{q}j)\xi^*(\mathbf{q}j') + \cdots,$$

where M_κ is the mass of the κth atom in the unit cell, $a_\beta(\kappa; \mathbf{q}j)$ is one of the coefficients of the transformation equations relating the $3rN$ normal coordinates $\xi(\mathbf{q}j)$ to the displacements u_n, $\mathbf{q}$ is an allowed wave vector in the Brillouin zone, j is one of the $3r$ branches of the phonon spectrum, and $\mathbf{x}$ is a primitive translation vector of the lattice. As indicated or implied in Equation (3.4), the second term has values only for $\mathbf{q} = 0$, and the third term has values only for $\mathbf{q}' = -\mathbf{q}$. Substituting Equation (3.4) into Equation (3.1), we obtain

$$p_{\rho, fi} = [\int \Psi_f^* \Psi_i \, dV] \, p_\rho^{(0)} +$$

$$+ \sum_{\beta\kappa, j} [\int \Psi_f^* \xi(j, \mathbf{q} = 0)\Psi_i \, dV] \frac{p_\rho^{(1)}(\beta\kappa)}{\sqrt{M_\kappa}} a_\beta(\kappa; j, \mathbf{q} = 0) +$$

$$+ \sum_{\substack{\beta\kappa l, \\ \mathbf{q}j}} \sum_{\substack{\beta'\kappa' \\ j'}} [\int \Psi_f^* \xi(\mathbf{q}j)\xi^*(\mathbf{q}j')\Psi_i \, dV] \frac{p_\rho^{(2)}(\beta\kappa l; \beta'\kappa')}{\sqrt{M_\kappa M_{\kappa'}}}$$

$$e^{i\mathbf{q}\cdot\mathbf{x}} a_\beta(\kappa; \mathbf{q}j) \, a_{\beta'}^*(\kappa'; \mathbf{q}j') + \cdots. \tag{3.5}$$

The integrals in the last two terms of Equation (3.5) determine the selection rules for first-order and second-order infrared absorption.

When the lattice vibrations are harmonic, the wave functions Ψ_i and Ψ_f represent assemblies of $3rN$ oscillators, each of which is specified by a normal coordinate, frequency, and energy quantum number. To a good approximation, Ψ_i and Ψ_f can be expressed as the product functions

$$\Psi_i = \prod_{n=1}^{3rN} \Psi_n(\xi_n, \omega_n, v_n^{(i)}) \tag{3.6}$$

and

$$\Psi_f = \prod_{n=1}^{3rN} \Psi_n(\xi_n, \omega_n, v_n^{(f)}), \tag{3.7}$$

where v_n is the energy quantum number $(0, 1, 2\ldots)$ of the nth oscillator. It is well-known that the harmonic-oscillator wave functions [56] are either symmetric or antisymmetric, depending on whether the quantum number v_n is even or odd. Thus the first term in Equation (3.5) vanishes unless $v_n^{(f)} = v_n^{(i)}$. Since this term is independent of time, it represents the permanent electric moments of the crystal for $u_n = 0$. First-order transitions are given by the second term, which vanishes unless $v_j^{(f)} = v_j^{(i)} \pm 1$. As $\mathbf{q} = 0$ for these vibrational states, the atoms in all of the unit cells vibrate in phase. The matrix elements vary with time as $e^{\pm i\omega_j t}$, and the transitions involved are funda-mentals. For a particular vibrational mode to be infrared-active, the wave function of the vibration must transform as the dipole term in Equation (3.2), that is, as linear combinations of the coordinates x, y, and z. These transformation properties of the normal modes are conveniently given in the character tables for the crystallographic point groups [57] and are easily identified (see Sections 2.2 and 2.3).

Overtones and combination modes are accounted for by the third and higher terms in Equation (3.5). For example, second harmonics are permitted when the relations $n = n'$ and $v_n^{(f)} = v_n^{(i)} + 2$ are satisfied; binary combinations are permitted when $n \neq n'$, $v_n^{(f)} = v_n^{(i)} + 1$, and $v_{n'}^{(f)} = v_{n'}^{(i)} + 1$. The major contributions to the second-order spec-trum will come from those regions of the Brillouin zone where the phonon density of states is high – regions where the dispersion curves are relatively flat. Although little is known about the general, second-order selection rules for layered materials, the properties of the combinations of phonons located at the center of the Brillouin zone (or at other high symmetry points) are listed in the tables compiled by Herzberg [58]. Moreover, if the lattice vibrations are anharmonic, second-order absorption can be mediated by the cubic and higher terms in the crystal potential. This complementary mechanism for two-phonon absorption has been discussed by Burstein [59].

3.2. ANALYSIS OF INFRARED SPECTRA

The frequency-dependent, complex dielectric function describes the response of a material to infrared radiation. This function can be determined from reflection and transmission measurements by fitting an oscillator model to the experimental data or by employing the dispersion (Kramers-Kronig) relations for complex functions. The oscillator method of analysis was developed by Huang [60] and Born and Huang [61], who applied it to the alkali halides. Barker [62] and Chang et al. [63] have extended the oscillator model to crystals with several infrared-active phonons. The general dielectric function can be written as

$$\varepsilon = \varepsilon_\infty + \sum_j \frac{4\pi\rho_j\omega_j^2}{\omega_j^2 - \omega^2 + i\gamma_j\omega_j\omega}, \tag{3.8}$$

where ε_∞ is the high-frequency dielectric constant ($\omega \gg \omega_j$); and ρ_j, ω_j, and γ_j are the oscillator strength, angular frequency, and damping constant, respectively, of the jth infrared-active mode. Equation (3.8) is the vibrational analogue of the electronic dielectric function, derived in the Drude-Lorentz electron theory of solids [64]. Uniaxial materials have two dielectric functions, one parallel and the other perpendi-cular to the axis of highest symmetry. Many of the layered crystal structures described in Section 2.1 have this type of symmetry.

Since the reflectivity at normal incidence is related to ε through the equation

$$R = \left| \frac{\sqrt{\varepsilon} - 1}{\sqrt{\varepsilon} + 1} \right|^2, \tag{3.9}$$

where the vertical bars denote the magnitude of the complex quantity inside, the infrared vibrational spectrum consists of a series of resonances for each polarization direction of the incident radiation. The parameters ρ_j, ω_j, and γ_j are in practice adjusted to obtain the best oscillator fit to the reflectance data. If the material is transparent at a given infrared frequency, supplementary transmission measurements are usually made to eliminate the errors in the fitting procedure due to reflections from more than one surface of the crystal. For $\omega \ll \omega_j$ Equation (3.8) becomes

$$\varepsilon_0 = \varepsilon_\infty + \sum_j 4\pi\rho_j, \tag{3.10}$$

where ε_0 is the low-frequency dielectric constant.

It is instructive to plot the frequency of a transverse wave, travelling through a dielectric material, as a function of the wave vector q. We choose as an example the layered crystal ε–GaSe, whose vibrational spectrum will be discussed at much greater length in Section 5.1. Although GaSe has three infrared-active lattice vibrations for $E \perp c$ (Table 2.10), we shall limit the dielectric function to the E'^3 resonance at 213.9 cm^{-1} (Table 5.3). The other two E' modes are relatively weak and contribute very little to the dielectric function (see Section 4 and Section 5.1). Since ε in Equation (3.8) is complex, q will also be complex, according to the equation

$$q^2 = \frac{\omega^2}{c^2}\varepsilon, \tag{3.11}$$

where c is the velocity of light in a vacuum. A plot of ω versus q for GaSe is shown in Figure 3.1, where we have used Equations (3.11) and (3.8) and the dispersion

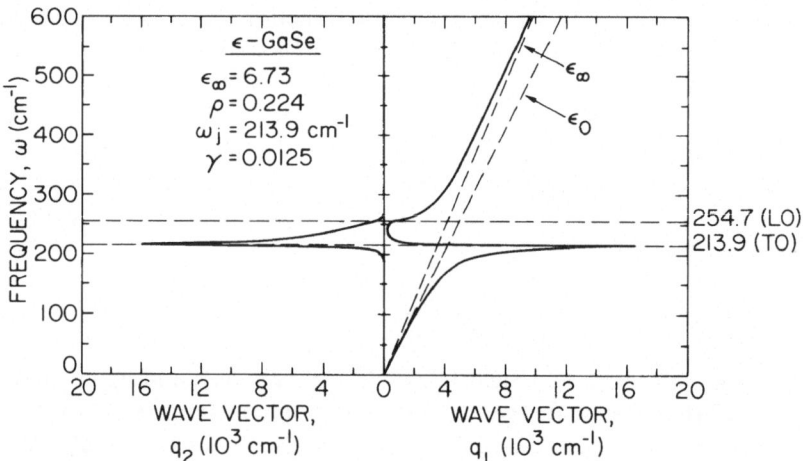

Fig. 3.1.   Polariton dispersion curve for the E'^3 resonance in ε-GaSe.

parameters for the E'^3 mode in Table 5.1 [65]. The real and imaginary parts of the wave vector $q = q_1 - iq_2$, which are related to the complex dielectric function $\varepsilon = \varepsilon_1 - i\varepsilon_2$ through Equation (3.11), are plotted along the abscissa. For values of ω between 200 and 300 cm^{-1}, the wave interacts strongly with the E'^3 phonon. The coupled photon-phonon mode (a polariton) is indicated by the departure of the curve of ω versus q_1 from the straight lines labelled ε_0 and ε_∞. When $\omega \ll \omega_j$ the slope of ω versus q_1 is $c/\sqrt{\varepsilon_0}$; when $\omega \gg \omega_j$ the slope is $c/\sqrt{\varepsilon_\infty}$. As ω increases toward the resonance frequency ω_j, q_1 becomes very large. If γ were equal to zero (no damping), q_1 and q_2 would approach the resonance frequency asymptotically from opposite sides and become infinitely large at ω_j. Between 213.9 and 254.7 cm^{-1} (the longitudinal optical (LO) phonon frequency), ε would be real and negative, q would be imaginary, and wave propagation would cease. Since γ is not zero in our example, the maximum values of q_1 and q_2 occur 0.8 cm^{-1} below and above, respectively, the TO frequency. The wave vector q also maintains a real part throughout the region bounded by the TO and LO phonon frequencies.

In calculating the LO phonon frequency in Figure 3.1, we made use of the Lyddane-Sachs-Teller relation (LST) [66] for a single undamped oscillator ($\gamma = 0$). The LO frequency could also have been obtained by finding the frequency for which the real part of the dielectric function is zero – this is the Drude rule. However, both procedures are theoretically incorrect when $\gamma \neq 0$. Chang et al. [63] have pointed out that, for a crystal with an arbitrary number of damped oscillators, the TO and LO frequencies are the complex values of ω in Equation (3.8) when $\varepsilon = \infty$ and $\varepsilon = 0$, respectively. The TO frequencies are given by*

$$\omega_{\text{TO}}^{(j)} = \omega_j \left[\pm \left(1 - \frac{\gamma_j^2}{4} \right)^{1/2} + i \frac{\gamma_j}{2} \right], \tag{3.12}$$

and the LO frequencies are the roots of the equation

$$\prod_{j=1}^{n} (\omega_j^2 - \omega_{\text{LO}}^2 + i\gamma_j\omega_j\omega_{\text{LO}}) + \sum_{j=1}^{n} \frac{4\pi\rho_j\omega_j}{\varepsilon_\infty} \prod_{\substack{k=1 \\ k \neq j}}^{n} (\omega_k^2 - \omega_{\text{LO}}^2 + i\gamma_j\omega_j\omega_{\text{LO}}) = 0, \tag{3.13}$$

where n is the number of oscillators in the dielectric function. The LST relation becomes

$$\frac{\varepsilon_0}{\varepsilon_\infty} = \prod_{j=1}^{n} \frac{|\omega_{\text{LO}}^{(j)}|^2}{\omega_j^2}. \tag{3.14}$$

When $\gamma_j = 0$, $\omega_{\text{TO}}^{(j)}$ reduces to ω_j, as expected, and the roots $\omega_{\text{LO}}^{(j)}$ are somewhat easier to obtain from Equation (3.13). When the damping constant is small, Chang et al. proposed an alternative approximate procedure to Equation (3.13) for obtaining the LO frequencies. If the real and imaginary parts of the dielectric function are known, the modulus

$$|\varepsilon| = (\varepsilon_1^2 + \varepsilon_2^2)^{1/2} \tag{3.15}$$

can be computed. The maxima and minima of the modulus yield the TO and LO frequencies, respectively. The procedure amounts to replacing the exact conditions

---

* The damping constant in the paper of Chang et al. [63] has the same dimensions as the angular frequency and is equal to $\gamma_j\omega_j$ in Equation (3.12). We have used a dimensionless damping constant and positive frequencies for the plane-wave solutions of the dielectric response (compare Chang et al's Equation (19) with our Equation (3.8)).

for the TO and LO frequencies, $\varepsilon = \infty$ or $\varepsilon = 0$, with the approximate conditions that $|\varepsilon|$ be either a maximum or a minimum.

The second method of analysis of infrared data employs the Kramers-Kronig (KK) dispersion relations. Although this method normally requires a digital computer in order to evaluate the integrations, the results complement the oscillator analysis previously described. The most useful dispersion relations involve the real and imaginary parts of the Fresnel reflection coefficient for normal incidence. The coefficient r is the complex ratio of the amplitudes of the reflected and incident waves. In complex polar form

$$r = \sqrt{R}\, e^{i\theta}, \tag{3.16}$$

where $R = rr^*$ and θ is the phase difference between the waves. From Equations (3.9) and (3.16) we have

$$\varepsilon_1 = \frac{(1 - R)^2 - 4R \sin^2 \theta}{(1 + R - 2\sqrt{R} \cos \theta)^2} \tag{3.17}$$

and

$$\varepsilon_2 = \frac{-4(1 - R)\sqrt{R} \sin \theta}{(1 + R - 2\sqrt{R} \cos \theta)^2}. \tag{3.18}$$

Equations (3.17) and 3.18) are functions of both R and θ, and θ is not easily measured, at least not over a wide range of frequencies; however, a dispersion relation exists between $\ln(\sqrt{R})$ and θ, which can be written as

$$\theta(\omega) = \frac{\omega}{\pi} \int_0^\infty \frac{\ln[R(\omega')] - \ln[R(\omega)]}{(\omega^2 - \omega'^2)}\, d\omega'. \tag{3.19}$$

By measuring $R(\omega)$, $\theta(\omega)$ can be evaluated through Equation (3.19), and the dielectric function follows from Equations (3.17) and (3.18). The KK method of analysis is straightforward, except for occasional problems associated with the convergence of the integral in Equation (3.19). If $R(\omega)$ has been measured between the frequencies ω_1 and ω_2, the contributions to the integral for $\omega < \omega_1$ and $\omega > \omega_2$ must be shown to be negligibly small or else calculated by assuming a functional dependence for $R(\omega)$. One of the most common assumptions is that R is a constant when $\omega < \omega_1$ or $\omega > \omega_2$, but this may lead to significant errors in $\theta(\omega)$ for certain crystals. These and other considerations have been discussed by Stern [67] and Robinson and Price [68], and we refer the reader to their articles.

3.3. Effective charge

The macroscopic effective charge of a polar lattice vibration is defined by the equation [69]

$$e^* = \left(\frac{\partial p}{\partial u}\right)_E, \tag{3.20}$$

where p is the electric dipole moment of the vibration, u is the relative atomic displacement coordinate, and E is the macroscopic electric field of the vibration. For a TO phonon the macroscopic electric field is identically zero. Polar vibrations that are described by the Lorentzian dielectric function of Equation (3.8) have transverse effective charges given by

$$e_T^* = 2\pi c \left(\frac{M\rho}{N}\right)^{1/2} \bar{v}_{TO}.$$ (3.21)

In this equation M is the mass of the oscillator, N is the number of oscillators per unit volume, and $\bar{v}_{TO}$ is the TO frequency (in wave numbers). M and N are model-dependent parameters, that is, they depend on the specific atomic displacements of the polar mode. In diatomic crystals with one TO phonon, M is the reduced mass of the two atoms. In the structurally more complicated layered materials, M and N can be determined only through a detailed knowledge of the atomic displacements. ε–GaSe, for example, has a primitive unit cell that contains eight atoms, and the dominant E''^3 TO mode has the Ga and Se sublattices vibrating 180° out of phase. The equivalent Lorentz oscillator is a Ga–Se dipole with a mass equal to the reduced mass of the Ga and Se atoms. Four oscillators occupy the volume of the unit cell. For most of the layered materials that we shall discuss in Section 5, the equivalent Lorentz oscillators and the parameters M and N are easily determined.

Other effective charges commonly used in analyzing polar modes are the macroscopic, longitudinal effective charge

$$e_L^* = \frac{e_T^*}{\varepsilon_\infty},$$ (3.22)

and the Szigeti effective charge [70]

$$e_s^* = \left(\frac{3}{\varepsilon_\infty + 2}\right) e_T^*.$$ (3.23)

Equations (3.21) through (3.23) were first derived for the alkali halides, and when restricted to diatomic cubic crystals, the charges of Equations (3.21) and (3.22) are called the Born [71] and Callen [72] effective charges, respectively. The Szigeti charge was derived by calculating the effective electric field acting on the ions of a lattice that has tetrahedral site symmetry. This local effective field differs from the macroscopic electric field, which is the total field averaged over the volume of the unit cell. The correction factor $3/(\varepsilon_\infty + 2)$ in Equation (3.23) measures the effect of placing point charges or ions at specific lattice sites. Although the Szigeti charge is often calculated for crystals that do not have tetrahedral site symmetry, the assumptions involved in these calculations are misleading. In Section 5.6 we shall use only the macroscopic effective charge.

The macroscopic charge for diatomic cubic crystals has been decomposed in various ways, according to the nature of the crystal bonding and the lattice dynamical model. We mention only two: the division of e_T^* into rigid (or static) and dynamic components [73], and the division into local and nonlocal components [74]. The first decomposition reflects the distinction between the charge transferred during bonding and the charge produced by deformable vibrating ions. The second decomposition distinguishes the charge localized on the atomic sites from the charge distributed throughout the unit cell. In general, the localized and nonlocalized charges include both rigid and dynamic components.

3.4. EXPERIMENTAL CONFIGURATIONS IN RAMAN SCATTERING

The two experimental configurations commonly employed in Raman investigations of opaque samples are illustrated in Figure 3.2. In back-scattering the incident light is normal to the surface of the crystal, and the scattered light is collected within a cone

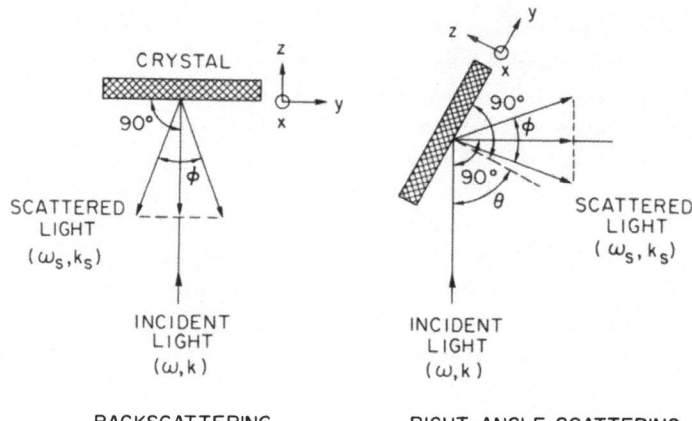

Fig. 3.2.   Two experimental configurations for Raman investigations of opaque samples.

of angle φ, whose axis coincides with the direction of the incident light. The specularly reflected light is either blocked or redirected away from the entrance slits of the spectrometer. Conservation of energy and momentum requires that

$$\omega = \omega_s \pm \omega_j \tag{3.24}$$

and

$$\mathbf{k} = \mathbf{k}_s - \mathbf{q}, \tag{3.25}$$

where k and k_s are the wave vectors of the incident and scattered light, respectively, $\mathbf{q}$ is the wave vector of the phonon, and the plus or minus sign refer to the creation or annihilation of a phonon during the scattering process. Since the wave vector of the incident (visible) light is a very small fraction of the width of the Brillouin zone, the phonons involved in first-order Raman scattering come from the center of the zone; thus $\mathbf{k} = \mathbf{k}_z$, $\mathbf{k}_s \simeq -\mathbf{k}_z$, and $\mathbf{q} \simeq 2\mathbf{k}_z$. The electric vector of the incident and back-scattered light may have components along both the x- and y-axes of the crystal. In right-angle scattering the direction of the incident light is at an angle θ with respect to the normal, and the scattered light is collected within a cone of angle φ. The axis of the cone and the direction of the incident light are perpendicular to each other. In this configuration the wave vector and electric vector of the incident and scattered light may have components along all three of the crystal axes; consequently, the polarization selection rules for scattering are more difficult to establish. The index of refraction of the crystal also changes the direction of the incident ray inside the crystal and further complicates the analysis of the incident and scattered light. Right-angle scattering, however, is the easier of the two experimental configurations to use, and if the incident ray is polarized along y or if the crystal has a large index of refraction, the analysis can be considerably simplified.†

3.5. RAMAN SELECTION RULES AND SCATTERED LIGHT INTENSITY

The scattering of light by lattice vibrations requires the nonlinear mixing of the photon and phonon frequencies. This is accomplished by the electric moment induced in the crystal by the radiation field. In classical terms the Raman effect occurs because the induced moment, which generates the scattered light, is the product of the crystal polarizability (a symmetric tensor) and the electric field intensity:

$$m_\rho = \sum_\sigma \alpha_{\rho\sigma} E_\sigma, \tag{3.26}$$

where the indices ρ and σ denote the x, y, or z axes of the crystal. Since the electronic masses are much lighter than the nuclear masses and ω is normally much greater than ω_j, the dominant contribution to $\alpha_{\rho\sigma}$ in Equation (3.26) comes from the electronic system of the crystal. In the Born-Oppenheimer approximation the electronic polarizability varies with the instantaneous positions of the vibrating nucleii; thus, the polarizability oscillates with the frequencies of the lattice, and the product $\alpha_{\rho\sigma} E_\sigma$ creates the sum and difference Raman frequencies, $\omega \pm \omega_j$.

The selection rules and scattered light intensity are governed by the matrix elements of the induced electric moment

$$m_{\rho, fi} = \Sigma [\int \Psi_f^* \alpha_{\rho\sigma} \Psi_i \, dV] E_\sigma. \tag{3.27}$$

When the integrand in Equation (3.27) is antisymmetric, $\alpha_{\rho\sigma, fi}$ is identically zero; when the integrand is symmetric, a Raman transition is allowed, but the scattered light intensity may nevertheless be low, depending on the specific functions involved in the integral. For the integrand to be symmetric, $\Psi_f^* \Psi_i$ must transform in the same way as $\alpha_{\rho\sigma}$. This symmetry requirement defines the general selection rule for Raman scattering. If $\hat{\mathbf{e}}_s$ is a unit vector perpendicular to the direction of observation, the light scattered in that direction is given by

$$S = C\omega_s^4 |\mathbf{m} \cdot \hat{\mathbf{e}}_s|^2$$
$$= C\omega_s^4 |\sum_{\rho\sigma} E_\sigma \alpha_{\rho\sigma} e_s^\rho|^2, \tag{3.28}$$

where C is a constant and e_s^σ is a component of the unit vector. In this equation we see that the polarizability links the polarization directions of the incident and scattered light.

Since lattice vibrations introduce small oscillations in the electronic polarizability, $\alpha_{\rho\sigma}$ can be expanded [55] in a Taylor series in the atomic displacements. This expansion is exactly analogous to Equation (3.2) for the electric moment p_ρ. By substituting the $\alpha_{\rho\sigma}$ expansion into Equation (3.27) and using harmonic-oscillator wave functions, we

† Forward scattering, a third experimental configuration, is sometimes used for crystals that are transparent to the exciting light. If the light is normal to the surface, then $\mathbf{k} = \mathbf{k}_z$, $\mathbf{k}_s \simeq \mathbf{k}_z$, and $\mathbf{q} \simeq 0$.

obtain the same quantum number relations for Raman scattering as for infrared absorption. First-order Raman scattering is given by the second term in the expansion, and the quantum number relation is $v_j^{(f)} = v_j^{(i)} \pm 1$, which limits the transitions to fundamentals. Since the second term varies jointly with the vector components m_ρ and E_σ in Equation (3.26), this term will transform as the product of the translations T_ρ and T_σ. Hence Raman-active vibrational modes transform as bilinear combinations of the coordinates x, y, and z. This is a particular selection rule for first-order Raman scattering, but it is widely used in analyzing spectra. Raman-active modes can be identified from the character tables of the point groups [57], where the transformation properties of $\rho\sigma$ are listed opposite the irreducible representations.

The semi-classical treatment of Raman scattering that we have just summarized was first given by Born and Bradburn [55]. However, their theory does not account for scattering from polar lattice vibrations, which possess long-range electric fields. Poulet [75] has incorporated polar scattering in the Born theory by expanding the polarizability in a power series in the phonon electric field as well as in the atomic displacements. Loudon [76], on the other hand, has carried through a direct calculation of the polar electron-lattice interaction, using third-order perturbation theory. His work also describes the Raman effect in noncubic crystals, which do not in general have the phonon displacement and the electric field of the polar vibration pointing in the same direction. In uniaxial crystals the scattered light is given by the following formula of Loudon [77]:

$$
S = \left[\sum_{\substack{\sigma,\rho,\tau= \\ x,y,z}} e_i^\sigma \, R_{\sigma\rho}^\tau (\alpha\zeta^\tau + \beta q^\tau) \, e_s^\rho \right]^2,
\tag{3.29}
$$

where e_i^σ is a component of the incident electric field, $R_{\sigma\rho}^\tau$ is the Raman tensor, α is a constant, β is proportional to the magnitude of the phonon electric field, and ζ^τ and q^τ are the components ($\tau = x$, y, or z) of unit vectors in the directions of the phonon displacement and phonon wave vector, respectively. An interesting result of Loudon's theory is that $R_{\sigma\rho}$ is not a symmetric tensor unless $\omega_j \ll \omega$. Since $R_{\sigma\rho}$ ($R_{\sigma\rho}^\tau$ summed over τ) is equal to $\alpha_{\rho\sigma}$, Equation (3.29) reduces to Equation (3.28) for nonpolar phonon scattering. The components of the Raman tensor $R_{\sigma\rho}^\tau$ are tabulated in Loudon's review paper [77].

Very few resonant Raman and second-order Raman studies have as yet been made on layered materials, and we shall not extend the foregoing discussion to cover these effects. The reader is referred to the articles by Loudon [77], Burstein [59], and Martin and Falicov [78].

4. Normal Vibrations in Layered Crystals

Layered crystal structures are characterized by large directional variations in the crystal bonding forces, which result in easy mechanical cleavage of the crystals along the basal plane. Long-wavelength phonons are directly affected by the magnitude and anisotropy of these forces. In this chapter we shall explore some of the important consequences of mechanical anisotropy and symmetry on the normal vibrations of layered materials.

4.1. GENERAL PROPERTIES

The lattice vibrations of a layered structure are conditioned by the number of layers in the primitive unit cell. If the cell contains only one layer, the modes are determined primarily by the strong restoring forces present within each layer. The forces between the layers usually can be neglected because of their relative weakness.* If the cell contains m layers ($m > 1$), the interlayer force constants, whether large or small, control not only the frequencies of $3(m - 1)$ optical modes but also the magnitude of the correlation splittings observed in infrared and Raman spectra. The $3(m - 1)$ modes are the rigid-layer vibrations of the crystal, which we shall discuss presently. Correlation splittings will be taken up in the following section.

Figure 4.1 shows the counter-phase modes of a two-layer polymorph of MoS_2, which Zvyagin and Soboleva [14] list as their polytypal structure No. 1. Although

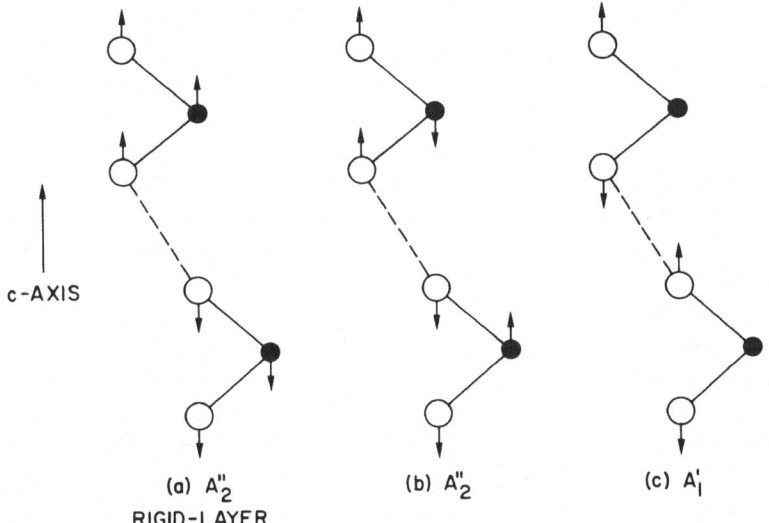

(a) A_2'' (b) A_2'' (c) A_1'
RIGID-LAYER

Fig. 4.1. Counterphase optical modes of ε-MoS_2 polarized along the c-axis. The black circles are the Mo atoms.

this polytype has not been found in nature or synthesized as yet in the laboratory, its lattice vibrations illustrate several fundamental ideas about normal modes in layered materials. We shall refer to this structure as ε-MoS_2 [18], because of its similarity to ε-GaSe. The irreducible representations of the eighteen long-wavelength modes of ε-MoS_2 are

$$4A_2'' + 2A_1' + 4E' + 2E''. \tag{4.1}$$

We have obtained this decomposition from the correlation diagram of Figure 2.7, using the fact that the symmetries of the pseudolattice and real crystal are the same. The displacement vectors shown in the upper halves of the unit cells of Figure 4.1

* $MoSe_2$ and possibly $MoTe_2$ are notable exeptions (see Section 5.2); their lattice vibrations cannot be separated into intralayer (or internal) and interlayer (or external) modes, because of the strength of the interlayer coupling. Other materials, such as GaSe, which have moderate electrostatic forces linking their layers together, will also have altered internal-mode frequencies.

characterize the internal modes of a single layer polarized along the c-axis. (The analogous, doubly degenerate E' and E'' modes are polarized at right angles to the c-axis.) Since ε–MoS_2 belongs to the space group D_{3h}^1 (P$\bar{6}$m2), there are no symmetry operations that transform the atoms of the upper layer in Figure 4.1 into the atoms of the lower layer; therefore, on the basis of symmetry the amplitudes of the vibrating atoms in the two layers are unrelated. To establish a connection between the layers apart from symmetry, we invoke the basic property of mechanical anisotropy. Since the interlayer interaction in ε–MoS_2 is weak, the internal vibrations of the two layers are almost independent; however, the presence of a small interlayer restoring force compels the layers to vibrate either in phase or 180° out of phase.* The displacement vectors of the corresponding atoms in the two layers will then be approximately equal, except possibly for the sign of the displacement. The absolute values of these displacements are strictly equal only when the interlayer interaction is vanishingly small. In the case of 2H–MoS_2 (Figure 5.5) symmetry arguments alone are sufficient to relate the vibrational amplitudes of the two layers.

The rigid-layer vibrations are shown in Figure 4.1(a). These are vibrations in which the layers move as nearly rigid units [81] in counterphase. Their conjugates are the acoustical modes. Another term, 'quasi-acoustical' [82], was originally used for the rigid-layer modes in MoS_2, in order to emphasize their relation to the acoustical branch of the phonon spectrum. In the reduced-zone scheme the acoustical branch of the MoS_2 pseudocrystal has a width of π/c in the z direction. By comparison, the zone widths of 2H–MoS_2 and 3T–MoS_2 are one half and one third as large, respectively. This means that the 'acoustical' branches of the 2H and 3T polytypes consist of two or three subbranches, pieced together as shown in Figure 4.2.† The frequency ω_R in Figure 4.2(a), which lies at the zone boundary in the MoS_2 pseudocrystal (or in 3R–MoS_2), is located at the zone center in 2H–MoS_2. In 3T–MoS_2 (Figure 4.2(b)) the rigid-layer frequency has a smaller value, because of the division of the zone into an odd number of parts. If the layers are perfectly rigid and the interactions are limited to nearest-neighbor layers, the acoustical branch of the pseudocrystal has the form of a sine curve. ω_R in 3T–MoS_2 is then equal to $\sin[(2\pi/3c)(c/2)] = 0.87$ times the maximum value of the curve, which is the rigid-layer frequency of the two-layer polytype. However, this simple reduced-zone picture does not take into account the variation in the stacking sequence. There is some evidence, based on the stacking fault energies of ε–GaSe and β–GaS [12, 83], that the rigid-layer frequency of 2H–MoS_2 should be larger than that of ε–MoS_2. Thus the dispersion curves of the various polytypes cannot be superimposed and compared directly as in Figure 4.2. We shall discuss this point further in the following section and quantify the frequency differences by means of the linear-chain model.

* Vibrations that differ only by the phases of the layers are known as conjugate modes or as Davydov doublets or multiplets [79, 80]. Counterphase vibrations have the atoms on opposite sides of the interlayer gap vibrating 180° out of phase.

† In this figure we have divided the width of the Brillouin zone into equal parts. A material that has nonuniform stacking of the layers or nonidentical coordinations of the atoms will have different interlayer distances and layer thicknesses. (By nonuniform stacking we mean that the relative orientation and position of any two nearest-neighbor layers is the same throughout the crystal.) These differences, however, have a relatively small effect on the subbranches shown in Figure 4.2.

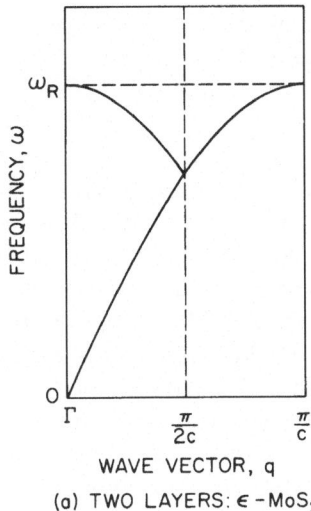
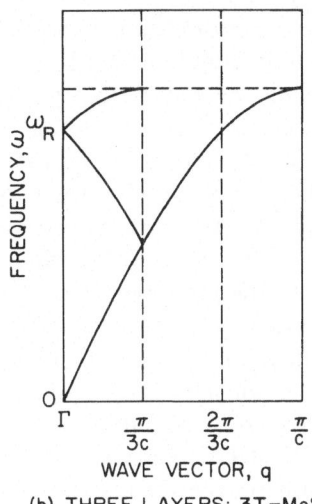

Fig. 4.2. Acoustical branch of the pseudolattice and the reduction of the Brillouin zone in 2H– and 3T–MoS$_2$ for $q \| c$. Similar reductions occur in other layered materials with more than one layer in the primitive unit cell.

The A_2'' mode in Figure 4.1(b) and its conjugate provide examples of infrared and Raman activity. The figure shows that the atomic displacements within the layers generate dipole moments. These moments either reinforce or cancel each other, depending on whether the layers vibrate in phase or out of phase. If there were a center of inversion between the layers, as in 2H–MoS$_2$, the net dipole moment would be either twice the layer moment or identically zero. Since the corresponding amplitudes of vibration in the two layers are only approximately equal in ε–MoS$_2$, both A_2'' modes have a dipole moment, although they are very different in size. From the character table for D_{3h}, we find that A_2'' transforms as the coordinate z; hence these modes are infrared-active for $E \| c$. (In 2H–MoS$_2$ only the antisymmetric mode is infrared-active, as the symmetric or counterphase mode lacks a net dipole moment.) The oscillator strength of the counterphase vibration in Figure 4.1(b) will clearly be much smaller than the oscillator strength of its conjugate. Because of the near degeneracy of the two vibrations, the in-phase mode will dominate the infrared spectrum in the neighborhood of the resonance frequency. Similar arguments apply to the infrared-active E' modes. Finally, the oscillator strength of the A_2'' rigid-layer mode in Figure 4.1(a) is expected to be very small, not only because of the slight relative displacements of the Mo and S atoms within each layer [81], but also because of the counterphase movement of the layers.

For the A_2'' mode in Figure 4.1(b) to be Raman-active, the polarizability of the atoms in the unit cell must change during the vibration. The Mo and S atoms within each layer vibrate against one another, and the polarizability of the layer is the same at opposite phases of the vibration. The polarizability of the layer therefore lacks a first-order term in the atomic displacements (compare Equation (3.2)). Moreover, since the polarizability due to the interlayer interaction is exactly the same at opposite phases of the vibration, the net polarizability of the atoms in the unit cell is invariant.

The A_2'' mode is therefore Raman-inactive. If we now rotate the displacement vectors of the A_2'' vibration in Figure 4.1(b) by 90°, we obtain the analogous E' mode. Our remarks concerning the polarizability of the layer apply to this vibration as well, but the interlayer phasing this time produces a change in the net polarizability of the unit cell; accordingly, Table 2.1 indicates that the E' vibration is Raman-active. The conjugate of the E' vibration has the two layers of the unit cell moving in phase. Since the polarizability of the layers is invariant during the vibration, the Raman activity of the conjugate has its origin in the small relative displacement of the sulfur atoms on opposite sides of the interlayer gap. The symmetry of ε–MoS_2 does not require the amplitudes of these sulfur atoms to be identical. (In $2H$–MoS_2 the vibrational amplitudes are identical, and the net polarizability of the E_{1u}^2 mode vanishes (Figure 5.5).) Although both E' vibrations are Raman-active, the counterphase mode will be much stronger than the in-phase mode in Raman scattering. Furthermore, because of the near degeneracy of conjugate modes, it is unlikely that the in-phase mode will be observed in Raman experiments, unless the resolution and sensitivity of the measurements are extremely good.

Another example of Raman activity is provided by the A_1' vibration in Figure 4.1(c). The Mo atoms are at rest in this mode, and the sulfur atoms in each layer vibrate against one another. The stretching and compression of the intralayer Mo–S bonds produce a first-order term in the polarizability. When the layers vibrate out of phase, the net polarizability of the unit cell is composed of positive contributions from the interlayer interaction as well as from the two layers. When the layers vibrate in phase, the layer polarizabilities cancel each other, and the contribution from the interlayer interaction is greatly reduced because of the layer phasing. The symmetry of ε–MoS_2 again prevents the net polarizability of the in-phase mode from vanishing. The analogous basal-plane modes, the E'' conjugate pair, have very different net polarizabilities. The coordination of the layer and the direction of the atomic displacements lead in this case to an invariant polarizability for the layer. Thus the Raman activities of the E'' pair are due to the interlayer interaction. The in-phase E'' mode has a smaller polarizability than the counterphase E'' mode; this mode in turn has a small polarizability than the counterphase A_1' mode. Raman experiments on $2H$–MoS_2 [65] have shown that the strongest and weakest lines in the spectrum correspond to the A_{1g} and E_{1g} vibrations, respectively. Since the displacement vectors for the A_{1g} and E_{1g} vibrations (Figure 5.5) are similar to those for the A_1' and E'' counterphase modes, these data corroborate our qualitative description of Raman activity in MoS_2.

The foregoing analysis of rigid-layer modes and infrared and Raman activities applies also to the GaSe, CdI_2, and graphite families. In general, the lattice vibrations of these materials can be assigned to $3r/m$ sets of m conjugate modes, where r and m are the number of atoms and number of layers, respectively, in the primitive unit cell. Three of these sets will contain an acoustical mode and $(m - 1)$ rigid-layer modes. Because of the weak interlayer coupling, the atomic displacements within each layer will be approximately the same, and the layers will vibrate either in phase or out of phase with their nearest neighbors. The infrared and Raman spectra, moreover, will be dominated by particular modes within each conjugate set. These modes will tend to mask the absorption or light scattering of their infrared- or Raman-active conjugate. The mode in each set with all of the layers vibrating in phase will tend to control the infrared spectrum; the mode with all (or most) of the layers vibrating out of phase will tend to control the Raman spectrum. These conclusions will be of particular value in Section 5, where we will review and assess the experimental investigations of layered materials. More complicated layered structures, such as As_2S_3, Sb_2S_3, and Bi_2Te_3, will have rigid-layer vibrations similar to those in the first four families; but the above rules for infrared and Raman activity should not be applied to these structures, as the displacement vectors for their normal modes have not been established.

4.2. LINEAR-CHAIN MODEL

The long-wavelength phonons in several families of layered materials can be treated theoretically by means of the linear-chain model [81]. In order to apply this model to a particular crystal, three requirements must be satisfied. First, the atoms of the primitive unit cells must lie on a series of parallel planes, and each plane must contain no more than one atom in each cell. Figure 4.3 illustrates this requirement in the two-layer

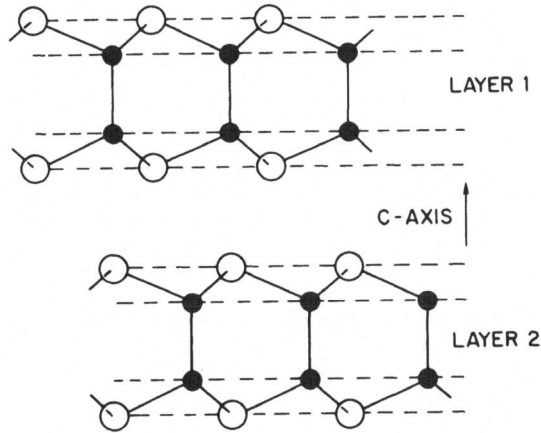

Fig. 4.3. Parallel planes of atoms in ε–GaSe, illustrating the first requirement of the linear-chain model.

polytype ε–GaSe. The planes run perpendicular to the c–axis in ε–GaSe and intersect one Ga atom or one Se atom in each unit cell. Second, the normal coordinates of the long-wavelength vibrations must be either parallel or perpendicular to the atomic planes. This requirement is also satisfied in ε–GaSe, as the displacements of the atoms are either along the basal plane or along the c-axis. Finally, the interactions between the planes must be restricted, in general, to nearest neighbors, since otherwise the number of interplanar force constants may exceed the number of measured phonon frequencies. The applicability of the linear-chain model depends upon an over-determination of the force constants, so that the assumptions concerning the number of near-neighbor interactions can be tested. The first two requirements are structural in nature, while the third is physical.

In order to demonstrate the formal results of the linear-chain model, we consider the crystal ε–GaSe. Since the structural requirements are satisfied, the normal vibrations of the atoms in the unit cell can be viewed as rigid vibrations of the atomic planes comprising the crystal. In the long-wavelength limit, corresponding atoms in all primitive unit cells vibrate in phase; hence, the relative distances between these atoms remain constant. Shear and compressional restoring forces between pairs of planes define the interplanar interactions. The linear chain representation of the normal vibrations of the atomic planes is shown in Figure 4.4, where the axis of the chain is

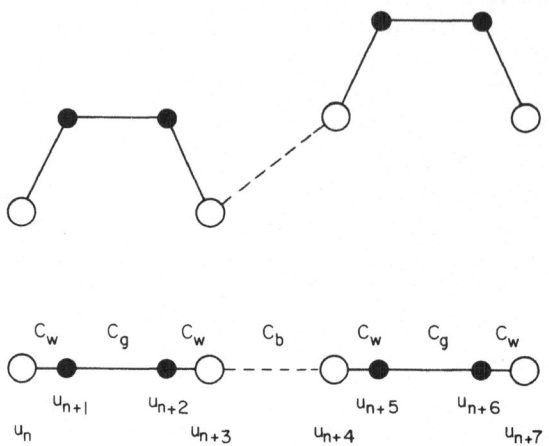

Fig. 4.4.   Linear-chain representation of the long-wavelength lattice vibrations of ε-GaSe. The atoms of the primitive unit cell are shown in the upper half of the figure (compare Figure 2.2), and the linear chain is shown in the lower half. The c-axis coincides with the chain axis.

parallel to the c-axis of the crystal. The dynamical equations for the vibrating planes can be expressed as

$$-M_n\omega^2 u_n = \sum_{n'} C_{nn'}(u_{n'} - u_n),\tag{4.2}$$

where M_n is the mass of the atom lying in plane n, ω is the angular frequency of the normal mode, u_n and $u_{n'}$ are the displacements of the planes n and n' from their equilibrium positions, and $C_{nn'}$ is the shear or compressional force constant between the planes n and n'. The size of the force constant $C_{nn'}$ is obviously linked to the number of atomic masses chosen to represent the plane. In Equation (4.2) $C_{nn'}$ is scaled to the mass of one atom. The frequencies of the modes are solutions of the $3r$ linear equations in Equation (4.2), where r is the number of atoms in the primitive unit cell.

For each symmetry direction in ε–GaSe, the secular determinant for ω is of eighth order, since there are eight atoms in the unit cell and the basal-plane modes are doubly degenerate. The nondegenerate A_2'' and A_1' modes (Table 2.10) have their atomic displacements along the axis of the chain, and the doubly degenerate E' and E'' modes have their displacements perpendicular to the chain. The order of the secular determinant for both symmetry directions can be reduced by making certain assumptions about the eigenvectors of the vibrations. Because of the weak coupling between the two layers in the unit cell, the normal modes are as shown in Figure 4.5. In constructing and identifying these modes, we have used Figure 2.8 and Table 2.10. If there were

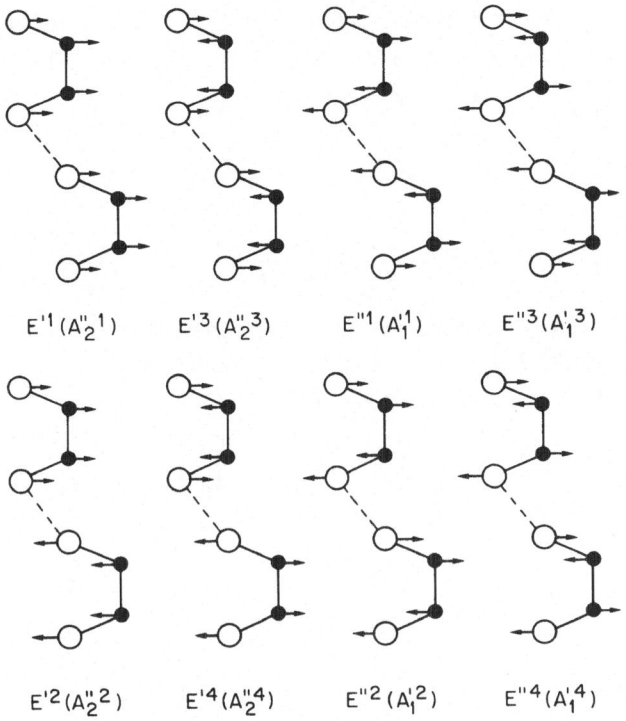

$E'^1 (A_2''^1)$ $E'^3 (A_2''^3)$ $E''^1 (A_1'^1)$ $E''^3 (A_1'^3)$

$E'^2 (A_2''^2)$ $E'^4 (A_2''^4)$ $E''^2 (A_1'^2)$ $E''^4 (A_1'^4)$

Fig. 4.5.   Displacement vectors for the twenty-four normal modes of ε-GaSe and their irreducible representations. The displacement vectors for the c-axis modes ($4A_2'' + 4A_1'$) are obtained by rotating the vectors for the basal-plane modes ($4E' + 4E''$) by 90°.

an inversion center between the layers, as in β–GaS (Figure 2.2), the modes in Figure 4.5 would be either symmetric or antisymmetric. The atoms in the two halves of the unit cell, coupled by the inversion operation, would then have the same absolute amplitudes of vibration. In ε–GaSe the vibrational amplitudes are only approximately equal, due to the lower symmetry.

The approximate amplitude relations for the E'^1 (or $A_2''^1$) and E'^3 (or $A_2''^3$) vibrations in Figure 4.5 are given by $u_n = u_{n+3} = u_{n+4} = u_{n+7}$ and $u_{n+1} = u_{n+2} = u_{n+5} = u_{n+6}$. These relations reduce the secular determinant of Equation (4.2) to second order. By restricting the interplanar interactions to nearest neighbors (see Figure 4.4), the frequencies of the modes become

$$\omega^2(A_2''^1, E'^1) = 0 \quad \text{(acoustical modes)} \tag{4.3}$$

and

$$\omega^2(A_2''^3, E'^3) = \frac{C_w}{M}, \tag{4.4}$$

where C_w is the shear or compressional force constant between the Ga and Se planes, $M = M_g M_s / (M_g + M_s)$, and M_g and M_s are the masses of the Ga and Se atoms, respectively. Similarly, for the E''^1 $(A_1'^1)$ and E''^3 $(A_1'^3)$ modes, $u_n = -u_{n+3} = -u_{n+4} = u_{n+7}$ and $u_{n+1} = -u_{n+2} = -u_{n+5} = u_{n+6}$. Their frequencies are

$$\omega^2(A_1'^1, E''^1) = \frac{C_w}{2M} + \frac{C_g}{M_g} - \left[\left(\frac{C_w}{2M} + \frac{C_g}{M_g} \right)^2 - \frac{2C_w C_g}{M_g M_s} \right]^{1/2} \tag{4.5}$$

and

$$\omega^2(A_1'^3, E''^3) = \frac{C_w}{2M} + \frac{C_g}{M_g} + \left[\left(\frac{C_w}{2M} + \frac{C_g}{M_g} \right)^2 - \frac{2C_w C_g}{M_g M_s} \right]^{1/2}, \tag{4.6}$$

where C_g is the shear or compressional force constant between the two Ga planes. Proceeding next to the E'^2 $(A_2''^2)$ and E'^4 $(A_2''^4)$ modes, we have $u_n = u_{n+3} = -u_{n+4} = -u_{n+7}$, $u_{n+1} = u_{n+2} = -u_{n+5} = -u_{n+6}$, and

$$\omega^2(A_2''^2, E'^2) = \frac{C_w}{2M} + \frac{C_b}{M_s} - \left[\left(\frac{C_w}{2M} + \frac{C_b}{M_s} \right)^2 - \frac{2C_w C_b}{M_g M_s} \right]^{1/2}$$

$$\simeq \frac{2C_b}{(M_g + M_s)} \tag{4.7}$$

and

$$\omega^2(A_2''^4, E'^4) = \frac{C_w}{2M} + \frac{C_b}{M_s} + \left[\left(\frac{C_w}{2M} + \frac{C_b}{M_s} \right)^2 - \frac{2C_w C_b}{M_g M_s} \right]^{1/2}$$

$$\simeq \frac{C_w}{M} \left[1 + \frac{2M^2 C_b}{M_s^2 C_w} \right], \tag{4.8}$$

where C_b is the shear or compressional force constant between the selenium planes. The approximate expressions for the frequencies in Equations (4.7) and (4.8) are obtained by assuming that the second term inside the square brackets is much smaller than the first term (a condition roughly equivalent to $C_b \ll C_w$). Lastly, the E''^2 $(A_1'^2)$ and E''^4 $(A_1'^4)$ modes have the amplitude relations, $u_n = -u_{n+3} = u_{n+4} = -u_{n+7}$ and $u_{n+1} = -u_{n+2} = u_{n+5} = -u_{n+6}$. Their frequencies are

$$\omega^2(A_1'^2, E''^2) = \frac{C_w}{2M} + \frac{C_b}{M_s} + \frac{C_g}{M_g} - \left[\left(\frac{C_w}{2M} + \frac{C_b}{M_s} + \frac{C_g}{M_g} \right)^2 - \right.$$

$$\left. - \frac{2}{M_g M_s} (C_w C_b + C_w C_g + 2C_g C_b) \right]^{1/2} \tag{4.9}$$

and

$$\omega^2(A_1'^4, E''^4) = \frac{C_w}{2M} + \frac{C_b}{M_s} + \frac{C_g}{M_g} + \left[\left(\frac{C_w}{2M} + \frac{C_b}{M_s} + \frac{C_g}{M_g} \right)^2 - \right.$$

$$\left. - \frac{2}{M_g M_s} (C_w C_b + C_w C_g + 2C_g C_b) \right]^{1/2} \tag{4.10}$$

Equations (4.3)–(4.10) give the vibrational frequencies of all 24 modes of ε–GaSe (eight are doubly degenerate) in terms of the six nearest-neighbor force constants.

If in each symmetry direction the number of experimental frequencies for ε–GaSe is greater than three, the linear-chain model can be applied and evaluated; however, we shall defer the discussion of the results of this model for ε–GaSe (and other layered materials) until Section 5.

The model just described can be applied to other members of the GaSe family (Section 2.1.1), including different polytypes of the same crystal. For example, the linear-chain expressions for the frequencies of β–GaS [81] are the same as those for ε–GaSe, except for the irreducible representations appearing on the left-hand sides of Equations (4.3)–(4.10). These symmetry species can be obtained from the correlations between the representations of the pseudolattice and the real crystal (compare Figures 2.8 and 2.9). Although β–GaSe has not as yet been synthesized, its vibrational frequencies are undoubtedly slightly different from those of ε–GaSe, because of the different stacking sequences of the layers. Both polytypes have a uniform stacking of the layers, which implies that the interlayer coupling constants in each polytype are equal; but the value of the coupling constant in β–GaSe, as mentioned previously [12, 83], is probably larger than that in ε–GaSe.

In the hypothetical three-layer polytype, 3T–GaSe (analogous to 3T–MoS$_2$), the stacking is nonuniform and includes interlayer coupling constants appropriate to ε–, β–, and α–GaSe (analogous to 2H–NbSe$_2$ in Figure 2.3). Equations (4.3)–(4.6) give the frequencies of the modes in 3T–GaSe that have all three layers in the unit cell vibrating in phase. The counterphase vibrations of this material have two of the layers in the unit cell vibrating against the third; within each conjugate set there is a splitting between the frequencies of the two counterphase modes. We can determine the magnitude of the splitting in the following way. We shall only consider the three conjugate sets composed of an acoustical (in-phase) mode and two rigid-layer (counterphase) modes. The three layers of the unit cell of 3T–GaSe are assumed to be perfectly rigid and coupled together by nearest-neighbor interlayer force constants, C_1, C_2, and C_3, which correspond to ε, β, and α stacking. Since the layers have identical masses, Equation (4.2) leads to the simultaneous equations

$$[(C_1 + C_3) - M\omega^2]u_1 - C_1 u_2 - C_3 u_3 = 0,$$

$$-C_1 u_1 + [(C_1 + C_2) - M\omega^2]u_2 - C_2 u_3 = 0,$$

and

$$-C_3 u_1 - C_2 u_2 + [(C_2 + C_3) - M\omega^2]u_3 = 0. \tag{4.11}$$

Solving the above secular equation in the coefficients, we obtain for the eigenfrequencies

$$\omega^2 = 0 \quad \text{(acoustical mode)} \tag{4.12}$$

and

$$\omega^2 = \frac{C_1 + C_2 + C_3}{M}\left[1 \pm \sqrt{1 - \frac{3(C_1 C_2 + C_1 C_3 + C_2 C_3)}{(C_1 + C_2 + C_3)^2}}\right], \tag{4.13}$$

where the splitting between the rigid-layer modes is given by the square-root term. If the stacking of the layers in 3T–GaSe were uniform, the interlayer force constants would be equal. The square-root term would then vanish, and the two degenerate rigid-layer modes would have the frequency

$$\omega^2 = \frac{3C_b}{M} = \frac{2C_b}{(2M/3)},$$ (4.14)

where we have set $C_1 = C_2 = C_3 \equiv C_b$ and replaced the layer mass by a reduced oscillator mass $(2M/3)$. For the two-layer polytype, ε–GaSe, we can rewrite the rigid-layer frequency in Equation (4.7) as

$$\omega^2(\varepsilon\text{–GaSe}) = \frac{2C_b(\varepsilon\text{–GaSe})}{(M/2)}.$$ (4.15)

Dividing Equation (4.14) by Equation (4.15), we have for the ratio of the rigid-layer frequencies in the two polytypes

$$\frac{\omega}{\omega(\varepsilon\text{–GaSe})} = \frac{\sqrt{3}}{2}\left[\frac{C_b}{C_b(\varepsilon\text{–GaSe})}\right]^{1/2}$$ (4.16)

If the interlayer force constants in 3T– and ε–GaSe are equal, then the ratio is $\sqrt{3}/2 = 0.87$, which is the same as the ratio given by the reduced-zone model of the rigid-layer vibrations. Our main point, however, is that the nonuniform stacking of the layers in 3T–GaSe removes the degeneracy between the two counterphase modes in each conjugate set. Furthermore, all three-layer materials are subject to this rule: a uniform stacking of the layers requires that the counterphase modes in each set be degenerate; conversely, a nonuniform stacking removes the degeneracy.

Similar equations can be derived from the linear-chain model for the MoS_2 and CdI_2 families. These have already been given in the literature and will not be reproduced here; however, in the case of CdI_2, where the bonding is more ionic and long-range electrostatic interactions between atoms cannot be neglected, the linear-chain model must be supplemented by a calculation of the local effective field in the crystal. The procedure for combining these two model calculations is described in Section 5.3.

5. Experimental Investigations

In this section we shall examine the experimental investigations of the various families of layered materials presented in Section 2. The family that has generated the most controversy, and consequently the largest number of papers, is GaSe. Much of the disagreement in the GaSe literature is linked to the different polytypes obtained by X-ray diffraction and the related symmetries, activities, and assignments of the normal modes. Since most of these problems have only recently been resolved, the GaSe family provides an engaging starting point for the discussion of experimental results.

5.1. GaSe FAMILY

The infrared optical properties of GaSe were first investigated in 1966 by Leung *et al.* [84]. Their crystals were grown by the Bridgman method, and they were large enough to permit the measurement of the reflectance and transmittance for $E \| c$ as well as $E \perp c$. One phonon band was observed between 216 and 1200 cm^{-1} for each polarization direction. By fitting Lorentzian oscillators to the experimental data, they obtained the dielectric constants and dispersion parameters for the two TO bands. These are given in Tables 5.1 and 5.2, together with earlier values of the dielectric constants. Additional measurements on capacitor-shaped specimens of GaSe, which were thin layers of the material with Au films deposited on the two faces, gave the low-frequency dielectric constant for $E \| c$.

These authors also noted the uncertainty in the literature [12] concerning the crystal structure of GaSe. In the original study of Schubert et al. [47], two polytypes of GaSe were identified, ε (hexagonal) and γ (rhombohedral), illustrated in Figure 2.2. Because of the nearly identical, X-ray scattering powers of the Ga and Se atoms, Schubert et al. were not able to establish the ordering of the atomic planes in each layer; however, the sequence Se–Ga–Ga–Se was thought to be the most probable. Tatarinova et al. [88] then proposed an alternative ordering in which the two inner planes of each layer were composed of Se rather than Ga atoms. This GaS anti-type ordering was criticized by Semiletov [89], who obtained the sequence Se–In–In–Se for the isomorphic material InSe. Jellinek and Hahn [11] followed with an X-ray study of powdered specimens, in which they showed that the structures of GaSe and β–GaS were the same (Figure 2.2).

TABLE 5.1

Dielectric constants for $E \perp c$ and oscillator parameters for the E'^3 lattice vibration of ε-GaSe (room-temperature data).

Authors	Method of determination	ε_∞	ε_0	ρ	γ	$\bar{\nu}_{TO}$ (cm^{-1})	$\bar{\nu}_{LO}$[a] (cm^{-1})
MacDonald [85]	Unreported	4.7	7 (± 1)	0.183[b]			
Brebner [86]	Interference fringes ($\simeq 1$ μm)	7.3					
Brebner and Déverin [87]	Reflectance (0.8–1.1 μm)	7.1 (± 0.2)					
Leung et al. [84]	Lorentz oscillator	8.4	10.2[b]	0.143	0.0087	230.7	254.2
	Interference fringes (7–33 μm)	7.45	9.80[b]	0.187		230.7	264.6
Kuroda et al. [94]	Lorentz oscillator	7.3	10.6[b]	0.263	0.047	211.0	254.3
Wieting and Verble [65]	Lorentz oscillator	6.73 (± 0.1)	9.54[b] (± 0.1)	0.224 (± 0.02)	0.0125 (± 0.001)	213.9 (± 1)	254.7 (± 5)
	Reflectance and transmittance	6.71 (± 0.1)					
Finkman and Rizzo [96]	Lorentz oscillator			0.21 (± 0.06)	0.02 (± 0.006)	215.0 (± 2)	

[a] Calculated from Equation (3.14) for a single oscillator.
[b] Calculated from Equation (3.10) for a single socillator.

TABLE 5.2

Dielectric constants for $E\|c$ and oscillator parameters for the $A_2''^3$ lattice vibration of ε-GaSe (room-temperature data).

Authors	Method of determination	ε_∞	ε_0	ρ	γ	$\bar{\nu}_{TO}$ (cm^{-1})	$\bar{\nu}_{LO}$[a] (cm^{-1})
Leung et al. [84]	Lorentz oscillator	7.1	7.6[b]	0.043	0.0076	237.0	245.8
	Capacitance		8.0 ($\pm$0.3)				
Finkman and Rizzo [96]	Lorentz oscillator			0.044 ($\pm$0.004)	0.01 ($\pm$0.001)	236 ($\pm$2)	

[a] Calculated from Equation (3.14) for a single oscillator.
[b] Calculated from Equation (3.10) for a single oscillator.

The disagreement over the polytype of GaSe was finally resolved in the recent papers of Terhell and Lieth [90, 91]. At this point, however, we shall only remark that Leung et al.'s infrared data are consistent with all three polytypes of GaSe (β, ε, and γ), due to the weakness of the interlayer interaction and the near degeneracies of certain pairs of vibrational modes. These physical effects and their relation to polymorphism will be discussed shortly.

The frequency given by Leung et al. for the TO phonon for $E \perp c$ has not been verified by other investigators [92–94, 65, 95, 96], who have agreed on a lower value for this mode (Table 5.3). However, Leung et al.'s TO and LO frequencies for $E\|c$ have been confirmed by Finkman and Rizzo [96] and Hoff and Irwin [97]. If the erroneous data for $E \perp c$ were the result of a spectrometer miscalibration, which is the most likely explanation, the spectral results for both polarization directions would have been affected equally. It is surprising that one TO frequency has been verified and the other has not.

Later in 1966 Wright and Mooradian [98] reported the first-order Raman spectra of GaS and GaSe, using helium-neon (6328 Å) and Nd:YAG (1.06 μm) lasers. Five lines were observed for each material, but no vibrational symmetry assignments for the frequencies were made. The assignments appearing in Tables 5.3 and 5.4 are those of the present authors. Wright and Mooradian stated that the symmetry of GaS is D_{3h}; however, Hahn and Frank's original paper [46] on GaS showed that the space group was D_{6h}^4 and that the layers were stacked as in the β modification. All of the frequencies of the Raman-active modes of GaS were obtained by Wright and Mooradian, except the frequency of the E_{2g}^2 rigid-layer mode, which was found later and independently by several authors [99–101].

Three related infrared studies of β–GaS and ε–GaSe* were reported next by Kuroda, Nishina, and Fukuroi [92–94]. Absorption spectra for $E \perp c$ were obtained between 50 and 4000 cm^{-1}, and several of the features were identified as one- and

* Kuroda et al. consistently mislabel ε-GaSe as β-GaSe. The structure appearing in the text of [94] is the ε and not the β polytype.

TABLE 5.3

Raman and infrared vibrational frequencies of ε-GaSe at room temperature. The frequencies are compared with those derived from the linear-chain model ($C_w^s = 9.98 \times 10^4$, $C_b^s = 1.61 \times 10^3$, and $C_g^s = 1.53 \times 10^4$ dyn cm^{-1}; $C_w^c = 1.23 \times 10^5$, $C_b^c = 6.03 \times 10^3$, and $C_g^c = 1.75 \times 10^5$ dyn cm^{-1}).

Vibrational frequencies (cm^{-1})

ε-GaSe normal mode	Leung et al. [84]	Wright and Mooradian [98]	Kuroda et al. [92–94]		Wieting and Verble [65]		Hayek et al. [99]		Irwin et al. [95, 97, 107]		Yoshida et al. [101]		Mercier and Voitchovsky [104]	Finkman [96]	Linear-chain Rizzo and model [81]	Equivalent mode in the β polytype
	Infrared	Raman	Infrared	Raman	Infrared	Raman	Infrared	Raman	Raman	Infrared	Raman	Infrared	Raman	Infrared		
A_1''															0	A_{2u}^1
A_2''				40.0[a]						36.7[f]			40		36.7[d]	B_{2g}^1
A_3'' TO	237.0													236	237.0[d]	A_{2u}^2 TO
A_3'' LO	245.8							247	247.0				248			A_{2u}^2 LO
A_4''															239.6	B_{2g}^2
A_1'		133.8				134.6		135	134.3		134[e]		136		135.0[d]	B_{1u}^1
A_2'								141					152		143.6	A_{1g}^1
A_3'															351.0	B_{1u}^2
A_4'		308.6		309.4[a]		307.8		307.5	308.0		309[e]		309		351.2	A_{1g}^2
E_1'															0	E_{1u}^1
E_2'						19.1		18	19.5	20.1[f]	20[e]	20[f]	18		19.1[d]	E_{2g}^1
E_3' TO	230.7		211.0		213.9					212.1				215.0	213.9[d]	E_{1u}^2 TO
E_3' LO	254.2	253.8				249[b]		254	252.1		250[e]		254[a]			E_{1u}^2 LO
E_4' TO							214.5		215.0						214.7	E_{2g}^2
E_1''		59.6				60.1		58.5	60.1		60[e]				56.5	E_{2u}^1
E_2''													59		60.1[d]	E_{1g}^1
E_3''															223.6	E_{2u}^2
E_4''		209.5				213.1[c]		210.5	211.9		211[c,e]		210		224.2	E_{1g}^2

[a] Obtained through a multiphonon analysis.
[b] Probably an average value for the (E_3' LO, A_3'' LO) doublet.
[c] Probably an average value for the (E_4'', E_4') doublet.
[d] Frequencies matched to the data, in order to determine the force constants (Section 4.2).
[e] Data taken at liquid-nitrogen temperature.
[f] Data taken at liquid-helium temperature.

TABLE 5.4

Raman and infrared vibrational frequencies of β-GaS at room temperature. The frequencies are compared with those derived from the linear chain model ($C_w^s = 1.12 \times 10^5$, $C_b^s = 1.47 \times 10^3$, and $C_g^s = 1.58 \times 10^4$ dyn cm^{-1}).

β-GaS normal mode	Wright and Mooradian [98] Raman	Kuroda et al. [92, 94] Infrared	van der Ziel et al. [100] Raman	Hayek et al. [99] Raman	Irwin et al. [95] Raman	Yoshida et al. [101] Raman	Mercier and Voitchovsky [104] Raman	Finkman and Rizzo [96] Infrared	Linear-chain model [81]	Equivalent mode in the ε polytype
A_{2u}^1									0	$A_1''^1$
B_{2g}^1										$A_1''^2$
A_{2u}^2								318		$A_1''^3$
B_{2g}^2										$A_1''^4$
B_{1u}^1										$A_1'^1$
A_{1g}^1	183.9	186.2[a]	188.0	188	187.9	188[c]	191			$A_1'^2$
B_{1u}^2										$A_1'^3$
A_{1g}^2	357.6	363.8[a]	359.9	363	360.7	363[c]	362			$A_1'^4$
E_{1u}^1									0	E'^1
E_{2g}^1			22.0	22	22.0	24[c]	24		22.0[b]	E'^2
E_{2g}^2		310							294.0[b]	E'^3
E_{1u}^2	289.3	296.3[a]	295.2	296	295.0	295[c]	294	294	295.8	E'^4
E_{1g}^1									71.5	E''^1
E_{2u}^1	69.6	58.7[a]	74.2	74	75.2	75[c]	76		75.2[b]	E''^2
E_{2u}^2									298.3	E''^3
E_{1g}^2	285.7		291.4	291.5	290.5				300.0	E''^4

[a] Obtained through a multiphonon analysis.
[b] Frequencies matched to the data, in order to determine the force constants (Section 4.2).
[c] Data taken at liquid-nitrogen temperature.

two-phonon absorption bands. The results of their oscillator analysis of the strong band in GaSe at 211 cm^{-1}, which neglects the small contribution of the nearly degenerate conjugate mode in the ε structure (see Section 4.2), are given in Table 5.1. Group theoretical analyses of the normal modes were also carried out, and a simple model of the long-wavelength rigid-layer modes was proposed. By combining their infrared absorption and electroluminescence data with the Raman results of Wright and Mooradian [98], they determined a number of the phonon frequencies in the two materials, which are listed in Tables 5.3 and 5.4. They specified the irreducible representations only for the basal-plane infrared-active modes in these tables.

Although Kuroda et al. assigned several minima in the transmission spectra of GaS and GaSe to overtones and combination modes, the frequencies of many critical-point phonons were unknown to them, and they did not employ any selection rules in their analysis. We offer the following comments on their assignments. The transmission minima at 632.5 cm^{-1} in GaS and 420 cm^{-1} in GaSe were identified as the first over-tones of the E_{1u}^2 and E'^3 modes, respectively (Tables 5.3 and 5.4). This identification is probably correct for GaSe, as the first overtone has $A_1' + E'$ symmetry ($E' \times E' = A_1' + E'$ [58]) and is infrared-active. However, the first overtone of the E_{1u}^2 mode in GaS should appear closer to the minimum at 592.5 cm^{-1}, rather than 632.5 cm^{-1}; Kuroda et al.'s 310 cm^{-1} frequency for the E_{1u}^2 mode is probably 16 cm^{-1} too high (compare Table 5.5 and Figure 5.2). The first overtone of E_{1u}^2 is also inactive in the infrared ($E_{1u} \times E_{1u} = A_{1g} + E_{2g}$ [58]). We propose a reassignment of the minimum at 592.5 cm^{-1} to the combination mode, E_{1u}^2 and its conjugate E_{2g}^2, which is infrared-active ($E_{1u} \times E_{2g} = B_{1u} + B_{2u} + E_{1u}$ [58]) and has a compatible frequency. The minimum at 618.8 cm^{-1} in GaSe cannot be the first overtone of the $A_1'^4$ mode at 308.6 cm^{-1} (Wright and Mooradian [98]), because the overtone is inactive ($A_1' \times A_1' = A_1'$ [58]). While it is possible that the 618.8 cm^{-1} minimum is the third over-tone of the E'^3 TO mode, the agreement between the frequencies is rather poor. In summary, apart from the four localized modes in their spectra and the overtones just mentioned, the multiphonon analysis of Kuroda et al. is unconvincing. In this family, as in all families of layered materials, much work remains to be done in the areas of multiphonon absorption and higher-order scattering.

The next investigation of GaSe by Wieting and Verble [65] combined both Raman and infrared experiments. Reflectivity measurements for $E \perp c$ were reported between 175 and 4100 cm^{-1}, and the data were analyzed by classical oscillator theory (see Table 5.1). In addition to the five lines observed by Wright and Mooradian [98], a sixth line was identified by symmetry and bonding arguments as the basal-plane rigid-layer mode. The crystal used in this investigation was grown by the Bridgman method, and its structure was established through Debye-Scherrer studies of finely chopped fragments taken from one edge of the crystal. The X-ray studies confirmed the earlier work of Jellinek and Hahn [11] that GaSe has the β structure. The group theoretical analysis, correlation diagram, and normal-mode assignments of the phonon frequencies were worked out by these authors on the basis of the β polytype. This proved, however, to be the wrong polytype for GaSe, as subsequent and more complete structural investigations have shown [90, 91]. These structural developments deserve a brief discussion.

GaSe crystals have been grown from the vapor phase [102] and by the Bridgman

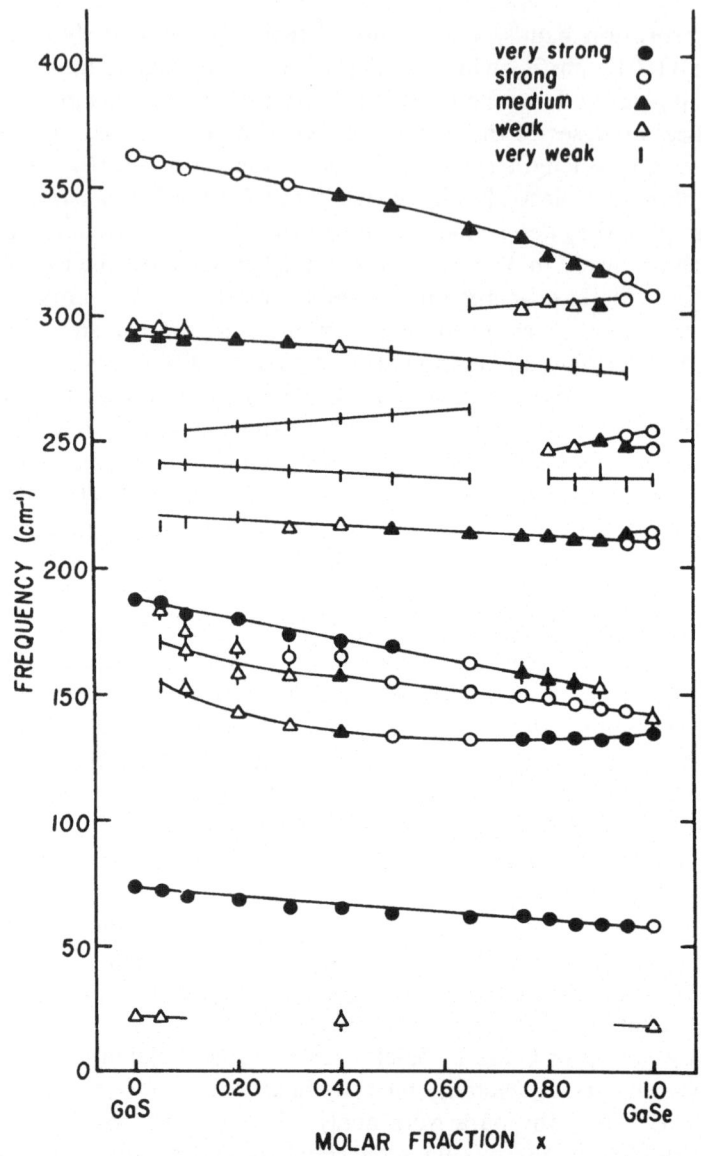

Fig. 5.1.   Frequencies of the Raman-active lattice vibrations of $GaS_{1-x}Se_x$ as a function of the molar fraction x (figure reproduced from Hayek *et al.* [99] by permission).

method [103]. The platelets produced by vapor transport are structurally superior to Bridgman crystals, which have many stacking faults due to displacements of the layers along the basal plane. Terhell and Lieth [91] have shown that Se-poor platelets of GaSe have the ε structure, when examined by a single-crystal X-ray technique, and that they exhibit several lines characteristic of the β structure, when examined by the Debye-Scherrer technique. These authors also showed that numerous stacking faults are introduced during powdering. Since the relatively stable β polytype [12, 83] cannot be created by powdering ε-GaSe, Terhell and Lieth concluded that Debye-Scherrer pictures of GaSe are misleading and that the β polytype of Jellinek and Hahn [11] is an artifact of the powdering process. In response to the work of Terhell and Lieth, we have re-examined the Bridgman crystal in our Raman and infrared investigation [65]. The single-crystal X-ray pictures* definitely show a hexagonal phase, without trace of γ–GaSe, but the ε and β structures cannot be differentiated, because of the degree of stacking disorder. We therefore concur with Terhell and Lieth. Our X-ray results, however, do not support the assertion of Aulich et $al.$ [103] and Mercier and Voitchovsky [104] that Bridgman crystals have some admixture of the γ polytype.

The phonon frequencies of GaSe obtained by Wieting and Verble [65] are reassigned in Table 5.3, according to the irreducible representations of the ε structure. Because of the lower crystal symmetry, the E_{1u}^2 and E_{2g}^2 lattice vibrations in the β polytype become a nearly degenerate pair of E' infrared-active modes in the ε polytype (Figure 2.8). The TO phonon band at 213.9 cm^{-1} should therefore be analyzed as a nearly degenerate pair. We have already discussed this point in Sections 3.2 and 4.1, where it was shown that the in-phase mode (E'^3 or E_{1u}^2 in GaSe) dominates the infrared spectrum. Thus the 213.9 cm^{-1} TO band is assigned to the E'^3 mode in Table 6.3, instead of to the E'^4 mode. Wherever there is a choice in the assignments in Table 5.3, we attribute the infrared frequency to the in-phase mode and the Raman frequency to the counter-phase mode. The displacement vectors and phases of all 24 normal modes of GaSe are illustrated in Figure 4.5. Lastly, the Raman line at 249 cm^{-1} has been reassigned to the (E'^3 LO, $A_2''^3$ LO) doublet for reasons we now take up.

A Raman study of the mixed-crystal system GaS$_{1-x}$Se$_x$ has been reported by Hayek et $al.$ [99]. These authors looked for correlation (or Davydov) splittings among the Raman-active modes and traced the movement of the phonon frequencies as the molar fraction x was varied between 0 and 1. Figure 5.1 summarizes their results. One-mode behavior [105, 107] is clearly observed for the $E_{2g}^2 \rightarrow E'^2$ and $E_{1g}^1 \rightarrow E''^2$ phonons; that is, the GaS E_{2g}^2 and E_{1g}^1 modes at 22 and 74 cm^{-1}, respectively, move continuously as x is varied and become the GaSe E'^2 and E''^2 modes at 18 and 58.5 cm^{-1}. Two-mode behavior, on the other hand, is observed for the $A_{1g}^1 \rightarrow A_1'^2$ phonons, since each end-member mode in the mixed-crystal system retains its separate identity for all values of x. In addition, Hayek et $al.$ concluded that the doublet at 210.5 and 214.5 cm^{-1} and the doublet at 247 and 254 cm^{-1} were also two-mode bands, but subsequent work has not supported their claim. According to Hayek et $al.$ the line at 210.5 cm^{-1} has E' symmetry; however, Hoff, Clayman, and Bromley [95, 97] have shown that the line has E'' symmetry. The two lines at 247 and 254 cm^{-1}, moreover, have been identified by Yoshida et $al.$ [101] and Hoff and Irwin [97] as the $A_2''^3$ and E'^3 LO modes. Table 5.3 reflects these changes in the assignments of the 210.5, 247, and 254 cm^{-1} lines.

The strongest case for the observation of correlation splittings in GaS$_{1-x}$Se$_x$ is based upon the two doublets at (135, 141) cm^{-1} and 307.5 cm^{-1} in GaSe. The doublets were well-resolved by Hayek et $al.$ only when 5 mole percent of S was added to GaSe. In the pure material ($x = 1$) the 135 cm^{-1} line was much stronger than the 141 cm^{-1} line, and under these conditions the polarization selection rules for the weaker line

---

* The X-ray study was conducted by F. L. Carter.

cannot be verified. We note that the stronger component of the (135, 141) doublet has the *lower* frequency. In Section 4 we argued that the stronger component should correspond to the counterphase mode, which presumably has a *higher* frequency than the in-phase mode, due to the interlayer interaction (compare Equations (4.5) and (4.9)). The frequencies of the E'^3 and E'^4 modes at 213.9 and 214.5 cm^{-1} are in agreement with the arguments of Section 4, and we expect similar agreement for the $(A_1'^1, A_1'^2)$ conjugate pair. Hayek *et al.* observed that the components of the doublet repelled each other and transferred intensity for $x \gtrsim 0.7$, and this may explain why the weaker component has the higher frequency. Although Hoff *et al.* [107] have suggested that the shoulder at 141 cm^{-1} in ε–GaSe may be due to second-order scattering or impurities, the magnitude of the doublet splitting (6 cm^{-1}) matches that of the linear-chain model (Table 5.3), and the comportment of the doublet is consistent with that of a conjugate pair, apart from the intensity interchange. By contrast, the behavior of the doublet at 307.5 cm^{-1} is considered normal. The linear-chain model predicts a very small splitting (0.2 cm^{-1}) for the $A_1'^3$ and $A_1'^4$ modes, and the counterphase $A_4'^4$ mode, which has the higher frequency, moves continuously as the molar fraction is varied and joins the A_{1g}^2 mode in GaS. The $A_1'^3$ mode weakens and finally vanishes for $x \lesssim 0.6$; the conjugate of the A_{1g}^2 mode was unobserved. Although Mercier and Voitchovsky's data [104] parallels and supports the results of Hayek *et al.*, it would be useful to have further experimental evidence of conjugate-mode behavior in GaS$_{1-x}$Se$_x$.

According to Irwin *et al.* [95, 97], the frequency of the E''^3 mode in ε–GaSe is lower than that of the E'^3 mode. In the linear-chain approximation the fundamental difference between the two modes is the participation of the Ga–Ga bond in the E''^3 mode, which contributes additional positive terms to the expression for the eigenfrequency (see Figure 4.5 and Equations (4.4) and (4.6)). The predicted frequency of the E''^3 mode, 223.6 cm^{-1} (Table 5.3), lies well above the E'^3 frequency. There is little doubt that the polarization data of Irwin *et al.* are correct, since similar results [100, 95] have been obtained on GaS (Table 5.4, the E_{1g}^2 and E_{2g}^2 modes); thus, the assumption of only nearest-neighbor interactions between the atomic planes of GaSe needs to be revised. Hoff *et al.* [107] have studied the variation of the phonon frequencies of the polar modes in ε–GaSe with propagation direction. Their data was found to satisfy the first limiting case of Loudon [77], in which electrostatic forces dominate over the anisotropy of the force constants in a uniaxial crystal. A lattice dynamical model for ε–GaSe should therefore include effective charges and long-range Coulomb forces. Bromley and Irwin [108] have proposed a model for ε–GaSe that contains both short-range forces, extending out to one unit-cell distance a, and Coulomb forces with two effective charges, one for atomic displacements perpendicular to the c-axis and another for displacements parallel to the c-axis. Their calculated dispersion curves agree quite well with Brebner, Jandl, and Powell's neutron data [109, 110] for $q \| c$.*

---

* Brebner *et al.* calculated the dispersion curves for ε-GaSe, using an axially symmetric Born-von Karman model and neglecting Coulomb forces. Although their curves generally fit the neutron data, the large LO–TO splittings in ε-GaSe ($\simeq 40$ cm^{-1} in the case of the E'^3 mode) and the wealth of experimental detail between 210 and 260 cm^{-1} in the phonon spectrum for $q = 0$ can only be accounted for by using effective charges.

TABLE 5.5

Dielectric constants and oscillator parameters for the A_{2u}^2 and E_{1u}^2 lattice vibrations of β-GaS (room-temperature data).

Authors	Method of determination	ε_∞	ε_0	ρ	γ	$\bar{\nu}_{TO}$ (cm^{-1})	$\bar{\nu}_{LO}$ (cm^{-1})
Brebner [86]	Interference fringes ($\simeq 1$ μm) $E\perp c$	9.0	11.5[a]				
Finkman and Rizzo [96]	Lorentz oscillator $E\perp c$			0.20 (± 0.06)	0.005 (± 0.002)	294 (± 2)	332.5[b]
	Lorentz oscillator $E\|c$			0.056 (± 0.006)	0.03 (± 0.003)	318 (± 2)	

[a] Calculated from Equation (3.10) for a single oscillator, using Finkman and Rizzo's value of ρ.
[b] Calculated from Equation (3.14) for a single oscillator, using Brebner's value of ε_∞.

Finkman and Rizzo [96] have investigated the far-infrared reflection spectra of β–GaS and ε–GaSe between 50 and 500 cm^{-1}. Their oscillator fits to the data for $E\|c$ and $E\perp c$ were carried out by trial and error (Tables 5.1, 5.2, and 5.5). As starting points for the oscillator fits, they used Kramers-Kronig analyses of the reflection spectra, taken at a 45° angle of incidence. Two of the frequencies obtained for $E\perp c$, 247 cm^{-1} (GaSe) and 340 cm^{-1} (GaS), were assigned to the E'''^3 and E_{2u}^2 vibrations. Since these vibrations are inactive in the infrared (Tables 2.9 and 2.10), Finkman and Rizzo proposed that the selection rules were altered by the stacking faults, which reduce the crystal symmetry to C_{3v} (γ–GaSe). We do not agree with this interpretation of the data. Stacking faults are much less of a problem in β–GaS than in ε–GaSe, and

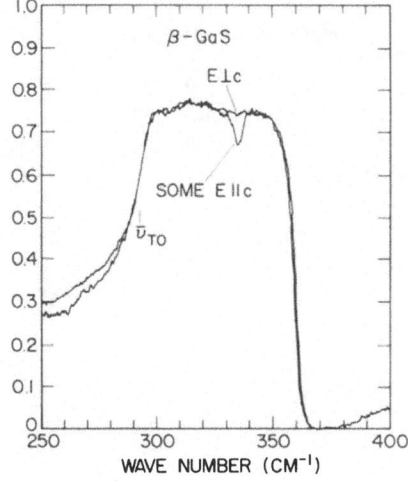

Fig. 5.2. Spectral reflectance of β-GaS at room temperature and near normal incidence.

the small dips in their reflection spectra for $E \perp c$ can be more easily explained as the $A_2''^3$ and A_{2u}^2 TO modes leaking through because of incomplete polarization. We have obtained very similar data on β–GaS (Figure 5.2),* when a small component of the electric vector was parallel to the c-axis. The 247 and 340 cm^{-1} features have been omitted from Tables 5.3 and 5.4. Finkman and Rizzo's rigid-layer frequencies at 37, 19.5, and 21.5 cm^{-1} were apparently meaured by a.c. techniques.

Hoff *et al.* [107] have compared the frequencies of the Raman-active modes in the two polytypes, ε–GaSe and γ–GaSe. Needle-shaped crystals having the γ structure were grown by the sublimation technique [111]. Their Raman data at room temperature is given in Table 5.6, and the atomic displacements for each mode are shown in Figure 5.3. These displacements are different from those of Hoff *et al.*, as they based their analysis on a nonprimitive hexagonal unit cell for γ–GaSe. Figure 2.2 indicates that the primitive unit cell of γ–GaSe contains only one molecule; consequently, all of the normal modes are internal modes, and no rigid-layer vibrations exist. Although these authors observed a Raman line in γ–GaSe at 20.9 ± 2 cm^{-1} and identified it as a

TABLE 5.6

Comparison of the long-wavelength lattice vibrations of ε- and γ-GaSe at room temperature [107]. The 134.3 cm^{-1} Raman line may belong to $A_1'^1$, if an intensity interchange occurs between conjugate modes (see text).

ε-GaSe irreducible representation	εGaSe frequency (cm^{-1})	γ-GaSe frequency (cm^{-1})	γ-GaSe irreducible representation
$A_2''^1$	0	0	A_1^1
$A_2''^2$	36.7^a		
$A_2''^3$ TO	237.0^b	235.7	A_1^2 TO
$A_2''^3$ LO	247.0^c	246.5	A_1^2 LO
$A_2''^4$			
$A_1'^1$		133.5	A_1^3
$A_1'^2$	134.3		
$A_1'^3$		307.4	A_1^4
$A_1'^4$	308.0		
E'^1	0	0	E^1
E'^2	19.5	(20.9 ± 2)	
E'^3 TO	212.1^a	213.9	E^2 TO
E'^3 LO	252.1 ± 2	253.2	E^2 LO
E'^4 TO	214.0 ± 2		
E''^1		59.5	E^3
E''^2	60.1		
E''^3		208.7	E^4
E''^4	211.9		

^{a} Reference [95].
^{b} Reference [84].
^{c} Reference [97].

* The crystals were supplied by Robert F. Shaw.

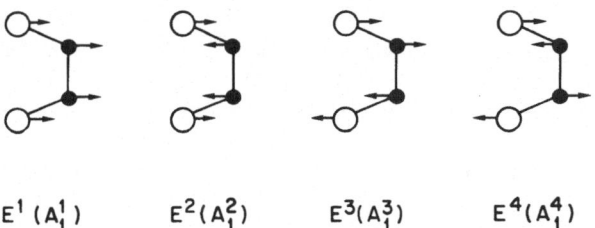

$$E^1 (A_1^1) \qquad E^2(A_1^2) \qquad E^3(A_1^3) \qquad E^4(A_1^4)$$

Fig. 5.3. Displacement vectors for the twelve normal modes of γ-GaSe and their irreducible representations. The displacement vectors for the c-axis modes ($4A_1$) are obtained by rotating the above vectors for the basal-plane modes ($4E$) by 90°.

rigid-layer mode, the line was very weak (much weaker than the E'^2 vibration in ε–GaSe) and probably stems from a small admixture of ε stacking in their crystal. In Table 5.6 we assign the Raman lines in ε–GaSe to the counterphase modes (as also do Hoff *et al.*) and list the normal vibrations of γ–GaSe opposite the in-phase modes of ε–GaSe. Since the interactions between nearest-neighbor layers are virtually the same in the two polytypes, the differences in the Raman frequencies should reflect the strength of the interlayer coupling. There are six pairs of modes that can be used to test the linear-chain model: $(A_1^3, A_1'^2)$, $(A_1^4, A_1'^4)$, (E^1, E'^2), $(E^2, E'^4 \text{ TO})$, (E^3, E''^2), and (E^4, E''^4). All of the in-phase modes (γ–GaSe irreducible representations) have smaller frequencies than their counterphase analogues in ε–GaSe, but the splittings of the crucial pairs, $(A_1^3, A_1'^2)$ and (E^3, E''^2), do not agree very well with the values, 8.6 cm^{-1} and 3.6 cm^{-1}, predicted by the linear-chain model (Table 5.3). Some of the discrepancy may be due to the experimental error in the Raman frequencies, which is typically ± 1 cm^{-1}, but it also seems likely that second-order interactions between more distant atomic planes are required in the model.

The only investigation of resonant Raman scattering in ε–GaSe has been reported by Hoff and Irwin [97]. They measured the wavelength dependence of the scattering efficiency below the energy gap and the angular variation of the scattering. Martin's theory [113] was used to interpret the data, but only partial agreement was obtained. For the polar modes the $1s$ or ground state of the free exciton in GaSe was established as the intermediate state in the scattering process. The scattered light, however, was observed to be independent of the scattering angle – a result at variance with Martin's theory. The scattering direction of the $A_2''^3$ LO phonon also apparently violated momentum conservation. Stacking faults in GaSe were tentatively invoked by the authors to explain both the angular independence of the scattering and the lack of $\mathbf{q}$ conservation. Other experimental results of this study were the breaking of Raman inactivity near resonance for the $A_2''^3$ LO mode, a two-phonon line at 509 cm^{-1}, and a pair of unexplained antiresonances for the A_1' nonpolar phonons.

Mushinskii and Kobolev [114] have reported an infrared study of β–InSe between 200 and 4000 cm^{-1}, in which they analyzed the reflectance data for $E \perp c$ by the Kramers-Kronig and classical oscillator methods. Their results are given in Table 5.7. The TO mode at 194 cm^{-1} satisfies the simple mass-scaling relation of Equation (4.4), if the mass of the In atom is substituted for that of the Ga atom and the force constant C_w^s is assumed to be invariant. Equation (4.4) predicts that the E_{1u}^2 TO frequency of InSe should be 190 cm^{-1}. An absorption spectrum was also obtained by Mushinskii

TABLE 5.7

Kramers-Kronig and classical oscillator analyses of the E_{1u}^2 lattice vibration of β-InSe (room-temperature data).

Author	Method of determination	ε_∞	ε_0	ρ	γ	$\bar{\nu}_{TO}$ (cm^{-1})	$\bar{\nu}_{LO}$ (cm^{-1})
Mushinskii and Kobolev [114]	Kramers-Kronig					194 (± 2)	212 (± 2)
	Lorentz oscillator	6.7	7.8[a]	0.088[b]	0.015		

[a] Calculated from Equation (3.14) for a single oscillator.
[b] Calculated from Equation (3.10) for a single oscillator.

and Kobolev; several features were interpreted as combination and overtone bands of the zone-boundary optical and acoustical phonons. The frequencies of these phonons were found to agree with Brout's sum rule [115, 116] to within 7%, but they differed from the linear expressions of Keyes [117] for zone-boundary phonons. However, these comparisons are questionable, since Brout's sum rule requires the inclusion of all phonon frequencies for a given $\mathbf{q}$, not just the infrared and acoustical phonons. Furthermore, Keyes developed his linear expressions for the diamond and zinc-blend structures, which are isotropic. The complex phonon spectrum of β-InSe also allows much greater flexibility in the analysis of the multiphonon absorption spectrum.

5.2. MoS$_2$ FAMILY

Wilson and Yoffe [16] reported the first observation of TO phonons in the MoS$_2$ family. They obtained the transmittance spectrum of 2H–MoS$_2$ between 250 and 4000 cm^{-1} and identified the absorption band at 374 cm^{-1} as the E_{1u} vibration (Table 2.5). Wieting and Verble [112] followed with a study of the reflectances for $E \perp c$ and $E \| c$, which they analyzed by the Kramers-Kronig and classical oscillator methods. The E_{1u} and A_{2u} infrared frequencies were also measured by Agnihotri et al. [118] and Lucovsky et al. [119]. All of these data are collected in Tables 5.8 and 5.9, including several determinations of the dielectric constants for both polarization directions.

The high-frequency dielectric constants in Table 5.8 are clearly incompatible: the experimental data fall into two broad groups, which have representative values of about 7.5 and 15, and the spectral regions overlap in the various investigations. Since there is no obvious way of reconciling the data, we show in Figure 5.4 three hitherto unreported spectra between 2000 and 18 000 cm^{-1}. The spectra for $E \perp c$ were obtained on a bright natural specimen, whose basal plane was perpendicular to the direction of the incident radiation. Multiple reflections from the lower surface of the specimen are negligible, except in the region between 3000 and 10 000 cm^{-1}; however, the correction to R^* in Figure 5.4 is seen to be small. At 2500 cm^{-1} (4 μm) the reflectivity is 0.349 $\pm$ 0.003. Thus $\varepsilon_\infty = 15.1 \pm 0.2$ (from Equation (3.9)), which agrees with our earlier

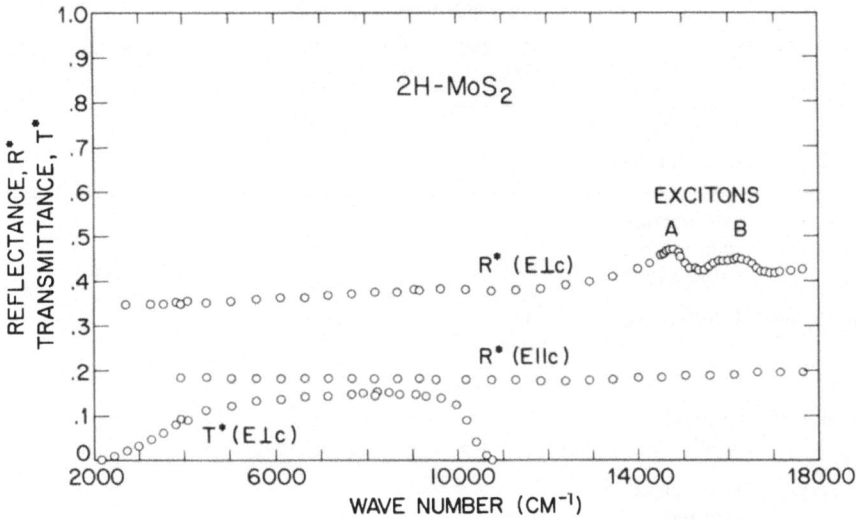

Fig. 5.4. Reflectance and transmittance spectra of 2H–MoS₂ at room temperature.

TABLE 5.8

Dielectric constants for $E \perp c$ and oscillator parameters for the E^2_{1u} lattice vibration of 2H-MoS₂ (room-temperature data, unless otherwise noted).

Authors	Method of determination	ε_∞	ε_0	ρ	γ	$\bar{\nu}_{TO}$ (cm^{-1})	$\bar{\nu}_{LO}$[a] (cm^{-1})
Bailly [120]	Transmittance ($\simeq$0.85 μm)	18.80 ($\pm$0.06)					
Lagrenaudie [121]	($\simeq$1.1 μm)	7.6[b]					
Frindt and Yoffe [122]	Interference fringes ($\simeq$2.5 μm)	7.0 ($\pm$1.4)					
Evans and Young [123]	Interference fringes ($\simeq$2 μm, 13.5–19 μm)	7.34 ($\pm$0.16)					
Wilson and Yoffe [16]	Transmittance					374	
Wieting and Verble [112]	Lorentz oscillator	15.2 ($\pm$0.4)	15.4[c] ($\pm$0.4)	0.016 ($\pm$0.001)	0.0025 ($\pm$0.0002)	384 ($\pm$1)	387 ($\pm$1)
	Interference fringes (2.5–20 μm)	15.2 ($\pm$1.6)					
Agnihotri et al. [118]	Absorptance					377 ($\pm$2)	
Lucovsky et al. [119]	Reflectance (pressed pellet)					384	

[a] Calculated from Equation (3.14) for a single oscillator.
[b] Data obtained at liquid-nitrogen temperature.
[c] Calculated from Equation (3.10) for a single oscillator.

TABLE 5.9

Dielectric constants for $E\|c$ and oscillator parameters for the A_{2u}^2 lattice vibration of 2H–MoS$_2$ (room-temperature data).

Authors	Method of determination	ε_∞	ε_0	ρ	γ	$\bar{\nu}_{TO}$ (cm^{-1})	$\bar{\nu}_{LO}$ (cm^{-1})
Bailly [120]	Polarization microscope ($\simeq 0.85$ μm)	4.12					
Evans and Young [123]	Interference rings (0.71–1.5 μm)	3.10 (± 0.04)					
Wieting and Verble [112]	Lorentz oscillator	6.25 (± 0.2)	6.28[a] (± 0.2)	0.0024 (± 0.0002)	0.005 (± 0.0005)	470 (± 1)	471[b] (± 1)
Lucovsky et al. [119]	Reflectance (pressed pellet)					470	

[a] Calculated from Equation (3.10) for a single oscillator.
[b] Calculated from Equation (3.14) for a single oscillator.

value [112]. At 1176 cm^{-1} (0.85μ), $\varepsilon_\infty = 18.0$ ($R = 0.382$), which agrees with Bailly's value [120]. We think these results confirm the larger dielectric constant, 15.2 ± 0.4, given in Table 5.8. The reflectance data for $E\|c$ in Figure 5.4 was obtained on a large thick specimen, cut parallel to the c-axis with a diamond saw and polished on a chamois lap with alcohol and Linde B. Between 4000 and 13 000 cm^{-1} the dielectric constant is 6.1 ($R = 0.18$), which is in agreement with our previous value in Table 5.9 but differs in varying amounts from those of Bailly [120] and Evans and Young [123]. When the electric vector was polarized perpendicular to the c-axis, this specimen exhibited the A and B exciton bands, but the reflectance was somewhat lower than that of the first or basal-plane specimen.

Raman investigations of 2H–MoS$_2$ have been reported by Verble and Wieting [82, 112, 124], Agnihotri et al. [118], and Chen and Wang [125]. Their data are listed in Table 5.10 and compared with the frequencies predicted by the linear-chain model [81]. Except for some of the Raman frequencies of Agnihotri et al., principally the E_{2g}^1, the agreement is generally good. The near degeneracy of the E_{1u}^2 and E_{2g}^2 modes and the low frequencies of the E_{2g}^1 and B_{2g}^1 rigid-layer modes are the physical effects of the weak interlayer coupling. In Figure 5.5 we show the atomic displacements for the normal vibrations of 2H–MoS$_2$, which illustrate the origin of these effects. Chen and Wang also identified several second-order features of the Raman spectrum, using the neutron-scattering data of Wakabayashi et al. [126].

The interlayer interaction in MoS$_2$ and other layered materials has often been assumed to be of the van der Waals type. A model for the rigid-layer lattice vibrations of 2H–MoS$_2$, based on central forces and a van der Waals potential, has been proposed by Verble et al. [124]. By resolving the nearest-neighbor forces into components parallel and perpendicular to the c-axis, they obtained the following expression for the

TABLE 5.10

Raman and infrared vibrational frequencies of 2H–MoS$_2$ at room temperature. The frequencies are compared with those derived from the linear-chain model ($C_w^s = 1.67 \times 10^5$, $C_b^s = 2.71 \times 10^3$, $C_w^c = 2.50 \times 10^5$, and $C_b^c = 7.56 \times 10^3$ dyn cm^{-1}).

2H–MoS$_2$ normal mode	Vibrational frequencies (cm^{-1})						
	Wilson and Yoffe [16]	Wieting and Verble [112]	Wakabayashi et al. [126]	Agnihotri et al. [118]	Lucovsky et al. [119]	Chen and Wang [125]	Linear-chain model [81]
A_{2u}^1							0
B_{2g}^1			56[a]				56[b]
A_{2u}^2		470			470		470[b]
B_{2g}^2							475.1
B_{1u}							363.9
A_{1g}		409		395		408.3	374.7
E_{1u}^1							0
E_{2g}^1		33.7	36[a]	118		32	33.7[b]
E_{1u}^2	374	384		377	384		384[b]
E_{2g}^2		383		377		383	386.2
E_{2u}							297.3
E_{1g}		287		298		286	302.1

[a] Data obtained by inelastic neutron scattering.
[b] Frequencies matched to the data, in order to determine the force constants (Section 4.2).

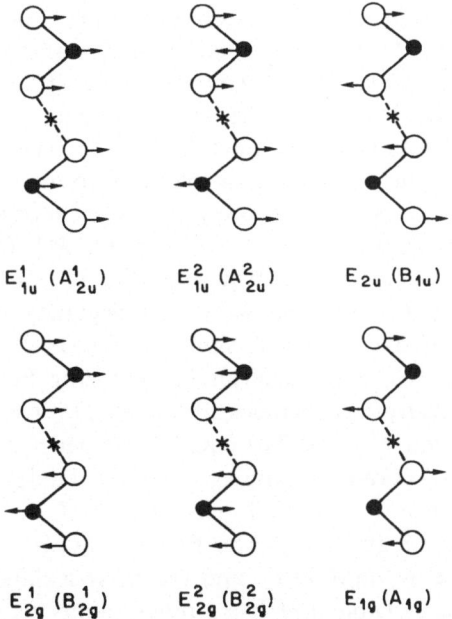

$E_{1u}^1 \ (A_{2u}^1)$ $E_{1u}^2 \ (A_{2u}^2)$ $E_{2u} \ (B_{1u})$

$E_{2g}^1 \ (B_{2g}^1)$ $E_{2g}^2 \ (B_{2g}^2)$ $E_{1g} \ (A_{1g})$

Fig. 5.5. Displacement vectors for the eighteen normal modes of 2H–MoS$_2$ and their irreducible representations. The displacement vectors for the c-axis modes ($2A_{2u} + 2B_{2g} + B_{1u} + A_{1g}$) are obtained by rotating the above vectors for the basal-plane modes ($2E_{1u} + 2E_{2g} + E_{2u} + E_{1g}$) by 90°.

ratio of the compressional, interlayer force constant to the shear, interlayer force constant:

$$\frac{C_b^c}{C_b^s} = \tfrac{3}{2} \tan \theta, \tag{5.1}$$

where θ is the angle between the sulfur-sulfur interlayer bond and the basal plane. For 2H–MoS$_2$, $\theta = 60.1°$ [15] and $C_b^c/C_b^s = 2.61$. The linear-chain model [81] predicts that the ratio of the force constants is given by the square of the ratio of the rigid-layer frequencies (see Equation (4.7) for ε–GaSe)

$$\frac{C_b^c}{C_b^s} = \frac{\omega^2(B_{2g}^1)}{\omega^2(E_{2g}^1)}. \tag{5.2}$$

Using the values in Table 5.10, we find that $C_b^c/C_b^s = 2.76$, in good agreement with the central-force value. However, Webb et al. [127] have shown that Equation (5.1) for central forces is incorrect and that it should be replaced by

$$\frac{C_b^c}{C_b^s} = 2 \tan^2 \theta. \tag{5.3}$$

This equation increases the ratio C_b^c/C_b^s from 2.61 to 6.0, and hence the central-force model of the interlayer coupling is inadequate. In the following article in this volume by Wakabayashi and Nicklow, the coupling between the layers is satisfactorily modelled with axially symmetric, instead of central, forces. The interatomic potential proposed by Verble et al. [124] also requires modification. The handbook value of the melting point of 2H–MoS$_2$ (1185 °C) was used by these authors to determine the depth of the potential well. Cannon [128] has demonstrated that the melting point is much higher, at least 1800 °C. If MoS$_2$ obeys the Tammann rule for the sintering temperature, the melting point would be approximately 2375 °C. The higher melting point of MoS$_2$ substantially alters the parameters of the interatomic potential.

Infrared and Raman studies of 2H–MoSe$_2$ have been reported by Agnihotri et al. [129, 118, 130], Lucovsky et al. [119], and Smith et al. [131]. Their data are presented in Tables 5.11 and 5.12. The large oscillator strength and LO–TO splitting (80 cm^{-1}) given by Agnihotri et al. are at variance with the reflectance data of Lucovsky et al., which show a relatively narrow resonance, similar to that in 2H–MoS$_2$. The oscillator parameters of Agnihotri et al. were determined by fitting Equation (3.8) to: (i) the experimentally derived absorption coefficient between 275 and 305 cm^{-1}, and (ii) the index of refraction between 250 and 310 cm^{-1}, obtained from interference fringes. Although these two sets of parameters agree, it is improbable that the substitution of Se for S in MoS$_2$ should lead to a 30-fold increase in both the oscillator strength and LO–TO splitting (Tables 5.8 and 5.11). The Raman data of Agnihotri also disagree fundamentally with those of Smith et al. and the mass-scaling results of Table 5.14. The rigid-layer E_{2g}^1 mode in Table 5.11 has a much higher frequency than the corresponding mode in 2H–MoS$_2$; this implies that the coupling between the layers is much stronger in MoSe$_2$. If the E_{2g}^1 frequency in 2H–MoSe$_2$ is correct, a parallel increase should be anticipated in 2H–MoTe$_2$.

TABLE 5.11

Dielectric constants for $E \perp c$ and oscillator parameters for the E_{1u}^2 lattice vibration of 2H–MoSe$_2$ (room-temperature data, unless otherwise noted).

Authors	Method of determination	ε_∞	ε_0	ρ	γ	$\bar{\nu}_{TO}$ (cm^{-1})	$\bar{\nu}_{LO}$[a] (cm^{-1})
Agnihotri et al. [129, 130]	Lorentz oscillator[b]	10.24	16.81	0.52[c]	0.004, 0.003	283	362.6
	Interference fringes (32–40 μm)	10.24	16.81	0.52[c]		277	354.9
Lucovsky et al. [119]	Reflectance (pressed pellet)					288	
Smith et al. [131]						286	

[a] Calculated from Equation (3.14) for a single oscillator.
[b] Data obtained at liquid-nitrogen temperature.
[c] Calculated from Equation (3.10) for a single oscillator.

TABLE 5.12

Raman and infrared vibrational frequencies of 2H–MoSe$_2$ at room temperature. The frequencies are compared with those derived from the linear-chain model ($C_w^s = 1.44 \times 10^5$, $C_b^s = 3.40 \times 10^4$, $C_w^c = 2.18 \times 10^5$, and 2.96×10^4 dyn cm^{-1}).

2H–MoSe$_2$ normal mode	Vibrational frequencies (cm^{-1})			
	Agnihotri et al. [118]	Lucovsky et al. [119]	Smith et al. [131]	Linear-chain model [81]
A_{2u}^1				0
B_{2g}^1				87.1
A_{2u}^2		350	352	352[a]
B_{2g}^2				359.2
B_{1u}				216.4
A_{1g}	361		244	244[a]
E_{1u}^1				0
E_{2g}^1	112		92	92[a]
E_{1u}^2	283	288	286	286[a]
E_{2g}^2	285		302	296.6
E_{2u}				175.8
E_{1g}	217		150	213.4

[a] Frequencies matched to the data, in order to determine the force constants (Section 4.2).

Studies reported on other members of the MoS_2 family are summarized in Tables 5.13 and 5.14. We have selected $2H-MoS_2$ as the reference material in Table 5.14, in order to calculate the frequencies of the remaining materials by mass scaling. For the B_{2g}^1 and E_{2g}^1 modes the reduced mass [81] is $(M_m + 2M_s)$; for the A_{2u}^2, E_{1u}^2, and E_{2g}^2 modes it is $M_m M_s/(M_m + 2M_s)$; and for the A_{1g} and E_{1g} modes it is M_s. (M_m is the

TABLE 5.13

Dielectric constants and TO phonon frequencies of $2H-MoTe_2$, $2H-WS_2$, $3R-WS_2$, and $2H-WSe_2$ (room-temperature data, unless otherwise noted).

Material	Method of determination	ε_∞ $(E \perp c)$	ε_∞ $(E \| c)$	$\bar{\nu}_{TO}$ (E_{1u}^2)	$\bar{\nu}_{TO}$ (A_{2u}^2)
$2H-MoTe_2$	Kramers-Kronig [132]	14			
	Absorptance [118]			240 (± 3)	
	Reflectance [133] (1.6 μm)	15[a]	8.4[a,b]		
$2H-WS_2$	Reflectance [119] (pressed pellet)			356	435
$3R-WS_2$	Reflectance [133] (0.83 μm)	19[a]	6.8[a,b]		
$2H-WSe_2$	Reflectance [119] (pressed pellet)			245	305

[a] Dielectric constant calculated from Equation (3.9).
[b] Reflectance data at liquid-nitrogen temperature.

mass of the metal atom and M_s is the mass of the chalcogen atom.) The agreement between the experimental data and the scaled frequencies is generally good, even though the rigid-layer frequencies of $2H-MoSe_2$ and $2H-MoTe_2$ are conspicuous exceptions. Moreover, the rigid-layer frequencies of $2H-NbSe_2$ agree very closely with the scaled frequencies. This result reinforces our previous remarks concerning $2H-MoSe_2$. If the interlayer coupling is primarily determined by the selenium-selenium bonds across the gap, the rigid-layer frequencies of $2H-MoSe_2$ and $2H-NbSe_2$ should be nearly equal. Since the Raman data indicate that the frequencies differ sharply, future investigations should be directed toward an explanation of these effects.

5.3. CdI_2 FAMILY

Apart from the hydroxides, the early studies [136, 137] of this family were limited to polycrystalline specimens and various mulls and solutions of powder specimens. Carabatos [138] has reported the first single-crystal Raman results on $2H-CdI_2$ and $2H-PbI_2$; Smith et al. [139] have reported the first infrared measurements on HfS_2. The oscillator parameters for HfS_2 and subsequent infrared results on the halides and chalcogenides are gathered in Table 5.15. All of the materials in this table have large splittings between the LO and TO frequencies (the LO frequency is typically twice the TO frequency). Comparing the three families mentioned thus far, we see that the

TABLE 5.14

Raman and infrared vibrational frequencies of the 2H polytypes of MoS_2, $MoSe_2$, $MoTe_2$, WS_2, WSe_2, and $NbSe_2$ at room temperature. The experimental data are compared with the frequencies obtained by mass-scaling the data for MoS_2.

Symmetry of normal mode	Vibrational frequencies (cm^{-1})										
	2H-MoS₂ [112] Data	2H-MoSe₂ Data	[131] Mass scaling	2H-MoTe₂ Data	[118] Mass scaling	2H-WS₂ Data	[119] Mass scaling	2H-WSe₂ Data	[119] Mass scaling	2H-NbSe₂ Data	[134] Mass scaling
A_{2u}^1	56		44		38		45		38	45.0[a]	45
B_{2g}^1	470	352	377		349	435	423	305	316		381
A_{2u}^2											
B_{2g}^2											
B_{1u}											
A_{1g}	409	244	261	321	205		409		261	230.9[b]	261
E_{1u}^1	33.7	92	27	108	23		27		23	29.6[b]	27
E_{2g}^1	384	286	308	240	285	356	345	245	258	238.3[b]	311
E_{1u}^2	383	302	307	237	284		344		258		310
E_{2g}^2											
E_{2u}											
E_{1g}	287	150	183	207	144		287		183		183

[a] Data obtained by inelastic neutron scattering [135].
[b] Data obtained at liquid-nitrogen temperature.

TABLE 5.15

Dielectric constants for $E \perp c$ and oscillator parameters for several members of the CdI_2 family (room-temperature data).

Material	Method of determination	ε_∞	ε_0	ρ	γ	$\bar{\nu}_{TO}$ (cm^{-1})	$\bar{\nu}_{LO}$ (cm^{-1})
$CdBr_2$	Reflectance [119] (pressed pellet)					$\simeq 95$	$\simeq 170$
CdI_2	Reflectance [119] (pressed pellet)					$\simeq 60$	$\simeq 135$
ZrS_2	Reflectance [119]	9.23	34.5	2.01	0.09	181	350
HfS_2	Reflectance [139]	5.3	20.5	1.21	0.025	160	315
	Reflectance [119]	6.20	23.1	1.34	0.17	166	318
HfSSe	Reflectance [119]	9.80	37.0	1.79	0.12	122	
					0.374	0.22	219
$HfSe_2$	Reflectance [119]	8.05	38.2	2.40	0.27	98	215
SnS_2	Reflectance [119] (pressed pellet)					$\simeq 200$	$\simeq 320$
PbI_2	Reflectance [119] (pressed pellet)					$\simeq 40$	$\simeq 110$

splittings and oscillator strengths are smallest in the MoS_2 family (Table 5.1), intermediate in the GaSe family (Table 5.8), and largest in the CdI_2 family. These trends will be discussed in Section 5.6 in terms of effective charges and chemical bonding. Figure 5.6 illustrates the displacement vectors for the normal modes of $2H–CdI_2$ (or $CdCl_2$). The Raman data on the halides and chalcogenides are collected in Table 5.16. The frequencies derived from mass scaling are also shown in this table, and the agreement is seen to be generally fair, except for the infrared data of Lockwood [140].

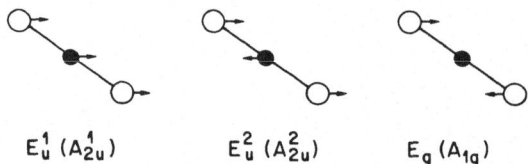

$E_u^1 (A_{2u}^1)$ $E_u^2 (A_{2u}^2)$ $E_g (A_{1g})$

Fig. 5.6. Displacement vectors for the nine normal modes of $2H–CdI_2$ (or $CdCl_2$) and their irreducible representations. The displacement vectors for the c-axis modes ($2A_{2u} + A_{1g}$) are obtained by rotating the above vectors for the basal-plane modes ($2E_{1u} + E_g$) by 90°.

Lockwood's A_{2u}^2 and E_u^2 TO frequencies were obtained on powder specimens dispersed in polyethylene; because of the surface modes in particles of small size [142], his frequencies fall somewhere in between the TO and LO frequencies in the bulk material. Smith et al. [139] have also compared the E_u^2 and E_g vibrations of HfS_2 by computing the local field for the polar modes and mass-scaling the E_u^2 frequency. Their model calculations demonstrate that the interlayer interaction has a negligible effect upon the vibrational frequencies of HfS_2.

TABLE 5.16

Raman and infrared vibrational frequencies of several members of the CdI_2 family (room-temperature data).

Material	Reference	A_{2u}^2	A_{1g}		E_u^2		E_g	
		Data	Data	Mass scaling	Data	Mass scaling	Data	Mass scaling
CdCl₂	[140]	164 (±2)ᵃ	233 (±1)	211	210 (±2)ᵃ	80	131 (±1)	82
	[141]		233.5				133.5	
	[140]	102 (±2)ᵃ	148 (±1)		161 (±2)ᵃ		77 (±1)	55
CdBr₂	[141]		146	141			74.5	
	[119]				≃95	65		
	[138]		112				28.2	
CdI₂	[141]		111.5	111.5ᵇ			43.5	43.5ᵇ
	[119]				≃60	60ᵇ		
MnCl₂	[140]	180 (±5)ᵃ	234.5 (±1)		230 (±5)ᵃ		144 (±1)	
CoCl₂	[140]	190 (±5)ᵃ	250 (±1)		235 (±5)ᵃ		152 (±0.5)	
TiS₂	[139]		335	337			232	253
TiSe₂	[139]		195	215			134	161
ZrS₂	[139]		333	337			235	253
	[119]				181	179		
ZrSe₂	[139]		194	215			148	161
HfS₂	[139]		337	337ᶜ	160	160ᶜ	253	253ᶜ
	[119]				166			
HfSSe	[119]				122, 219			
HfSe₂	[139]		198	215			155	161
	[119]				98	120		
SnS₂	[119]				≃200			

ᵃ Transmittance data on powder samples dispersed in polyethylene.
ᵇ Reference value for mass scaling in the cadmium dihalides.
ᶜ Reference value for mass scaling in the IVA dichalcogenides.

Ghosh [143] has used the linear-chain model to interpret the long-wavelength lattice vibrations of $CdBr_2$, $CdCl_2$, $MnCl_2$, and $CoCl_2$. His major conclusion, however, is weakened by several errors. The reduced masses in Ghosh's expressions for the frequencies of the infrared-active modes ($A_{2u} + E_u$) should be replaced by $mM/(M + 2m)$ [81], where m and M are the masses of the halogen and metal atoms, respectively. His analysis also neglects the effect of the polarization field of an infrared mode, which in ionic materials substantially alters the vibrational energy. Third, he compares the calculated frequencies of the A_{2u}^2 and E_u^2 modes with the infrared data of Lockwood [140] and invokes second-neighbor interactions to account for the discrepancies. Since Lockwood's frequencies are demonstrably too high (Table 5.16), Ghosh's conclusion concerning second-neighbor interactions in $CdCl_2$ and $CdBr_2$ lacks experimental support. The correct procedure for applying the linear-chain model to ionic materials is to calculate the local field for the polar modes, as Smith *et al.* [139] have done for HfS_2. The shift in the frequency of the polar mode from its value for zero polarization is a function of the dielectric constants and Lorentz factor [144, 145]. Because of the ionic character of the bond in all four of these materials (see Section 5.6), electrostatic interactions between distant atoms should be expected.

Nakashima [146] and Zallen and Slade [147] have investigated by means of Raman scattering the effects of polytypism in PbI_2. The earlier work of Carabatos [138] on $2H–PbI_2$ yielded three Raman lines at 76, 94.3, and 101 cm^{-1}; the latter pair were tentatively identified as the E_g and A_{1g} modes, respectively. Table 5.17 indicates, however, that the E_g and A_{1g} vibrations at room temperature lie at 74 and 96 cm^{-1} [147]

TABLE 5.17

Comparison of the long-wavelength lattice vibrations of $2H–PbI_2$ and $4H–PbI_2$ (room-temperature data, unless otherwise noted).

2H–PbI$_2$ normal mode	Vibrational frequencies (cm^{-1})					4H–PbI$_2$ normal mode
	2H–PbI$_2$			4H–PbI$_2$		
	[138]	[119]	[147]	[146]^a	[147]	
A_{2u}^1	(acoustical)			(acoustical)		A_1^1
						B_1^1
						A_1^2
A_{2u}^2						B_1^2
				97	96	A_1^3
A_{1g}	101		96			B_1^3
				(acoustical)		E_1^1
E_u^1	(acoustical)			25.0	14	E_2^1
						E_1^2
E_u^2		$\simeq 40$				E_2^2
				78.2	73.9^b	E_1^3
E_g	94.3		74	75.2	77.2^b	E_2^3

^{a} Data taken at 36 K.
^{b} Data taken at 9 K.

and that the broad feature near 101 cm^{-1} is a second-order band [146]. Two other discrepancies appear in the table, and to resolve them we refer to the 4H–PbI$_2$ displacement vectors in Figure 5.7. Although Nakashima attributed a very weak Raman line at 25 cm^{-1} to the rigid-layer E_2^1 mode,* Zallen and Slade's spectra clearly show a lower frequency for this mode. Using a valence force model for 4H–PbI$_2$, Zallen and Slade have calculated that the B_1^1 rigid-layer frequency should be approximately

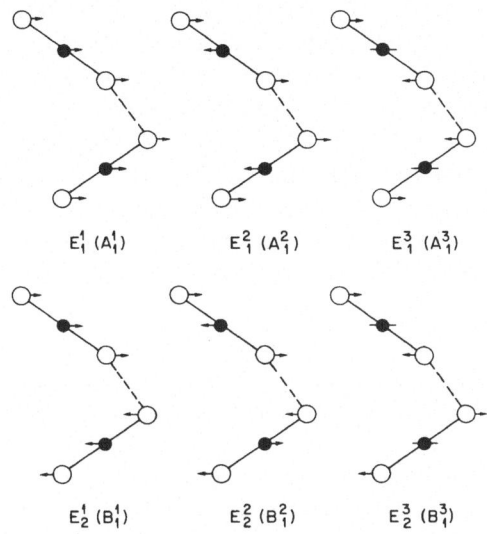

$$E_1^1 \; (A_1^1) \qquad E_1^2 \; (A_1^2) \qquad E_1^3 \; (A_1^3)$$

$$E_2^1 \; (B_1^1) \qquad E_2^2 \; (B_1^2) \qquad E_2^3 \; (B_1^3)$$

Fig. 5.7.   Displacement vectors for the eighteen normal modes of 4H–PbI$_2$ and their irreducible representations. The displacement vectors for the c-axis modes ($3A_1 + 3B_1$) are obtained by rotating the above vectors for the basal-plane modes ($3E_1 + 3E_2$) by 90°. The horizontal bars through the Pb atoms indicate that the directions of the displacements are undetermined.

27 cm^{-1}. This is close enough to Nakashima's Raman frequency to suggest that the silent B_1^1 mode (Table 2.14) may have become weakly active, perhaps through disorder in the layer stacking. The second discrepancy concerns the ordering of the frequencies of the E_1^3 and E_2^3 conjugate modes. In the E_g vibration of 2H–PbI$_2$, the layers vibrate 180° out of phase (as defined in Section 4.1), and the corresponding counterphase vibration in the 4H polytype has E_2 symmetry. Zallen and Slade have measured almost identical frequencies for the E_g and E_2^3 modes at 9 K (77.8 as compared with 77.2 cm^{-1}); predictably, the in-phase E_1^3 mode exhibited a somewhat lower frequency, 73.9 cm^{-1}. Since Nakashima apparently did not determine the polarizability components for his 75.2 and 78.2 cm^{-1} lines, the experimental data support the assignments of Zallen and Slade.

The hydroxides of Mg and Ca form a special 2H stacking group within the CdI$_2$ family. The axes of the hydroxyl ions are oriented parallel to the c-axis of the crystal, and the oxygen atoms are directed inward toward the center of the layer. In addition

* The superscripts on the irreducible representations of the normal modes in Figure 5.7 differ from those of Nakashima [146]. The superscripts are, of course, arbitrary and serve only as labels for modes having the same symmetry.

to the nine normal modes of 2H–CdI$_2$ (Equation (2.13) and Table 2.13), six more internal and external modes are associated with the hydroxyl ions. Their vibrational symmetries are given by

$$A_{2u} + A_{1g} + E_u + E_g. \tag{5.4}$$

The A representations in Equation (5.4) label the O–H bond-stretching internal vibrations, which are either symmetric or antisymmetric with respect to inversion through the metal atom. The E_u and E_g librations are doubly degenerate wagging motions of the hydroxyl ions, with displacements perpendicular to the c-axis. A number of infrared and Raman investigations, dating back to the original work of Coblentz [1], have been reported on Mg(OH)$_2$ and Ca(OH)$_2$. Much of the effort and interest has centered on the complex series of single-phonon and multiphonon resonances between 2500 and 4000 cm^{-1}, where the internal hydroxyl vibrations are observed. A review of the literature, supplemented by a careful reinvestigation of the infrared and Raman frequencies of Mg(OH)$_2$ and Ca(OH)$_2$, has recently been published by Dawson et al. [148]. We summarize their results in Table 5.18. These authors

TABLE 5.18

Infrared and Raman frequencies of Mg(OH)$_2$, Mg(OD)$_2$, Ca(OH)$_2$, and Ca (OD)$_2$ at room temperature [148]. The hydroxyl modes are $A_{2u}^3 + A_{1g}^2 + E_u^3 + E_g^2$.

| Symmetry of normal mode | Vibrational frequencies (cm^{-1}) | | | | | |
| | Mg(OH)$_2$ | Mg(OD)$_2$ | | Ca(OH)$_2$ | Ca(OD)$_2$ | |
	Data	Data	Mass scaling	Data	Data	Mass scaling
A_{2u}^1						
A_{2u}^2	461			334		
A_{1g}	443	434	431	357	350	349
E_u^1						
E_u^2	361			287.5		
E_g	280	277	272	254	252	247
A_{2u}^3	3688			3640		
A_{1g}^2	3652	2696	2655	3620	2661	2632
E_u^3	416			373		
E_g^2	725	506	527	680	475	495

also studied the deuterated specimens of the two materials and compared their data with the Raman frequencies calculated by mass scaling. The good agreement indicated in Table 5.18 confirms their mode assignments. The two E_u vibrations, which have similar infrared frequencies, were distinguished by their oscillator strengths: the stronger infrared resonance should be the E_u^2 mode, which has the metal and hydroxyl sublattices vibrating out of phase. Correlation or Davydov (A_{2u}^3, A_{1g}^2) splittings of 36 cm^{-1} and 20 cm^{-1} were observed in Mg(OH)$_2$ and Ca(OH)$_2$, respectively.

5.4. GRAPHITE

A Raman study of hexagonal graphite has been reported by Tuinstra and Koenig [149]. Although they obtained the same normal-mode decomposition as in Equation (2.15), they attributed the Raman line at 1575 cm^{-1} to a pair of unresolved E_{2g} phonons. In their analysis of the vibrational displacements, the E_{2g} phonons differ only by the phasing of the two layers in the unit cell; since the interlayer coupling is relatively weak, the two vibrations are nearly degenerate. However, hexagonal graphite has an inversion center between the layers (Figure 2.4), and the normal vibrations occur as symmetric and antisymmetric pairs, which reflect the two possible phasings of the layers. Thus the conjugate of the Raman-active E_{2g}^2 mode in Figure 5.8 is the infrared-active E_{1u}^2 mode, and the other E_{2g} mode is the basal-plane rigid-layer mode. The

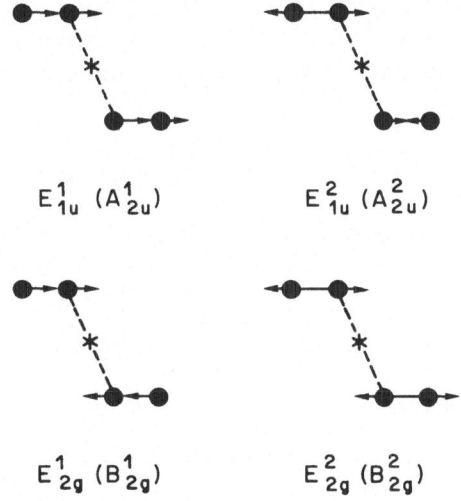

$$E_{1u}^1 \ (A_{2u}^1) \qquad\qquad E_{1u}^2 \ (A_{2u}^2)$$

$$E_{2g}^1 \ (B_{2g}^1) \qquad\qquad E_{2g}^2 \ (B_{2g}^2)$$

Fig. 5.8. Displacement vectors for the twelve normal modes of hexagonal graphite and their irreducible representations. The displacement vectors for the c-axis modes ($2A_{2u} + 2B_{2g}$) are obtained by rotating the above vectors for the basal-plane modes ($2E_{1u} + 2E_{2g}$) by 90°.

1575 cm^{-1} Raman line belongs to the E_{2g}^2 vibration [21]. These authors also observed a second Raman line at 1355 cm^{-1} in pyrolytic graphite, commercial graphite, and activated charcoal, which was assigned to an A_{1g} phonon located at the K point in the Brillouin zone. Because of the structural disorder in graphitic materials, Tuinstra and Koenig reasoned that the zone-boundary phonon becomes active through a particle-size effect similar to that found in polymers [150]. We propose an alternative interpretation. Structural investigations [151, 152] of powdered and commercial graphites have shown that the rhombohedral and hexagonal polytypes are often mixed. The space group of rhombohedral graphite is D_{3d}^5 (R$\bar{3}$m), and a correlation analysis yields the following correspondences between the irreducible representations of the hexagonal and rhombohedral polymorphs:

$$A_{2u} \rightarrow A_{2u}(z)$$
$$B_{2g} \rightarrow A_{1g}(x^2 + y^2, z^2)$$

$$E_{1u} \to E_u(x, y)$$

$$E_{2g} \to E_g(x^2 - y^2, yz; xy, zx), \tag{5.5}$$

where we have indicated the infrared and Raman activities in parentheses. The B_{2g}^2 mode is silent in the hexagonal structure but becomes Raman-active as an A_{1g} vibration in rhombohedral graphite. Nicklow et $al.$ [153], using an axially symmetric Born-von Kármán model, have calculated that the frequency of the B_{2g}^2 mode is approximately 1400 cm^{-1}. This agrees rather well with the 1355 cm^{-1} of Tuinstra and Koenig. Furthermore, Tuinstra and Koenig found that the 1355 cm^{-1} line was weakest in stress-annealed pyrolytic graphite and strongest in activated charcoal. The latter material probably has the greatest admixture of the rhombohedral polytype.

Brillson et $al.$ [21] have obtained both the infrared and Raman spectra of hexagonal graphite. A small peak in the reflectance at 1588 $\pm$ 5 cm^{-1} was attributed to the E_{1u}^2 mode, and a single Raman line at 1574 $\pm$ 1 was attributed to the E_{2g}^2 mode. The rigid-layer E_{2g}^1 and infrared-active A_{2u}^2 modes were unobserved.* Table 5.19 compares

TABLE 5.19

Infrared and Raman vibrational frequencies of hexagonal graphite at room temperature.

Graphite normal mode	Vibrational frequencies (cm^{-1})			
	Dolling and Brockhouse [154]	Tuinstra and Koenig [149]	Brillson et $al.$ [21]	Nicklow et $al.$ [153]
A_{2u}^1				
B_{2g}^1	128 $\pm$ 2			126
A_{2u}^2				1390[b]
B_{2g}^2		(1355[a])		1400[b]
E_{1u}^1				
E_{2g}^1	43 $\pm$ 10			45
E_{1u}^2			1588 $\pm$ 5	1575[b]
E_{2g}^2		1575	1574 $\pm$ 1	1575[c]

[a] An A_{1g} phonon in rhombohedral graphite or zone-boundary A_{1g} phonon in hexagonal graphite (see text).
[b] Calculated using an axially symmetric force-constant model.
[c] Matched in the calculation of [153] to Tuinstra and Koenig's frequency.

their measurements with Tuinstra and Koenig's data and several other frequencies obtained from inelastic neutron scattering. Bellodi et $al.$ [155] also measured the reflectance spectrum of graphite for $E \perp c$, but they did not observe the E_{1u}^2 peak reported by Brillson et $al.$

5.5. As_2S_3, Sb_2S_3, AND Bi_2Te_3 FAMILIES

Although the sesquichalcogenides of As, Sb, and Bi have different crystal structures, we shall discuss them together in this section, as few studies have been made of their

* The A_{1u} and B_{1g} representations in the group-theoretical analysis of Brillson et $al.$ [21] should be replaced by A_{2u} and B_{2g}.

long-wavelength lattice vibrations. As_2S_3 and As_2Se_3 have been investigated by Zallen et al. [24], who obtained infrared spectra for $E \perp b$ and Raman spectra in both backscattering and right-angle scattering configurations. Because of the low crystal symmetry and large number of atoms in the primitive unit cell (10 within each layer of the cell), the spectra are more difficult to interpret than the other families we have discussed. According to Table 2.16, there are 13 infrared-active modes with B_u symmetry for $E \perp b$. Zallen et al. have pointed out that the B_u vibrations can be divided into two groups, with moments along the a- or c-axes. This follows from the dominant orthorhombic symmetry of the layer (or pseudolattice). Referring to Table 2.17, we find that seven A_1 modes and six B_2 modes should be observed for $E \| c$ and $E \| a$, respectively. Table 5.20 lists the LO and TO frequencies that were obtained from KK

TABLE 5.20

Classical oscillator parameters and Raman vibrational frequencies of As_2S_3 and As_2Se_3 at room temperature [24].

Material	Infrared Polarization	ρ	γ	$\bar{\nu}_{LO}$ (cm^{-1})	$\bar{\nu}_{TO}$ (cm^{-1})	Raman frequency (cm^{-1})
	c-axis	0.003	0.03	386	383	382
	c-axis	0.02	0.03	363	354	355
As_2S_3	c-axis	0.06	0.03	324	311	311
$\varepsilon_0^c = 10.9$	c-axis	0.04	0.17	205	198	204
$\varepsilon_\infty^c = 7.3$	c-axis	0.04	0.08	166	159	154
$\varepsilon_0^a = 14.3$	c-axis	0.12	0.05	148	140	136
$\varepsilon_\infty^a = 11.0$	a-axis	0.008	0.02	378	375	
	a-axis	0.008	0.01	349	345	
	a-axis	0.19	0.03	327	299	
	a-axis	0.06	0.03	185	181	
	c-axis				268[a]	273
	c-axis	0.008	0.02	250	248	248
	c-axis	0.10	0.04	232	217	216
As_2Se_3	c-axis	0.05	0.24	136	132	132
$\varepsilon_0^c = 12.4$	c-axis	0.03	0.05	108	105	104
$\varepsilon_\infty^c = 8.8$	c-axis	0.10	0.03	97	94	90
$\varepsilon_0^a = 13.9$	a-axis	0.0008	0.01	249	249	
$\varepsilon_\infty^a = 10.5$	a-axis	0.07	0.04	238	224	
	a-axis	0.12	0.03	209	201	
	a-axis	0.03	0.14	134	132	
	a-axis	0.05	0.04	108	106	

[a] Obtained from transmittance data.

and oscillator analyses of the reflectance data. The strongest modes for each symmetry direction were assumed to have their As and S (or Se) sublattices vibrating 180° out of phase. The As_2S_3 high-frequency dielectric constants are similar to those of Evans and Young [156]: $\varepsilon_\infty^c = 7.0$ and $\varepsilon_\infty^a = 8.9$ (at $\simeq 2 \ \mu m$).

Since the A_1 and B_2 modes of the pseudolattice become B_u and A_g modes in the crystal (Figure 2.16), one also expects to observe two groups of Raman-active A_g modes with selection rules derived from the A_1 and B_2 vibrations. Zallen et al., however, were able to identify only the modes having moments along the c-axis, which in backscattering exhibit xx and yy components of the Raman tensor. The Raman frequencies of these modes are given in Table 5.20. Even though the displacement vectors for the (B_u, A_g) pairs in this table have not been established, they are undoubtedly Davydov doublets or conjugate modes. A simple scaling relation was also determined for the frequencies in Table 5.20. This relation, which is surprisingly independent of the specific eigenvectors of the modes, can be expressed as

$$r = \frac{\bar{v}(\text{As}_2\text{Se}_3)}{\bar{v}(\text{As}_2\text{S}_3)} = 0.70 \pm 0.03. \tag{5.6}$$

Most of the difference $(1 - r)$ between the As_2S_3 and As_2Se_3 frequencies was accounted for by the heavier Se mass. Finally, Zallen et al. observed two Raman lines in As_2S_3 at 25 and 36 cm^{-1}, which they attributed to the rigid-layer vibrations.

Petzelt and Grigas [53] have measured the infrared reflectance spectra of Sb_2S_3, Sb_2Se_3, and Bi_2S_3 between 25 and 400 cm^{-1}. Most of their data were obtained at liquid-nitrogen and room temperature with the electric vector lying in the easy cleavage plane of the material, which is the xz plane in the structural illustration of Figure 2.5. Their group-theoretical analysis of the long-wavelength phonons is summarized in Table 2.18. Four B_{1u}, nine B_{2u}, and nine B_{3u} modes are infrared-active for $E\|c$, $E\|a$, and $E\|b$, respectively. The B_{2u} and B_{3u} modes are generated by the correlation splitting of the B_u modes in the pseudocrystal (Figure 2.17). Since the interlayer coupling is relatively weak, these modes should have nearly the same vibrational frequencies; thus, apart from accidental degeneracies, a total of 13 first-order resonances (four $\|c$ and nine $\|a, b$) should appear in the infrared spectrum of each material. The data of Petzelt and Grigas are unusually complex, and through KK analyses they succeeded only in setting wide limits for the number of observed first-order modes. With $E\|c$, 3–13 peaks in ε_2 were found for Sb_2S_3, 3–5 for Sb_2Se_3, and 3–10 for Bi_2S_3; with $E\|a$, 9–13 peaks for Sb_2S_3, about 8 for Sb_2Se_3, and 9–12 for Bi_2S_3. All of the stronger modes appeared below 300 cm^{-1}, and several modes were observed near the lowest measurable frequency, 25 cm^{-1}. If the symmetry of these materials is D_{2h}, all of the preceding peaks in ε_2 cannot be first-order transitions; however, Petzelt and Grigas suggested that the structure of Sb_2S_3 may be C_{2v}^9 (Pn2$_1$a), which would increase the number of infrared-active modes to 14 for each of the three symmetry directions. (C_{2v} symmetry is produced by displacing the atoms slightly from their positions along the z-axis of the crystal.) As they themselves noted, the small distortion of the lattice required to change the symmetry to C_{2v} would introduce only very weak, additional infrared resonances. It seems more likely that the extra peaks are second-order bands. In this regard complementary Raman experiments, which have not as yet been made, could help to establish the crystal structure. Moreover, the A_g and B_{1g} Raman-active modes are conjugate pairs; by measuring the small shifts in frequency between the Raman lines with diagonal and off-diagonal (xy) components of the scattering tensor (Table 2.10), the correlation splittings could be determined.

The infrared reflectances of Bi_2Te_3 and Bi_2Se_3 have been obtained by Unkelbach et al. [157] and by Köhler and Becker [158]. Their classical-oscillator and KK analyses of the data yielded the parameters listed in Table 5.21. The oscillator model included two Lorentz terms for the two infrared-active E_u modes (Table 2.19) and an additional Drude term to account for free-carrier absorption. Carrier concentrations are $\geqslant 10^{17}$ cm^{-3} in these materials. The E_u vibrations, which have all of the atoms in motion, are characterized as follows: E_u^1 is the acoustical mode, E_u^2 is an optical mode in which the Bi and chalcogen $(C_{3v} + D_{3d})$ sublattices vibrate out of phase, and E_u^3 is an

TABLE 5.21

Oscillator parameters for the infrared-active modes of Bi_2Se_3 and Bi_2Te_3 at room temperature.

Oscillator parameter	Bi_2Se_3 [158]		Bi_2Te_3 [157]	
	E_u^2	E_u^3	E_u^2	E_u^3
ε_∞	29		80	
ε_0	96		360 ± 50	
ρ	3.4	1.9	19.1	3.2
γ	0.050	0.095	0.24	1.0
$\bar{\nu}_{TO}(cm^{-1})$	69.5	92	49	101

optical mode in which the chalcogen (C_{3v}) and chalcogen (D_{3d}) sublattices vibrate out of phase. (The Bi atoms in E_u^3 are almost at rest [159]). We have attributed the stronger infrared resonances of Bi_2Te_3 and Bi_2Se_3 to the E_u^2 modes.

The E_u frequencies for Bi_2Te_3 in Table 5.21 do not agree with the earlier lattice dynamical calculations of Jenkins *et al.* [159]. These authors used a Born-von Kármán model and limited the interactions to nearest and next nearest neighbors. Long-range electrostatic forces were also neglected. The discrepancies between the data and Jenkins *et al.*'s dispersion curves indicate that the polarization field of the polar modes and the contributions of more distant neighbors need to be included in the model. In both Bi_2Te_3 and Bi_2Se_3 the oscillator strengths are similar to those in the CdI_2 family (Table 5.15). Although the LO–TO splittings were not determined by Unkelbach *et al.* and Köhler and Becker, they should be approximately the same as the splitting, for example, in $HfSe_2$.

5.6. CHEMICAL BONDING

In this section we compare the effective charges and force constants of layered materials with the proposed bonding schemes. Table 5.22 lists the macroscopic effective charges and the linear-chain force constants for most of the materials discussed in the foregoing sections. The effective charges were calculated by means of Equation (3.21) and the appropriate reduced masses of the infrared oscillators. For the GaSe, MoS_2, and CdI_2 families, we used the reduced masses of the linear-chain model: $M = M_m M_n/(M_m + M_n)$ for GaSe and $M = M_m M_n/(M_m + 2M_n)$ for MoS_2 and CdI_2, where M_m and M_n are the masses of the metal and nonmetal atoms, respectively. For As_2S_3 the choice of a mass is less clear, but if we assume with Zallen *et al.* [24] that the infrared mode with the largest LO–TO splitting corresponds approximately to a rigid counterphase vibration of the As and S sublattices, then a resonable choice for the reduced

TABLE 5.22

Macroscopic effective charges and bonding anisotropies in six families of layered materials (room-temperature data).

Material	e_T^*/e $(E \perp c)$	e_T^*/e $(E \| c)$	C_w^s ($\times 10^5$ dyn cm^{-1})	C_b^s/C_w^s	C_w^c ($\times 10^5$ dyn cm^{-1})	C_b^c/C_w^c
β-GaS [96, 95]	2.0	1.2	1.1	0.013	1.3	
ε-GaSe [65, 84]	2.2	1.0	1.0	0.016	1.2	0.075
γ-GaSe [107]			1.0ᵃ		1.2	
β-InSe [114]	1.5		1.0			
2H–MoS₂ [112]	0.78	0.37	1.7	0.016	2.5	0.030
2H–MoSe₂ [129, 131]	4.4		1.4	0.22	2.4	
2H–MoTe₂ [118]			1.2	0.56		
2H–WS₂ [119]			1.8		2.7	
2H–WSe₂ [119]			1.5		2.3	
2H–NbSe₂ [134]			0.98ᵇ	0.033		
CdBr₂ [119]			0.18			
CdI₂ [119]			0.083			
ZrS₂ [119]	4.6		0.36			
HfS₂ [139]	3.5		0.36			
HfSe₂ [119]	4.3		0.24			
SnS₂ [119]			0.49			
2H–PbI₂ [119]			0.054			
4H–PbI₂ [147]			0.054ᶜ	0.28		
Mg(OH)₂ [148]			0.54		0.89	
Ca(OH)₂ [148]			0.45		0.60	
Graphite [21, 153]			8.9	0.0008		
As₂S₃ [24]	1.1ᵈ		1.1ᵈ	0.014–		
	1.8ᵉ		1.0ᵉ	0.030ᶠ		
As₂Se₃ [24]	1.3ᵈ		0.85ᵈ			
	1.1ᵉ		0.90ᵉ			
Bi₂Se₃ [158]	5.5		0.14			
Bi₂Te₃ [157]	11		0.094			

ᵃ Assume $C_w^s(\gamma\text{-GaSe}) \simeq C_w^s(\varepsilon\text{-GaSe})$.
ᵇ Assume $\omega(E_{1u}^2) \simeq \omega(E_{2g}^2)$.
ᶜ Assume $\omega(E_1^2) \simeq \omega(E_u^2)$ for 4H– and 2H–PbI₂.
ᵈ For $E\|c$.
ᵉ For $E\|a$.
  Rigid-layer displacement vectors undetermined.

mass is $M = (\frac{2}{3})M_m M_n/[(\frac{2}{3})M_m + M_n]$. Similar considerations apply to Bi₂Te₃. In most cases the force constants within (C_w) and between (C_b) the layers differ only slightly from the simple coupled-oscillator values given by $C = M\omega^2$. The differences are more substantial for MoSe₂ and MoTe₂, as the interlayer coupling is stronger. To determine the force constants of the remaining materials, the simple coupled-oscillator equation was used.

The effective charges of the transition-metal dichalcogenides in Table 5.22 are smallest in MoS_2 and largest in ZrS_2. Since these materials are chemically similar, Lucovsky et al. [119, 160] concluded that the bonding is predominantly ionic in ZrS_2, HfS_2, and $HfSe_2$ and covalent in MoS_2, $MoSe_2$, WS_2, and WSe_2.† Their inclusion of $MoSe_2$, WS_2, and WSe_2 in the covalent group is based on the weakness and narrowness of the infrared reflection bands observed in pressed pellets of the materials. These authors also point out that the interatomic distances and cation-anion radius ratios in the two groups favor different bonding ionicities. In the GaSe family we observe that e_T^* is also relatively large; however, this does not necessarily imply that the materials are ionic, since the charge may contain a large dynamic component. Alternatively, Lucovsky et al. argue that some charge transfer is required in order to create the sp^3 bonds [161] between each gallium atom and its four nearest neighbors. As the outer electron configuration of atomic Ga is $4s^2 4p^1$, each chalcogen atom must contribute an additional electron to form the tetrahedral bond; thus, a charge of $-1e$ resides on each gallium site. Although Lucovsky et al. state that the static component is approximately $-2e$, which leaves only a very small dynamic component in e_T^*, the static charge is probably much smaller. Schluter's [162] energy-band and charge-density calculations show that the localized charge on the gallium sites is $+0.4e$ (note the reversal of sign, which indicates that the covalent charge has been redistributed) and the bonding is predominantly covalent. If we use Schluter's value, about 80% of e_T^* in GaSe originates in the deformation of the electronic charge during the lattice vibration.

In the sesquichalcogenides the proposed schemes vary according to the crystal structure and atomic valence. Charge transfer is not required in As_2S_3, because of the three-fold coordination of the As atoms and the two-fold coordination of the chalcogen atoms. (As and S or Se come from groups V and VI of the periodic table, respectively.) Thus the static charges are very small, and the values of e_T^* in Table 5.22 are probably dynamic. In Bi_2Te_3 the Bi and $Te^{(2)}$ (D_{3d}) atoms are located at octahedral sites (see Section 2.1.7). Drabble and Goodman [163] have proposed that the six bonding orbitals are $sp^3 d^2$ hybrids, with each Bi atom ($6s^2 6p^3$) receiving one electron from a $Te^{(1)}$ atom. The additional ionic component shortens the $Bi-Te^{(1)}$ bond length (3.12 Å) by comparison with the $Bi-Te^{(2)}$ bond length (3.22 Å). However, charge transfer cannot account for the large effective charges of Bi_2Se_3 and Bi_2Te_3 in Table 5.22. The doubling of e_T^* in Bi_2Te_3 indicates that the tellurium atom has a greater polarizability than the selenium atom [158]. This observation and Drabble and Goodman's bonding scheme imply that the macroscopic effective charges for Bi_2Se_3 and Bi_2Te_3 are mostly dynamic.

Several trends are apparent in the tabulated values of the force constants within and between the layers. The force constants for the shear vibrations, which have atomic displacements parallel to the basal plane, are always smaller than the force constants for the analogous compressional vibrations. Within each family of materials, C_w^s and C_w^c (the superscripts denote 'shear' and 'compressional', respectively) tend to decrease as one moves to the left in the periodic table or down the periodic table (smaller

---

† The large value of e_T^* ($E \perp c$) for $MoSe_2$ in Table 5.22 is an anomaly, as we have already mentioned in Section 5.2.

group number, larger row number). Differences are also observed between the intra-layer force constants of covalent and ionic materials: the largest value of C_w^s is found in graphite and the smallest in PbI_2 – a variation of more than two orders of magnitude. The relatively small intralayer force constants in Bi_2Se_3 and Bi_2Te_3 suggest a greater degree of ionicity than the bonding scheme of Drabble and Goodman [163] seems to allow. Finally, the ratio C_b/C_w varies widely and indicates the degree to which the lattice vibrations of a particular material may be considered layered or 'two-dimensional'. The lower the values of C_b/C_w the more the material behaves as an isolated single layer. The sharp increase of C_b^s/C_w^s in the molybdenum dichalcogenides, when Se or Te is substituted for S, is at variance with an exclusively van der Waals bond between the layers. Since the interlayer coupling is stronger in $MoSe_2$ and $MoTe_2$, one expects a shortening of the Se–Se or Te–Te bond length across the van der Waals gap. The structural data [16], however, shows that the chalcogen-chalcogen bond lengths in MoS_2, $MoSe_2$, and $MoTe_2$ are roughly consistent with $3s^2$, $4s^2$, and $5s^2$ van der Waals interactions between nearest-neighbor atoms. In $NbSe_2$ the Se–Se bond length (3.52 Å) is actually shorter than the bond length in $MoSe_2$ (3.75 Å), even though its rigid-layer frequency (29.6 cm^{-1}) is much smaller than $MoSe_2$'s (92 cm^{-1}). If the experimental evidence for the stronger interlayer interactions in $MoSe_2$ and $MoTe_2$ is reliable, it is of fundamental importance to establish a force-constant and bonding model for the rigid-layer vibrations of these materials.

Appendix

The nomenclature for the irreducible representations of the 32 point groups has unfortunately never been standardized. While infrared and Raman spectroscopists prefer the notation adopted by the molecular chemists [164, 165, 57], lattice dynamical theorists and experimentalists working in inelastic neutron scattering employ a variety of notations [166–168]. Warren [44] has discussed the question of nomenclature in his review paper on lattice vibrations; although he made no specific recommendations, he stressed the need for efficient communication among workers in related fields. In this spirit we provide the following table (Table A.1), comparing the notation of the molecular spectroscopists [57] with the notation of Koster *et al.* [169].

Acknowledgments

The authors wish to express their thanks to M. Hass, H. B. Rosenstock, and F. L. Carter for critical discussions and suggestions, and to Mrs G. M. LaRochelle for typing the manuscript.

TABLE A.1

Equivalent irreducible representations for the thirty-two point groups [57, 168].

Point group	Equivalent irreducible representations

C_1 $A = \Gamma_1$

C_i $A_u = \Gamma_1^-$
$A_g = \Gamma_1^+$

C_2 $A = \Gamma_1$ $B = \Gamma_2$

C_s $A' = \Gamma_1$ $A'' = \Gamma_2$

C_{2h} $A_u = \Gamma_1^-$ $B_u = \Gamma_2^-$
$A_g = \Gamma_1^+$ $B_g = \Gamma_2^+$

D_2 $A = \Gamma_1$ $B_1 = \Gamma_3$ $B_2 = \Gamma_2$ $B_3 = \Gamma_4$

C_{2v} $A_1 = \Gamma_1$ $A_2 = \Gamma_3$ $B_1 = \Gamma_4$ $B_2 = \Gamma_2$

D_{2h} $A_u = \Gamma_1^-$ $B_{1u} = \Gamma_3^-$ $B_{2u} = \Gamma_2^-$ $B_{3u} = \Gamma_4^-$
$A_g = \Gamma_1^+$ $B_{1g} = \Gamma_3^+$ $B_{2g} = \Gamma_2^+$ $B_{3g} = \Gamma_4^+$

C_4, S_4 $A = \Gamma_1$ $B = \Gamma_2$ $E = \Gamma_3 + \Gamma_4$

C_{4h} $A_u = \Gamma_1^-$ $B_u = \Gamma_2^-$ $E_u = \Gamma_3^- + \Gamma_4^-$
$A_g = \Gamma_1^+$ $B_g = \Gamma_2^+$ $E_g = \Gamma_3^+ + \Gamma_4^+$

$D_4, C_{4v},$
$D_{2d}(V_d)$ $A_1 = \Gamma_1$ $A_2 = \Gamma_2$ $B_1 = \Gamma_3$ $B_2 = \Gamma_4$ $E = \Gamma_5$

D_{4h} $A_{1u} = \Gamma_1^-$ $A_{2u} = \Gamma_2^-$ $B_{1u} = \Gamma_3^-$ $B_{2u} = \Gamma_4^-$ $E_u = \Gamma_5^-$
$A_{1g} = \Gamma_1^+$ $A_{2g} = \Gamma_2^+$ $B_{1g} = \Gamma_3^+$ $B_{2g} = \Gamma_4^+$ $E_g = \Gamma_5^+$

C_3 $A = \Gamma_1$ $E = \Gamma_2 + \Gamma_3$

$C_{3i}(S_6)$ $A_u = \Gamma_1^-$ $E_u = \Gamma_2^- + \Gamma_3^-$
$A_g = \Gamma_1^+$ $E_g = \Gamma_2^+ + \Gamma_3^+$

D_3, C_{3v} $A_1 = \Gamma_1$ $A_2 = \Gamma_2$ $E = \Gamma_3$

D_{3d} $A_{1u} = \Gamma_1^-$ $A_{2u} = \Gamma_2^-$ $E_u = \Gamma_3^-$
$A_{1g} = \Gamma_1^+$ $A_{2g} = \Gamma_2^+$ $E_g = \Gamma_3^+$

C_6 $A = \Gamma_1$ $B = \Gamma_4$ $E_1 = \Gamma_5 + \Gamma_6$ $E_2 = \Gamma_2 + \Gamma_3$

C_{3h} $A' = \Gamma_1$ $A'' = \Gamma_4$ $E' = \Gamma_2 + \Gamma_3$ $E'' = \Gamma_5 + \Gamma_6$

C_{6h} $A_u = \Gamma_1^-$ $B_u = \Gamma_4^-$ $E_{1u} = \Gamma_5^- + \Gamma_6^-$ $E_{2u} = \Gamma_2^- + \Gamma_3^-$
$A_g = \Gamma_1^+$ $B_g = \Gamma_4^+$ $E_{1g} = \Gamma_5^+ + \Gamma_6^+$ $E_{2g} = \Gamma_2^+ + \Gamma_3^+$

D_6 $A_1 = \Gamma_1$ $A_2 = \Gamma_2$ $B_1 = \Gamma_3$ $B_2 = \Gamma_4$ $E_1 = \Gamma_5$ $E_2 = \Gamma_6$

C_{6v} $A_1 = \Gamma_1$ $A_2 = \Gamma_2$ $B_1 = \Gamma_4$ $B_2 = \Gamma_3$ $E_1 = \Gamma_5$ $E_2 = \Gamma_6$

D_{3h} $A_1' = \Gamma_1$ $A_2' = \Gamma_2$ $A_1'' = \Gamma_3$ $A_2'' = \Gamma_4$ $E' = \Gamma_6$ $E'' = \Gamma_5$

D_{6h} $A_{1u} = \Gamma_1^-$ $A_{2u} = \Gamma_2^-$ $B_{1u} = \Gamma_3^-$ $B_{2u} = \Gamma_4^-$ $E_{1u} = \Gamma_5^-$ $E_{2u} = \Gamma_6^-$
$A_{1g} = \Gamma_1^+$ $A_{2g} = \Gamma_2^+$ $B_{1g} = \Gamma_3^+$ $B_{2g} = \Gamma_4^+$ $E_{1g} = \Gamma_5^+$ $E_{2g} = \Gamma_6^+$

T $A = \Gamma_1$ $E = \Gamma_2 + \Gamma_3$ $F = \Gamma_4$

T_h $A_u = \Gamma_1^-$ $E_u = \Gamma_2^- + \Gamma_3^-$ $F_u = \Gamma_4^-$
$A_g = \Gamma_1^+$ $E_g = \Gamma_2^+ + \Gamma_3^+$ $F_g = \Gamma_4^+$

O, T_d $A_1 = \Gamma_1$ $A_2 = \Gamma_2$ $E = \Gamma_3$ $F_1 = \Gamma_4$ $F_2 = \Gamma_5$

O_h $A_{1u} = \Gamma_1^-$ $A_{2u} = \Gamma_2^-$ $E_u = \Gamma_3^-$ $F_{1u} = \Gamma_4^-$ $F_{2u} = \Gamma_5^-$
$A_{1g} = \Gamma_1^+$ $A_{2g} = \Gamma_2^+$ $E_g = \Gamma_3^+$ $F_{1g} = \Gamma_4^+$ $F_{2g} = \Gamma_5^+$

List of Symbols

a	coefficients in the transformation equations for the atomic displacements
a_j	multiplicity of the jth irreducible representation in a decomposition
a, b, c	lengths of the unit-cell edges
c	velocity of light in a vacuum; superscript on force constant denoting 'compressional'
$C_{nn'}$	interplanar force constant
C_b	force constant between adjacent layers
C_w	force constant within each layer
e	electron charge
e^σ, e^ρ	components of the electric field intensity
$e*$	effective charge
e_T^*	macroscopic, transverse effective charge
E	electric field intensity
h	order of a symmetry group
j	subscript or index for the $3r$ branches of the phonon spectrum
$\mathbf{k}$	wave vector of a photon
$\mathbf{k}_s$	wave vector of a scattered photon
l	index for the primitive unit cells
LO, TO	abbreviation or subscript denoting 'longitudinal optical' or 'transverse optical'
m	induced electric moment in a crystal; number of layers in the primitive unit cell
M	reduced mass of an oscillator; mass of an atom
n	subscript or index for the atoms or atomic planes of a crystal; number of oscillators in a dielectric function
N	number of primitive unit cells
p	electric moment of a lattice vibration
$\mathbf{q}$	wave vector of a phonon
r	number of atoms in the primitive unit cell
R	symmetry operation; reflectivity of a surface
$R*, T*$	reflectance or transmittance of a crystal
$R_{\sigma\rho}$	Raman tensor
s	superscript on force constant denoting 'shear'
$\mathbf{S}$	set of rotational symmetry operations
u	displacement of an atom from its equilibrium position
v	vibrational, energy quantum number
$\mathbf{v}(S)$	nonprimitive translation combined with a rotation
x, y, z	crystallographic axes; Cartesian axes
$\mathbf{x}(m)$	primitive translation vector of a lattice
$\mathbf{X}(\kappa)$	position vector of κth atom, referred to origin of the primitive unit cell
$\mathbf{X}(l)$	position vector of lth primitive unit cell
α	polarizability of a crystal
β	subscript denoting the $x, y,$ or z component
γ	damping constant of an oscillator

Γ	a reducible representation
$\Gamma^{(j)}$	an irreducible representation
ε	complex dielectric function of a crystal
ε_1	real part of the dielectric function
ε_2	imaginary part of the dielectric function
ε_0	low-frequency dielectric constant
ε_∞	high-frequency dielectric constant
$\eta(R)$	number of atoms left invariant by an operation of the factor group
θ	phase difference between the incident and reflected waves; angle of incidence
κ	subscript or index for the atoms of the primitive unit cell
$\bar{\nu}$	wave number of a photon
ξ	normal coordinates of the atomic displacements
ρ	oscillator strength
ρ, σ, τ	subscript or superscript denoting the x, y, or z component
τ_0	nonprimitive translation vector in a symmetry operation
φ	angle of an n-fold axis of symmetry; cone angle of scattered radiation
$\chi(R)$	character of a reducible representation
$\chi^{(j)}(R)$	character of an irreducible representation
Ψ	time-dependent wave function of a vibrational state
ω	angular frequency of a photon
ω_s	angular frequency of a scattered photon
ω_R	angular frequency of a rigid-layer vibration
$*$	superscript indicating the complex conjugate
$\hat{}$	denotes a unit vector

References

[1] W. W. Coblentz: *Phys. Rev.* **20** (1905), 252.

[2] D. Krishnamurti: *Proc. Ind. Acad. Sci.* **A50** (1959), 223, 232, 247.

[3] A. C. Menzies: *Rep. Prog. Phys.* **16** (1953), 83.

[4] S. Bhagavantam: *Proc. Ind. Acad. Sci.* **A37** (1953), 350.

[5] J.-P. Mathieu: *J. Phys. Radium* **16** (1955), 220; *Physical Society (London) Year Book*, 1956, p. 23: in *Optik und Spektroskopie aller Wellenlängen*, ed. by P. Görlich, Academie-Verlag, Berlin 1962, p. 476.

[6] W. Vedder and D. F. Hornig: *Adv. Spectry.* **2** (1961), 189.

[7] S. S. Mitra: *Solid State Phys.* **13** (1962), 1; in *Optical Properties of Solids*, ed. by S. Nudelman and S. S. Mitra, Plenum, New York 1969, p. 333.

[8] A. D. Yoffe: *Ann. Rev. Mat. Sci.* **3** (1973), 147; *Adv. Solid State Phys.* **13** (1973), 1.

[9] R. Huisman and F. Jellinek: *J. Less-Common Metals* **17** (1969), 111.

[10] F. J. di Salvo, B. G. Bagley, J. M. Voorhoeve, and J. V. Waszczak : *J. Phys. Chem. Solids* **34** (1973), 1357.

[11] F. Jellinek and H. Hahn: *Z. Naturforsch.* **16b** (1961), 713.

[12] Z. S. Basinski, D. B. Dove, and E. Mooser: *Helv. Phys. Acta* **34** (1961), 373.

[13] R. W. G. Wyckoff: *Crystal Structures*, Interscience, New York 1963, Vol. 1, p. 144.

[14] B. B. Zvyagin and S. V. Soboleva : *Kristallografiya*, **12** (1967), 57; [*Soviet Physics-Crystallography*, **12** (1967), 46].

[15] Wyckoff: *op. cit.*, Vol. 1, p. 280.

[16] J. A. Wilson and A. D. Yoffe: *Adv. Phys.* **73** (1969), 193.

[17] B. E. Brown and D. J. Beerntsen: *Acta. Cryst.* **18** (1965), 31.

[18] F. Jellinek: *Acta Cryst.* **13** (1960), 1021.

[19] Wyckoff: *op. cit.*, Vol. 1, p. 274.

[20] Ibid., p. 26.

[21] L. J. Brillson, E. Burstein, A. A. Maradudin, and T. Stark: in *The Physics of Semimetals and Narrow-Gap Semiconductors*, ed. by D. L. Carter and R. T. Bate, Pergamon, Oxford 1971, p. 187.

[22] R. W. G. Wyckoff: *Crystal Structures*, Interscience, New York 1964, Vol. 2, p. 26.

[23] A. A. Vaipolin: *Kristallografiya*, **10** (1965), 596; [*Soviet Physics-Crystallography* **10** (1966), 509].

[24] R. Zallen, M. L. Slade, and A. T. Ward: *Phys. Rev.* **B3** (1971), 4257; R. Zallen: in *Lattice Dynamics and Intermolecular Forces*, ed. by S. Califano, Academic, New York 1975, p. 159.

[25] Wyckoff: *op. cit.*, Vol. 2, p. 27; S. Ščavničar: *Z. Krist.* **114** (1960), 85.

[26] Wyckoff: *op. cit.*, Vol. 2, p. 30.

[27] J. R. Wiese and L. Muldawer: *J. Phys. Chem. Solids* **15** (1960), 13.

[28] F. Seitz: *Ann. Math.* **37** (1936), 17.

[29] E. B. Wilson, J. C. Decius, and P. C. Cross: *Molecular Vibrations*, McGraw-Hill, New York 1955, Appendix X-8.

[30] S. Bhagavantum and T. Venkatarayudu: *Proc. Indian Acad. Sci*, **A9** (1939), 224.

[31] S. Bhagavantum and T. Venkatarayudu: *The Theory of Groups and Its Application to Physical Problems*, Academic, New York 1969, p. 140.

[32] D. F. Hornig: *J. Chem. Phys.* **16** (1948), 1063.

[33] H. Winston and R. S. Halford: *J. Chem. Phys.* **17** (1949), 607.

[34] E. A. Wood: *Bell System Tech. J.* **43** (1964), 541; Bell System Monograph 4680, 1964.

[35] M. C. Tobin: *J. Chem. Phys.* **23** (1955), 891.

[26] M. C. Tobin: *J. Mol. Spectry.* **4** (1960), 349.

[37] R. Zbinden: *Infrared Spectroscopy of High Polymers*, Academic, New York 1964.

[38] A. U. Macrae: *Science* **139** (1963), 379.

[39] E. Alexander and K. Herrmann: *Z. Krist.* **69** (1928), 285.

[40] *International Tables for X-Ray Crystallography*, Vol. I, ed. by N. F. M. Henry and K. Lonsdale, Kynoch, Birmingham, England 1969.

[41] W. G. Fateley, F. R. Dollish, N. T. McDevitt, and F. F. Bentley: *Infrared and Raman Selection Rules for Molecular and Lattice Vibrations: The Correlation Method*, Wiley-Interscience, New York 1972, p. 201.

[42] *International Tables for X-Ray Crystallography*, *op. cit.*, p. 294.

[43] M. Tinkham: *Group Theory and Quantum Mechanics*, McGraw-Hill, New York 1964, p. 30.

[44] J. L. Warren: *Rev. Mod. Phys.* **40** (1968), 38.

[45] *International Tables for X-Ray Crystallography*, *op. cit.*, p. 304.

[46] H. Hahn and G. Frank: *Z. Anorg. Allgem. Chem.* **278** (1955), 340.

[47] K. Schubert, E. Dörre, and M. Kluge: *Z. Metallk.* **46** (1955), 216.

[48] Wyckoff: *op. cit.*, Vol. 1, p. 267.

[49] Ibid., p. 272.

[50] Tinkham: *op. cit.*, p. 326.

[51] *International Tables for X-Ray Crystallography*, *op. cit.*, p. 99.

[52] Tinkham: *op. cit.*, p. 325.

[53] J. Petzelt and J. Grigas: *Ferroelectrics* **5** (1973), 59.

[54] H. Goldstein: *Classical Mechanics*, Addison-Wesley, Reading, Massachusetts 1950, p. 318.

[55] M. Born and M. Bradburn: *Proc. Roy. Soc. London* **A188** (1947), 161.

[56] L. Pauling and E. B. Wilson: *Introduction to Quantum Mechanics*, McGraw-Hill, New York 1935, p. 77.

[57] Wilson, Decius, and Cross: *op. cit.*, Appendix X-5.

[58] G. Herzberg: *Infrared and Raman Spectra of Polyatomic Molecules*, Vol. II of *Molecular Spectra and Molecular Structure*, Van Nostrand Rheinhold, New York 1945, II 3(e).

[59] E. Burstein: in *Phonons and Phonon Interactions*, ed. by T. A. Bak and W. A. Benjamin, New York 1964, p. 276.

[60] K. Huang: *Proc. Roy. Soc. London* **A208** (1951), 352.

[61] M. Born and K. Huang: *Dynamical Theory of Crystal Lattices*, Clarendon, Oxford 1954, Sections 7 and 8.

[62] A. S. Barker, Jr.: *Phys. Rev.* **136** (1964), A1290.

[63] I. F. Chang, S. S. Mitra, J. N. Plendl, and L. C. Mansur: *Phys. Status Solidi* **28** (1968), 663.

[64] J. C. Slater: *Insulators, Semiconductors, and Metals*, Vol. 3 of *Quantum Theory of Molecules and Solids*, McGraw-Hill, New York 1963, p. 95.

[65] T. J. Wieting and J. L. Verble: *Phys. Rev.* **B5** (1972), 1473.

[66] R. H. Lyddane, R. G. Sachs, and E. Teller: *Phys. Rev.* **59** (1941), 673.

[67] F. Stern: in *Solid State Physics*, ed. by F. Seitz and D. Turnbull, Academic, New York 1963, Vol. 15, p. 299.

[68] T. S. Robinson: *Proc. Roy. Soc. London* **B65** (1952), 910; T. S. Robinson and W. C. Price: *Proc. Phys. Soc. London* **B66** (1953), 969.

[69] E. Burnstein, A. Pinczuk and R. F. Wallis: In *The Physics of Semimetals and Narrow-Gap Semiconductors*, ed. by D. L. Carter and R. T. Bate, Pergamon, Oxford 1971, p. 251.

[70] B. Szigeti: *Trans. Faraday Soc.* **45** (1949), 155.

[71] Born and Huang: *op. cit.*, Section 9.

[72] H. B. Callen: *Phys. Rev.* **76** (1949), 1394.

[73] W. Cochran: *Nature* **191** (1961), 60.

[74] E. Burstein: *J. Phys. Chem. Solids Suppl.* **21** (1965), 315; G. Lucovsky, R. M. Martin, and E. Burstein: *Phys. Rev.* **B4** (1971), 1367.

[75] H. Poulet: *Ann. Phys. Paris* **10** (1955), 908.

[76] R. Loudon: *Proc. Roy. Soc.* **A275** (1963), 218.

[77] R. Loudon: *Adv. Phys.* **13** (1964), 423.

[78] R. M. Martin and L. M. Falicov: in *Light Scattering in Solids*, ed. by M. Cardona, Springer-Verlag, New York 1975, p. 79.

[79] A. S. Davydov: *J. Exptl. Theoret. Phys.* U.S.S.R. **18** (1948), 210.

[80] L. C. Kravitz, J. D. Kingsley, and E. L. Elkin: *J. Chem. Phys.* **49** (1968), 4600.

[81] T. J. Wieting: *Solid State Commun.* **12** (1973), 931.

[82] J. L. Verble and T. J. Wieting: *Phys. Rev. Letters* **25** (1970), 362.

[83] Z. S. Bazinski, D. B. Dove, and E. Mooser: *Phys. Rev.* **34** (1963), 469.

[84] P. C. Leung, G. Andermann, W. G. Spitzer, and C. A. Mead: *J. Phys. Chem. Solids* **27** (1966), 849.

[85] R. H. Bube and E. L. Lind: *Phys. Rev.* **115** (1959), 1159.

[86] J. L. Brebner: *J. Phys. Chem. Solids* **25** (1964), 1427.

[87] J. L. Brebner and J.-A. Deverin: *Helv. Phys. Acta* **38** (1965), 650.

[88] L. I. Tartarinova, Yu K. Auleitner, and Z. G. Pinsker: *Kristallografiya* **1** (1956), 537; [*Soviet Physics-Crystallography* **1** (1956), 426].

[89] S. A. Semiletov: *Kristallografiya* **3** (1958), 288; [*Soviet Physics-Crystallography* **3** (1958), 292].

[90] J. C. J. M. Terhell and R. M. A. Lieth: *Phys. Status Solidi* **5** (1971), 719.

[91] J. C. J. M. Terhell and R. M. A. Lieth: *Phys. Status Solidi* **10** (1972), 529.

[92] N. Kuroda, Y. Nishina, and T. Fukuroi: *J. Phys. Soc. Japan* **24** (1968), 214.

[93] Y. Nishina, N. Kuroda, and T. Fukuroi: in *Proceedings of the Eleventh International Conference on the Physics of Semiconductors*, Nauka, Leningrad 1968, Vol. II, p. 1024.

[94] N. Kuroda, Y. Nishina, and T. Fukuroi: *J. Phys. Soc. Japan* **28** (1970), 981.

[95] J. C. Irwin, R. M. Hoff, B. P. Clayman, and R. A. Bromley: *Solid State Commun.* **13** (1973), 1531.

[96] E. Finkman and A. Rizzo: *Solid State Commun.* **15** (1974), 1841.

[97] R. M. Hoff and J. C. Irwin: *Phys. Rev.* **10** (1974), 3464.

[98] G. B. Wright and A. Mooradian: *Bull. Am. Phys. Soc.* **11** (1966), 812.

[99] M. Hayek, O. Brafman, and R. M. A. Lieth: *Phys. Rev.* **B8** (1973), 2772.

[100] J. P. van der Ziel, A. E. Meixner, and H. M. Kasper: *Solid State Commun.* **12** (1973), 1213.

[101] H. Yoshida, S. Nakashima, and A. Mitsuishi: *Phys. Status Solidi* **B59** (1973), 655.

[102] R. M. A. Lieth, C. W. M. van der Heyden, and J. W. M. van Kessel: *J. Crystal Growth* **5** (1969), 251.

[103] E. Aulich, J. L. Brebner, and E. Mooser: *Phys. Status Solidi* **31** (1969), 129.

[104] A. Mercier and J. P. Voitchovsky: *Solid State Commun.* **14** (1974), 757.

[105] I. F. Chang and S. S. Mitra: *Adv. Phys.* **20** (1971), 359; *Phys. Rev.* **172** (1968), 924.

[106] G. Lucovsky, M. Brodsky, and E. Burstein: in *Localized Excitations in Solids*, ed. by R. F. Wallis, Plenum, New York 1968, p. 592.

[107] R. M. Hoff, J. C. Irwin, and R. M. A. Lieth: *Can. J. Phys.* **53** (1975), 1606.

[108] R. A. Bromley and J. C. Irwin: private communication.

[109] J. L. Brebner, S. Jandl, and B. M. Powell: *Solid State Commun.* **13** (1973), 1555.

[110] S. Jandl, J. L. Brebner, and B. M. Powell: *Phys. Rev.* **B13** (1976), 686.

[111] J. C. J. M. Terhell and R. M. A. Lieth: *J. Crystal Growth* **16** (1972), 54.

[112] T. J. Wieting and J. L. Verble: *Phys. Rev.* **B3** (1971), 4286.

[113] R. M. Martin: *Phys. Rev.* **B4** (1971), 3676.

[114] V. P. Mushinskii and V. I. Kobolev: *Fiz. Tverd. Tela* **14** (1972), 1275; [*Soviet Physics – Solid State* **14** (1972), 1098].

[115] R. Brout: *Phys. Rev.* **113** (1959), 43.

[116] H. B. Rosenstock: *Phys. Rev.* **129** (1963), 1959; H. B. Rosenstock and G. Blanken: *Phys. Rev.* **145** (1966), 546.

[117] R. W. Keyes: *J. Chem. Phys.* **37** (1962), 72.

[118] O. P. Agnihotri, H. K. Sehgal, and A. K. Garg: *Solid State Commun.* **12** (1973), 135.

[119] G. Lucovsky, R. M. White, J. A. Benda, and J. F. Revelli: *Phys. Rev.* **B7** (1973), 3859.

[120] R. Bailly: *Am. Mineralogist* **33** (1948), 519.

[121] J. Lagrenaudie: *J. Phys. Radium* **15** (1954), 299.

[122] R. F. Findt and A. D. Yoffe: *Proc. Roy. Soc. London* **A273** (1963), 69.

[123] B. L. Evans and P. A. Young: *Proc. Roy. Soc. London* **A284** (1965), 402.

[124] J. L. Verble, T. J. Wieting, and P. R. Reed: *Solid State Commun.* **11** (1972), 941.

[125] J. M. Chen and C. S. Wang: *Solid State Commun.* **14** (1974), 857.

[126] N. Wakabayashi, H. G. Smith, and R. M. Nicklow: *Phys. Rev* **B12** (1975), 659; *Bull. Am. Phys. Soc.* **17** (1972), 292.

[127] A. W. Webb, J. L. Feldmann, E. F. Skelton, L. C. Towle, C. Y. Liu, and I. L. Spain: *J. Phys. Chem. Solids* **37** (1976), 329.

[128] P. Cannon: *Nature* **183** (1959), 1612.

[129] O. P. Agnihotri and H. K. Sehgal: *Phil. Mag.* **26** (1972), 753.

[130] A. K. Garg, H. K. Sehgal, and O. P. Agnihotri: *Solid State Commun.* **12** (1973), 1261.

[131] J. E. Smith, Jr., J. B. Torrance, and M. W. Shafer: *Bull. Am. Phys. Soc.* **18** (1973), 396; J. E. Smith, Jr.: private communication.

[132] V. Grasso, G. Mondo, and G. Saitta: *J. Phys. C: Solid State Phys.* **5** (1972), 1101.

[133] W. Y. Liang: *J. Phys. C: Solid State Phys.* **6** (1973), 551.

[134] C. S. Wang and J. M. Chen: *Solid State Commun.* **14** (1974), 1145.

[135] N. Wakabayashi, H. G Smith, and R M. Nicklow: *Bull. Am. Phys. Soc.* **17** (1972), 292.

[136] D. M. Adams, M. Goldstein, and E. F. Mooney: *Trans. Farad. Soc.* **59** (1963), 2228.

[137] J.-P. Mon: *Compt. Rend. Acad. Sci. Paris* **B262** (1966) 493.

[138] C. Carabatos: *Compt. Rend. Acad. Sci. Paris* **B272** (1971), 465.

[139] J. E. Smith, Jr., M. I. Nathan, M. W. Shafer, and J. B. Torrance: in *Proceedings of the Eleventh International Conference on the Physics of Semiconductors*, Nauka, Leningrad 1968, Vol. II, p. 1306.

[140] D. J. Lockwood: *J. Opt. Soc. Am.* **63** (1973), 374; J. H. Christie and D. J. Lockwood: in *Proceedings of the Second International Conference on Light Scattering in Solids*, ed. by M. Balkanski, Flammarion Sciences, Paris 1971, p. 145 ; D. J. Lockwood: in *Light Scattering Spectra of Solids*, ed. by G. B. Wright, Springer, Berlin 1969, p. 75.

[141] S. Nakashima, H. Yoshida, T. Fukumoto, and A. Mitsuishi: *J. Phys. Soc. Japan* **31** (1971), 1847.

[142] R. Fuchs and K. L. Kliewer: *J. Opt. Soc. Am.* **58** (1968), 319.

[143] P. N. Ghosh: *Solid State Commun.* **16** (1975), 811.

[144] Slater: *op. cit.*, p. 187.

[145] H. Mueller: *Phys. Rev.* **47** (1935), 947; *Phys. Rev.* **50** (1936), 547.

[146] S. Nakashima: *Solid State Commun.* **16** (1975), 1059.

[147] R. Zallen and M. L. Slade: *Solid State Commun.* **17** (1975), 1561.

[148] P. Dawson, C. D. Hadfield, and G. R. Wilkinson: *J. Phys. Chem. Solids* **34** (1973), 1217.

[149] F. Tuinstra and J. L. Koenig: *Bull. Am. Phys. Soc.* **15** (1970), 296; *J. Chem. Phys.* **53** (1970), 1126.

[150] R. F. Shaufele and T. Shimanouchi: *J. Chem. Phys.* **47** (1967), 3605.

[151] H. Lipson and A. R. Stokes: *Proc. Roy. Soc. London* **A181** (1942), 101.

[152] F. Laves and Y. Baskin: *Z. Krist.* **107** (1956), 337.

[153] R. Nicklow, N. Wakabayashi, and H. G. Smith: *Phys. Rev.* **B5** (1972), 4951.

[154] G. Dolling and B. N. Brockhouse: *Phys. Rev.* **128** (1962), 1120.

[155] G. Bellodi, A. Borghese, G. Guizetto, L. Nosenzo, E. Reguzzoni, and G. Samoggia: *Phys. Rev.* **B12** (1975), 5951.

[156] B. L. Evans and P. A. Young: *Proc. Roy. Soc. London* **A297** (1966), 230.

[157] K. H. Unkelbach, Ch. Becker, H. Köhler, and A. von Middendorff: *Phys. Status Solidi* **B60** (1973), K41.

[158] H. Köhler and C. R. Becker: *Phys. Status Solidi* **B61** (1974), 533.

[159] J. O. Jenkins, J. A. Rayne, and R. W. Ure, Jr.: *Phys. Rev.* **B5** (1972), 3171.

[160] R. M. White and G. Lucovsky: *Solid State Commun.* **11** (1972), 1369.

[161] G. Fischer and J. L. Brebner: *J. Phys. Chem. Solids* **23** (1962), 1363.

[162] M. Schlüter: *Il Nuovo Cimento* **13B** (1973), 313.

[163] J. R. Drabble and C. H. L. Goodman: *J. Phys. Chem. Solids* **5** (1958), 142.

[164] R. S. Mulliken: *Phys. Rev.* **43** (1933), 279.

[165] G. Herzberg: *op. cit.*, p. 104.

[166] L. Bouckaert, R. Smoluchowsky, and E. Wigner: *Phys. Rev.* **50** (1936), 58.

[167] S. L. Altman and A. P. Cracknell: *Rev. Mod. Phys.* **37** (1965), 19.

[168] J. L. Warren, J. L. Yarnell, G. Dolling, and R. A. Cowley: *Phys. Rev.* **158** (1967), 805.

[169] G. F. Koster, J. O. Dimmock, R. G. Wheeler, and H. Statz: *Properties of the Thirty-Two Point Groups*, MIT Press, Cambridge 1963.

NEUTRON SCATTERING AND LATTICE DYNAMICS
OF MATERIALS WITH LAYERED STRUCTURES

N. WAKABAYASHI and R. M. NICKLOW

Solid State Division, Oak Ridge National Laboratory*, Oak Ridge, Tenn. 37830, U.S.A.

1. Lattice Dynamics and Inelastic Neutron Scattering

1.1. LATTICE DYNAMICS IN THE HARMONIC APPROXIMATION

There are numerous textbooks and review articles [1–3] on lattice dynamics, and, in particular, the microscopic approach to the theory of lattice dynamics has been a subject of considerable activity in recent years [4–6]. Here a brief summary of fundamental equations for the more conventional Born-von Kármán formalism is presented in order to define notations that will be used later.

A crystal is regarded as a periodic array of N unit cells each of which contains r atoms. The equilibrium position of the κth atom in the lth unit cell is given by

$$\mathbf{X}\begin{pmatrix} l \\ \kappa \end{pmatrix} = \mathbf{X}(l) + \mathbf{X}(\kappa)$$

where $\mathbf{X}(l)$ is the position vector for the lth unit cell and $\mathbf{X}(\kappa)$ is the location of the κth atom within a unit cell. The total potential energy of a crystal, Φ, is assumed to be a function of the instantaneous positions of all atoms. When each atom is displaced from its equilibrium position by $\mathbf{u}\begin{pmatrix} l \\ \kappa \end{pmatrix}$, the potential energy measured with respect to that for the equilbrium configuration may be expanded as

$$\Delta\Phi = \tfrac{1}{2} \sum_{\substack{l\kappa\alpha \\ l'\kappa'\beta}} \Phi_{\alpha\beta}\begin{pmatrix} ll' \\ \kappa\kappa' \end{pmatrix} u_\alpha\begin{pmatrix} l \\ \kappa \end{pmatrix} u_\beta\begin{pmatrix} l' \\ \kappa' \end{pmatrix} + \cdots,$$

where $\alpha, \beta = x, y,$ or z, and

$$\Phi_{\alpha\beta}\begin{pmatrix} ll' \\ \kappa\kappa' \end{pmatrix} = \frac{\partial^2\Phi}{\partial X_\alpha\begin{pmatrix} l \\ \kappa \end{pmatrix} \partial X_\beta\begin{pmatrix} l' \\ \kappa' \end{pmatrix}} = \Phi_{\beta\alpha}\begin{pmatrix} l'l \\ \kappa'\kappa \end{pmatrix}. \tag{1.1}$$

In the harmonic approximation, higher order terms are neglected. $\Phi_{\alpha\beta}\begin{pmatrix} ll' \\ \kappa\kappa' \end{pmatrix}$ is a force constant that represents the force exerted in the α-direction on the atom $(l\kappa)$ when the atom $(l'\kappa')$ is displaced by a unit distance in the β-direction. It should depend only on the relative position of the cells l and l' and, hence, can be written as

$$\Phi_{\alpha\beta}\begin{pmatrix} ll' \\ \kappa\kappa' \end{pmatrix} = \Phi_{\alpha\beta}\begin{pmatrix} l - l' \\ \kappa\kappa' \end{pmatrix}.$$

* Operated by Union Carbide Corporation for the United States Energy Research and Development Administration.

T. J. Wieting and M. Schlüter (eds.), Electrons and Phonons in Layered Crystal Structures. 409–464.
All Rights Reserved.
Copyright © 1979 by D. Reidel Publishing Company, Dordrecht, Holland.

Due to the translational and rotational invariance of the potential energy of a crystal, force constants must satisfy the relations

$$\sum_{l'\kappa'} \Phi_{\alpha\beta}\begin{pmatrix} l-l' \\ \kappa \quad \kappa' \end{pmatrix} = 0. \tag{1.2}$$

and

$$\sum_{l'\kappa\beta\gamma} \left\{ \Phi_{\alpha\beta}\begin{pmatrix} l-l' \\ \kappa \quad \kappa' \end{pmatrix} X_\gamma\begin{pmatrix} l'-l \\ \kappa' \quad \kappa \end{pmatrix} - \Phi_{\alpha\gamma}\begin{pmatrix} l-l' \\ \kappa \quad \kappa' \end{pmatrix} X_\beta\begin{pmatrix} l'-l \\ \kappa' \quad \kappa \end{pmatrix} \right\} = 0, \tag{1.3}$$

where

$$X_\alpha\begin{pmatrix} l-l' \\ \kappa \quad \kappa' \end{pmatrix} = X_\alpha\begin{pmatrix} l \\ \kappa \end{pmatrix} - X_\alpha\begin{pmatrix} l' \\ \kappa' \end{pmatrix}.$$

The force constants may be considered to be the elements of a 3×3 force constant matrix

$$\Phi\begin{pmatrix} l-l' \\ \kappa \quad \kappa' \end{pmatrix} = \left\{ \Phi_{\alpha\beta}\begin{pmatrix} l-l' \\ \kappa \quad \kappa' \end{pmatrix} \right\},$$

the form of which must be consistent with the symmetry of the crystal. Since the matrix may also be regarded as a tensor, the interatomic forces expressed in this manner are sometimes called general tensor forces. The equations of motion for the atoms in the lattice are

$$M_\kappa \ddot{u}_\alpha\begin{pmatrix} l \\ \kappa \end{pmatrix}, t \end{pmatrix} = -\sum_{l'\kappa'\beta} \Phi_{\alpha\beta}\begin{pmatrix} l-l' \\ \kappa \quad \kappa' \end{pmatrix} u_\beta\begin{pmatrix} l' \\ \kappa' \end{pmatrix}, t \end{pmatrix}, \tag{1.4}$$

where M_κ is the mass of the κth atom. As a consequence of the lattice periodicity, these $3rN$ equations can be reduced to a set of $3r$ linear homogeneous equations for $3r$ unknowns. Substituting the form

$$u_\alpha\begin{pmatrix} l \\ \kappa \end{pmatrix}, t \end{pmatrix} = u_\alpha(\mathbf{q}; \kappa) e^{i(\mathbf{q}\cdot\mathbf{X}\begin{pmatrix} l \\ \kappa \end{pmatrix} - \omega t)}$$

into Equation (1.4), one obtains the equations for $u_\alpha(\mathbf{q}; \kappa)$,

$$M_\kappa \omega^2 u_\alpha(\mathbf{q}; \kappa) = \sum_{l'\kappa'\beta} \Phi_{\alpha\beta}\begin{pmatrix} l-l' \\ \kappa \quad \kappa' \end{pmatrix} e^{i\mathbf{q}\cdot\mathbf{X}\begin{pmatrix} l-l' \\ \kappa \quad \kappa' \end{pmatrix}} u_\beta(\mathbf{q}; \kappa) \tag{1.5}$$

or

$$\mathbf{M}\omega^2\mathbf{u} = \mathbf{D}\mathbf{u},$$

where $\mathbf{M}$ and $\mathbf{D}$ are $3r \times 3r$ matrices with elements

$$M_{\alpha\beta}(\kappa\kappa') = M_\kappa \delta_{\alpha\beta}\delta_{\kappa\kappa'},$$

where $\delta_{\alpha\beta}$ and $\delta_{\kappa\kappa'}$ are Kronecker deltas and

$$D_{\alpha\beta}(\mathbf{q}; \kappa\kappa') = \sum_{l'} \Phi_{\alpha\beta}\begin{pmatrix} l-l' \\ \kappa \quad \kappa' \end{pmatrix} e^{i\mathbf{q}\cdot\mathbf{X}\begin{pmatrix} l-l' \\ \kappa \quad \kappa' \end{pmatrix}}. \tag{1.6}$$

The matrix **D** is the dynamical matrix. One can eliminate the mass matrix **M** by writing

$$u_\alpha(\mathbf{q}; \kappa) = \frac{1}{\sqrt{M_\kappa}}\, v_\alpha(\mathbf{q}; \kappa).$$

Then Equation (1.5) is transformed to

$$\omega^2 \mathbf{v} = \mathbf{D}' \mathbf{v} \tag{1.7}$$

where

$$D'_{\alpha\beta}(\mathbf{q}; \kappa\kappa') = \frac{1}{\sqrt{M_\kappa M_{\kappa'}}}\, D_{\alpha\beta}(\mathbf{q}; \kappa\kappa').$$

Equation (1.7) has a solution if

$$\det |D'_{\alpha\beta}(\mathbf{q}; \kappa\kappa') - \omega^2 \delta_{\kappa\kappa'} \delta_{\alpha\beta}| = 0. \tag{1.8}$$

For each value of $\mathbf{q}$ this equation will give $3r$ solutions of ω which are denoted by $\omega_j(\mathbf{q})$ ($j = 1, 2, \ldots, 3r$). The relationship between the vibrational frequency ω and the wave vector $\mathbf{q}$

$$\omega = \omega_j(\mathbf{q})$$

is the dispersion relation for the jth branch of the normal modes of vibration of the lattice; or the phonon dispersion relation. For each $\omega_j(\mathbf{q})$ there exists a $3r$ dimensional vector

$$\mathbf{e}(\mathbf{q}j) = \{e_\alpha(\mathbf{q}j; \kappa)\} \tag{1.9}$$

which satisfies the equation

$$\omega_j^2(\mathbf{q})\, e_\alpha(\mathbf{q}j; \kappa) = \sum_{\kappa'\beta} D'_{\alpha\beta}(\mathbf{q}; \kappa\kappa')\, e_\beta(\mathbf{q}j; \kappa') \tag{1.10}$$

and is orthonormalized as

$$\sum_{\kappa,\alpha} e_\alpha^*(\mathbf{q}j; \kappa)\, e_\alpha(\mathbf{q}j'; \kappa) = \delta_{jj'}$$

and

$$\sum_j e_\alpha^*(\mathbf{q}j; \kappa)\, e_\beta(\mathbf{q}j; \kappa') = \delta_{\alpha\beta}\delta_{\kappa\kappa'}.$$

$\mathbf{e}(\mathbf{q}j)$ is the eigenvector or the polarization vector of the phonon $(\mathbf{q}, j)$.

The relations (1.2) on the force constants have the consequence that the frequencies of three branches of $\omega_j(\mathbf{q})$ ($j = 1, 2, 3$) vanish as $\mathbf{q} \to 0$ corresponding to the translation of the crystal as a whole along three mutually perpendicular directions. These three branches are called acoustic branches because, in the limit of very long wavelengths, i.e., $\mathbf{q} \to 0$, they give the frequencies of sound waves. The propagation of sound waves can also be described by the macroscopic equations of motion of an elastic medium,

$$\rho \ddot{u}_\alpha(\mathbf{X}) = \sum_{\gamma\beta\lambda} C_{\alpha\gamma,\beta\lambda} \frac{\partial^2 u_\beta(\mathbf{X})}{\partial x_\lambda \partial x_\gamma}, \qquad (\alpha, \beta, \gamma, \lambda = 1, 2, 3), \tag{1.11}$$

where ρ is the density, $\mathbf{u}(\mathbf{X})$ is the strain at $\mathbf{X}$, and the $C_{\alpha\gamma,\beta\lambda}$ are elastic constants.

Conventional abbreviations for the subscripts on the elastic constants $\alpha\gamma \to i$ are: $11 \to 1, 22 \to 2, 33 \to 3, 23 \to 4, 13 \to 5$, and $12 \to 6$. Assuming wavelike solutions gives, in analogy to Equation (1.5),

$$\rho\omega^2(\mathbf{q})\mathbf{u}(\mathbf{q}) = \mathbf{A}(\mathbf{q})\mathbf{u}(\mathbf{q}), \tag{1.12}$$

where

$$A_{\alpha\beta}(\mathbf{q}) = \sum_{\gamma\lambda} C_{\alpha\gamma,\beta\lambda} q_\gamma q_\lambda. \tag{1.13}$$

The matrix $\mathbf{A}(\mathbf{q})$ is the analogue of the dynamical matrix $\mathbf{D}(\mathbf{q})$ discussed above. Since $\mathbf{A}(\mathbf{q})$ is quadratic in $\mathbf{q}$, $\omega_j(\mathbf{q})$ is a linear function of $|\mathbf{q}|$. This linear dependence can be written $\omega_j(\mathbf{q}) = C_j(\hat{q}) |\mathbf{q}|$ with $C_j(\hat{q})$ denoting the velocity of acoustic waves or initial slope as $q \to 0$) corresponding to the jth branch $(j = 1, 2, 3)$ in the direction $\hat{q} = \mathbf{q}/|\mathbf{q}|$. For directions of high symmetry in the crystal, $C_j(\mathbf{q})$ is usually determined by a simple linear combination of the $C_{\alpha\gamma,\beta\lambda}$ (or C_{ij}). For example in the [001] or c-direction of a crystal with the *hcp* structure, the longitudinal and transverse acoustic dispersion curves have initial slopes given by $\sqrt{C_{11}/\rho}$ and $\sqrt{C_{44}/\rho}$, respectively. Further, if the small $\mathbf{q}$ series expansion of Equation (1.5) is compared to Equation (1.12), it is possible to obtain relationships between the force constants and the elastic constants. If the elastic constants are known, such relationships can be useful in the development of a model for the interatomic forces provided the range of the forces is not too great. Throughout the remaining sections v and ω will be used interchangeably with the definition $\omega = 2\pi v$.

1.2. MODEL OF LATTICE DYNAMICS

In order to perform actual calculations of frequencies and eigenvectors for a crystal and compare them with experimental results, one must construct a force model in which certain parameters are used to characterize the interatomic interactions. The most straightforward and obvious model is the one in which each atomic force constant defined by Equation (1.1) is regarded as a parameter; the so-called general tensor force model. This model is, in principle, the most general since it involves no assumptions about the nature of the interatomic interactions, except that the force constant matrices must be consistent with the crystal symmetry. However, the number of adjustable parameters is then so large that one must restrict the range of the interactions to a few near-neighbors in order to carry out actual calculations. In practice interatomic forces are very often rather long range, and hence a large number of distant neighbors must be included in the model. Therefore, further assumptions are usually made about the nature of the interactions. One of the assumptions commonly made is that of axially symmetric forces [7]. The potential energy of a crystal is assumed to consist of two-body potentials $\phi^{(ij)} (\mathbf{X}_i - \mathbf{X}_j)$ summed over all pairs of atoms i, j,

$$\Phi = \sum_{i>j} \phi^{(ij)}(\mathbf{X}_i - \mathbf{X}_j).$$

Furthermore, the two-body potential is assumed to be a central potential, i.e., it is a function only of the distance between the atoms i and j,

$$\phi^{(ij)}(\mathbf{X}_i - \mathbf{X}_j) = \phi^{(ij)}(|\mathbf{X}_i - \mathbf{X}_j|).$$

Thus,

$$
\Phi_{\alpha\beta}\begin{pmatrix} l \, l' \\ \kappa\kappa' \end{pmatrix} = \frac{\partial^2 \phi}{\partial X_\alpha\begin{pmatrix} l \\ \kappa \end{pmatrix} \partial X_\beta\begin{pmatrix} l' \\ \kappa' \end{pmatrix}} = -\frac{\partial^2 \phi^{(l\kappa, l'\kappa')}}{\partial X_\alpha\begin{pmatrix} l - l' \\ \kappa \; \kappa' \end{pmatrix} \partial X_\beta\begin{pmatrix} l - l' \\ \kappa \; \kappa' \end{pmatrix}}
$$

$$
= -\left\{ \left(\phi''(R) - \frac{\phi'(R)}{R} \right) \frac{X_\alpha\begin{pmatrix} l - l' \\ \kappa \; \kappa' \end{pmatrix} X_\beta\begin{pmatrix} l - l' \\ \kappa \; \kappa' \end{pmatrix}}{R^2} + \frac{\phi'(R)}{R} \delta_{\alpha\beta} \right\}
$$

$$
= -\left\{ (\phi_r - \phi_t) \frac{X_\alpha\begin{pmatrix} l - l' \\ \kappa \; \kappa' \end{pmatrix} X_\beta\begin{pmatrix} l - l' \\ \kappa \; \kappa' \end{pmatrix}}{R^2} + \phi_t \delta_{\alpha\beta} \right\}, \tag{1.14}
$$

where

$$
\phi_t = \frac{\phi'(R)}{R} = \frac{1}{R} \frac{d\phi^{(l\kappa, l'\kappa')}}{dR} \quad \text{and} \quad \phi_r = \phi''(R) = \frac{d^2\phi^{(l\kappa, l'\kappa')}}{dR^2}
$$

with

$$
R = \left| \mathbf{X}\begin{pmatrix} l \\ \kappa \end{pmatrix} - \mathbf{X}\begin{pmatrix} l' \\ \kappa' \end{pmatrix} \right| = \left| \mathbf{X}\begin{pmatrix} l - l' \\ \kappa \; \kappa' \end{pmatrix} \right|.
$$

The ϕ_r and ϕ_t parameters represent radial (stretching or compressional) and transverse (shearing) components of the force constants, respectively. This axially symmetric model has only these two parameters for each pair of interacting atoms.

An even simpler model is the central force model in which all pairs of atoms are assumed to be located at the equilibrium positions of central pair potentials. This means $\phi'(R) = 0$ or $\phi_t = 0$ and, hence, there are no shearing forces. The model has only one parameter for each pair of atoms and potential to the second order in atomic displacements is

$$
\tfrac{1}{2} \sum \phi_r^{(l\kappa, l'\kappa')} \left| \left\{ \mathbf{u}\begin{pmatrix} l \\ \kappa \end{pmatrix} - \mathbf{u}\begin{pmatrix} l' \\ \kappa' \end{pmatrix} \right\} \cdot \frac{\mathbf{X}\begin{pmatrix} l - l' \\ \kappa \; \kappa' \end{pmatrix}}{R} \right|^2.
$$

These models are, in a certain sense, extensions of pair potentials (such as Lennard-Jones, Morse, or Buckingham potentials) used to describe interactions between atoms in molecules [8]. In fact, sometimes the Born-Mayer potential $\phi = Ae^{-Br}$ has been used to represent the part of the interatomic interaction due to the electron shell overlap in ionic crystals [9].

Another approach used to characterize interatomic interactions is that of the valence force model [10, 11], which may be regarded as a natural extension of the valence forces, used to describe molecular spectra [12], to the study of lattice vibrations for covalent crystals. Parameters in this force model can be regarded as force constants associated with changes in covalent bond lengths and bond angles. Thus one advantage of the model is that the potential energy of the lattice is expressed in terms of variables which are invariant under translations and rigid body rotations of the lattice as a whole [13] and the relations imposed on parameters such as Equations (1.2) and (1.3) are automatically satisfied. For example, in the case shown in Figure 1.1 the potential energy may have terms such as

$$\tfrac{1}{2}K_r^{(1)}(\Delta r_1)^2,\ \tfrac{1}{2}K_\theta^{(2)}(\Delta\theta_2)^2,\ K_{rr}^{(12)}(\Delta r_1)(\Delta r_2),$$

and

$$K_{r\theta}^{(13)}(\Delta r_1)(\Delta\theta_3). \tag{1.15}$$

In this general form, the model was first applied to the analysis of the phonon dispersion curves for diamond measured by neutron inelastic scattering techniques [13]. A simpler version of the model, slightly modified to allow for the planar structure of

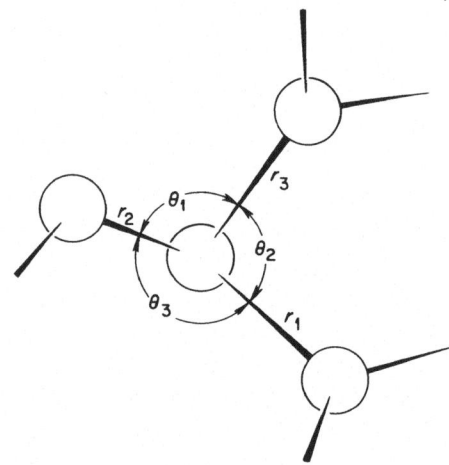

Variables for Valence Force Model.

Fig. 1.1.   Typical bond and angle variables of the valence force model.

graphite, has been used to calculate the specific heat in terms of valence force constants which were estimated from those of benzene and other organic molecules [14].

The valence force model should be most appropriate to the lattice dynamics of crystals in which interatomic force interactions are limited to a few near neighbors interacting through strong covalent bonds. In applying the model to calculations on actual crystals, one must choose the most significant terms in the potential energy function in order to limit the number of parameters in the model. Some physical picture of bonding (e.g. hybridization of orbitals) is desirable in order to make such a choice.

For all of the models discussed above, one assumed implicitly that the ranges of the interatomic forces are relatively short so that only interactions between fairly close neighbors are included. Thus the Coulomb interaction between charges in ionic crystals cannot be represented by such models. However, since the form of the potential is explicitly known for this interaction, the total contribution to the dynamical matrix from this part of the central pair potential can be obtained directly by summation over the entire crystal. The result is usually expressed as

$$D_{\alpha\beta}^{(\text{Coul})}(q \mid \kappa\kappa') = \frac{Z_\kappa Z_{\kappa'}}{v_a} C_{\alpha\beta}(q \mid \kappa\kappa'),$$

where Z_κ and $Z_{\kappa'}$ are the charges on atoms κ and κ', respectively, and v_a is the volume of the unit cell. $C_{\alpha\beta}(q \mid \kappa\kappa')$ is a Coulomb coefficient and is determined uniquely by the atomic arrangement of a crystal [15, 16]. A rigid ion model which includes Coulomb interactions between all ions and overlap repulsive forces between near neighbors has been applied to various ionic crystals [15, 17], but has met with little success in reproducing the experimental data, mainly because the ions were assumed to be non-polarizable.

One method of introducing the dipole moments on ions in calculations of phonon frequencies is that incorporated in the shell model [17]. In this model an atom or ion in a crystal is divided into two dynamical entities, the core and the shell as shown in Figure 1.2. With **u** denoting a $3r$-dimensional column vector consisting of atomic

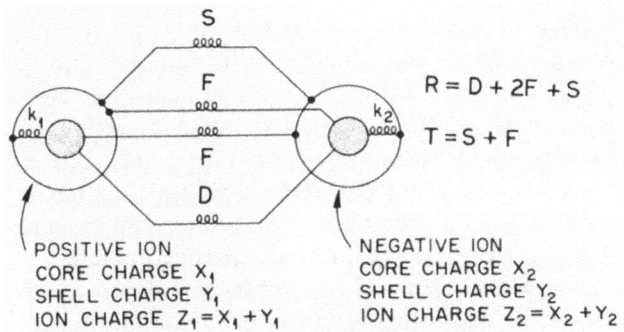

Fig. 1.2. Interactions represented by the parameters of the shell model.

(ionic displacements) from equilibrium positions and **w** denoting a similar vector of displacements of the shells from the instantaneous positions of atoms (ions) to which they belong, one can write the equations of motion as

$$\omega^2 \mathbf{Mu} = (\mathbf{R} + \mathbf{ZCZ})\mathbf{u} + (\mathbf{T} + \mathbf{ZCY})\mathbf{w} \tag{1.16}$$

$$0 = (\mathbf{T}^\dagger + \mathbf{YCZ})\mathbf{u} + (\mathbf{S}' + \mathbf{YCY})\mathbf{w} \tag{1.17}$$

with

$$S'_{\alpha\beta}(\mathbf{q} \mid \kappa\kappa') = k_\kappa \delta_{\kappa\kappa'}\delta_{\alpha\beta} + S_{\alpha\beta}(\mathbf{q} \mid \kappa\kappa'),$$

$$Z_{\alpha\beta}(\kappa\kappa') = Z_\kappa \delta_{\kappa\kappa'}\delta_{\alpha\beta}$$

$$Y_{\alpha\beta}(\kappa\kappa') = Y_\kappa \delta_{\kappa\kappa'}\delta_{\alpha\beta}$$

where Y_κ is the charge of the shell which is bound to the core of the κth atom (ion) by the spring constant k_κ. R, T and S are dynamical matrices corresponding to atom-atom, atom-shell, and shell-shell interactions, respectively. They include only short range forces and can be treated by the models described before. **C** is a matrix of Coulomb coefficients. The electronic dipole moment at an atomic site can be expressed by the relative displacement of the core and shell, **w**, as

$$\rho = \mathbf{Y}\mathbf{w}.$$

By eliminating **w** from Equations (1.16) and (1.17), the dynamical matrix becomes

$$\mathbf{D} = \mathbf{R} + \mathbf{Z}C\mathbf{Z} - (\mathbf{T} + \mathbf{Z}\mathbf{C}\mathbf{Y})(\mathbf{S}' + \mathbf{Y}\mathbf{C}\mathbf{Y})^{-1}(\mathbf{T}^\dagger + \mathbf{Y}\mathbf{C}\mathbf{Z}). \qquad (1.18)$$

Thus, even when $Z = 0$, as in the case of Si and Ge, there is a contribution

$$-\mathbf{T}(\mathbf{S}' + \mathbf{Y}\mathbf{C}\mathbf{Y})^{-1}\mathbf{T}^\dagger \qquad (1.19)$$

to the dynamical matrix due to the polarization of the electrons. This term is, of course, long range in behavior. In the simplest form of the shell model, the short range interactions are taken to be due only to shell-shell interactions. In this case, $\mathbf{R} = \mathbf{T} = \mathbf{S}$ and $\mathbf{T}^\dagger = \mathbf{T}$.

1.3. INELASTIC SCATTERING OF NEUTRONS

The inelastic scattering of neutrons has proved to be an invaluable experimental technique for studying, in detail, the dynamics of crystalline solids. The reason lies in the fact that the energy and wavevector of a thermal neutron is comparable in magnitude to that of a phonon; so that changes in these quantities resulting from the scattering of the neutron by the vibrating crystal lattice are relatively large and easily measured. Since there are numerous detailed theoretical treatments of neutron scattering already published [18–20] we will give here only a brief summary of the main results which are used in the interpretation of experimental data.

We begin with a brief discussion of the experimental procedure. The instrument predominantly used in the neutron study of lattice dynamics is the three-axis crystal spectrometer. A schematic diagram of such an instrument which is located at the High Flux Isotope Reactor (HFIR) of the Oak Ridge National Laboratory is shown in Figure 1.3. The neutron beam from the reactor, which has a Maxwellian energy spectrum peaked at about 40 meV, is monochromatized by Bragg diffraction at the monochromator crystal located on axis-1. The energy selected, usually between 13 and 150 meV for lattice dynamics studies, depends on the d-spacing of the diffracting planes and on the Bragg angle, $2\theta_M$, which is continuously and automatically variable in the instrument illustrated [21]. This monochromatic beam of neutrons each with an initial, preselected energy E_0 and momentum $\hbar k_0$ ($|k_0| = \sqrt{2mE_0}$, where m is the neutron mass) is incident on the sample located on axis-2. The orientation of the sample with respect to the incident beam is important and is given here by the angle ψ. The energy E' (and momentum $\hbar \mathbf{k}'$) of the neutrons which are scattered through the angle ϕ is measured, again by Bragg diffraction, by the analyzing crystal located on axis-3.

The cross section or intensity appropriate to a certain energy change $\hbar\omega$ and momentum change $\hbar\mathbf{Q}$ of the scattered neutrons, where

$$\hbar\omega = E_0 - E' \qquad (1.20)$$

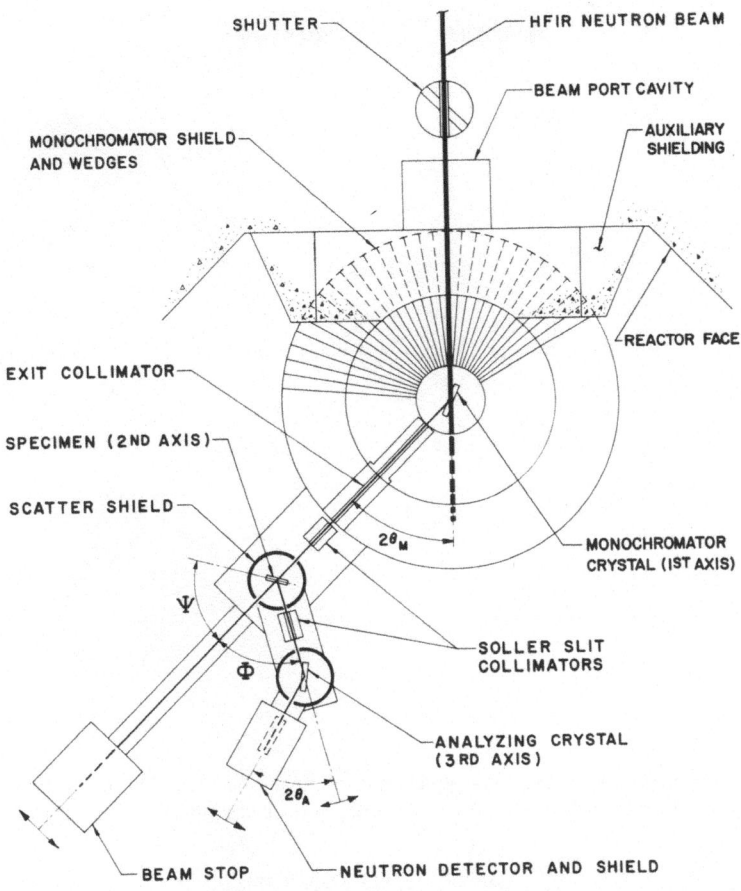

Fig. 1.3. Schematic of the triple-axis neutron spectrometer located at the High Flux Isotope Reactor, Oak Ridge National Laboratory.

and
$$\hbar\mathbf{Q} = \hbar(\mathbf{k}_0 - \mathbf{k}'),$$

is determined by energy and momentum conservation conditions which depend on the dynamical properties of the sample under study in a manner to be discussed in more detail later. Basically, however, a peak in the distribution of scattered neutrons will occur when ω and $\mathbf{Q}$ coincide with the dispersion relation $\omega(\mathbf{Q})$ of the sample. A simple illustration of the scattering experiment as viewed, or constructed, in the reciprocal lattice of an fcc structure is given in Figure 1.4a. Each reciprocal lattice vector τ corresponds to an equivalent origin for the unique set of normal mode wavevectors $\mathbf{q}$ ($\mathbf{Q} = 2\pi\tau + \mathbf{q}$), within the Brillouin zone shown surrounding each point in the figure. The particular τ or zone investigated experimentally depends on the angles ψ and ϕ of Figure 1.3 as well as on $|\mathbf{k}_0|$ and $|\mathbf{k}'|$ and can be varied at will.

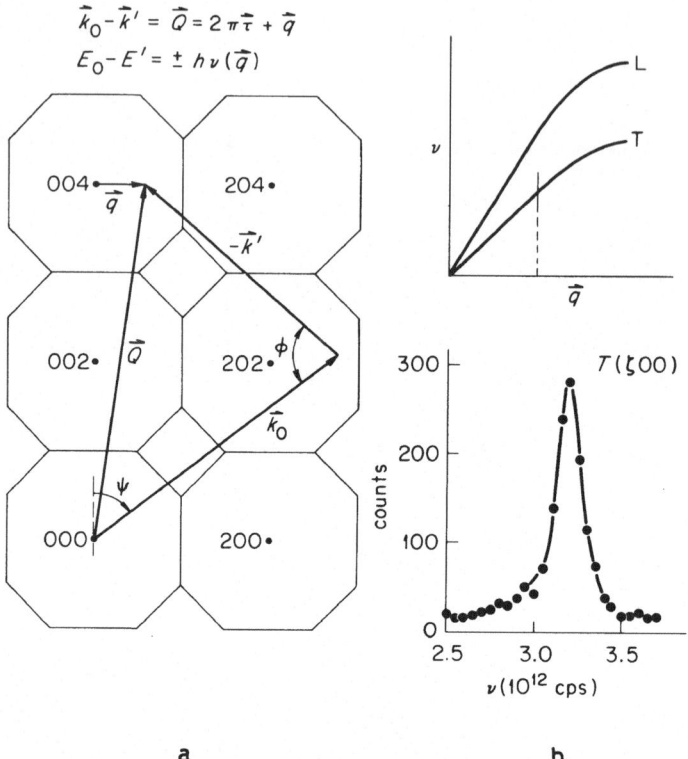

$$\vec{k_0} - \vec{k}' = \vec{Q} = 2\pi\vec{\tau} + \vec{q}$$
$$E_0 - E' = \pm h\nu(\vec{q})$$

a   **b**

Fig. 1.4.   (a) Schematic illustration of a typical scattering geometry in reciprocal-space.   (b) Example of typical Constant-Q measurement of one point on the phonon dispersion relation.

    With the computer control of present day spectrometers, it is possible to search for peaks in the intensity distribution of the scattered neutrons by scanning along any line in $(\omega, \mathbf{Q})$. However, it is often most convenient to carry out scans in ω for constant $\mathbf{Q}$, or scans in $\mathbf{Q}$ for constant ω. A simple example of a constant $\mathbf{Q}$ measurement is illustrated in Figure 1.4b. The location of the peak in the neutron intensity determines one point on $\omega(\mathbf{Q})$. The experiment is then repeated for many $\mathbf{Q}$ to map out $\omega(\mathbf{Q})$ throughout the zone. Other experimental details of the neutron inelastic scattering technique can be found elsewhere [22].

    The differential cross section for neutron scattering, or, the number of neutrons scattered per unit energy E', per unit solid angle Ω by a crystal, depends on the vibrational atomic displacements according to [20]

$$\frac{\mathrm{d}^2\sigma}{\mathrm{d}\Omega\,\mathrm{d}E'} = \frac{k'}{k_0}\frac{1}{2\pi h}\int_{-\infty}^{\infty} \mathrm{d}t \exp(-i\omega t) \sum_{\substack{ll' \\ \kappa\kappa'}} \exp\left[-i\mathbf{Q}\cdot\left\{\mathbf{X}\begin{pmatrix}l\\\kappa\end{pmatrix} - X\begin{pmatrix}l'\\\kappa'\end{pmatrix}\right\}\right] \times$$

$$\times \left\langle \mathbf{Q}\cdot\mathbf{u}\begin{pmatrix}l\\\kappa\end{pmatrix},0 \right) \mathbf{Q}\cdot\mathbf{u}\begin{pmatrix}l'\\\kappa'\end{pmatrix},t \right\rangle, \tag{1.21}$$

where $\langle \ldots \rangle$ is the thermal average of the displacement-displacement correlation function and $\mathbf{u}\begin{pmatrix} l' \\ \kappa', \end{pmatrix} t$ is the displacement of atom (l', κ') at time t and $\mathbf{u}\begin{pmatrix} l \\ \kappa, \end{pmatrix} 0$ is the corresponding quantity for atom (l, κ) at $t = 0$.

In terms of the normal mode variables discussed in previous sections the cross section for energy loss of the neutron (phonon creation) is given by

$$\frac{d^2\sigma}{d\Omega\, dE_f} = \frac{\hbar}{4\pi} \frac{k_f}{k_i} (n_j + 1) \left| \sum_\kappa \mathbf{Q} \cdot \mathbf{e}(\mathbf{q}j \mid \kappa)\, b_\kappa \exp[2\pi i \boldsymbol{\tau} \cdot \mathbf{X}(\kappa)]/(\omega_j M_\kappa)^{1/2} \right|^2,$$

(1.22)

where $n_j = [\exp(\hbar\omega_j/kT) - 1]^{-1}$, $\omega_j = \omega_j(q)$, and the summation is over all the atoms in the unit cell, having neutron scattering lengths b_κ. It is desirable to note several properties of this cross section which are important for the interpretation of experimental results and for the development of a satisfactory force model for the material under investigation.

The $\mathbf{Q} \cdot \mathbf{e}$ factor provides a means for investigating the polarization character (longitudinal or transverse) of each mode, since for a given $\mathbf{q}$ it is often possible by selecting different $\boldsymbol{\tau}$ to achieve scattering geometries with $\mathbf{Q}$ alternately nearly parallel or nearly perpendicular to $\mathbf{q}$. Also the phases between the displacement of the atoms in the unit cell vibrating in the mode $(\mathbf{q}, j)$ are described by the eigenvector $\mathbf{e}(\mathbf{q}j \mid \kappa)$. This fact together with the phase factors $\exp(2\pi i \boldsymbol{\tau} \cdot \mathbf{X}(\kappa))$ means that each mode is characterized by a neutron scattering structure factor (the quantity within the absolute signs of Equation (1.22)) which varies in a distinct way with $\mathbf{q}$ and $\boldsymbol{\tau}$. The dependence of the neutron scattering intensity on polarization character, $\mathbf{q}$ and $\boldsymbol{\tau}$ facilitates both the identification of the different branches of the dispersion relation in a complex crystal structure and the investigation of individual branches which are in close proximity without the need for excessively good energy resolution.

Of particular importance, of course, is that the structure factor may be calculated for any mode if we know the eigenvectors $\mathbf{e}(\mathbf{q}j \mid \kappa)$ which may in turn be obtained from a theoretical model. Thus the neutron scattering intensity is important experimental information which, together with the measured frequencies, may be essential in arriving at a valid model for the interatomic forces. Often, however, an adequate reproduction of the measured $\omega_j(\mathbf{q})$ alone is difficult. We will be content with this goal for the materials reported in this article, using the measured intensities only for purposes of branch and polarization identifications.

2. Atomic Vibrations in Layered Compounds

Special features of the lattice dynamics of layered compounds are mainly related to the extreme anisotropy in the interatomic forces. Interactions between atoms in different layers are very weak compared with those between atoms in the same layer. Hence, the lattice vibrations of such materials can be described approximately by those of an isolated single layer except for very low frequency modes of vibration in which the weak interlayer interactions are significant in determining their frequencies. For a network of atoms forming a single layer phonons propagate in the layer, and dispersion curves are defined only for wave vectors in a plane parallel to the layer. It should be pointed out, however, that dynamics of such a layer is not that of a truly two-dimensional lattice, since a layer may have a finite dimension in the direction perpendicular to the layer. Even if the layer consists of a single sheet of atoms with only two-dimensional extent, the atoms have three-dimensional motion. When such layers are stacked together to form a three-dimensional crystal having a unit cell consisting of n-layers between which the interactions are zero, all the phonon dispersion curves have n-fold degeneracies. If small but finite interactions are introduced between these layers, phonons in different layers will be coupled and the degeneracies will be lifted. As a result the dispersion relation will consist of several sets of n closely spaced branches with spacings that depend mainly on the interlayer interactions.

For example, in the case of a hcp crystal with a unit cell consisting of two layers ($n = 2$), as shown schematically in Figure 2.1, the dynamical matrix may be written as

$$\mathbf{D} = \mathbf{D}_0 + \mathbf{D}', \tag{2.1}$$

where

$$\mathbf{D}_0 = \begin{pmatrix} \mathbf{d}_0 & 0 \\ 0 & \mathbf{d}_0 \end{pmatrix} \quad \text{and} \quad \mathbf{D}' = \begin{pmatrix} \mathbf{d}_{11} & \mathbf{d}_{12} \\ \mathbf{d}_{12}^+ & \mathbf{d}_{11}^* \end{pmatrix}. \tag{2.2}$$

$\mathbf{d}_0$ is the dynamical matrix for a layer and has a dimension $3p \times 3p$ where p is the number of atoms in the unit cell of a layer (thus $r = 2p$ in the present example). $\mathbf{D}'$ is the part of the dynamical matrix arising from the extension in the third dimension and

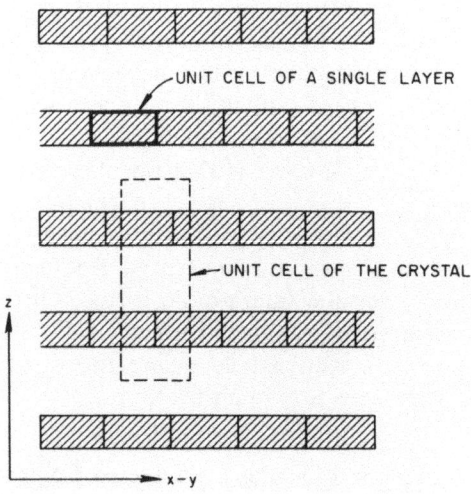

Fig. 2.1.   Illustration of the unit cell for a layered structure with $n = 2$

it includes the interlayer interactions. The eigenvector $\mathbf{e}^0$ $(\mathbf{q}_p j)$ for the jth mode in a single layer and the phonon frequency ω_j^0 $(\mathbf{q}_p)$ satisfy the equation

$$\mathbf{d}_0(\mathbf{q}_p) \, \mathbf{e}^0(\mathbf{q}_p j) = \{\omega_j^0(\mathbf{q}_p)\}^2 \, \mathbf{e}^0(\mathbf{q}_p j), \tag{2.3}$$

where $\mathbf{q}_p$ is a two-dimensional wave vector in a layer. The eigenvectors for $\mathbf{D}_0$ may be constructed in terms of $e^0 (\mathbf{q}_p j)$. However, because of the two-fold degeneracy, there are two linearly independent vectors, such as

$$\mathbf{e}^{(1)}(\mathbf{q}_p j) = \begin{pmatrix} e^0(\mathbf{q}_p j) \\ 0 \end{pmatrix} \quad \text{and} \quad \mathbf{e}^{(2)}(\mathbf{q}_p j) = \begin{pmatrix} 0 \\ e^{0*}(\mathbf{q}_p j) \end{pmatrix}$$

or

$$\mathbf{e}^{(+)}(\mathbf{q}_p j) = \frac{1}{\sqrt{2}} \begin{pmatrix} e^0(\mathbf{q}_p j) \\ e^{0*}(\mathbf{q}_p j) \end{pmatrix} \quad \text{and} \quad \mathbf{e}^{(-)}(\mathbf{q}_p j) = \frac{1}{\sqrt{2}} \begin{pmatrix} e^0(\mathbf{q}_p j) \\ -e^{0*}(\mathbf{q}_p j) \end{pmatrix} \qquad (2.4)$$

Thus

$$\mathbf{D}_0(\mathbf{q}_p) \, \hat{e}^{(\pm)}(\mathbf{q}_p j) = \{\omega_j^0(\mathbf{q}_p)\}^2 \, \mathbf{e}^{(\pm)}(\mathbf{q}_p j). \qquad (2.5)$$

If $\mathbf{D}'$ is treated as a small perturbation, frequencies of normal modes for $\mathbf{D}$ satisfying

$$\mathbf{D}(\mathbf{q}) \, \mathbf{e}(\mathbf{q} j) = \omega_j^2(\mathbf{q}) \, \mathbf{e}(\mathbf{q} j) \qquad (2.6)$$

may be obtained by the well-known perturbation theory for degenerate systems. The result is given by

$$\{\omega_j^{(\pm)}(\mathbf{q})\}^2 = \{\omega_j^0(\mathbf{q}_p)\}^2 +$$
$$+ [e^0(\mathbf{q}_p j) \cdot \mathbf{d}_{11}(\mathbf{q}) \, e^0(\mathbf{q}_p j)] \pm \text{Re}[e^0(\mathbf{q}_p j) \cdot \mathbf{d}_{12}(\mathbf{q}) \, e^{0 *}(\mathbf{q}_p j)],$$
$$(2.7)$$

where $\text{Re}[A]$ is the real part of A and $\mathbf{q} = (\mathbf{q}_p, q_z)$. It should be noted that although $\omega_j^0(\mathbf{q}_p)$ is determined by the force constants within a single layer, these force constants may be affected by the introduction of the interlayer interactions. As can be seen from the double sign in Equation (2.7), there are two nearly equal frequencies, corresponding to vibrations of adjacent layers being in and out of phase with each other. These modes are represented by $\mathbf{e}^{(+)}(\mathbf{q}_p j)$ and $\mathbf{e}^{(-)}(\mathbf{q}_p j)$.

It should be noted that the difference in the squares of frequencies $\{\omega^{(+)}\}^2 - \{\omega^{(-)}\}^2$ is of first order in the interlayer interactions; therefore,

$$\frac{\omega^{(+)} - \omega^{(-)}}{\omega^0}$$

is of first order in the ratio interlayer-force/intralayer-force. In general it can be extremely difficult to measure by neutron scattering methods the difference in the two frequencies for the optic modes.

For wave vectors perpendicular to the plane of the layer (the c^* direction in the above example for an hcp structure), vibrational modes may be classified into two types: those in which atoms within a layer move as an almost (in the case of a planar layer, exactly) rigid unit, and those in which the atoms of a layer vibrate relative to one another almost independently of those of neighboring layers. The former modes may be loosely called *rigid-layer modes* and the latter *intralayer modes*. These may be compared with the external and internal modes of molecular crystals, the molecule being the entire layer in the present case. In fact, for the above example of an *hcp* crystal, from Equation (2.7) with $\mathbf{q} = (00q_z)$,

$$\omega_j^2(q_z) = \{\omega_j^0(0)\}^2 + [e^0(oj) \cdot \mathbf{d}_{11}(q_z) \, e^0(oj)] \pm \text{Re}[e^0(oj) \cdot \mathbf{d}_{12}(q_z) \, e^{0 *}(oj)].$$
$$(2.8)$$

For the jth mode that corresponds to an acoustic mode in the original single layer, $\omega_j^0(0) = 0$. Thus, for such a mode, there exists two vibrational modes whose frequencies are given by

$$\omega_j^2(q_z) = [e^0(oj) \cdot d_{11}(q_z) \, e^0(oj)] \pm \text{Re}[e^0(oj) \cdot d_{12}(q_z) \, e^0 \,^*(oj)]. \qquad (2.9)$$

These are the rigid layer modes, and one of them which gives $\omega_j(q_z \to 0) \to 0$ is the acoustic mode and the other is the low frequency optic mode that corresponds strictly to the out-of-phase vibration of rigid layers at $q_z = 0$. Other frequencies given by Equation (2.8) may be approximated by

$$\omega_j(q_z) \simeq \omega_j^0(0), \qquad (2.10)$$

corresponding to intralayer modes.

Thus, rigid layer modes whose frequencies are given by Equation (2.9) are determined by weak interlayer forces and have very low frequencies, whereas intralayer modes, whose frequencies are approximately equal to the frequencies of vibrations in a single layer as given by Equation (2.10), have no dispersion to first order in the interlayer interactions.

Another interesting feature of the dynamics of layered compounds is that dispersion curves for certain transverse acoustic modes with wave vectors within a layer may have extremely small initial slopes arising from weak interlayer interactions. It can be shown quite generally that the macroscopic elastic constants of the type $C_{3\alpha3\beta}$ vanish for all α and β if there is no interaction between layers stacked along the z-direction (X_α with $\alpha = 3$) as a result of the rotational invariance condition given by Equation (1.3). For example, in the case of an hcp crystal, the velocity of the transverse elastic wave propagating in the basal plane with polarization vector perpendicular to the plane (TA$_\perp$ mode) is given by $\sqrt{C_{44}/\rho}$ where ρ is the density of the crystal. Since

$$C_{44} = C_{3131} = 0$$

for noninteracting layers, the phonon dispersion curves for these TA$_\perp$ modes must have zero initial slopes (see Appendix).

Even when interlayer interactions are introduced, the slopes will remain small since they must be the same as the initial slopes of the phonon dispersion curves for the transverse acoustic modes propagating along the z-direction, the latter being rigid layer modes with low frequencies. However, phonons propagating in the basal plane are essentially intralayer modes and are expected to have high frequencies for large wave vector q. Therefore, the dispersion curves for $TA_\perp$ modes in layered compounds may have convex (almost quadratic) shapes with very small initial slopes.

Finally, the microscopic approach to the phonon theory for quasi-two dimensional lattices should be mentioned. Often a microscopic theory can be formulated naturally in reciprocal space rather than in real space as is done for the Born-von Kármán formalism. This is especially suitable for metals since the interatomic interactions are mediated by conduction electrons, and consequently, they are very long range in real space. Thus, a large number of force constants are required in the Born-von Kármán type of models. Microscopic calculations of phonon dispersion curves were first performed for metals [23–25] but more recently formulations for semiconductors [26,

28] and ionic crystals [29, 30, 31] have been studied and results of actual calculations give reasonable agreement with experimental results. A microscopic theory is generally based on the pseudo-potential function for the smooth part of the valence electron wave function and on the dielectric function $\varepsilon(\mathbf{Q}, \mathbf{Q'})$ [32, 33] which represents the response of the electron system to an external electric field. The dielectric function may be regarded as an infinite matrix whose rows and columns are designated by $\mathbf{Q}$ and $\mathbf{Q'}$. The potential energy of the lattice may be expressed to second order in the atomic displacements in terms of the Fourier transform of the pseudo-potential for atom κ, $W_\kappa(Q)$ [34], and the dielectric function matrix. The contribution from the valence electrons to the dynamical matrix is given by [34]

$$D_{\alpha\beta}^{bs}(\mathbf{q} \mid \kappa\kappa') = \sum_{\mathbf{G},\mathbf{G'}} (\mathbf{q} + \mathbf{G})_\alpha (\mathbf{q} + \mathbf{G'})_\beta \frac{1}{v(\mathbf{q} + \mathbf{G})} \{\varepsilon^{-1}(\mathbf{q} + \mathbf{G}, \mathbf{q} + \mathbf{G'}) -$$

$$- \delta_{\mathbf{G},\mathbf{G'}}\} W_\kappa(\mathbf{q} + \mathbf{G}) W_{\kappa'}(\mathbf{q} + \mathbf{G'}) \, e^{i[\mathbf{G}\cdot\mathbf{X}(\kappa) - \mathbf{G'}\cdot\mathbf{X}(\kappa')]}, \qquad (2.11)$$

where G and G' are reciprocal lattice vectors and $v(\mathbf{Q})$ is the Fourier transform of the electron-electron interaction which is given by

$$\frac{4\pi e^2}{V_a} \frac{1}{Q^2}$$

in the case of the Coulomb interaction where V_a is the volume of the unit cell.

One of the major difficulties in carrying out actual calculations for a microscopic theory is the inversion of the matrix ε. This presents no trouble if ε is diagonal as in the case of a spherical Fermi surface for a free electron system. However, in general, one must limit the dimension to a tractable number in order to invert the matrix. To avoid this problem, Sinha *et al.* [17] made a factorization Ansatz in which ε is assumed to be of the form $\varepsilon(Q, Q') \propto f(Q)f(Q')$ and they applied the method to calculations for semiconductors and ionic crystals with reasonable success [27, 29, 30].

For the two-dimensional lattices, the inversion of the dielectric function matrix may be more feasible because the lower dimensionality should reduce the computation time. Also the sum over G and G' in Equation (2–11) may be performed in less time. Thus, the layered compounds seem to offer a promising opportunity for calculating phonon dispersion curves from first principles without involving drastic approximations; however, the question of the validity of pseudo-potentials for transition metal compounds presents a difficulty [35, 36].

3. Neutron Scattering Experiments on Layered Compounds

3.1. INTRODUCTION

There are several difficulties encountered in obtaining phonon dispersion curves experimentally which are particularly serious in the study of layered compounds. One is the lack of single crystal samples large enough for neutron scattering experiments. There are some exceptions in this respect, such as MoS_2 which is found in nature in the form of large single crystals. However, most of the crystals available have rather large mosaic spreads both within and perpendicular to layers. Generally, a large mosaic distribution leads to an increase in the number of various spurious scattering processes which can contaminate or obscure the peaks in the distribution of scattered neutrons due to one-phonon processes. In layered compounds, owing to the generally large separation of the layers (large c-lattice constant), the reciprocal lattice points along the c^* direction are very closely spaced. Consequently the probability for spurious processes is unusually high in layered compounds.

Another difficulty is associated with the sample shape. Measurements of the transverse vibrational modes which have both wave vectors and polarization vectors parallel to the plane of a layer requires sample orientations in which the c-axis is perpendicular to the scattering plane (the plane containing both the incident and scattered neutron beams). However, because of extremely anisotropic mechanical properties, the sample shape is often in the form of a thin plate with the c-axis perpendicular to the surface. Hence it is almost impossible to measure the modes mentioned above, because for a sample with the c-axis perpendicular to the scattering plane, the path of neutrons in the sample becomes very long and the absorption may be quite high.

Finally, a word about energy resolution. Compared to optical methods, the energy resolution of neutron inelastic scattering is rather poor in the range of neutron energies required for the observation of the phonon dispersion curves for layered compounds. It is, in general, 0.5 meV $\sim$ 3 meV which typically amounts to $2 \sim 5\%$ energy resolution. Thus, as was pointed out previously, it is impossible to distinguish the two nearly degenerate frequencies corresponding to a pair of modes which differ from each other only by the vibrational phase of adjacent layers. As a result, crudely speaking, only half the number of possible branches can be clearly separated, except for the low frequency rigid layer modes.

These problems combined make neutron scattering experiments on these substances very difficult at present, and the amount of data accumulated so far is very small. In the following, experimental data are presented with some emphasis on experimental details.

3.2. GRAPHITE

The graphite structure consists of two-dimensional hexagonal networks of carbon atoms which are stacked together to form a three-dimensional crystal. A perspective illustration which shows the positions of the atoms in graphite is given in Figure 3.1. The atoms labeled A, B, C, and D are all in the same unit cell. The structure belongs to the space group D_{6h}^4. Since the unit cell contains four atoms, there are in general 12 branches to the dispersion relation. It is convenient to classify the various branches in terms of their group theoretical representations. The results of a group theoretical analysis are given in Table 3.1. Since graphite is the simplest layered compound, its mechanical and electronic properties including the electron energy band structure have been studied extensively. However, as is usually the case with a layered compound, it is not possible to obtain a large single crystal, and only pyrolytic graphite samples of appreciable size are available. Pyrolytic graphite consists of very thin ($\sim 1\mu$) graphite crystallites randomly oriented about a common c-axis. Since this presents experimental difficulties which may possibly be encountered in measurements on other layered compounds, the experimental technique used to measure the phonon spectra for graphite will be discussed somewhat in detail.

TABLE 3.1

Irreducible representations and eigenvectors for graphite

	Nb[a]	Nd[b]	Eigenvectors											
Γ_2^-	2	1	(o	o	u	o	o	v	o	o	u	o	o	v)
Γ_3^+	2	1	(o	o	u	o	o	v	o	o	$-u$	o	o	$-v$)
Γ_5^+	2	2	linear combinations of											
			(u	o	o	v	o	o	$-u$	o	o	$-v$	o	o) and
			(o	u	o	o	v	o	o	$-u$	o	o	$-v$	o).
Γ_6^-	2	2	linear combinations of											
			(u	o	o	v	o	o	u	o	o	v	o	o) and
			(o	u	o	o	v	o	o	u	o	o	v	o)
Σ_1	4	1	(u_1	o	o	v_1	o	o	u_2	o	o	v_2	o	o)
Σ_3	4	1	(o	o	u_1	o	o	v_1	o	o	u_2	o	o	v_2)
Σ_4	4	1	(o	u_1	o	o	v_1	o	o	u_2	o	o	v_2	o)
M_1^+	2	1	(u	o	o	v	o	o	$-u$	o	o	v	o	o)
M_2^-	2	1	(u	o	o	v	o	o	u	o	o	$-v$	o	o)
M_3^+	2	1	(o	o	u	o	o	v	o	o	$-u$	o	o	v)
M_3^-	2	1	(o	u	o	o	v	o	o	u	o	o	$-v$	o)
M_4^+	2	1	(o	u	o	o	v	o	o	$-u$	o	o	$-v$	o)
M_4^-	2	1	(o	o	u	o	o	v	o	o	u	o	o	$-v$)
T_1	4	1	(u_1	u_2	o	v_1	v_2	o	$-u_1$	u_2	o	$-v_1$	v_2	o)
T_2	2	1	(o	o	u	o	o	v	o	o	$-u$	o	o	$-v$)
T_3	2	1	(o	o	u	o	o	v	o	o	u	o	o	v)
T_4	4	1	(u_1	u_2	o	v_1	v_2	o	u_1	$-u_2$	o	v_1	$-v_2$	o)
Δ_1	2	1	(o	o	u	o	o	v	o	o	u	o	o	v)
Δ_2	2	1	(o	o	u	o	o	v	o	o	$-u$	o	o	$-v$)
Δ_5	2	2	linear combinations of											
			(u	o	o	v	o	o	$-u$	o	o	$-v$	o	o) and
			(o	u	o	o	v	o	o	$-u$	o	o	$-v$	o)
Δ_6	2	2	linear combinations of											
			(u	o	o	v	o	o	u	o	o	v	o	o) and
			(o	u	o	o	v	o	o	u	o	o	v	o)

	Nb	Nd			Nb	Nd
K_1	1	1		A_1	2	2
K_2	1	1		A_3	2	4
K_3	1	1				
K_4	1	1				
K_5	3	2				
K_6	1	2				

[a] Nb: number of modes
[b] Nd: degree of degeneracy

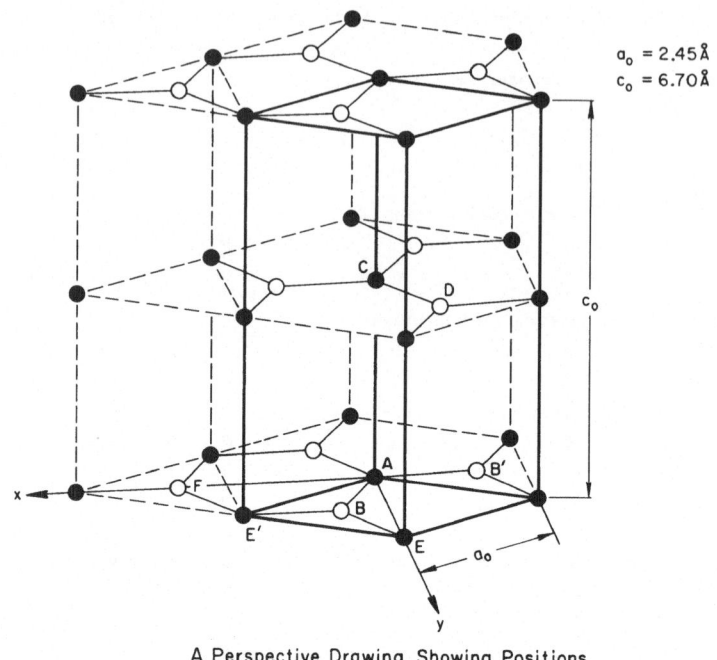

A Perspective Drawing Showing Positions
of the Atoms in Ordinary Graphite.

Fig. 3.1. Perspective drawing showing positions of the atoms in graphite.

The reciprocal space appropriate for neutron scattering experiments on graphite may be visualized as shown in Figure 3.2. It consists of the reciprocal lattice of the ideal graphite lattice rotated about the c^* axis to produce smeared out rings of reciprocal lattice points except along the c^*-axis direction itself. The rings shown in Figure 3.2 represent the [110] and [100]-type reciprocal lattice points after such a rotation. The only portion of the phonon dispersion relation that can be studied unambiguously by coherent neutron scattering techniques is that corresponding to the longitudinal modes propagating along the [001] direction, (Δ_1, Δ_2), because the one-phonon neutron scattering cross section given by Equation (1.22) contains a factor $\mathbf{Q} \cdot \mathbf{e}$, where $\mathbf{Q} = 2\pi\tau + \mathbf{q}$ and the reciprocal lattice vector τ is well defined only along the [001] direction. The first observation of the phonon dispersion curves of graphite were made by Dolling and Brockhouse [37] on a pyrolytic graphite sample which had about 5° angular spread of the c-axis. They could observe the longitudinal acoustic and the lowest optic modes along the [001] direction, and also were able to obtain limited information about low frequency transverse modes along the same direction. However, because of the large angular spread of the sample, it was very difficult for them to extract accurate phonon frequencies for these modes. Since the time of this work, the quality of pyrolytic graphite had improved considerably, and it became possible to obtain large samples having less than 0.4° angular spread of the c-axis. New measurements were performed [38] on such a sample to extend the work of Dolling and Brockhouse.

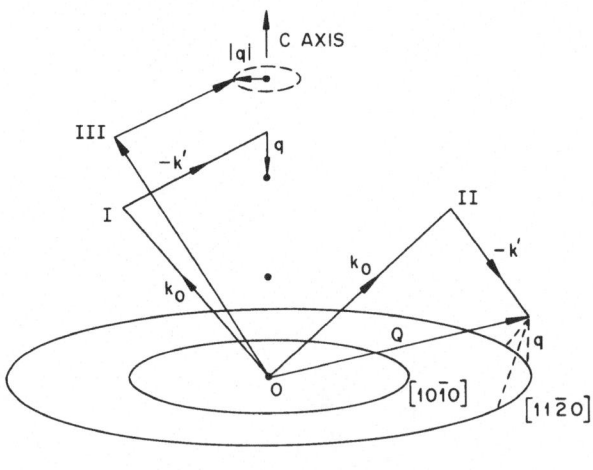

Reciprocal Space Diagram
for Pyrolitic Graphite

Fig. 3.2.   Reciprocal-space diagram for pyrolytic graphite showing the three neutron scattering configurations described in the text.

The sample crystal was oriented with the c-axis in the plane of scattering. Four different scattering configurations were used in the measurements, three of which are shown schematically in Figure 3.2. The scattering plane contains the c-axis, the incident neutron wave vector, and the scattered neutron wave vector.

For configuration I constant-$\mathbf{Q}$ measurements were carried out. In this configuration the frequencies of the longitudinal modes propagating along the c^* direction can be measured unambiguously as a function of wave vector. The results obtained for the acoustic (LA) and the lowest frequency optic (LO) modes are shown in Figure 3.3. The measurements give frequencies which are slightly lower than those obtained previously [37]. This difference is undoubtedly due to the larger mosaic spread of the sample used in the earlier work which gave rise to contributions to the observed neutron groups from the higher frequency modes that have wave vectors slightly off the [001] direction.

With configuration II constant-$\mathbf{Q}$ measurements were carried out to determine the frequencies of the transverse acoustic modes and the lowest frequency optic modes propagating along the [001] direction. Hereafter these will be called the $TA(00\zeta)$ and $TO(00\zeta)$ modes. As indicated by the dashed arrows in Figure 3.2, the wave vector component perpendicular to the scattering plane is not well defined in this experiment because any point on the (110) ring is an origin for a wave vector. In addition the finite q-resolution of the instrument permits the observation of scattering by phonons with wave vectors in the scattering plane but which are also inclined to the [001] direction. However, one expects the frequencies of the transverse (00ζ) modes to be lower than those of all other modes observable in this configuration, since these other modes involve a distortion of the very stiff basal plane. It should be pointed out that the structure factors (Section 1.3) for the $TA(00\zeta)$ and $TO(00\zeta)$ modes are zero in

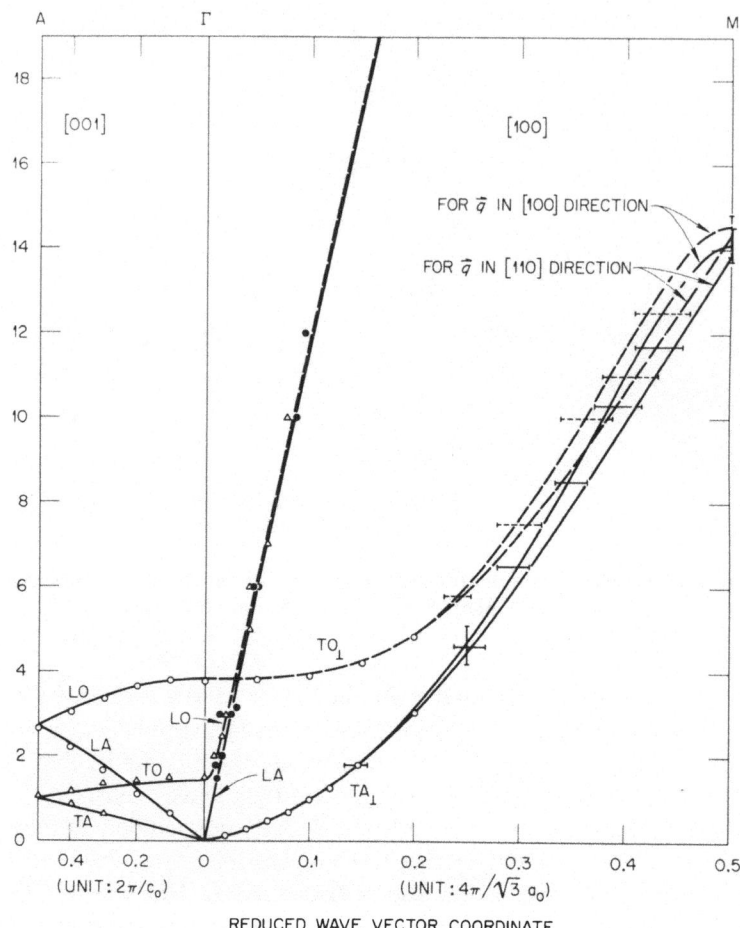

Fig. 3.3. Phonon frequencies measured for pyrolytic graphite. All measurements for **q** in the basal plane have been plotted with **q** expressed in units appropriate for the [100] direction. The lines represent calculations with the model discussed in the text. For the $TA_\perp$ and $TO_\perp$ branches calculations for **q** in both the [100] and [110] directions are shown.

alternate Brillouin zones along c, such that only one of the branches is visible in a given zone. Thus, as the frequency is scanned in a constant-**Q** experiment one should observe no phonon scattering below $v_T(\zeta)$, the frequency corresponding to the visible transverse mode with the wave vector $\mathbf{q} = (00\zeta)2\pi/c$. Then as $v_T(\zeta)$ is reached in the scan, a rapid increase in the scattering is expected. Above $v_T(\zeta)$ the scattering by phonons having wave vectors inclined to the [001] direction dominates. However, the intensity of this scattering will decrease (slowly) with increasing v, primarily because of the frequency dependent factors in the neutron scattering cross section, viz.

$$\frac{1}{v}\{1 + [e^{hv/kT} - 1]^{-1}\},$$

although the polarizations of the phonons involved and the magnitude of the instrumental q-resolution in the scattering plane also are important. The net result is that the maximum of the intensity distribution of neutrons obtained in a configuration 11 experiment always occurs at a frequency which is slightly higher than $v_T(\zeta)$.

A detailed numerical calculation of the intensity distribution for $TO(00\zeta)$ at $\zeta = 0$ was carried out and is described in [38]. The result of this calculation is compared to the experimental data in Figure 3.4.

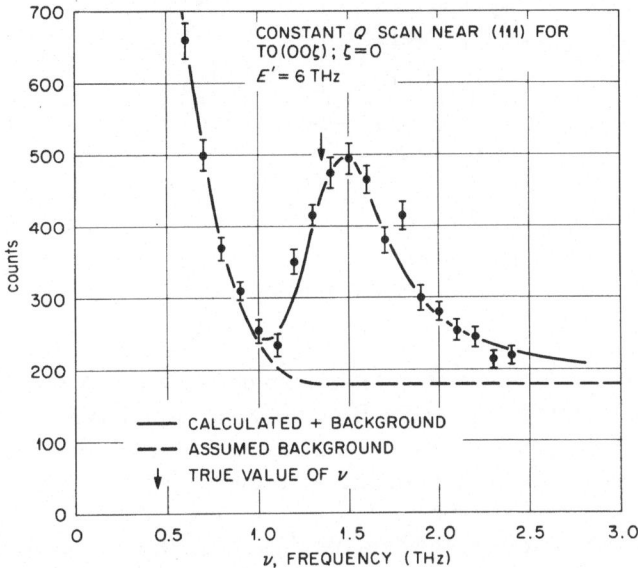

Fig. 3.4.   Comparison of the calculated and measured neutron intensity distributions for a Constant-Q scan of the $TO(00\zeta)$ branch for $\zeta = 0$ of graphite.

On the basis of the excellent agreement obtained between the calculated and experimental results as shown in Figure 3.4, it was concluded that the true frequency of the transverse optic phonon with zero wave vector is very nearly 1.35 THz. This frequency, indicated by the arrow in the figure, corresponds to an intensity which is about 80% of the maximum intensity (above the assumed background) observed near 1.48 THz. It was therefore assumed that a reasonable value for the true frequency at other values of ζ is that appropriate to an intensity which is approximately 80% of the maximum observed in each scan. The (uncorrected) frequencies corresponding to the positions of the intensity maxima are plotted in Figure 3.3.

In configuration III the transverse modes ($TA_\perp$ and $TO_\perp$), having wave vectors in and polarizations perpendicular to the basal plane, were investigated. For this configuration only the magnitude $|\mathbf{q}|$, but not the direction of the wave vector is well defined by the scattering geometry. However, very well defined neutron scattering peaks were observed for small $|\mathbf{q}|$, indicating that the dispersion relation for these modes is quite isotropic for wave vectors in the basal plane. Even at rather large $|\mathbf{q}|$ well defined, though broadened, peaks were observed as indicated in Figure 3.5a where a constant energy scan is shown, and in Figure 3.5b where a constant-Q scan

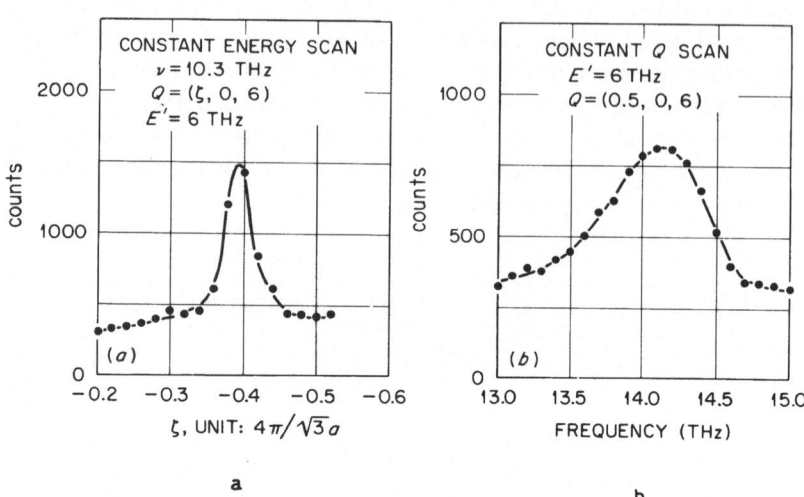

Fig. 3.5. (a) Constant-energy scan of $TA_\perp$ branch of graphite. (b) Constant-Q scan of $TA_\perp$ branch of graphite.

for $|\mathbf{q}| = 0.5 \, (4\pi/\sqrt{3}a)$ is shown. This latter scan corresponds to a wave-vector magnitude equal to that appropriate for the Brillouin zone boundary in the [100] direction, i.e., the M point. Constant-Q scans were carried out at the M point because the dispersion surfaces at that point for both the acoustic and optic modes have horizontal tangents.

The general conclusion from the data obtained in configuration III is that for small $|\mathbf{q}|$, e.g. $< 0.2 \, (4\pi/\sqrt{3}a)$, the $TA_\perp$ and $TO_\perp$ dispersion surfaces are very isotropic so that the widths of the measured neutron peaks were mainly due to the instrumental resolution. As $|\mathbf{q}|$ increased the observed peak width increased, indicating a small departure from perfect isotropy. In this case a high resolution constant-Q scan would measure the distribution of frequencies for those $TA_\perp$ (or $TO_\perp$) modes which all have the same $|\mathbf{q}|$.

Calculations of such distribution functions were carried out for several preliminary force models with a modification of the frequency-distribution calculation method outlined by Raubenheimer and Gilat [39] in order to investigate the shapes to be expected for them. One result which is typical of those obtained is shown in Figure 3.6. Most of the scans actually obtained in configuration III near the $|\mathbf{q}|$ shown in Figure 3.6 were constant-energy scans. Nevertheless, the calculations did at least indicate that the distribution of frequencies between the extreme values, corresponding here to wave vectors along [100] and [110] directions, was reasonably uniform. When smeared out by the instrumental resolution, such a frequency distribution may be expected to give rise to a neutron peak whose position would be intermediate between the values of these extreme frequencies and whose width would be proportional to the departure from isotropy of the dispersion surface.

As $|\mathbf{q}|$ is increased the neutron peaks continue to be rather well defined up to $|\mathbf{q}| \simeq 0.5 \, (4\pi/\sqrt{3}a)$. Any $|\mathbf{q}|$ larger than this value is beyond the Brillouin zone boundary

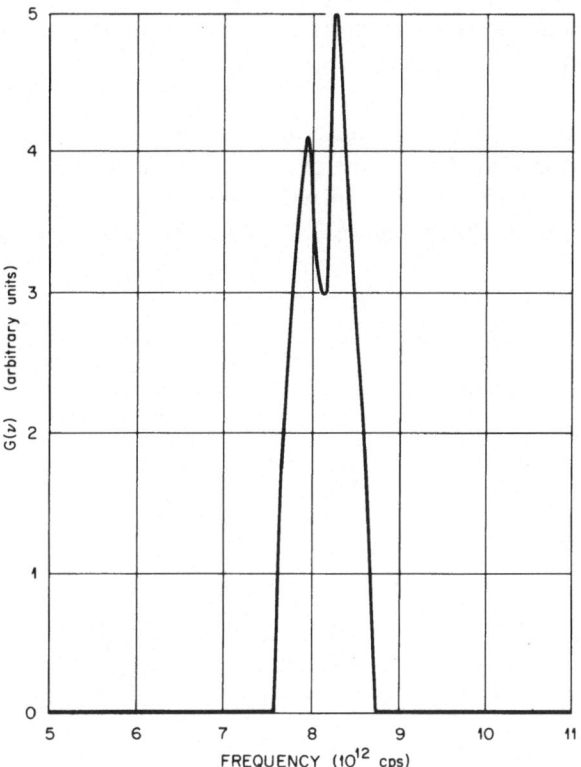

Fig. 3.6.   Calculated frequency distribution for $TA_\perp$ branch for $|\mathbf{q}| = 0.35(4\pi/\sqrt{3}a)$.

and the dispersion curves for $\mathbf{q}$ along the [100] and [110] directions are no longer nearly superimposed. In fact, no easily interpretable neutron groups were obtained for $|\mathbf{q}| > 0.5 \ (4\pi/\sqrt{3}a)$.

Configuration IV is similar to II except that $\mathbf{Q}$ (and $\mathbf{q}$) is in the basal plane. In this configuration constant energy scans were carried out to study the longitudinal acoustic and lowest frequency longitudinal optic modes having wave vectors in the basal plane. For reasons already mentioned above for configuration II, the phonon wave vector is not well defined in configuration IV. However, similar to observations for configuration II, there are ranges for v and $\mathbf{q}$ where no phonon scattering is expected. For example, at a point $\mathbf{Q} = (1 + \zeta, \ 1 + \zeta, \ 0) \ 4\pi/\sqrt{3}a$, just beyond the (110) ring, the shortest wavevector of a phonon that can contribute to the scattering is $(\zeta, \ \zeta, \ 0)$ $4\pi/\sqrt{3}a$. Any other wave vector connecting this $\mathbf{Q}$ with a point on the (110) ring is longer and therefore belongs to a phonon having a frequency higher than that with the wave vector $(\zeta, \ \zeta, \ 0) \ 4\pi/\sqrt{3}a$. If the energy transfer of the neutron $(E_0 - E') = hv$ is chosen so that the frequency v is smaller than $v_L(\zeta)$, the frequency of the longitudinal $(\zeta\zeta0)$ branch at ζ, no scattering will be observed. Then as ζ is reduced, as in a constant energy scan, a rapid increase in the neutron scattering intensity will occur when ζ reaches a value so that $v = v_L(\zeta)$. For still smaller ζ the intensity may remain high, even though v will then be greater than $v_L(\zeta)$, because of the scattering from longitudinal and transverse modes with wavevectors out of the scattering plane. The intensity will then drop abruptly inside the (110) ring for $\mathbf{Q} = (1 - \zeta, 1 - \zeta, 0)\, 4\pi/\sqrt{3}a$ when again $v < v_L(\zeta)$.

Numerical calculations of the scattering expected for such constant energy scans are also described in [38]. The results of a calculation carried out for the $(\zeta\zeta 0)$ longitudinal optic mode which is degenerate with the $TO(00\zeta)$ branch at $\zeta = 0$ are compared to the experimental results in Figure 3.7. The calculated intensity distribution when

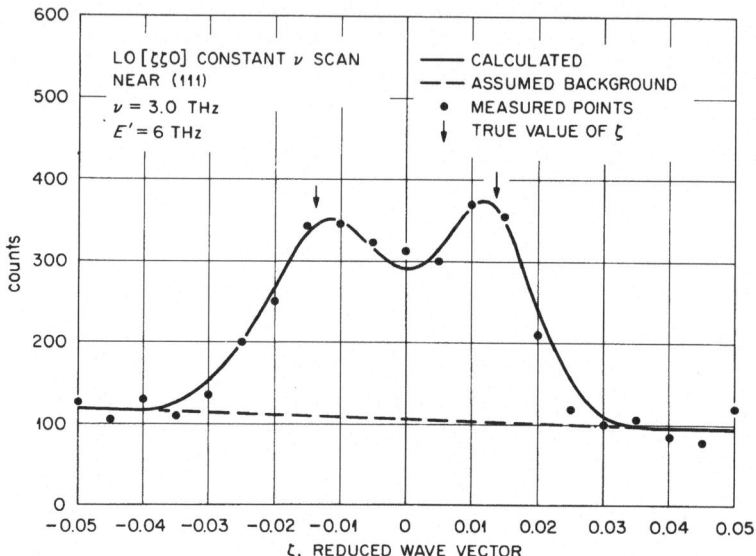

Fig. 3.7.   Calculated and measured constant-energy scan of the $LO(\zeta\zeta 0)$ branch for $v = 3.0$ THz.

added to a slightly sloping background is in excellent agreement with the experimental results. One common and expected feature of all such calculations is the small shift of the peak position toward values of ζ smaller than those which correspond to the true values defined by $v = v_L(\zeta)$. The corrected experimental results obtained for the longitudinal branches are shown in Figure 3.3 where ζ is expressed in units appropriate for the [100] direction. Well defined peaks were observed at low frequency, but the intensity decreased so rapidly with increasing frequency that reliable data could not be obtained above about 12 THz.

There have been numerous studies on the calculation of the phonon spectra and the specific heat of graphite. One of the first theoretical investigations is that of Komatsu and Nagamiya [40] who calculated the frequency distribution function by treating graphite as a system of thin elastic plates spaced at a constant distance. The lattice vibrations were assumed to be separable into two independent types of modes, the one corresponding to vibrations of a strictly two-dimensional layer with atomic displacements parallel to the layer and the other corresponding to the motions in the third direction, perpendicular to the layer. The latter vibrations are coupled to the motions of neighboring layers through a compressional force between them. They estimated Young's modulus and Poisson's ratio of a single layer of graphite from the carbon-carbon bond stretching forces of C_2H_6, C_2H_4, and benzene. Since there was no shearing force between layers in this model, the frequency distribution function was proportional to v for low frequencies instead of to v^2 which leads to a T^2 dependence of the specific heat at low temperatures. However, at very low temperature, the specific heat must have a T^3 dependence in accordance with the three dimensional structure of the crystal and, hence, the shearing force may not be neglected.

Komatsu [41] extended this work by including the shearing coupling between adjacent layers. He also investigated the variation of the specific heat due to stacking faults of layers using the same model. Although this model could give the correct T^3 dependence of the low temperature specific heat, the overall agreement with the experimental results for higher temperatures was not satisfactory.

Yoshimori and Kitano [14] and later Young and Koppel [42] extended the calculation beyond the semi-continuum approximation and constructed an atomic force constant model which may be regarded as a special type of the valence force model. The model included the force constants associated with the bond-stretching and bending within a layer and also the constant describing the force necessary to displace an atom along the c-axis relative to the plane formed by the instantaneous positions of its three nearest neighbor atoms in the same layer. For the interlayer interaction, only a compressional force constant between the layers was introduced through the nearest out-of-layer neighbor bond stretching force. Thus, this model also would not give the T^3 dependence for the low temperature specific heat. More recently Ahmadieh and Rafizadeh [43] calculated phonon dispersion curves using a four-parameter model based on two types of pair potentials, a strong force between a carbon atom and its three nearest neighbors and van der Waals forces for all the other interactions including that between the atoms of adjacent layers. Both pair potentials were taken to be of the Lennard-Jones 6–12 form and the model parameters were estimated from normal mode frequencies of a benzene molecule and from the nitrogen-nitrogen van der Waals potential of solid γ-nitrogen. The phonon dispersion curves calculated from the model do not agree with the experimental results described above, except for the low frequency modes along the [001]-direction. Especially, since the rotational invariance condition is not imposed on the parameters, the convex form of the $TA_\perp$ mode, discussed in II, is not reproduced by the model. Thus it is clear that a more general model is required to explain even the small amount of available data for phonon dispersion curves and optical measurements. In the following, several more general models are examined.

In Figure 3.1 atoms B, E (and E'), F, and C are, respectively, the first, second, third, and fourth nearest neighbors of A, and the force models to be considered are mostly limited to those which include interactions extending to the fourth nearest neighbor atoms. This is a natural limitation because the shortest distance between adjacent basal planes (i.e., between A and C) corresponds to the fourth nearest neighbor distance, and it is certainly desirable to include this interaction in any force model of graphite. On the other hand, with the small amount of data available, consideration of additional interactions between more distant neighbors either in the same basal plane or in adjacent planes would be fruitless.

The force constant matrices of a general tensor force model for graphite are given

N. WAKABAYASHI AND R. M. NICKLOW

in Table 3.2. Actually this model is not rigorously complete because the symmetry of the graphite lattice would allow certain second-neighbor force constants to be different, e.g., those between B–B' can be different from those between A–E'. However, if we could ignore the influence of the adjacent planes on the interactions within a plane, these two sets of force constants could be assumed to be identical. In any event, even with this assumption there are 12 independent force constants in the model and further approximations are desirable.

TABLE 3.2

General force-constant matrices for graphite

Neighbor relative to origin atom[a]	Coordinate of typical atom	Matrix elements ϕ_{xy}		
First	B': $(-a/\sqrt{3}, 0, 0)$	α_1	0	0
		0	β_1	0
		0	0	γ_1
Second	E: $(0, a, 0)$	α_2	ϵ_2	0
		$-\epsilon_2$	β_2	0
		0	0	γ_2
Third	F: $(2a/\sqrt{3}, 0, 0)$	α_3	0	0
		0	β_3	0
		0	0	γ_3
Fourth	C: $(0, 0, \frac{1}{2}c)$	α_4	0	0
		0	α_4	0
		0	0	γ_4

[a] Atom A in Figure 3.1.

The model of Yoshimori and Kitano is a special type of the valence force model (the bond-bending and bond-stretching model, BBS) and is equivalent to the following restrictions:

$$\alpha_1 = \kappa, \qquad \beta_1 = 6\mu, \qquad \gamma_1 = 2\mu'/3$$

$$\alpha_2 = -3\mu/4, \qquad \beta_2 = \mu/4, \qquad \gamma_2 = -\mu'/9, \qquad \epsilon_2 = \sqrt{3}\,\mu/4$$

$$\alpha_3 = \beta_3 = \gamma_3 = 0, \qquad \alpha_4 = 0, \qquad \gamma_4 = \kappa'. \tag{3.1}$$

Here κ is the bond-stretching force constant between nearest neighbors such as A–B, μ is a bond-bending force constant which describes the force necessary to change the angle formed by the A–B and A–B' bonds, μ' is a bond-bending force constant which describes the force necessary to displace an atom along the c-axis relative to its nearest neighbors in the same plane, and κ' is a bond stretching force constant between nearest neighbors on adjacent planes such as A–C. Note that for this model $\gamma_1 = -6\gamma_2$. The assumption of the axially symmetric model (AS) leads to the following restrictions:

$$\alpha_1 = \phi_r^1, \qquad \beta_1 = \phi_t^1, \qquad \gamma_1 = \phi_t^1$$

$$\alpha_2 = \phi_t^2, \qquad \beta_2 = \phi_r^2, \qquad \gamma_2 = \phi_t^2$$

$$\alpha_3 = \phi_r^3, \qquad \beta_3 = \phi_t^3, \qquad \gamma_3 = \phi_t^3$$

$$\alpha_4 = \phi_t^4, \qquad \gamma_4 = \phi_r^4 \tag{3.2}$$

where the ϕ_r's and ϕ_t's are defined in Equation (1.14). In this model the force constants for the B–B' and A–E' interactions are necessarily identical.

The BBS model is more appealing on physical grounds because it appears to be a natural consequence of the concept of strong covalent bonding between carbon atoms. However, this model as previously formulated neglects the third-neighbor interactions while a part of the fourth-neighbor interaction is included. Also for this model the elastic constant $C_{44} = 0$ because $\alpha_4 = 0$ and $\gamma_1 = -6\gamma_2$. Including bond-stretching forces between more distant neighbors does not remedy this nonphysical property. Furthermore, it was found that the $TA_\perp$ and $TO_\perp$ branches, which depend only on γ_1, γ_2, and γ_3, could not be satisfactorily reproduced with the restriction $\gamma_1 = -6\gamma_2$ even for $\gamma_3 \neq 0$.

The AS model, while not as physically appealing as the BBS model, is a somewhat more general model for vibrations perpendicular to the basal plane since no relation between the γ_n's (hereafter ϕ_t^n) is imposed; also $\phi_t^4 \neq 0$ thereby giving a finite C_{44}. Actually the ϕ_t^n constants are not independent because the initial slope at $\mathbf{q} \to 0$ of the $TA_\perp$ branch, which depends only on ϕ_t^1, ϕ_t^2, and ϕ_t^3, and that of the $TA(00\zeta)$ branch, which depends only on ϕ_t^4, are both determined by C_{44}. Thus one finds for the AS (and the general-tensor) model,

$$\phi_t^4 = (\phi_t^1/2 + 3\phi_t^2 + 2\phi_t^3)a^2/c^2. \tag{3.3}$$

The constants ϕ_t^4 and ϕ_r^4 were determined from a least squares analysis of the data obtained for the longitudinal and transverse branches along (00ζ) that were studied in configurations I and II. Then, keeping these constants fixed, ϕ_t^1, ϕ_t^2, and ϕ_t^3 were determined from an analysis of the TA and TO branches with the restriction expressed in Equation (3.3) included.

Finally the force constants ϕ_r^1, ϕ_r^2, and ϕ_r^3 were determined by a least squares fitting of the longitudinal modes studied in configuration IV with all the other constants held fixed at the values determined by the methods mentioned above. However, since only very limited information for these modes was available in the neutron measurements, the value of the Raman frequency [44] (1575 cm^{-1}) and the value of the elastic constant C_{66} measured by Seldin [45] were included in this stage of the analysis. The lines in Figure 3.3 show the fit to the neutron data obtained with this model.

The values obtained for all the force constants are given in Table 3.3. It is difficult to make a direct comparison of the values of these constants and those of the BBS model reported by Young and Koppel [42] since the two models are based on different general physical assumptions. About all that can be concluded generally is that those constants which describe similar atomic motions in the two models, i.e., that of one atom relative to another, have similar orders of magnitude. Moreover, the restrictions of the BBS model, $6\phi_t^2 = -\phi_t^1$ and $\phi_t^3 = 0$, as mentioned above are not compatible with the results obtained for the $TA_\perp$ branches and $TO_\perp$ branches. The restriction imposed by Equation (3.3) gives $6\phi_t^2 + 4\phi_t^3 \simeq -\phi_t^1$, since $\phi_t^4 \simeq 0$ (i.e., $\alpha_4 \simeq 0$) in comparison to the other constants in Equation (3.3). Thus if ϕ_t^3 were zero the restrictions of the two models on the z displacements would have been very similar. However, the relatively large magnitude of ϕ_t^3 indicates that there is a significant force in the z direction on atom A where atom F is displaced in the z direction. In terms of a bonding picture the existence of such a force could perhaps be interpreted to mean that a bond such as $E' - B$ possesses a considerable resistance to twisting.

TABLE 3.3

Force constants for graphite (10^5 dyn cm^{-1})

ϕ_r^1	3.62 ($\pm 25\%$)	ϕ_r^3	-0.037 ($\pm 100\%$)
ϕ_t^1	1.99 ($\pm 3\%$)	ϕ_t^3	0.288 ($\pm 3\%$)
ϕ_r^2	1.33 ($\pm 30\%$)	ϕ_r^4	0.058 ($\pm 2\%$)
ϕ_t^2	-0.520 ($\pm 3\%$)	ϕ_t^4	0.0077 ($\pm 5\%$)

A calculation of the entire dispersion relation for graphite in the [100], [110], and [001] directions is shown in Figure 3.8. In the [100] and [110] directions there are 12 distinct frequencies for each wave vector. However, because of the very small magnitude of ϕ_t^4, modes corresponding to atomic displacements parallel to the basal planes are nearly degenerate in pairs except at low frequencies near Γ. Each line in Figure 3.8 which goes to any one of the four highest frequencies at M represents such a pair.

Obviously the neutron measurements shown in Figure 3.3 correspond to a very small portion of the frequency range encompassed by the normal modes in graphite. Thus the absolute accuracy of the derived force constants and, hence, of the calculated high frequency branches is very difficult to assess. The optical data, which were used in the fitting analysis, accurately pin down the location of the highest frequency branch (longitudinal optic) at Γ. On the other hand, an AS model was found to be inadequate when used in an analysis of a rather complete measurement [46] of the dispersion relation of the hexagonal metal Tb, and there is no reason to expect such a model to be more reliable for graphite. Thus, while the qualitative features of the high frequency region of the calculated dispersion relation may be reliable, the details of these curves should be interpreted with caution.

As was pointed out in Section 2, one consequence of the extreme anisotropy of the forces in graphite is the almost quadratic form of the dispersion relation at small q for the $TA_\perp$ branch. For small q, this branch has the form

$$v^2 \propto Aq^2 + Bq^4, \tag{3.4}$$

as shown by Komatsu [41] using an elasticity theory for graphite which did include a finite C_{44}. At very small q, v varies linearly with q, as it must because of the term $\sqrt{A}q$. However the coefficient A which is proportional to C_{44} and hence proportional to the small constant ϕ_t^4 through the linear combination of the ϕ_t^n's given in Equation (3-3),

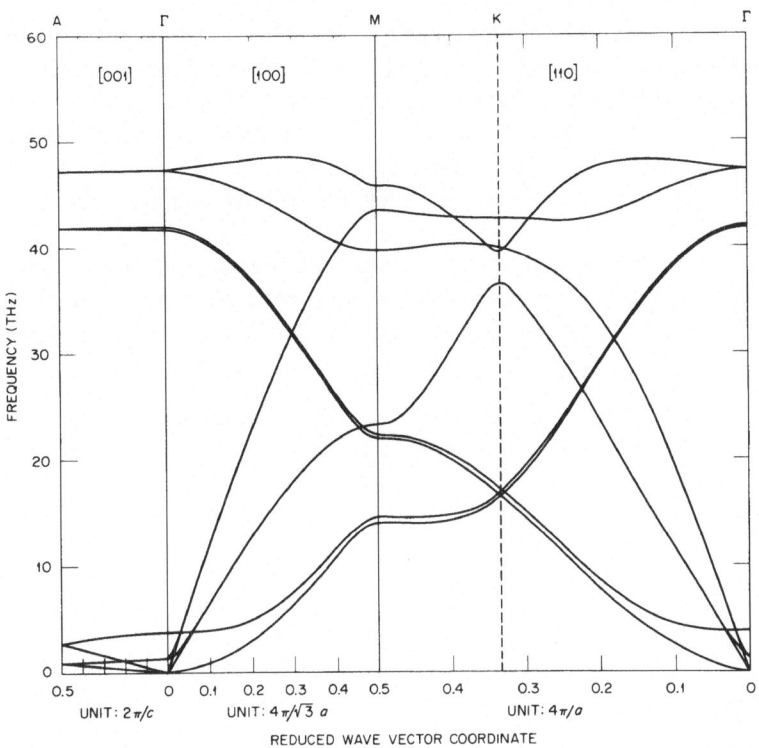

Fig. 3.8.   Phonon dispersion relation for graphite in the [001], [100], and [110] directions as calculated with the AS model.

is very much smaller than B which is given by a different linear combination of the very large constants ϕ_t^1, ϕ_t^2, and ϕ_t^3. Thus as q increases the Bq^4 term soon dominates in Equation (3.4) to give $v \propto q^2$. While this qualitative behavior is predicted by Komatsu's theory, the quantitative agreement between his calculations and the measurements is not good because he used values for C_{44} which apparently are not appropriate for the pyrolytic graphite used in the neutron scattering experiment.

The elastic constants of graphite which are calculated with the AS model are compared to the measurements of Seldin [45] in Table 3.4. The agreement between the

TABLE 3.4

Elastic constants for graphite (10^{11} dyn cm^{-2})

	Calculated from force model	Measured by velocity of sound
C_{11}	144 ± 20	106 ± 2
C_{33}	3.71 ± 0.05	3.65 ± 0.1
C_{44}	0.46 ± 0.02	0.40 ± 0.04
C_{66}	46	44 ± 2

two sets of constants is rather good except for C_{11}. This constant gives the initial slope of the longitudinal acoustic branch in the [100] direction. Since the neutron data for this branch were difficult to interpret unambiguously, the discrepancy may not be unreasonable. However, the excellent agreement obtained between the calculated and measured neutron intensity distributions, as illustrated in Figure 3.5 seems to indicate that the true value of C_{11} for graphite is probably intermediate between the values given in the table.

Figure 3.9 shows the frequency distribution function $g(v)$ that was calculated from the AS model with the method described by Raubenheimer and Gilat [39] appropriately modified for the graphite lattice. The general features of this $g(v)$ and that

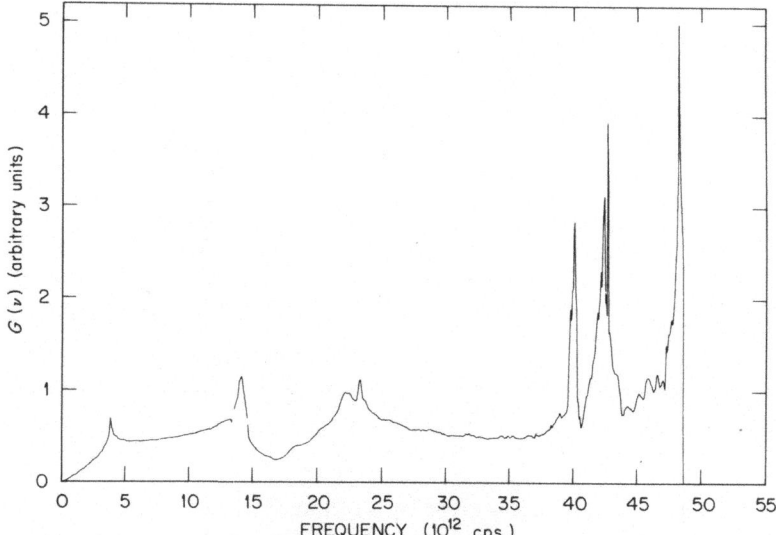

Fig. 3.9.   Frequency distribution function for graphite calculated with force model determined from neutron scattering data.

calculated by Young and Koppel [42] are very different, although both distributions have pronounced peaks near ~ 4 THz, ~ 14 THz, and ~ 48 THz.

The lattice specific heat at constant volume C_v was calculated with the $g(v)$ shown in Figure 3.9. One of the interesting features of the experimental results for C_v is the transition from a T^3 behavior expected for any solid at very low temperatures to an almost T^2 behavior at higher temperatures (10 K) expected for a two-dimensional lattice. The calculated C_v reproduces almost exactly the experimental results for natural graphite [47] as illustrated in Figure 3.10.

3.3. Transition metal dichalcogenides (TX$_2$)

The electronic properties of the transition metal dichalcogenides with layer type crystal structures have been investigated extensively in recent years. Of particular interest are the compounds MoS_2 and $NbSe_2$ which have physical properties that are fairly well described by a rigid band model for their electronic band structures [48], in spite of

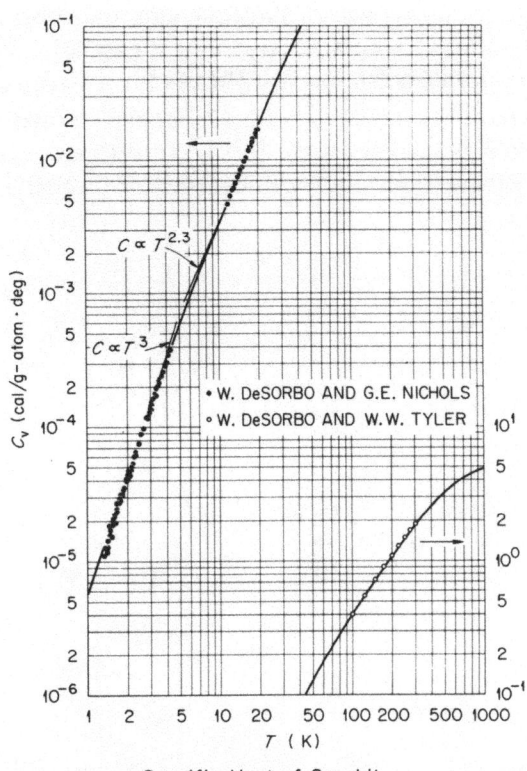

Specific Heat of Graphite.

Fig. 3.10. Comparison of the measured specific heat and that calculated from the frequency-distribution function shown in Figure 3.9.

their very different electronic conductivities. MoS_2 is a semiconductor with physical properties very similar to those of graphite. On the other hand, $NbSe_2$ is a metal with a rather high superconducting transition temperature (~ 7 K). The relationship between superconductivity, electron-phonon interaction and anomalous features in the phonon dispersion relations of high transition temperature superconductors [49–51], particularly those with d-electrons, has become a subject of widespread interest [52]. Such anomalies are more easily identified if a comparison can be made with the phonon spectra of a related material which is nonsuperconducting. Thus in addition to being of intrinsic interest, the joint study of the lattice dynamics of MoS_2 and $NbSe_2$ makes such a comparison possible.

Another transition metal dichalcogenide on which neutron measurements have been performed is $TiSe_2$ [53]. It possesses a crystal structure having only one layer per unit cell in contrast to those compounds discussed previously.

3.3.1. MoS_2 (2H)

For MoS_2 a layer consists of one plane of Mo atoms sandwiched by two planes of sulfur atoms. In hexagonal MoS_2 ($a = 3.1604$ Å, $c = 12.295$ Å) which belongs to the space group D_{6h}^4, two of these layers build the repeating unit in the c-direction so that the unit cell contains six atoms, two of molybdenum and four of sulfur. The basic structure of bonding within a layer is illustrated in Figure 3.11. There are in general, 18 branches of the phonon dispersion relation, but in high symmetry directions some branches are degenerate. The results of a group theoretical analysis are summarized in Tables 3.5–3.7. In Table 3.5 are listed the irreducible representations that should appear in the high symmetry directions and at high symmetry points in reciprocal

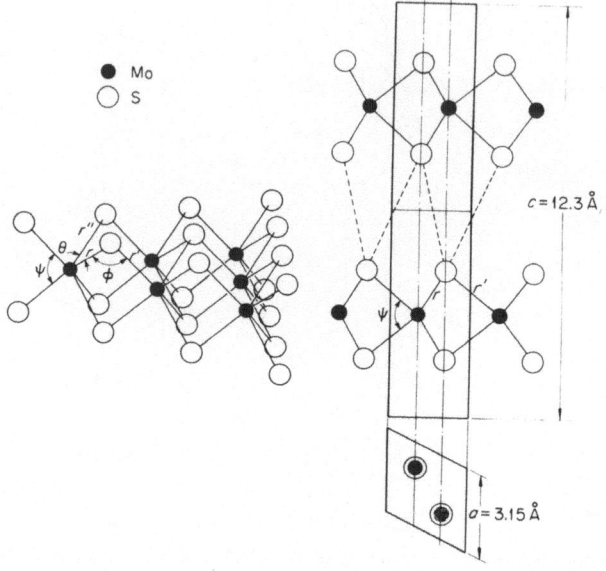

Fig. 3.11. Schematic of bonding within a layer of hexagonal MoS_2 (left) and relative configuration of neighboring layers (right). The basal projection of the atomic positions is shown below. The unit cell is outlined by thick lines.

space, along with the number of nondegenerate branches belonging to each representation. The forms of eigenvectors for some branches are also shown. For example, in the [001]-direction (Δ), there are four irreducible representations, Δ_1, Δ_2, Δ_5, and Δ_6. Δ_1 and Δ_2 are nondegenerate and each of them contains three branches which are purely longitudinal. Δ_5 and Δ_6 are doubly degenerate and six branches which are purely transverse belong to each of these representations. There are also four irreducible representations in the [100]-direction (Σ), Σ_1, Σ_2, Σ_3, and Σ_4, none of which is degenerate. There are six Σ_1's, six Σ_3's, two Σ_2's and four Σ_4's. Σ_2 and Σ_4 are purely transverse modes with polarization vectors parallel to the basal plane, and, in particular, in the modes belonging to Σ_2, Mo atoms are stationary with S atoms within a layer moving in opposite directions. Σ_1 and Σ_3 are neither purely longitudinal nor purely transverse. In Σ_1, motions of Mo atoms have only components perpendicular to the basal plane. Although these modes are of mixed character, the acoustic modes belonging to them may be regarded to be either mostly longitudinal (Σ_1) or mostly transverse (Σ_3). Thus the acoustic modes may be designated as LA, $TA_\perp$, or $TA_\parallel$ where $TA_\perp$ ($TA_\parallel$) signifies that the eigenvector is perpendicular (parallel) to the basal plane.

Vibrational modes at the Γ point have been discussed by Wieting and Verble [54], and the correspondence to the present notation is tabulated in Table 3.6. There are eight different irreducible representations and 12 different frequencies at Γ. But, since one of the Γ_2^- modes and one of the Γ_6^- modes are acoustic and have zero frequencies, there should exist 10 distinct nonzero frequencies, including two low-lying rigid layer modes (Γ_3^+ and Γ_5^+). Thus four pairs of intralayer modes are expected to be observed. In fact, these pairs should be $\Gamma_2^- - \Gamma_3^+$, $\Gamma_5^+ - \Gamma_6^+$, $\Gamma_1^+ - \Gamma_4^-$, and $\Gamma_5^- - \Gamma_6^+$. Similar pairs for other wave vectors can be obtained by inspection from the table of eigenvectors, and they are listed in Table 3.7.

The neutron measurements of the dispersion relation were performed [55] for wave vectors along the [001]- and [100]-directions, and the results are shown in Figure 3.12. The uncertainty in the measured frequencies is generally about 2% in Figure 3.12. Since the transverse Σ_2 and Σ_4 modes have polarization vectors parallel to the plane of the thin plate of the sample crystal, it was not possible to observe these modes. Except for the region of very low frequencies (2 THz), the frequencies of pair modes could not be obtained separately, and for the data shown no distinctions between the two branches in each pair were made.

The very low frequency branches in the [001]-direction are those corresponding to the rigid layer modes. The experimental data for the transverse acoustic mode in the [100] direction ($TA_\perp$) seem to indicate a very small initial slope for that branch, which is expected from the consideration of the rotational invariance discussed previously.

A few measurements of the low frequency branches were made along the [110]-direction. The frequencies of phonons with wave vectors having magnitudes less than $0.3\ (4\pi/a)$ are almost identical to those of phonons in the [100]-direction that have the same magnitude of wave vector. This near isotropy of the phonon dispersion curves in the basal plane has been found also in the case of pyrolytic graphite (see Section 3.1) as well as in various metals with the hcp crystal structure.

As discussed previously one of the simplest models that may be used to describe lattice dynamics is the axially-symmetric model. The use of this model may be justified in the present case for the interlayer interaction which is considered to be due to weak van der Waals forces. However, a valence force model should be more suitable to describe the intralayer interactions in view of their strong covalent character. Therefore, a mixed model was used to analyze the neutron data: the interlayer interaction was assumed to be due to an axially symmetric force between sulfur atoms of neighboring layers and the intralayer interactions were described by the stretching and bending of Mo–S bonds. The bonds and bond angles considered in the model are shown in Figure 3.11. In terms of the force constants, K_r, K_θ, K_ϕ, and K_ψ, the potential energy is written as

$$\tfrac{1}{2}K_r(\Delta r)^2 + \tfrac{1}{2}K_\theta(r_0\Delta\theta)^2 + \tfrac{1}{2}K_\phi(r_0\Delta\phi)^2 + \tfrac{1}{2}K_\psi(r_0\Delta\psi)^2,$$

where r_0 is the bond length. In addition, there should be the cross terms, such as $K_{rr'}^\phi(\Delta r)(\Delta r')$ and $K_{r\theta}(\Delta r)(r_0\Delta\theta)$. There are also two parameters, ϕ_r and ϕ_t, as defined in Equation (1.14), corresponding to the axially symmetric force between the sulfur atoms in different layers. Thus, the rigid-layer modes are determined mainly by ϕ_r and ϕ_t, and the other modes are determined mostly by the valence force constant K's.

The result of a least squares fit of the data to a model which includes six parameters,

TABLE 3.5

Irreducible representations and eigenvectors for MoS$_2$

	Nb[a]	Nd[b]	Eigenvectors
Γ_1^+	1	1	$(o\ \ o\ \ o\ \ o\ \ o\ \ o\ \ o\ \ o\ \ \tfrac{1}{2}\ \ o\ \ o\ \ -\tfrac{1}{2})$
Γ_4^-	1	1	$(o\ \ o\ \ o\ \ o\ \ o\ \ o\ \ o\ \ o\ \ \tfrac{1}{2}\ \ o\ \ o\ \ -\tfrac{1}{2})$
Γ_2^-	2	1	$(o\ \ o\ \ u\ \ o\ \ o\ \ o\ \ v\ \ o\ \ v\ \ o\ \ v\ \ o)$
Γ_3^+	2	1	$(o\ \ o\ \ u\ \ o\ \ o\ \ o\ \ -v\ \ o\ \ -v\ \ o\ \ -v\ \ o)$
Γ_5^+	2	2	linear combinations of
			$(u\ \ o\ \ o\ \ o\ \ v\ \ o\ \ o\ \ -u\ \ o\ \ -v\ \ o\ \ o)$
			$(o\ \ u'\ \ o\ \ o\ \ o\ \ v'\ \ -u'\ \ o\ \ o\ \ o\ \ -v'\ \ o)$ and
Γ_5^-	1	2	linear combinations of
			$(o\ \ o\ \ \tfrac{1}{2}\ \ o\ \ -\tfrac{1}{2}\ \ o\ \ o\ \ o\ \ -\tfrac{1}{2}\ \ o\ \ \tfrac{1}{2}\ \ o)$
			$(o\ \ o\ \ o\ \ \tfrac{1}{2}\ \ o\ \ -\tfrac{1}{2}\ \ o\ \ o\ \ o\ \ \tfrac{1}{2}\ \ o\ \ -\tfrac{1}{2})$ and
Γ_6^+	1	2	linear combinations of
			$(o\ \ o\ \ \tfrac{1}{2}\ \ o\ \ -\tfrac{1}{2}\ \ o\ \ o\ \ o\ \ \tfrac{1}{2}\ \ o\ \ \tfrac{1}{2}\ \ o)$
			$(o\ \ o\ \ o\ \ \tfrac{1}{2}\ \ o\ \ -\tfrac{1}{2}\ \ o\ \ o\ \ o\ \ -\tfrac{1}{2}\ \ o\ \ -\tfrac{1}{2})$ and
Γ_6^-	2	2	linear combinations of
			$(u\ \ o\ \ v\ \ o\ \ v\ \ o\ \ u\ \ o\ \ v\ \ o\ \ v\ \ o)$
			$(o\ \ u'\ \ o\ \ v'\ \ o\ \ v'\ \ o\ \ u'\ \ o\ \ v'\ \ o\ \ v')$ and
M_1^+	3	1	$(u\ \ o\ \ v_1\ \ o\ \ v_1\ \ o\ \ u\ \ o\ \ v_1\ \ o\ \ v_1\ \ o)$
M_1^-	1	1	$(o\ \ o\ \ \tfrac{1}{2}\ \ o\ \ -\tfrac{1}{2}\ \ o\ \ o\ \ o\ \ \tfrac{1}{2}\ \ o\ \ -\tfrac{1}{2}\ \ o)$
M_2^+	1	1	$(o\ \ o\ \ o\ \ \tfrac{1}{2}\ \ o\ \ -\tfrac{1}{2}\ \ o\ \ o\ \ o\ \ \tfrac{1}{2}\ \ o\ \ -\tfrac{1}{2})$
M_2^-	3	1	$(u\ \ o\ \ v_1\ \ o\ \ v_1\ \ o\ \ -u\ \ o\ \ -v_1\ \ o\ \ -v_1\ \ o)$
M_3^+	3	1	$(o\ \ u\ \ o\ \ v_1\ \ o\ \ v_1\ \ o\ \ -u\ \ o\ \ -v_1\ \ o\ \ -v_1)$
M_3^-	2	1	$(o\ \ u\ \ o\ \ v\ \ o\ \ v\ \ o\ \ u\ \ o\ \ v\ \ o\ \ v)$
M_4^+	2	1	$(o\ \ u\ \ o\ \ v\ \ o\ \ v\ \ o\ \ -u\ \ o\ \ -v\ \ o\ \ -v)$
M_4^-	3	1	$(o\ \ u\ \ o\ \ v_2\ \ o\ \ -v_1\ \ o\ \ -u\ \ o\ \ -v_2\ \ o\ \ v_1)$
Σ_1	6	1	$(u_1\ \ o\ \ w_1\ \ o\ \ w_1\ \ o\ \ u_2\ \ o\ \ w_2\ \ o\ \ w_2\ \ -w_2)$

	Nb	Nd																		
Σ_2	2	1	$(0$	0	0	v_1'	0	0	0	w_1	0	0	0	v_2'	0	0	0	$-v_2'$	$0)$	
Σ_3	6	1	$(0$	0	u_1	0	w_1	0	$-v_1$	0	0	u_2	v_2	0	0	w_2	0	$-v_2$	0	$w_2)$
Σ_4	4	1	$(0$	0	u_1'	0	0	0	v_1'	0	0	u_2'	v_2	0	0	0	v_2	0	v_2	$0)$
Δ_1	3	1	$(0$	0	0	0	v_1	0	0	0	u	0	v_2	0	0	0	v_1	0	0	$v_2)$
Δ_2	3	1	$(0$	0	u	0	v_1	0	0	v_2	$-u$	0	0	0	0	$-v_1$	0	0	0	$-v_2)$
Δ_5	3	2	linear combinations of																	
Δ_6	3	2	linear combinations of																	
T_1	5	1	$(u_1$	u_2	v_2	v_1	v_3	$-v_3$	$-u_1$	$-u_2$	0	$-v_1$	v_2	v_3	$-v_1$	v_3	$-v_1$	v_2	v_2	$-v_3)$
T_2	4	1	$(0$	0	v_2	v_1	v_3	$-v_2$	0	0	$-u$	v_1	$-v_2$	$-v_3$	v_1	$-v_1$	v_1	$-v_2$	v_2	$-v_3)$
T_3	4	1	$(0$	0	v_2	v_1	v_3	$-v_2$	0	0	u	$-v_1$	v_2	v_3	$-v_1$	v_1	v_1	$-v_2$	v_2	$v_3)$
T_4	5	1	$(u_1$	u_2	v_2	v_1	v_3	$-v_3$	u_1	u_2	0	v_1	$-v_2$	$-v_3$	v_1	v_1	v_1	$-v_2$	v_2	$v_3)$

(Rows Δ_5 and Δ_6: "and"; vectors given as linear combinations of two tuples each.)

	Nb	Nd		Nb	Nd
K_1	2	1	A_1	3	2
K_2	1	1	A_3	3	4
K_3	2	1			
K_4	1	1			
K_5	3	2			
K_6	3	2			

a Nb: number of modes,
b Nd: degree of degeneracy.

TABLE 3.6

Comparison between two group theoretical notations for MoS$_2$ at Γ

Γ_1^+	Γ_2^-	Γ_4^-	Γ_3^+	Γ_5^+	Γ_5^-	Γ_6^+	Γ_6^-
A_{1g}	A_{2u}	B_{1u}	B_{2g}	E_{2g}	E_{2u}	E_{1g}	E_{1u}

TABLE 3.7

Pair modes in MoS$_2$

Γ_1^+	Γ_2^-	Γ_5^+	Γ_5^-	M_1^+	M_1^-	M_3^+	M_3^-	Δ_1	Δ_5	T_1	T_2
Γ_4^-	Γ_3^+	Γ_6^-	Γ_6^+	M_2^-	M_2^+	M_4^-	M_4^+	Δ_2	Δ_6	T_4	T_3

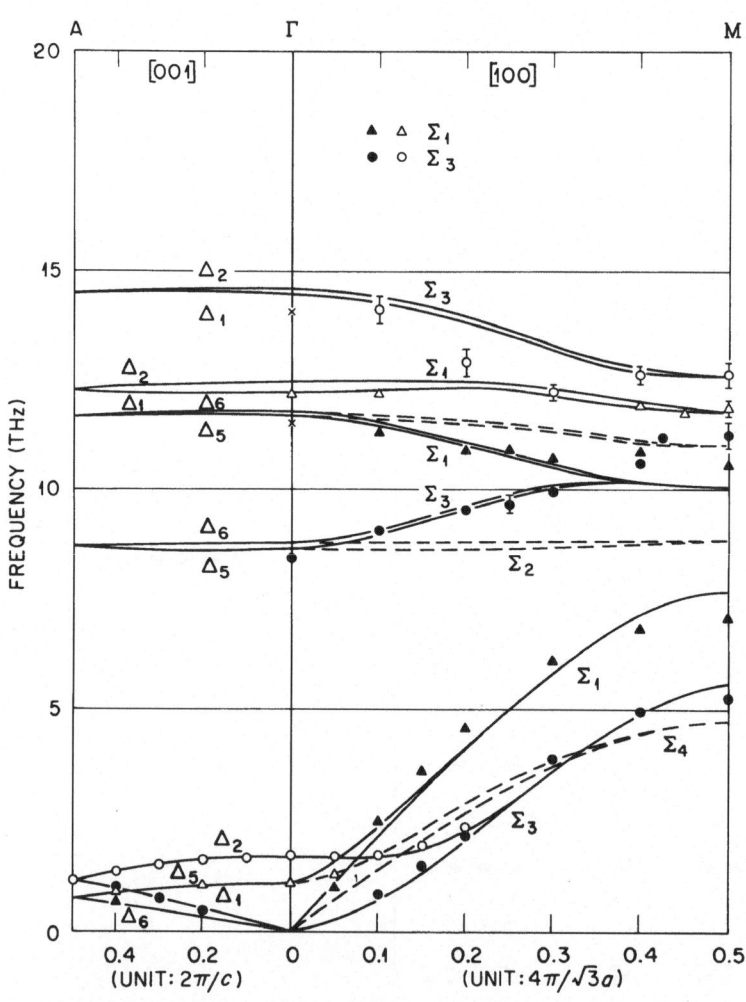

Fig. 3.12.   Phonon dispersion curves for hexagonal MoS$_2$ along the [001] and [100] directions. The lines are calculated curves from model described in the text.

ϕ_r, ϕ_t, K_r, K_θ, K_ϕ, and K_ψ was found to be inadequate to reproduce the data for large wave vectors. Since sulfur atoms are situated at positions of low symmetry, one might expect a large contribution to the potential function from the cross term $K_{rr'}^\phi$. A fit to a model that included this additional parameter was in significantly better agreement with the data. Force constants determined by the fitting calculation are given in Table 3.8 and the dispersion curves calculated from this model are shown in Figure

TABLE 3.8

Force constants determined for MoS_2 from analysis of dispersion-curve measurements.
Units: 10^5 dyn cm^{-1}.

K_r	1.3846
K_θ	0.1502
K_ϕ	0.1892
K_ψ	0.1381
$K_{rr'}^\phi$	−0.1722
ϕ_r	0.0311
ϕ_t	0.0072

3.12. The solid lines represent Σ_1 and Σ_3 branches and the broken lines Σ_2 and Σ_4 which were not observed experimentally. However, the restriction on the force constants imposed by the rotational invariance was not satisfied, and the elastic constant C_{44} calculated from ΔTA and $\Sigma TA_\perp$ differ almost by a factor of 2. Including one more parameter, $K_{rr''}^\theta$, in the model did not improve the quality of the fit, and, in fact, $K_{rr''}^\theta$ thus determined had a value smaller than that of $K_{rr'}^\phi$, by about an order of magnitude. Therefore, the seven parameter model was used to calculate the phonon density of states, $g(v)$, and the temperature dependence of the specific heat.

The $g(v)$ was calculated by the Raubenheimer-Gilat [39] method adapted for the present model. The result is shown in Figure 3.13. The peaks near 8.8 and 11.0 THz

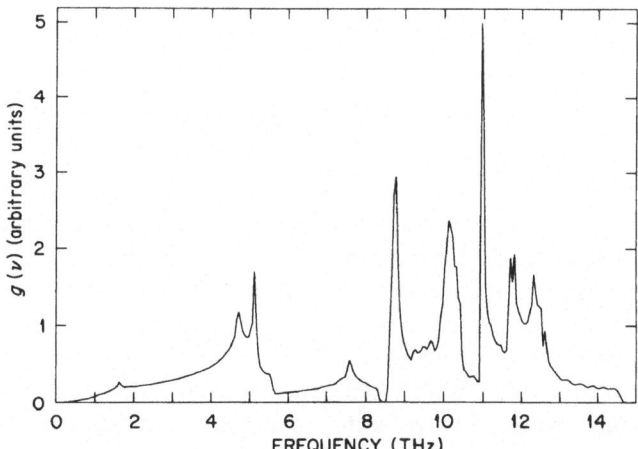

Fig. 3.13. Frequency distribution function for MoS_2 calculated with force model derived from neutron scattering data.

seem to correlate with several very flat branches such as those calculated for the Σ_2 and Σ_4 modes. Since it was impossible to observe these modes in the experiment, those prominent peaks may be quite model dependent. However, the overall feature may be accurate enough to allow a fairly reliable calculation of the temperature dependence of the specific heat. As shown in Figure 3.14 the calculated specific heat has the T^3 dependence only below 6 K, similar to the case of graphite (see Section 3.1).

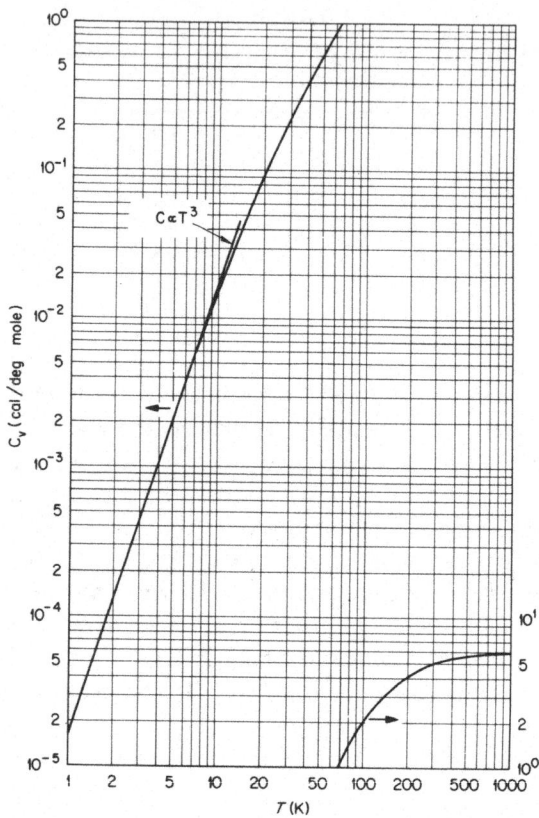

Fig. 3.14. Specific heat for MoS_2 calculated from the frequency distribution function shown in Figure 3.13.

From this result the temperature dependence of the Debye temperature was derived and is shown in Figure 3.15. An interesting point is that the Debye temperature is a monotonically increasing function of temperature in the region above 1 K, for which the calculation was carried out. This behavior seems to be common to those materials that have large c-lattice parameters and extremely weak interlayer interactions.

Although the valence force model seemed to reproduce the essential features of the experimental results, for large wave vectors there are discrepancies clearly outside of the experimental uncertainties. In order to investigate the possible effects of the polarizability of the atoms on the dispersion curve calculations, a very simple shell model (see Section 1.2) in which only Mo atoms were assumed to be polarizable was

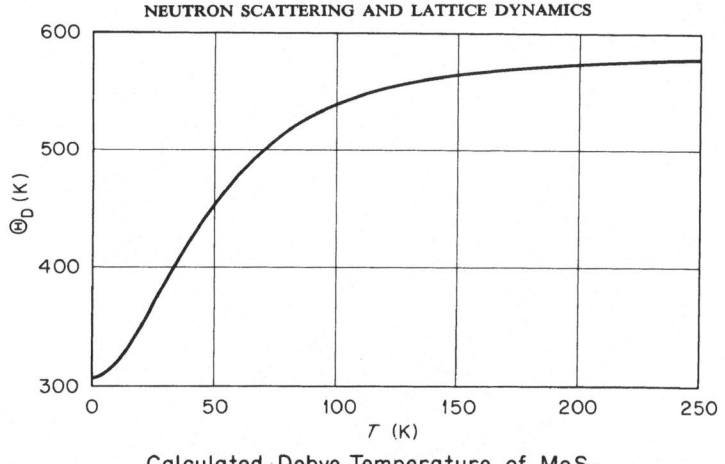

Calculated Debye Temperature of MoS$_2$

Fig. 3.15. Temperature dependence of the Debye temperature for MoS$_2$ as calculated from the specific heat shown in Figure 3.14.

constructed. Although the sulfur atoms are probably more polarizable, the calculations of the Coulomb interactions were very simple in this model and permitted an estimate of the possible influence of the polarizability to be made easily. It was found that if the polarizability was made large enough to affect the frequencies for large wave vectors, the splitting between the longitudinal and transverse optic modes became too large to be consistent with the results of the optical measurements [56, 57]. Also, the relevant Coulomb coefficients for wave vectors in the c-direction were found to be smaller than those in the a–b plane by more than three orders of magnitude due to the extremely large c/a ratio. The electrostatic interactions, therefore, will not appreciably affect the dispersion curves in the [001] direction. In order to improve the model, it will probably be necessary to include the polarizability of sulfur, and/or the covalent character of the bonding must be treated more accurately by including additional cross terms in the potential function. The choice of the most significant cross term was made rather arbitrarily in the above model. The only reasonable criterion should come from the detailed knowledge of the electronic configuration of the bonding between the molybdenum and sulfur atoms, and various efforts in recent years [58, 59] may eventually produce accurate information about the electronic wavefunctions of the ground and excited states in this and similar materials.

3.3.2. NbSe$_2$

As mentioned earlier, NbSe$_2$ is a superconductor with a moderately high transition temperature (~ 7 K), and one might expect large electron-phonon interactions in this material. Thus the phonon dispersion curves may show strong anomalies as in the cases of other high T_c superconductors [49–51]. Unfortunately, the samples available have been extremely small and in the form of very thin crystals (1 mm). The neutron inelastic scattering measurements [60] were obtained on a sample consisting of several thin crystals of NbSe$_2$ containing 5 % Mo impurities (total volume of about 0.02 cm^3). Figure 3.16 shows the experimentally observed phonon frequencies. Because of the small size of the sample, only very limited information was obtained.

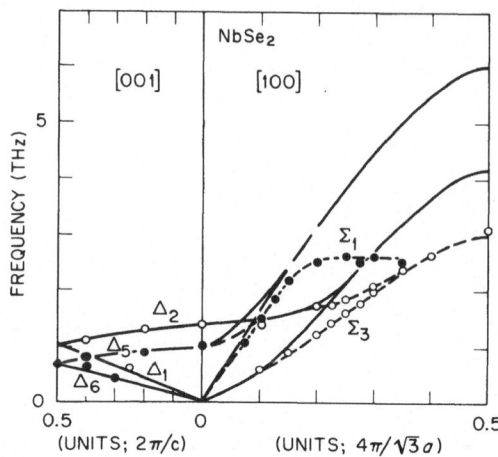

Fig. 3.16.   Phonon dispersion curves for the low frequency branches of $NbSe_2$. The solid lines are the results of the calculations described in the text.

Since $NbSe_2$ forms a crystal structure having the space group D_{6h}^4 which is essentially the same as that of MoS_2 group theoretical considerations for $NbSe_2$ are similar to those for MoS_2. The low frequency Σ_1 (nearly longitudinal) and Σ_3 (nearly transverse) branches, as well as the low-lying Δ_1, Δ_2, Δ_5, and Δ_6 branches have been observed. The solid lines in the figure represent the phonon dispersion curves calculated on the basis of the force constant model obtained for MoS_2 but with atomic masses appropriate for $NbSe_2$. The good agreement for the Δ-direction indicates that the effective interlayer interactions in the two materials are almost identical. However, rather drastic deviations from the calculation exist for both Σ_1 and Σ_3 branches. In particular, a sharp bend occurs in the Σ_1 branch near $q = 0.2$ (in units of $4\pi/\sqrt{3}a$, where a is the lattice constant in the layer). This branch is of special interest from the viewpoint of a microscopic theory for lattice dynamics. As was discussed before, the contribution to the dynamical matrix from the conduction electrons can be expressed in terms of the dielectric function $\varepsilon(\mathbf{Q})$. In a case where the Fermi surface is a cylinder with its axis along the z-direction, which may represent approximately the Fermi surface of a quasi-two dimensional system such as a layered compound, the dielectric function is given by [61, 62, 63]

$$\varepsilon(\mathbf{Q}) = \frac{m^* k_F k_C}{\pi^2 h^2} \, \mathrm{Re} \left\{ 1 - \left(1 - \frac{4k_F^2}{Q^2} \right)^{1/2} \right\}, \tag{3.5}$$

where k_F is the wave number of the Fermi surface, $k_C = 2\pi/c$, and m^* is the effective mass of the electron. $\varepsilon(\mathbf{Q})$ given by Equation (3.5) has a square-root singularity at $Q = 2k_F$, and, as can be seen from Equation (2.11), the phonon frequency should rise as q is increased through $2k_F$. For a hole Fermi surface, the singularity in $\varepsilon(\mathbf{Q})$ changes sign and accordingly the phonon frequency should fall as shown schematically in Figure 3.17. The APW calculation of the electron band structure for $NbSe_2$ [64] seems to indicate that there are two nearly cylindrical hole Fermi surfaces whose radii k_F are

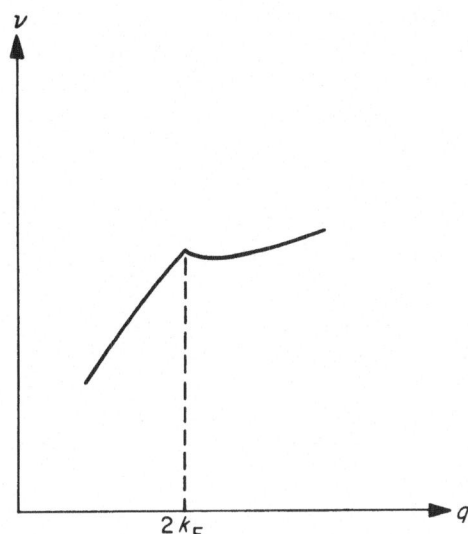

Fig. 3.17. The form of the Kohn anomaly expected for a cylindrical Fermi surface.

approximately 0.15 and 0.25 $(4\pi/\sqrt{3}a)$. This result implies that Kohn anomalies are possible at $q = 2k_F = 0.3$ and 0.5. Although the experimental data near $q = 0.5$ are not accurate enough to show a definite Kohn anomaly near that wave vector, the value of $2k_F = 0.3$ may correspond to the wave vector for the bend observed in the Σ_1 branch. Actually a value of $2k_F \simeq \frac{1}{3}$ has been deduced from measurements of the charge density wave periodicity by various diffraction methods [65, 66] in this material. A stronger Kohn anomaly as well as a charge density wave have also been observed in $TaSe_2$ [66]. The deviation of the measurements from the calculated curves not only for the Σ_1 branch but also for the Σ_3 branch may be an indication that contributions to the screening in Equation (2.11) arising from Umklapp processes may be important. Although the calculated band structure suggests the possibility of a Kohn anomaly for wave vectors along the c-direction, the experimental data show no anomaly and can be reproduced by the simple atomic forces used in the analysis of the data for the semi-conductor MoS_2. This is to be expected since the electron-phonon interaction in this direction should be very weak.

3.3.3. $TiSe_2$

$TiSe_2$ has the trigonal crystal structure of D_{3d}^3 symmetry with layers consisting of Se–Ti–Se stacked along the c-direction. It is convenient to use the Miller indices of the hcp structure to specify the reciprocal space coordinates with the c-axis of the hcp structure being coincident with that of the trigonal one. Thus, the hexagonal basal plane is parallel to the layers, and a direct comparison of phonon dispersion curves can be made with those for the layered compounds of the D_{6h}^4 group such as graphite, MoS_2 or $NbSe_2$.

Due to the small size of the available sample, only low frequency parts of acoustic modes could be measured in the experiment of Stirling *et al.* [53] (Figure 3.18). The slightly upward curvature of certain transverse modes were observed in [100] and [110] directions. These are the modes that involve the atomic motions parallel to the *c*-axis as in the case of graphite or MoS_2. The elastic constants C_{11}, C_{33}, C_{44}, and C_{66} $[=\frac{1}{2}(C_{11} - C_{12})]$ can be determined directly from the initial slopes of the acoustic modes (Table 3.9). Also the value of C_{14} can be calculated from slopes of certain

TABLE 3.9

Measured slopes at small q and elastic constants calculated for TiSe$_2$

Mode	Slope (m s^{-1})	Elastic constant (10^{10} dyne cm^{-2})
$[00\zeta]L$	2740 ± 100	$C_{33} = 39.0 \pm 3.0$
$[00\zeta]T$	1660 ± 60	$C_{44} = 14.3 \pm 1.0$
$[\zeta00]L$	4870 ± 200	
$[\zeta00]T_2$	2740 ± 60	$(C_{66} = 39.0 \pm 2.0)$
$[\zeta\zeta0]L$	4800 ± 200	$C_{11} = 120.0 \pm 10.0$
$[\zeta\zeta0]T_2$	2820 ± 30	
		$C_{12} = 42.0 \pm 10.0$
		$C_{14} = 12.0 \pm 10.0$

branches. The small value for C_{14} is indicative of the basically two-dimensional nature of the structure. The dispersion curves in the basal plane seem to be very isotropic for large ranges of q-values as is evidenced by the plot in the central part of Figure 3.18. No evidence for the Kohn type anomaly seen in NbSe$_2$ and TaSe$_2$ was detected in this work. However, a recent X-ray study by Woo *et al.* [67] has shown the existence of a superlattice reflection at the L-point in the reciprocal space below 150 K. This is considered to correspond to the formation of the charge density wave with the periodicity $(2a, 0, 2c)$ where a and c are the lattice constants.

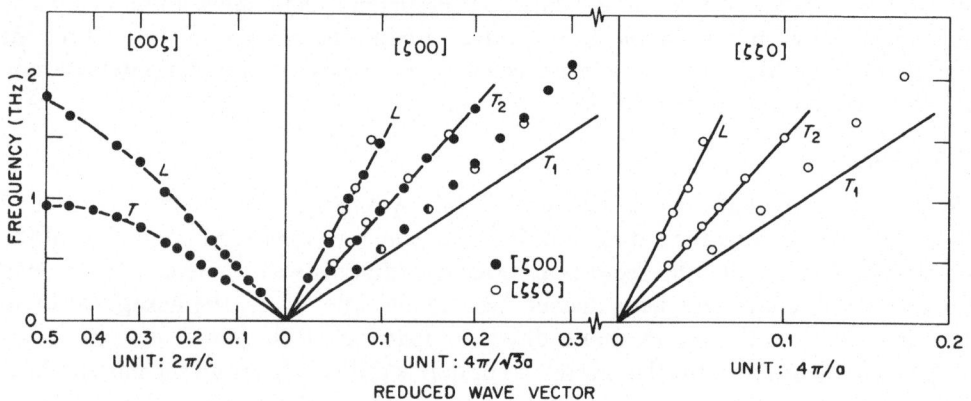

Fig. 3.18. Phonon dispersion curves for the low frequency branches of TiSe$_2$.

3.4. GaSe

In GaSe the tightly bound layers consist of 4 two-dimensional sheets of like atoms. Along the c-axis these sheets are in the sequence Se–Ga–Ga–Se, and the strong bonding both within and between the sheets of a particular layer is thought to be covalent in character. The complete four-fold layers are bound together by much weaker forces again, usually considered to be of the Van der Waals type.

Due to the weak character of the interlayer forces GaSe may occur in three different polytypes. Neutron measurements were carried out on a sample of the ε-polytype (a hexagonal structure with space group D_{3h}^1). The unit cell is shown in Figure 3.19. The group theoretical analysis has been carried out by Jandl and Brebner [68] and the result is given in Table 3.10.

The frequencies of phonons propagating along the Δ, Σ symmetry directions of the Brillouin zone were measured. All measurements were made at a temperature of 100 K [69]. The results are shown in Figure 3.20.

In the Σ direction only modes of the Σ_1, Σ_3 symmetry representations were observed; those of the Σ_2, Σ_4 representations have structure factors which are identically zero

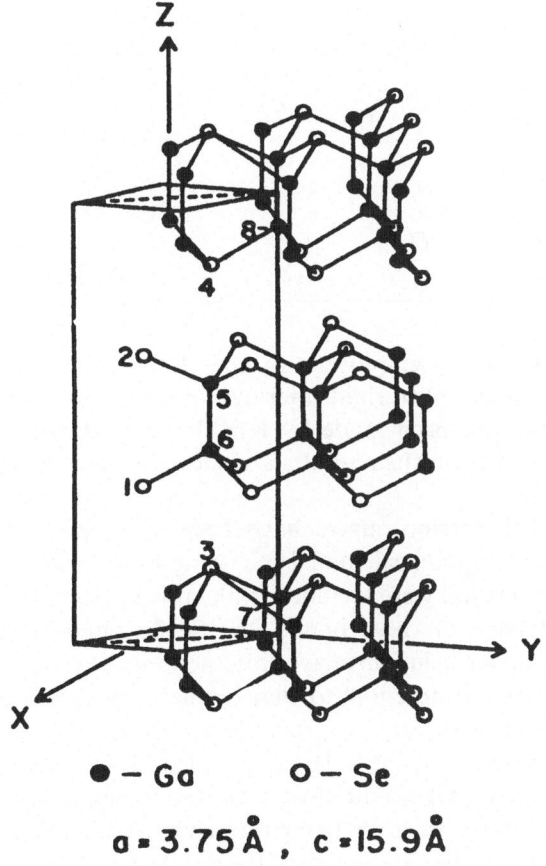

● – Ga   O – Se

a = 3.75 Å , c = 15.9 Å

Fig. 3.19.   Unit cell of ε-GaSe showing intralayer bonding. The numbered atoms (1–8) are contained in the unit cell.

TABLE 3.10

Mode symmetries of the ε polytype of GaSe

Position and group $G_0(q)$	I.R.	Vibration direction	Displacement of equivalent atoms	
$\Gamma, A;$ D_{3h}	A'_1	z	Equal and opposite	All Se atoms
	A''_2	z	Equal	same amplitude
$\Gamma = 4(A''_2 \oplus A'_1 \oplus E' \oplus E'')$	E'	x, y	Equal	All Ga atoms
	E''	x, y	Equal and opposite	same amplitude
$\Delta;$ C_{3v}	A_1	z		
$\Delta = 8(A_1 \oplus E_1)$	E_1	x, y		
	Σ_1	x, z	Equal along x Equal and opposite along z	
$\Sigma, M;$ C_{2v}	Σ_2	y	Equal and opposite	
$\Sigma = 4(2\Sigma_1 \oplus 2\Sigma_3 \oplus \Sigma_2 \oplus \Sigma_4)$	Σ_3	x, z	Equal and opposite along x Equal along z	
	Σ_4	y	Equal	
$T;$ C_3	A'	x, y, z	Equal along x and y Opposite along z	
$T = 12(A' \oplus A'')$	A''	x, y, z	Opposite along x and y Equal along z	
$K;$ C_{3h}	A'	z	Opposite	
	A''	z	Equal	
$K = 4(A'' \oplus A' \oplus E' \oplus E'')$	E'	x, y	Equal	
	E''	x, y	Opposite	

for the scattering geometry employed. In the Δ direction the four lowest energy branches represent almost pure rigid interlayer modes. Ultrasonic measurements of the elastic constants have been made by Khalilov and Rzaev [70]. The velocity of sound lines determined from their results are shown by the dashed lines through the origin for Δ direction.

The experimental dispersion curves have been analyzed in terms of an axially-symmetric (AS) lattice dynamical model including only short-range forces. The five shortest bonds in the crystal were included in the force system, and these are specified in Table 3.11. Only one of these bonds, the Se–Se bond, describes an interlayer interaction, the remainder being intralayer interactions. The Ga–Ga bond (number 4) has an interatomic separation equal to that of the Se–Se bond connecting atoms in adjacent unit cells. Consequently, in principle both interactions should be included. However, since the bonds are not only equal in length but also parallel, their corresponding parameters (ϕ_r, ϕ_t) would always appear in combination in the dynamical matrix. Consequently the sum of the two parameters was assigned to the Ga–Ga bond as a single parameter. The 10 parameters of the model are not independent. A relationship exists among the force constants owing to the symmetry requirement that the slopes of the acoustic Δ_3, Σ_3 modes as $|\zeta| \to 0$ are both determined by the elastic

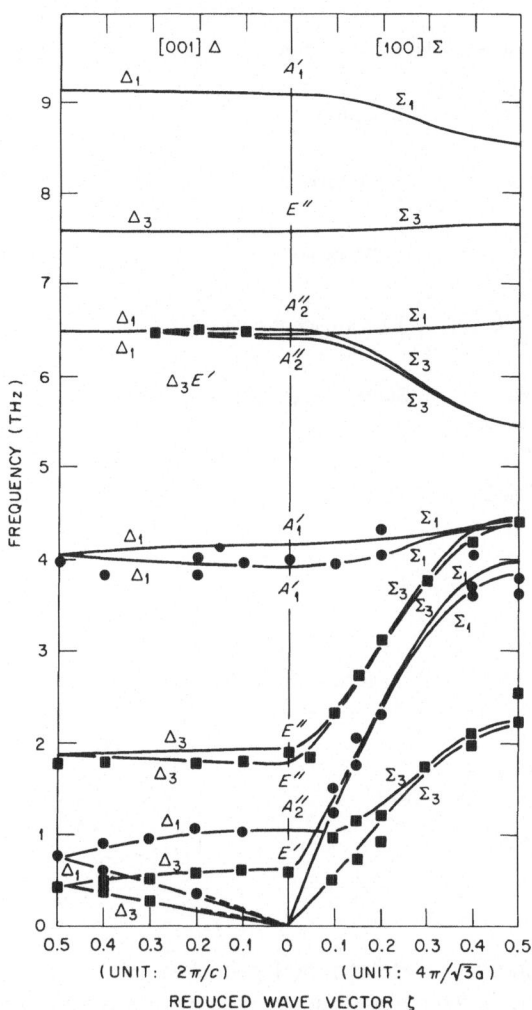

Fig. 3.20. Phonon dispersion curves for ε-GaSe. The lines are calculated from the model described in the text.

constant C_{44}. This relationship was used to define the parameter ϕ_t^5. The nine remaining parameters were determined by 'least-squares' fitting to the experimental frequencies, and the dispersion curves calculated from the 'best-fit' parameters are compared with experiment in Figure 3.20.

The model provides a fair description of the lower frequency dispersion curves and the higher frequency optical modes. The parameters obtained from the analysis are given in Table 3.12. It is evident from the magnitude of the parameters that the Ga–Ga, Ga–Se bonds (1, 2, respectively, of Table 3.11) are significantly stronger than any other bonds of the model. The interlayer Se–Se bond is much weaker than the dominant bonds, and their relative magnitudes support the contention of covalent intralayer and Van der Waals interlayer forces.

TABLE 3.11

Characteristics of the bonds included in the AS model (see Figure 3.19)

Bond	Location	$r(\text{Å})$
Ga–Ga (1)	Same layer	2.388
Ga–Se (2)	Same layer	2.473
Se–Se (3)	Two adjacent layers	3.850
Ga–Ga (Se–Se) (4)	Same layer, different unit cells	3.750
Ga–Se (5)	Same layer	4.186

TABLE 3.12

Force constants for GaSe (Nm^{-1})

$\phi_r^1 = 131 \pm 7$	$\phi_t^1 = 43 \pm 5$
$\phi_r^2 = 66 \pm 2$	$\phi_t^2 = 19 \pm 1$
$\phi_r^3 = 2.5 \pm 0.3$	$\phi_t^3 = 0.3 \pm 0.1$
$\phi_r^4 = 14 \pm 1$	$\phi_t^4 = -3.7 \pm 0.4$
$\phi_r^5 = 7 \pm 1$	$\phi_t^5 = -5$

The interlayer interaction is determined primarily from the four lowest frequency dispersion curves along Δ. The model then predicts the splitting of the conjugate modes (see Equation (2.7)) to be $4\,\text{cm}^{-1}$ and $7\,\text{cm}^{-1}$ for the lowest E'' and A_1' respectively. The latter value is consistent with the measurements reported by Hayek et al. [71], who observed a splitting for the lowest A_1' mode of $6\,\text{cm}^{-1}$. However, they did not observe any splitting for the lowest E'' mode.

The fitted dispersion curves agree well with the velocity of sound lines derived from the elastic constant measurements [70] and also generally agree with the optical measurements. The optic mode whose frequency is 7.5 THz is predicted to be of E'' symmetry representation while the mode with frequency 6.4 THz is predicted to have E' symmetry. These assignments agree with the group theory analysis given previously [68]. The most serious discrepancy between the model and optical measurement is in the frequency of the upper A_2'' mode. This mode is calculated to have a frequency of 6.5 THz while the IR measurements indicate a frequency of 7.11 THz.

The phonon frequency distribution function $g(\nu)$ was calculated using the method of Raubenheimer and Gilat [39] and the result is shown in Figure 3.21. The distribution shows four distinct bands whose origin can be seen from Figure 3.20.

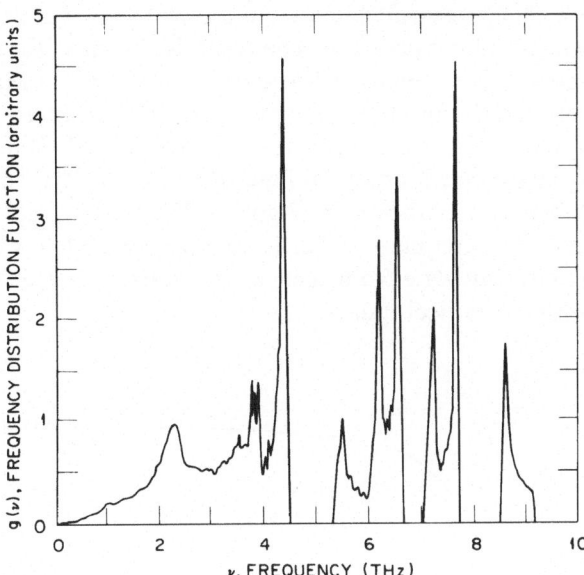

Fig. 3.21. Frequency distribution function for ε-GaSe calculated with force model derived from neutron scattering data.

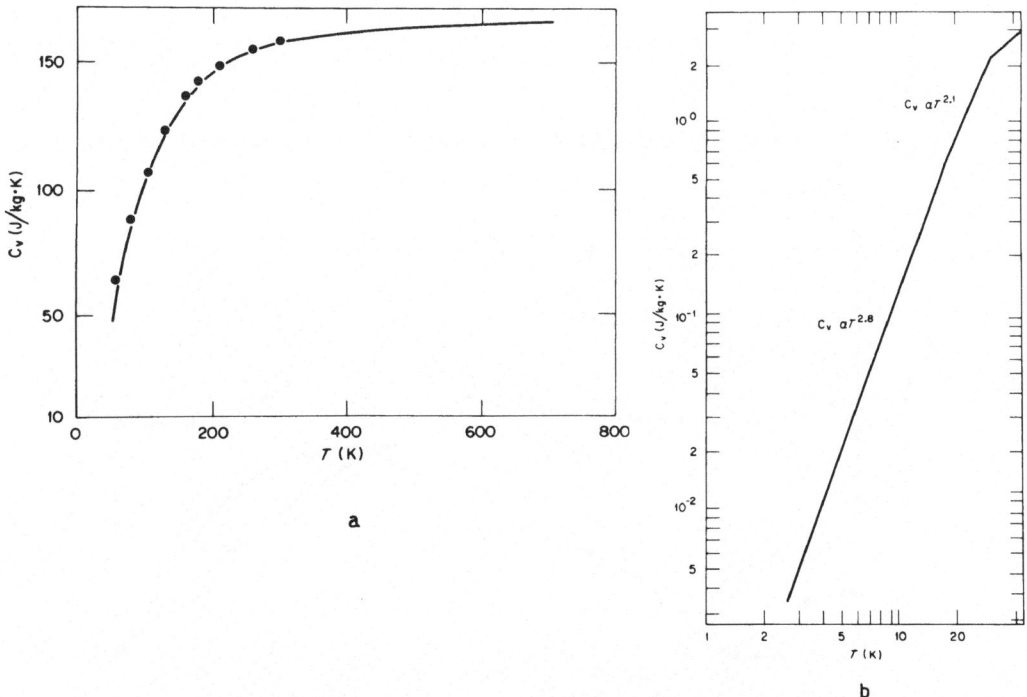

a

b

Fig. 3.22. (a) Temperature dependence of the specific heat for GaSe calculated from the frequency distribution function in Figure 3.21. The solid points are the experimental results [72]. (b) Log–log plot of the specific heat calculated for ε-GaSe.

The specific heat $C_v(T)$ was calculated from the $g(v)$ shown in Figure 3.21 assuming that $g(v)$ is independent of temperature. The result is shown as the solid line in Figure 3.22a. The solid points show the experimental C_v measured by Mamdeov *et al.* [72]. The agreement between the present calculation and the observed C_v is excellent. As in the case of graphite, there is a temperature range (16 ~ 30 K) in which C_v shows a behavior close to that expected for a two-dimensional solid. The temperature variation of the Debye temperature is shown in Figure 3.23. As for graphite and MoS_2 the curves do not show the minimum which is characteristic of this function for many solids. Jandl *et al.* also calculate from their model two-phonon distribution functions for comparison with the optical data.

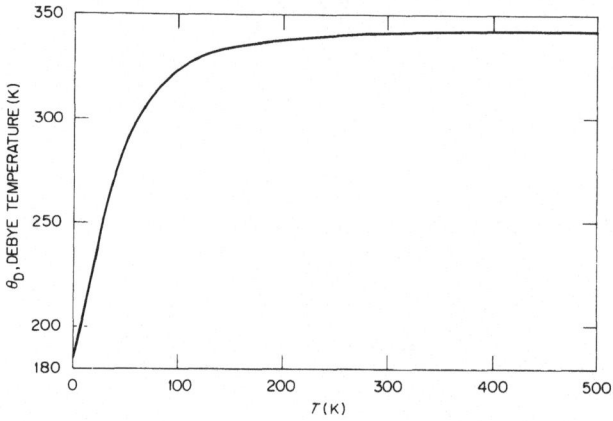

Fig. 3.23.   Temperature dependence of the calculated Debye temperature for ε-GaSe.

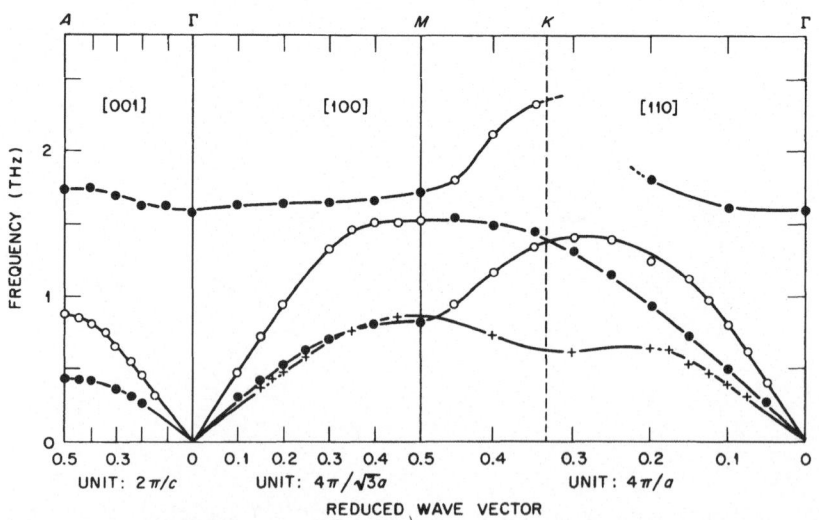

Fig. 3.24.   Phonon frequencies measured for PbI_2.

3.5. PbI_2

PbI_2 is a semi-conductor which possesses several polytypes (2H, 4H, 12R etc.). The nearest distances between Pb atoms in two neighboring layers and within a layer are 6.979 Å and 4.557 Å, respectively. The ratio of these distances is 1.53 for PbI_2 which is much smaller than corresponding values for MoS_2 (1.95) or $NbSe_2$ (1.82). Thus the anisotropy in the interatomic forces may be expected to be less pronounced in this material.

The phonon dispersion curves measured at room temperature by Dorner *et al.* [73] (Figure 3.24) indeed show little quasi-two dimensional features. The large dispersion in the [001] direction clearly indicates that the intra- and inter-layer forces have similar magnitudes. A few phonons in 12R type PbI_2 have been measured both at 50 K and room temperature [74]. Large temperature effects on the scattered neutron peak width as well as on the level of the background counts have been observed, indicating the existence of strong anharmonicity in this material.

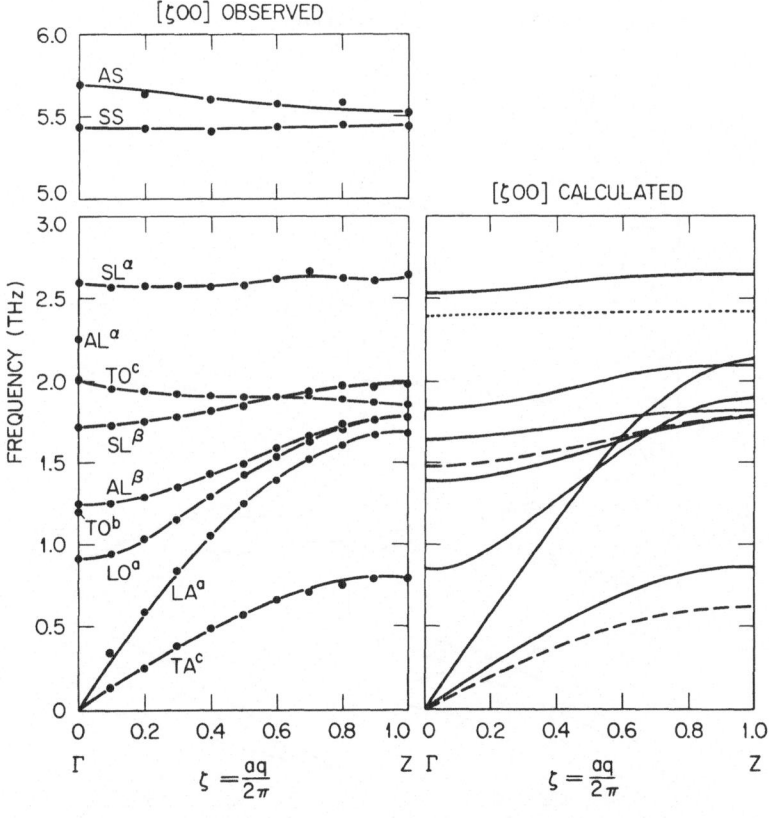

Phonon Dispersion Curves in I_2 at 77 K.

Fig. 3.25. Phonon frequencies of Iodine at 77 K in the [100] direction. *AS*: asymmetric stretching; *SS*: symmetric stretching; SL^α: symmetric libration, α component; etc.

3.6. IODINE

Iodine is another example of a crystal with a layered structure which does not exhibit properties characteristic of an anisotropic layered compound. This fact is reflected in the experimental phonon dispersion curves [75] shown in Figures 3.25 and 3.26, where the longitudinal acoustic modes in the *a*-direction (normal to the layers) have frequencies as high or higher as similar modes within the layers.

Figure 3.27 illustrates the two-dimensional network of molecules within a layer which are coupled to adjacent layers, presumably, by weak Van der Waals forces (the nearest I–I distances between layers, 4.269 Å, is the distance normally attributed to Van der Waals interactions between iodine atoms). This interaction and all more distant interactions are represented in the lattice dynamical calculations by a Buckingham-six potential energy function. Within each layer there are short I–I distances of 2.496 Å and these short contacts are usually considered to be due to weak intermolecular bonds of the charge-transfer type and are, perhaps, responsible for many of the bonding properties of iodine, such as the formation of I_5^- and I_7^--ions and I_2-benzene complexes, to cite a few examples. This interaction has been represented by another Buckingham-six potential function, although a Morse potential function has also been used with satisfactory results.

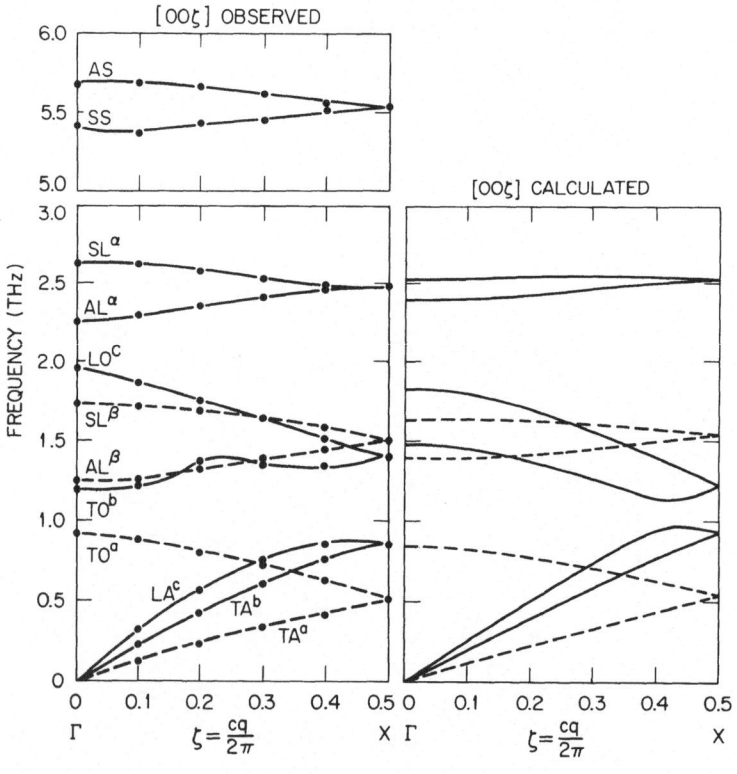

Phonon Dispersion Curves in I_2 at 77°K.

Fig. 3.26. Phonon frequencies of Iodine at 77 K in the [001] direction.

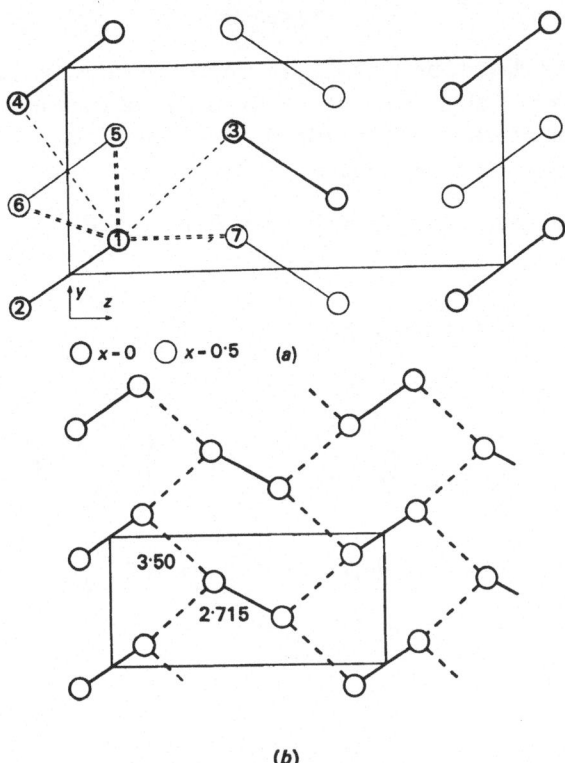

$\bigcirc x\text{-}0 \quad \bigcirc x\text{-}0\cdot5 \quad$ (a)

(b)

Fig. 3.27. (a) The structure of Iodine in (100) projection. (b) The two dimensional network of iodine molecules in the (100) plane.

There is a second I–I interaction (atom 1-atom 4) within the 2-d network of 3.97 Å which is intermediate between the two distances cited above. This could be a steric effect due to packing of non-spherical molecules and not represent a second, weak, charge-transfer type of bond. However, in order to get good agreement of the calculated $LA[100]$ branch with the observed neutron data it was necessary to postulate a separate potential function for this interaction (a third Buckingham-six function). Otherwise the calculated frequencies for this branch were much too high, higher even (at the zone boundary) than the librational modes.

The rather isotropic nature of the forces between the layers and within the layers in iodine probably reflects the fact that the Van der Waals forces between atoms with high atomic number are comparable to, or at least of the same order of magnitude as, the weak intermolecular charge-transfer forces. This is in contrast to the situation in graphite where the Van der Waals forces between layers of atoms of low atomic number are much smaller than the strong covalent forces within the layers.

In these calculations the iodine molecules have been considered to be rigid dumbbells. In view of the small splitting and dispersion of the internal stretching frequencies (5.68 to 5.42 THz at $q = 0$), this is probably not a bad assumption at this stage of the calculations. This condition should be relaxed in a more complete treatment of the lattice dynamics of iodine.

Appendix

In this appendix the elastic constant $C_{\alpha\gamma,\alpha3}$ is shown to vanish if there are no interlayer interactions. The $\kappa = 1$ atom is located at the origin of each unit cell, i.e., $X_\alpha(1) = 0$ for $\alpha = x, y, z$. The notation is that of Born and Huang [1].

The elastic constants are expressed as

$$C_{\alpha\gamma,\beta\lambda} = [\alpha\beta, \gamma\lambda] + [\beta\gamma, \alpha\lambda] - [\alpha\gamma, \beta\lambda] + (\alpha\gamma, \beta\lambda).$$

Thus

$$C_{\alpha\gamma,\alpha3} = [\alpha\alpha, \gamma3] + (\alpha\gamma, \alpha3)$$

But

$$[\alpha\alpha, \gamma3] = \frac{1}{2v} \sum_{\kappa\kappa'} \left\{ - \sum_{l'} \Phi_{\alpha\alpha}\begin{pmatrix} ol' \\ \kappa\kappa' \end{pmatrix} X_\gamma\begin{pmatrix} ol' \\ \kappa\kappa' \end{pmatrix} X_3\begin{pmatrix} ol' \\ \kappa\kappa' \end{pmatrix} \right\}.$$

However, since there are no interlayer interactions, $X_3\begin{pmatrix} ol' \\ \kappa\kappa' \end{pmatrix} = X_3\begin{pmatrix} oo \\ \kappa\kappa' \end{pmatrix} = X_3(\kappa) - X_3(\kappa')$ for l' with nonvanishing $\Phi_{\alpha\beta}\begin{pmatrix} ol' \\ \kappa\kappa' \end{pmatrix}$.

Thus

$$[\alpha\alpha, \gamma3] = -\frac{1}{v} \sum_{l'\kappa'\kappa} \Phi_{\alpha\alpha}\begin{pmatrix} ol' \\ \kappa\kappa' \end{pmatrix} X_\gamma\begin{pmatrix} ol' \\ \kappa\kappa' \end{pmatrix} X_3(\kappa).$$

Also

$$(\alpha\gamma, \alpha3) = -\frac{1}{v} \sum_{\kappa\kappa'} \sum_{\mu\nu} \left\{ \left(\sum_{l''\kappa''} \Phi_{\mu\alpha}\begin{pmatrix} ol'' \\ \kappa\kappa'' \end{pmatrix} X_\gamma\begin{pmatrix} ol'' \\ \kappa\kappa'' \end{pmatrix} \right) \times \right.$$

$$\left. \times \Gamma_{\mu\nu}(\kappa\kappa') \left(\sum_{l'''\kappa'''} \Phi_{\nu\alpha}\begin{pmatrix} ol''' \\ \kappa'\kappa''' \end{pmatrix} X_3\begin{pmatrix} ol''' \\ \kappa'\kappa''' \end{pmatrix} \right) \right\}$$

$$= \frac{1}{v} \sum_{\kappa\kappa'} \sum_{\mu\nu} \left\{ \left(\sum_{l''\kappa''} \Phi_{\mu\alpha}\begin{pmatrix} ol'' \\ \kappa\kappa'' \end{pmatrix} X_\gamma\begin{pmatrix} ol'' \\ \kappa\kappa'' \end{pmatrix} \right) \Gamma_{\mu\alpha}(\kappa\kappa') \times \right.$$

$$\left. \times \left(\sum_{\kappa'''} \left(\sum_{l'''} \Phi_{\nu\alpha}\begin{pmatrix} ol''' \\ \kappa'\kappa''' \end{pmatrix} \right) X_3(\kappa''') \right) \right\}$$

$$= \frac{1}{v} \sum_{\mu\kappa\kappa'''} \left(\sum_{l''\kappa''} \Phi_{\mu\alpha}\begin{pmatrix} ol'' \\ \kappa\kappa'' \end{pmatrix} X_\gamma\begin{pmatrix} ol'' \\ \kappa\kappa'' \end{pmatrix} \right) X_3(\kappa''') \sum_{\nu\kappa'} \left\{ \Gamma_{\mu\nu}(\kappa\kappa') \times \right.$$

$$\left. \times \left(\sum_{l'''} \Phi_{\nu\alpha}\begin{pmatrix} ol''' \\ \kappa'\kappa''' \end{pmatrix} \right) \right\}.$$

Using the definition of Γ, one obtains

$$(\alpha\gamma, \alpha3) = \frac{1}{v} \sum_{\mu\kappa\kappa'''} \left(\sum_{l''\kappa''} \Phi_{\alpha\alpha}\begin{pmatrix} ol'' \\ \kappa\kappa'' \end{pmatrix} X_\gamma \begin{pmatrix} ol'' \\ \kappa\kappa'' \end{pmatrix} \right) X_3(\kappa''')\delta_{\mu\alpha}\delta_{\kappa\kappa'''}$$

$$= -[\alpha\alpha, \gamma3].$$

Thus $C_{\alpha\gamma,\alpha3} = 0$.

In particular, for

$$C_{1313} = C_{44} = 0.$$

Similarly C_{14} can be shown to vanish.

Acknowledgment

The authors wish to thank H. G. Smith for many helpful discussions.

List of Symbols

A	Matrix of elastic constants
a, b, c	Lattice parameter
C	Matrix of Coulomb coefficients
C_i	Velocity of acoustic wave
C_{ij}	Elastic constant
C_v	Specific heat
$C_{\alpha\beta}(\mathbf{q} \mid \kappa\kappa')$	Coulomb coefficient
$C_{\alpha\gamma,\beta\lambda}$	Elastic constant
D	Dynamical matrix
D'	Mass reduced dynamical matrix
$\mathbf{D}_0$	Dynamical matrix for a two-dimensional crystal layer
$D_{\alpha\beta}$	Element of dynamical matrix
E	Neutron energy
e	Eigenvector
$g(v)$	Phonon frequency distribution function
k	Neutron wave vector
k_F	Wave number of Fermi surface
K_r	Stretching force constant in valence force model
$K_{r\theta}$	Force constant for simultaneous bond bending and bond stretching
K_θ	Bond bending force constant in valence force model
M	Matrix of atomic masses
m	Neutron mass
m^*	Effective mass of electron
M_κ	Mass of atom κ
Q	Change in neutron wave vector due to scattering by crystal
q	Wave vector of phonon

R, T, S	Matrices of certain combinations of core-core, core-shell, and shell-shell interaction parameters in shell model
u	Displacement vector from equilibrium position
v	Mass reduced displacement vector
$v(\mathbf{Q})$	Fourier transform of electron-electron interaction
w	Displacement of ionic shell
$W_\kappa(\mathbf{Q})$	Fourier transform of pseudo-potential for atom κ
$\mathbf{X}(l)$	Position vector of lth atom or lth unit cell
Y_κ	Charge of shell of atom κ
Z_κ	Charge on atom κ
$\delta_{\alpha\beta}$	Kronecker delta
$\Delta_i, \Sigma_i, \Gamma_i$	Irreducible representations
$\varepsilon(\mathbf{Q}, \mathbf{Q}')$	Dielectric function
ζ	Component of reduced phonon wave vector
θ_D	Debye temperature
ν	$\omega/2\pi$
ρ	Density
$\boldsymbol{\rho}$	Electronic dipole moment of ion
$\boldsymbol{\tau}, \mathbf{G}$	Reciprocal lattice vector of crystal
Φ	Potential energy of crystal
$\boldsymbol{\phi}$	Matrix of force constant
$\phi_{\alpha\beta}$	Force constant
ϕ_t, ϕ_r	Tangential and radial force constants in axially symmetric force model
ω	Frequency of phonon

References

[1] M. Born and K. Huang: *Dynamical Theory of Crystal Lattices*, Oxford University Press, Oxford 1954.

[2] A. A. Maradudin, E. W. Montroll, G. H. Weiss, and I. P. Ipatova: *Solid State Physics*, Suppl. 3, 2nd ed., Academic Press, New York 1971.

[3] G. Venkataraman, L. A. Feldkamp, and V. C. Sahni: *Dynamics of Perfect Crystals*, MIT Press, Cambridge 1975.

[4] W. Hanke and H. Bilz: *Neutron Inelastic Scattering* 1972, International Atomic Energy Agency, Vienna 1972, p. 3.

[5] S. K. Sinha: *CRC Crit. Rev. Solid State Sci.* **3** (1972), 273.

[6] *Dynamical Properties of Solids*, Vols. I and II, edited by G. K. Horton and A. A. Maradudin, North-Holland, Amsterdam 1975.

[7] G. W. Lehman, T. Wolfram, and R. E. DeWames: *Phys. Rev.* **128** (1962), 1593.

[8] See for example, Ian M. Torrens: *Interatomic Potentials*, Academic Press, New York 1972.

[9] D. C. Wallace: *Phys. Rev.* **176** (1968), 832; *Phys. Rev.* **187** (1969), 991.

[10] P. N. Keating: *Phys. Rev.* **145** (1966), 637.

[11] R. M. Martin: *Phys. Rev.* **B1** (1970), 4005.

[12] See for example, Gerhard Herzberg: *Molecular Spectras and Molecular Structure*, Van Nostrand, New York 1945.

[13] H. L. McMurry, A. W. Solbrig, Jr., and J. K. Boytor: *J. Phys. Chem. Solids* **28** (1967), 2359.

[14] A. Yoshimori and F. Kitano: *J. Phys. Soc. Japan* **11** (1958), 352.

[15] E. W. Kellerman: *Phil. Trans. Roy. Soc. London* **A238** (1940), 513.

[16] R. A. Cowley: *Acta Cryst.* **15** (1962), 687.

[17] A. D. B. Woods, W. Cochran, and B. N. Brockhouse: *Phys. Rev.* **119** (1960), 980.
[18] *Thermal Neutron Scattering*, ed. by P. A. Egelstaff, Academic Press, London, New York 1965.
[19] Mikhail Krivoglaz: *Theory of X-ray and Thermal Neutron Scattering by Real Crystals*, Plenum Press, New York 1969.
[20] W. Marshall and S. W. Lovesey: *Theory of Thermal Neutron Scattering*, Oxford University Press, London 1971.
[21] M. K. Wilkinson, H. G. Smith, W. C. Koehler, R. M. Nicklow, and R. M. Moon: *Neutron Inelastic Scattering*, International Atomic Energy Agency, Vienna 1968, Vol. II, p. 253.
[22] G. Dolling: in Vol. I of Ref. 6.
[23] L. J. Sham: *Proc. Roy. Soc.* **A283** (1965), 33.
[24] W. A. Harrison: *Pseudopotentials in the Theory of Metals*, Benjamin, New York 1966.
[25] Walter F. King, III and P. H. Cutler: *Phys. Rev.* **B3** (1971), 2485.
[26] R. M. Martin: *Phys. Rev.* **186** (1969), 871.
[27] S. K. Sinha, R. P. Gupta, and D. L. Price: *Phys. Rev. Letters* **26** (1971), 1324.
[28] L. J. Sham: in Vol. I of Ref. 6.
[29] R. Zeyher: *Phys. State Solid* **B48** (1971), 711.
[30] N. Wakabayashi and S. K. Sinha: *Phys. Rev.* **B10** (1974), 745.
[31] H. Bilz, B. Gliss, and W. Hanke: in Vol. I of Ref. 6.
[32] L. J. Sham and J. M. Ziman: *Solid State Phys.* **15** (1963), 221.
[33] D. Pines: *Elementary Excitations in Solids*, Benjamin, New York 1963.
[34] See for example, Marvin L. Cohen, Volker Heine, and D. Weair: *Solid State Phys.* **24** (1970).
[35] W. A. Harrison: *Phys. Rev.* **181** (1969), 1036.
[36] S. K. Sinha and B. W. Harmon: *Phys. Rev. Letters* **35** (1975), 1515.
[37] G. Dolling and B. N. Brockhouse: *Phys. Rev.* **128** (1962), 1120.
[38] R. M. Nicklow, N. Wakabayashi, and H. G. Smith: *Phys. Rev.* **B5** (1972), 4951.
[39] L. J. Raubenheimer and G. Gilat: *Phys. Rev.* **157** (1967), 586.
[40] K. Komatsu and T. Nagamiya: *J. Phys. Soc. Japan* **6** (1951), 438.
[41] K. Komatsu: *J. Phys. Soc. Japan* **10** (1955) 346; *J. Phys. Chem. Solids* **6** (1958), 381.
[42] James A. Young and Juan U. Koppel: *J. Chem. Phys.* **42** (1965), 357.
[43] Aziz A. Ahmadieh and Hamid A. Rafizadeh: *Phys. Rev.* **B7** (1973), 4527.
[44] F. Fuinstra and J. L. Koenig: *Bull. Am. Phys. Soc.* **15** (1970), 296.
[45] E. S. Seldin: *Proceedings of Ninth Biennial Conference on Carbon*, Chesnut Hill, Mass., June 18–20, 1969, p. 59. Compiled by Defence Ceramic Information Center, Columbus, Ohio.
[46] J. C. Gylden Houmann and R. M. Nicklow: *Phys. Rev.* **B1** (1970), 3943.
[47] W. Desorbo and G. E. Nichols: *J. Phys. Chem. Solids* **6** (1958), 352; W. Desorbo and W. W. Tyler: General Electric Technical Report No. RL-1202, 1954.
[48] See this volume, Schlüter and Fong: Section C.I(b).
[49] Y. Nakagawa and A. D. B. Woods: *Phys. Rev. Letters* **11** (1963), 271.
[50] B. M. Powell, P. Martel, and A. D. B. Woods: *Phys. Rev.* **171** (1968), 727.
[51] H. G. Smith and W. Gläser: *Phys. Rev. Letters* **25** (1970), 1611.
[52] *Superconductivity in d- and f-Band Metals*, ed. by David H. Douglass, AIP Conference Proceedings No. 4, American Institute of Physics, New York 1972.
[53] W. G. Stirling, B. Dorner, J. D. N. Cheeke, and Revelli: *Solid State Commun.* **18** (1976), 931.
[54] J. L. Verble and T. J. Wieting: *Phys. Rev. Letters* **25** (1970), 362.
[55] N. Wakabayashi, H. G. Smith, and R. M. Nicklow: *Phys. Rev.* **B12** (1975), 659.
[56] T. J. Wieting and J. L. Verble: *Phys. Rev.* **B3** (1971), 4286.
[57] J. L. Verble, T. J. Wieting, and P. R. Reed: *Solid State Commun.* **11** (1972), 941.
[58] J. C. Phillips: *Covalent Binding in Crystals, Molecules, and Polymers*, The University of Chicago Press, Chicago 1969.
[59] G. Lucousky and R. M. White: *Phys. Rev.* **B8** (1973), 660.
[60] N. Wakabayashi, H. G. Smith, and H. R. Shanks: *Phys. Letters* **50A** (1974), 367.
[61] A. M. Afanasev and Y. Kagon: *JETP* **16** (1963), 1030.
[62] C. Kittel: *Solid State Phys.* **22** (1968), 1.
[63] R. I. Sharp: *J. Phys.* **C2** (1969), 432.
[64] L. F. Mattheiss: *Phys. Rev. Letters* **30** (1973), 784.

[65] J. A. Wilson, F. J. DiSalvo, and S. Mahajan: *Advances Phys.* **24** (1975), 117.
[66] D. E. Moncton, J. D. Axe, and F. J. DiSalvo: *Phys. Rev. Letters* **34** (1975), 734.
[67] K. C. Woo, F. C. Brown, W. L. McMillan, R. J. Miller, M. J. Schaffman, and M. P. Sears: *Phys. Rev.* **B14** (1976), 3242.
[68] S. Jandl and J. L. Brebner: *Can. J. Phys.* **52** (1974), 2454.
[69] S. Jandl, J. L. Brebner, and B. M. Powell: *Phys. Rev.* **B13** (1976), 686.
[70] K. M. Khalilov and K. I. Rzaev: *Soviet Phys.-Crystallogr.* **11** (1957), 786.
[71] M. Hayek, O. Brafman, and R. M. A. Lieth: *Phys. Rev.* **B8** (1973), 2772.
[72] K. K. Mamedov, I. G. Kerimov, V. N. Kostryukov, and M. I. Mekhtiev: *Soviet Phys.-Semicond.* **1** (1967), 363.
[73] B. Dorner, R. E. Ghosh, and G. Harbeke: *Phys. Status Solidi* **B73** (1976), 655.
[74] N. Wakabayashi: Private communication.
[75] H. G. Smith, M. Nielsen, and C. B. Clark: *Chem. Phys. Letters* **33** (1975), 75.